# Fundamentals of
# NUMERICAL COMPUTATION

# Fundamentals of NUMERICAL COMPUTATION

TOBIN A. DRISCOLL
RICHARD J. BRAUN
UNIVERSITY OF DELAWARE
NEWARK, DELAWARE

Society for Industrial and Applied Mathematics
Philadelphia

Copyright © 2018 by the Society for Industrial and Applied Mathematics

10 9 8 7 6 5 4 3 2 1

All rights reserved. Printed in the United States of America. No part of this book may be reproduced, stored, or transmitted in any manner without the written permission of the publisher. For information, write to the Society for Industrial and Applied Mathematics, 3600 Market Street, 6th Floor, Philadelphia, PA 19104-2688 USA.

*Mathematica* is a registered trademark of Wolfram Research, Inc.

MATLAB is a registered trademark of The MathWorks, Inc. For MATLAB product information, please contact The MathWorks, Inc., 3 Apple Hill Drive, Natick, MA 01760-2098 USA, 508-647-7000, Fax: 508-647-7001, *info@mathworks.com*, *www.mathworks.com*.

Front cover photos courtesy of the Computer History Museum.

| | |
|---|---|
| *Publications Director* | Kivmars H. Bowling |
| *Acquisitions Editor* | Paula Callaghan |
| *Developmental Editor* | Gina Rinelli Harris |
| *Managing Editor* | Kelly Thomas |
| *Production Editor* | Louis R. Primus |
| *Copy Editor* | Louis R. Primus |
| *Production Manager* | Donna Witzleben |
| *Production Coordinator* | Cally A. Shrader |
| *Graphic Designer* | Lois Sellers |

**Library of Congress Cataloging-in-Publication Data**
Names: Driscoll, Tobin A. (Tobin Allen), 1969- author.
Title: Fundamentals of numerical computation / Tobin Driscoll, Richard Braun, University of Delaware, Newark, Delaware.
Description: Philadelphia : Society for Industrial and Applied Mathematics, [2018] | Series: Other titles in applied mathematics ; 154 | Includes bibliographical references and index.
Identifiers: LCCN 2017042102 (print) | LCCN 2017048879 (ebook) | ISBN 9781611975086 | ISBN 9781611975079
Subjects: LCSH: Numerical analysis--Data processing.
Classification: LCC QA297 (ebook) | LCC QA297 .D75 2018 (print) | DDC 518.0285/53--dc23
LC record available at https://lccn.loc.gov/2017042102

 is a registered trademark.

For our wives
and children

# Contents

Preface — xi

List of Functions — xvii

List of Terms — xix

List of MATLAB Commands — xxvii

Prologue — 1

## I — 7

**1 Numbers, problems, and algorithms** — 9
  1.1 Floating point numbers — 9
  1.2 Problems and conditioning — 15
  1.3 Stability of algorithms — 20

**2 Square linear systems** — 31
  2.1 Polynomial interpolation — 31
  2.2 Computing with matrices — 36
  2.3 Linear systems — 44
  2.4 LU factorization — 51
  2.5 Efficiency of matrix computations — 61
  2.6 Row pivoting — 68
  2.7 Vector and matrix norms — 74
  2.8 Conditioning of linear systems — 80
  2.9 Exploiting matrix structure — 84

**3 Overdetermined linear systems** — 95
  3.1 Fitting functions to data — 96
  3.2 The normal equations — 103
  3.3 The QR factorization — 107
  3.4 Computing QR factorizations — 113

**4 Roots of nonlinear equations** — 121
  4.1 The rootfinding problem — 121
  4.2 Fixed point iteration — 127
  4.3 Newton's method in one variable — 135

|     |                                          |     |
| --- | ---------------------------------------- | --- |
| 4.4 | Interpolation-based methods              | 143 |
| 4.5 | Newton for nonlinear systems             | 152 |
| 4.6 | Quasi-Newton methods                     | 159 |
| 4.7 | Nonlinear least squares                  | 166 |

## 5 Piecewise interpolation and calculus — 175

| | | |
| --- | --- | --- |
| 5.1 | The interpolation problem | 175 |
| 5.2 | Piecewise linear interpolation | 182 |
| 5.3 | Cubic splines | 189 |
| 5.4 | Finite differences | 196 |
| 5.5 | Convergence of finite differences | 202 |
| 5.6 | Numerical integration | 208 |
| 5.7 | Adaptive integration | 217 |

## 6 Initial-value problems for ODEs — 227

| | | |
| --- | --- | --- |
| 6.1 | Basics of initial-value problems | 227 |
| 6.2 | Euler's method | 235 |
| 6.3 | Systems of differential equations | 242 |
| 6.4 | Runge–Kutta methods | 249 |
| 6.5 | Adaptive Runge–Kutta | 255 |
| 6.6 | Multistep methods | 261 |
| 6.7 | Implementation of multistep methods | 266 |
| 6.8 | Zero-stability of multistep methods | 273 |

## II — 279

## 7 Matrix analysis — 281

| | | |
| --- | --- | --- |
| 7.1 | From matrix to insight | 281 |
| 7.2 | Eigenvalue decomposition | 286 |
| 7.3 | Singular value decomposition | 293 |
| 7.4 | Symmetry and definiteness | 299 |
| 7.5 | Dimension reduction | 304 |

## 8 Krylov methods in linear algebra — 311

| | | |
| --- | --- | --- |
| 8.1 | Sparsity and structure | 312 |
| 8.2 | Power iteration | 319 |
| 8.3 | Inverse iteration | 326 |
| 8.4 | Krylov subspaces | 332 |
| 8.5 | GMRES | 338 |
| 8.6 | MINRES and conjugate gradients | 344 |
| 8.7 | Matrix-free iterations | 349 |
| 8.8 | Preconditioning | 353 |

## 9 Global function approximation — 359

| | | |
| --- | --- | --- |
| 9.1 | Polynomial interpolation | 359 |
| 9.2 | The barycentric formula | 365 |
| 9.3 | Stability of polynomial interpolation | 369 |
| 9.4 | Orthogonal polynomials | 377 |
| 9.5 | Trigonometric interpolation | 384 |
| 9.6 | Spectrally accurate integration | 390 |

|  |  |  |
|---|---|---|
|  | 9.7 Improper integrals | 398 |
| 10 | **Boundary-value problems** | **409** |
|  | 10.1 Shooting | 410 |
|  | 10.2 Differentiation matrices | 418 |
|  | 10.3 Collocation for linear problems | 425 |
|  | 10.4 Nonlinearity and boundary conditions | 431 |
|  | 10.5 The Galerkin method | 439 |
| 11 | **Diffusion equations** | **449** |
|  | 11.1 Black–Scholes equation | 449 |
|  | 11.2 The method of lines | 455 |
|  | 11.3 Absolute stability | 463 |
|  | 11.4 Stiffness | 471 |
|  | 11.5 Method of lines for parabolic PDEs | 477 |
| 12 | **Advection equations** | **485** |
|  | 12.1 Traffic flow | 485 |
|  | 12.2 Upwinding and stability | 490 |
|  | 12.3 Absolute stability for advection | 495 |
|  | 12.4 The wave equation | 499 |
| 13 | **Two-dimensional problems** | **507** |
|  | 13.1 Tensor-product discretizations | 507 |
|  | 13.2 Two-dimensional diffusion and advection | 515 |
|  | 13.3 Laplace and Poisson equations | 524 |
|  | 13.4 Nonlinear elliptic PDEs | 531 |
| A | **Review of linear algebra** | **539** |
|  | **Bibliography** | **545** |
|  | **Index** | **551** |

# Preface

> I've developed an obscene interest in computation, and I'll be returning to the United States a better and impurer man.
> —John von Neumann

It might seem that computing should simply be a matter of translating formulas from the page to the machine. But even when such formulas are known, applying them in a numerical fashion requires care. For instance, rounding off numbers at the sixteenth significant digit can lay low such stalwarts as the quadratic formula! Fortunately, the consequences of applying a numerical method to a mathematical problem are quite understandable from the right perspective. In fact, it is our mastery of what can go wrong in some approaches that gives us confidence in the rest of them.

If mathematical modeling is the process of turning real phenomena into mathematical abstractions, then numerical computation is largely about the transformation from abstract mathematics to concrete reality. Many science and engineering disciplines have long benefited from the tremendous value of the correspondence between quantitative information and mathematical manipulation. Other fields, from biology to history to political science, are rapidly catching up. In our opinion, a young mathematician who is ignorant of numerical computation in the 21st century has much in common with one who was ignorant of calculus in the 18th century.

## To the student

Welcome! We expect that you have had lessons in manipulating power series, solving linear systems of equations, calculating eigenvalues of matrices, and obtaining solutions of differential equations. We also expect that you have written computer programs that take a nontrivial number of steps to perform a well-defined task, such as sorting a list. Even if you have rarely seen how these isolated mathematical and computational tasks interact with one another, or what they have to do with practical realities, you are part of the intended audience for this book.

Based on our experiences teaching this subject, our guess is that some rough seas may lie ahead of you. Probably you do not remember learning all parts of calculus, linear algebra, differential equations, and computer science with equal skill and fondness.

This book draws from all of these areas at times, so your weaknesses are going to surface once in a while. Furthermore, this may be the first course you have taken that does not fit neatly within a major discipline. Von Neumann's use of "impurer" in the quote above is a telling one: numerical computation is about solving problems, and the search for solution methods that work well can take us across intellectual disciplinary boundaries. This mindset may be unfamiliar and disorienting at times.

Don't panic! There is a great deal to be gained by working your way through this book. It goes almost without saying that you can acquire computing skills that are in much demand for present and future use in science, engineering, and mathematics—and, increasingly, in business, social science, and humanities, too. There are less tangible benefits as well. Having a new context to wrestle with concepts like Taylor series and matrices may shed new light on why they are important enough to learn. It can be rewarding to apply skills to solve relatable problems. Finally, the computational way of thought is one that complements other methods of analysis and can serve you well.

## MATLAB

All of the coded functions and most of the examples are presented in MATLAB. MATLAB was invented by an academic numerical analyst who wanted better tools for teaching the subject. Today MATLAB is a behemoth with endless bells and whistles, and it is crucial to major industrial sectors. It has both fans and detractors. At its core, though, MATLAB remains an environment that is excellent for lightweight, easily understood expression and exploration of the concepts that are in this text.

While a student should have one standard programming course under his or her belt before starting this book, prior knowledge of MATLAB should not be necessary. We encourage you to download the examples and execute them yourself (details on accessing them are given below). Not only should you see on the screen exactly what is in the book, but you should try to make changes and, in the words of Mark Zuckerberg, "move fast and break things." As with a natural language, the easiest path to fluency is by imitating a proficient speaker and gradually building your vocabulary as the need arises.

If you prefer a more to-the-point introduction only to MATLAB, there are lots of videos and tutorial resources online (for example, search for "matlab academy"), and we can shamelessly recommend at least one book: *Learning MATLAB*, by one of us (Driscoll).

## To the instructor

First of all, don't overlook the online content—you can find the details below.

The plausibly important introductory material on numerical computation for the majority of undergraduate students easily exceeds the capacity of two semesters—and of one textbook. As instructors and as authors, we face difficult choices as a result. We set aside the goal of creating an agreeable canon. Instead we hope for students to experience an echo of that "obscene interest" that von Neumann so gleefully described and pursued. For while there are excellent practical reasons to learn about numerical computing, it also stands as a subject of intellectual and even emotional relevance. We have seen students excited and motivated by applications of their newly found abilities to problems in mechanics, biology, networks, finance, and more—problems that are of unmistakable importance in the jungle beyond university textbooks, yet largely impenetrable using only the techniques learned within our well-tended gardens.

In writing this book, we have not attempted to be encyclopedic. We're sorry if some of your favorite topics don't appear or are minimized in the book. (It happened to us too; many painful cuts were made from prior drafts.) But in an information-saturated world, the usefulness of a textbook lies with teaching process, not content. We have aimed not for a cookbook but for an introduction to the principles of cooking.

Still, there *are* lots of recipes in the book—it's hard to imagine how one could become a great chef without spending time in the kitchen! Our language for these recipes is MATLAB, for a number of reasons: it is precise, it is executable, it is as readable as could be hoped for our purposes, it rewards thinking at the vector and matrix level, and (at this writing) it is widespread and worth knowing. There are 46 standalone functions and over 150 example scripts, all of them downloadable exactly as seen in the text. Some of our codes are quite close to production quality, some are serviceable but lacking, and others still are meant for demonstration only. Ultimately our codes are mainly meant to be clear, not ideal. We try to at least be explicit about the shortcomings of our implementations.

Just as good coding and performance optimization are secondary objectives of the book, we cut some corners in the mathematics as well. We state and in some cases prove the most essential and accessible theorems, but this is not a theorem-oriented book, and in some cases we are content with less precise arguments. We have tried to make clear where solid proof ends and where approximation, estimation, heuristics, and other indispensable tools of numerical analysis begin.

The examples and exercises are meant to show and encourage a numerical mode of thought. As such they often focus on issues of convergence, experimentation leading to abstraction, and attempts to build upon or refine presented ideas. Some exercises follow the premise that an algorithm's most interesting mode of operation is failure. We expect that any learning resulting from the book is likely to take place mostly from careful study of the examples and working through the problems.

## Contents

Chapter 1 explains how computers represent numbers and arithmetic, and what doing so means for mathematical expressions. Chapter 2 discusses the solution of square systems of linear equations and, more broadly, basic numerical linear algebra. Chapter 3 extends the linear algebra to linear least squares. These topics are the bedrock of scientific computing, because "everything" has multiple dimensions, and while "everything" is also nonlinear, our preferred starting point is to linearize.

Chapters 4 through 6 introduce what we take to be the rest of the most common problem types in scientific computing: roots and minimization of algebraic functions, piecewise approximation of functions, numerical analogs of differentiation and integration, and initial-value problems for ordinary differential equations. We also explain some of the most familiar and reliable ways to solve these problems, effective up to a certain point of size and/or difficulty. Chapters 1 through 6 can serve for a single-semester survey course. If desired, Chapter 6 could be left out in favor of one of Chapters 7, 8, or 9.

The remaining chapters are intended for a second course in the material. They go into more sophisticated types of problems (eigenvalues and singular values, boundary value problems, and partial differential equations), as well as more advanced techniques for problems from the first half (Krylov subspace methods, spectral approximation, stiff problems, boundary conditions, and tensor-product discretizations).

## Online content

We have created many downloadable materials for this book. There will be a maintained repository at `www.siam.org/books/ot154`. The materials include the following:

- All of the functions presented in the book.

- All of the computational examples, written as scripts that you can execute to reproduce the results.

- Slides and videos related to presenting the material in class.

- Laboratory exercises useful for hands-on problem solving with expert supervision. These draw from a wide variety of practical and mathematical applications. They have been used by the authors.

- Project ideas, typically motivated by real-world applications. These too have been used by the authors.

## Acknowledgments

We are, of course, deeply indebted to all who taught us and inspired us about numerical computation over the years. We are thankful to Rodrigo Platte, who used the book before it was fully baked and offered numerous suggestions. We thank an enthusiastic group of grad students for proofreading help: Samuel Cogar, Shukai Du, Kristopher Hollingsworth, Rayanne Luke, Navid Mirzaei, Nicholas Russell, and Osman Usta. We are also grateful to Paula Callaghan and the publishing team at SIAM, whose dedication to affordable, high-quality books makes a real difference in the field.

We thank our families for their support and endurance. Last but not least, we are grateful to the many students at the University of Delaware who have taken courses based on iterations of this book. Their experiences are what convinced us that the project was worth finishing.

## About the cover

- *(front cover, top left)* The Z4 digital computer was completed in Berlin in 1945 by Konrad Zuse, who hoped to sell it commercially. It could only perform 2.5 additions per second and used movie film punched with holes for input; the film could literally be looped for repeated instructions. The Z4 was hastily moved out of Berlin at the end of World War II to keep it out of Soviet hands, and it was stored in various houses and barns. In 1949 Eduard Stiefel (one of the co-discoverers of the conjugate gradients algorithm) had it brought to the ETH in Zürich.

- *(front cover, top right)* The UNIVAC 9400 was introduced by the Sperry Rand Corporation in 1967. It was programmed using then-standard 80-column punch cards. The 9400 had a maximum internal storage of 131,072 bytes and could perform over $10^5$ additions per second.

- *(front cover, middle right)* The SWAC was built by the U. S. National Bureau of Standards (now the National Institute of Standards and Technology) in 1950. It was the fastest computer in the world, able to do 15,000 additions per second. In the words of Alexandra Forsythe, who, with her husband George, was among the early users and legendary pioneers in computer science, "you ran everything twice and if the results agreed you figured it was ok." It played a key role in the Nobel Prize in Chemistry awarded to Dorothy Hodgkin for discovering the structure of vitamin $B_{12}$.

- *(front cover, bottom right)* The ENIAC, completed in 1946 at the University of Pennsylvania, was one of the first general-purpose computers. It weighed over 27,000 kg and could add 5,000 times per second. The ENIAC took weeks to program, requiring the manual manipulation of switches and cables. In the photo are Betty Jennings (left) and Frances Bilas (right), two of the six all-female programmers.

- *(front cover, bottom left)* The TRADIC computer, created at Bell Labs in 1954, was among the first in the world to rely on transistors (684 of them) rather than vacuum tubes for computation. Its lead developer was Jean Felker, who promised the U. S. Air Force "a computer as reliable as a hammer!" It could perform 62,000 additions per second. The TRADIC was programmed using removable boards that could each hold up to 128 machine instructions, including a reusable subroutine.

- *(rear cover)* The authors are standing (TAD at right) at the University of Delaware with a sculpture called *Reconnect*, by Ronald Longsdorf. The six-foot-plus steel and resin column of computer parts is shrouded by connecting cables. (Photo by Andrew Bernoff.)

# List of Functions

| | | |
|---|---|---|
| 1.3.1 | (`horner`) Evaluate a polynomial by Horner's rule. | 21 |
| 2.3.1 | (`forwardsub`) Solve a lower triangular linear system. | 48 |
| 2.3.2 | (`backsub`) Solve an upper triangular linear system. | 48 |
| 2.4.1 | (`lufact`) LU factorization for a square matrix. | 57 |
| 3.2.1 | (`lsnormal`) Solve linear least squares by normal equations. | 105 |
| 3.3.1 | (`lsqrfact`) Solve linear least squares by QR factorization. | 112 |
| 3.4.1 | (`qrfact`) QR factorization by Householder reflections. | 117 |
| 4.3.1 | (`newton`) Newton's method for a scalar rootfinding problem. | 141 |
| 4.4.1 | (`secant`) Secant method for scalar rootfinding. | 146 |
| 4.5.1 | (`newtonsys`) Newton's method for a system of equations. | 156 |
| 4.6.1 | (`fdjac`) Finite difference approximation of a Jacobian. | 160 |
| 4.6.2 | (`levenberg`) Quasi-Newton method for nonlinear systems. | 163 |
| 5.2.1 | (`hatfun`) Hat function/piecewise linear basis function. | 184 |
| 5.2.2 | (`plinterp`) Piecewise linear interpolation. | 186 |
| 5.3.1 | (`spinterp`) Cubic not-a-knot spline interpolation. | 193 |
| 5.4.1 | (`fdweights`) Fornberg's algorithm for finite difference weights. | 200 |
| 5.6.1 | (`trapezoid`) Trapezoid formula for numerical integration. | 211 |
| 5.7.1 | (`intadapt`) Adaptive integration with error estimation. | 220 |
| 6.2.1 | (`eulerivp`) Euler's method for a scalar initial-value problem. | 236 |
| 6.3.1 | (`eulersys`) Euler's method for a first-order IVP system. | 247 |
| 6.4.1 | (`ie2`) Improved Euler method for an IVP. | 251 |
| 6.4.2 | (`rk4`) Fourth-order Runge–Kutta for an IVP. | 252 |
| 6.5.1 | (`rk23`) Adaptive IVP solver based on embedded RK formulas. | 258 |
| 6.7.1 | (`ab4`) Fourth-order Adams–Bashforth formula for an IVP. | 267 |
| 6.7.2 | (`am2`) Second-order Adams–Moulton (trapezoid) formula for an IVP. | 269 |

| | | |
|---|---|---|
| 8.2.1 | (`poweriter`) Power iteration for the dominant eigenvalue. | 322 |
| 8.3.1 | (`inviter`) Shifted inverse iteration for the closest eigenvalue. | 327 |
| 8.4.1 | (`arnoldi`) Arnoldi iteration for Krylov subspaces. | 336 |
| 8.5.1 | (`arngmres`) GMRES for a linear system. | 341 |
| 9.2.1 | (`polyinterp`) Polynomial interpolation by the barycentric formula. | 367 |
| 9.5.1 | (`triginterp`) Trigonometric interpolation. | 385 |
| 9.6.1 | (`ccint`) Clenshaw–Curtis numerical integration. | 393 |
| 9.6.2 | (`glint`) Gauss–Legendre numerical integration. | 395 |
| 9.7.1 | (`intde`) Doubly exponential integration over $(-\infty, \infty)$. | 401 |
| 9.7.2 | (`intsing`) Integrate function with endpoint singularities. | 404 |
| 10.1.1 | (`shoot`) Shooting method for a two-point boundary-value problem. | 412 |
| 10.2.1 | (`diffmat2`) Second-order accurate differentiation matrices. | 420 |
| 10.2.2 | (`diffcheb`) Chebyshev differentiation matrices. | 422 |
| 10.3.1 | (`bvplin`) Solve a linear boundary-value problem. | 427 |
| 10.4.1 | (`bvp`) Solve a nonlinear boundary-value problem. | 434 |
| 10.5.1 | (`fem`) Piecewise linear finite elements for a linear BVP. | 444 |
| 11.2.1 | (`diffper`) Differentiation matrices for periodic end conditions. | 457 |
| 13.2.1 | (`rectdisc`) Discretization on a rectangle. | 518 |
| 13.3.1 | (`poissonfd`) Solve Poisson's equation by finite differences. | 529 |
| 13.4.1 | (`newtonpde`) Newton's method to solve an elliptic PDE. | 534 |

# List of Terms

**adjacency matrix**  A matrix whose nonzero entries represent the edges between nodes in a network. 283

**advection equation**  A linear partial differential equation describing pure transport; canonical example of a hyperbolic PDE. 486

**algorithm**  Complete set of instructions for mapping data to a result. 20

**Arnoldi iteration**  Stable iteration for generating a Krylov subspace. 335

**asymptotic**  Functions or sequences are asymptotic if their ratio is 1 in the limit of interest. 61

**backward error**  The distance to a problem for which a computed solution is the exact answer. 24

**backward substitution**  An explicit procedure for solving any nonsingular, square, upper triangular system of linear equations. 46

**bandwidth**  The number of nonzero diagonals in a banded matrix. 85

**barycentric formula**  Computationally useful expression for the interpolating polynomial as a ratio of rational terms. 366

**"big-O"**  Describes an upper bound on asymptotic behavior of a function. 61

**boundary-value problem**  A differential equation with which partial information about the solution is given at two values of the independent variable. 410

**cardinal functions**  A set of interpolation basis functions in which each is one at one interpolation node and zero at all the other nodes. 180

**CFL condition**  Requirement that the numerical domain of dependence of a PDE contain the exact domain of dependence, possibly dictating an upper bound on the time step. 491

**Cholesky factorization**  Factorization of a symmetric positive definite matrix $A$ into $R^T R$ for upper triangular $R$. 90

**collocation**  Solution of a differential equation by imposing it approximately at a set of nodes. 425

**condition number** The ratio of change in the output to perturbation of the input. The measurements might be in relative or absolute senses. The condition number measures the inherent sensitivity of an answer to all forms of perturbation, including floating point error. 16

**conjugate gradients** A Krylov subspace iterative method for solving linear systems, useful for large, symmetric, positive definite matrices. 345

**cubic spline** Interpolating function that is piecewise cubic and twice continuously differentiable. 190

**diagonalizable** A matrix having an eigenvalue decomposition; opposite terms are *nondiagonalizable* and *defective*. 287

**differentiation matrix** Matrix that maps a vector of point values of a function to a vector approximate point values of one of its derivatives. 419

**Dirichlet condition** Boundary condition specifying the value of the solution. 433

**dominant eigenvalue** The eigenvalue of largest complex magnitude (absolute value). 320

**eigenvalue decomposition** Similarity transformation of a square matrix to a diagonal matrix. 287

**Euler's method** Prototype of all numerical methods for solving initial-value problems, though of limited practical utility on its own. 235

**evolutionary PDE** A partial differential equation in which one of the independent variables is time or a close analog. 450

**extrapolation** Use of multiple estimates and knowledge of the error structure to derive asymptotically superior estimates. 213

**finite difference formula** Linear combination of values of a function at nodes to approximate the value of its derivative at a point. 196

**finite element method** Use of piecewise integration to pose a linear system of equations for the approximate solution of a boundary-value problem. 443

**fixed point iteration** Attempted solution of a fixed point problem by repeated application of its defining function. 131

**fixed point problem** Solution of the equation $g(x) = x$, or, in higher dimensions, $G(x) = x$. 127

**floating point numbers** An idealized subset of real numbers on which numerical computation can be performed. 9

**flop** Floating point operation. Used as an estimate of computational complexity. 62

**forward substitution** An explicit procedure for solving any nonsingular, square, lower triangular system of linear equations. 46

**Gauss–Newton method** Solution of a nonlinear least-squares problem by iterative solution of linear least squares problems. 167

**Gaussian elimination** Process of performing row operations in a triangular pattern in order to transform a linear system of equations into upper triangular form. 51

**global error** For a numerical method for initial-value problems, the error of the method at a fixed value of the independent variable (typically time). 239

**GMRES** A Krylov subspace iterative method for solving linear systems, useful for large matrices. 339

**hat functions** Cardinal basis functions for piecewise linear interpolation. 183

**heat equation** A linear partial differential equation describing pure diffusion; canonical example of a parabolic PDE. 451

**hermitian** The conjugate transpose of a matrix, or a matrix that is unchanged after taking its conjugate transpose. 286

**identity matrix** A diagonal matrix that has ones on the diagonal. Multiplying by the identity leaves any matrix unchanged. 37

**implicit** A formula that defines a quantity only indirectly, e.g., as the solution of a nonlinear equation. 261

**initial-value problem** A differential equation with which complete information about the solution is given at a value of the independent variable. 227

**interpolation** Construction of a function that passes through a given set of points. 31, 175

**inverse iteration** Power iteration with the (possibly shifted) inverse of a matrix, to find the smallest or a targeted eigenvalue. 326

**Jacobian matrix** Matrix of first partial derivatives, appearing in the linear term of a multidimensional Taylor series. 153

**Kronecker product** Alternative type of matrix multiplication, useful for functions on a tensor-product domain. 525

**Krylov subspace** Span of the vectors $u, Au, \ldots, A^{m-1}u$, useful in solving linear systems for large matrices. 332

**Lagrange formula** Theoretically useful expression for the interpolating polynomial. 361

**Laplace equation** A linear partial differential equation describing steady state; canonical example of an elliptic PDE. 524

**least squares** Minimizing the sum of squares of a residual, or, equivalently, the 2-norm of the residual. 95

**linear convergence** Convergence of a sequence such that the error is asymptotically reduced by a constant factor at each iteration. 132

**linear least squares problem** Solution of least squares in the overdetermined linear system $Ax \approx b$. 99

**Lipschitz condition** An upper bound on the amount by which a function can "spread apart" two different inputs. 134

**local truncation error** Truncation error for each individual step of a numerical method for initial-value problems. 238, 263

**LU factorization** Factorization of a matrix into the product of a unit lower triangular and an upper triangular matrix. 56

**machine epsilon** Relative error in the representation of floating point numbers and floating point arithmetic. Also, the smallest number that can be added to 1 and result in a different floating point number. 10

**matrix condition number** The product of the norm of a matrix and the norm of its (pseudo)inverse. Equals the condition number of the linear system problem in the square case. 80

**matrix inverse** The inverse of a square nonsingular matrix $A$, written $A^{-1}$, is the unique matrix such that $AA^{-1}$ and $A^{-1}A$ are the identity. 543

**method of lines** Solution technique for partial differential equations in which each independent variable is discretized separately. 457

**multistep** One of two major families of numerical methods for initial-value problems, using information from multiple time steps to boost accuracy. 261

**Neumann condition** Boundary condition specifying the derivative of the solution. 433

**Newton's method** Rootfinding method that uses a tangent line in one dimension, or the Jacobian matrix in higher dimensions, to approximate the function and define the next root estimate. 138

**Newton–Cotes formula** Numerical integration formula derived by exact integration of a piecewise polynomial interpolant. 209

**nonlinear least squares** Minimization of the 2-norm of a nonlinear system of $m$ constraints in $n$ variables, where $m > n$. 166

**nonsingular** A square matrix is nonsingular (or invertible) if it has an inverse. 543

**norm** A function that assigns a nonnegative "size" to every vector, matrix, or function. 74

**normal equations** A square linear system of equations equivalent to the linear least squares problem. 103

**ONC matrix** Matrix with orthonormal columns (not necessarily square). 108

**order of accuracy** Power of $h$ in the leading (lowest-order) term of the truncation error. 203, 240, 263

**orthogonal** For vectors, having inner product equal to zero. For matrices, see orthogonal matrix. 107

**orthogonal matrix** A matrix that is square and has orthonormal columns. 109

**orthogonal polynomials** Family of polynomials whose distinct members have an integral inner product equal to zero, as with Legendre and Chebyshev polynomials. 380

**orthonormal** Describes unit vectors that are pairwise orthogonal. 108

**overdetermined** Having more constraints than the available free variables can fulfill. 95

**permutation matrix** An identity matrix with its rows (equivalently, its columns) reordered. Used to reorder the rows (columns) of another matrix in the same manner. 70

**PLU factorization** LU factorization with row pivoting for stability. 70

**power iteration** Approximation of the dominant eigenvalue by means of repeated multiplication by a matrix. 321

**preconditioning** Use of an approximate inverse to accelerate the convergence of a Krylov subspace iterative method. 353

**pseudoinverse** A matrix that maps the data (right-hand side vector) of a linear least squares problem to the solution vector. 104

**QR factorization** Factorization of a matrix into the product of an orthogonal matrix and an upper triangular matrix. Frequently used to solve linear least squares problems. 109

**quadratic convergence** Convergence of a sequence such that the error is approximately squared at each iteration. 139

**quasi-Newton methods** Rootfinding methods that modify Newton's method by approximation of the Jacobian matrix and/or improving the likelihood of convergence. 159

**quasimatrix** Collection of functions that have algebraic parallels to columns of a matrix. 379

**Rayleigh quotient** Scalar-valued function whose gradient is zero at the eigenvectors of a symmetric matrix. 301

**residual** The amount by which an estimate of the solution is incorrect when compared to the data of the problem. In the linear system $Ax = b$, this is $b - A\tilde{x}$ for an estimate $\tilde{x}$; in a rootfinding problem, it's the value of the objective function at the root estimate. 82, 125

**rootfinding problem** Finding a solution to an algebraic equation or system of equations. 121

**row pivoting** Swapping of rows to get the largest possible pivot element during Gaussian elimination. 69

**Runge phenomenon** Manifestation of the instability of polynomial interpolation at equally spaced nodes as degree increases. 373

**Runge–Kutta** One of two major families of numerical methods for initial-value problems, using multiple stages at each time step to boost accuracy. 249

**secant method** Rootfinding method for a function of one variable that uses a secant line to approximate the function and define the next root estimate. 145

**shooting** Unstable technique for solving a boundary-value problem in which an initial value is searched for by a rootfinding algorithm. 411

**(significant) digits** In scientific notation, the number of digits appearing in the mantissa (nonexponential) part of the expression. 11

**simple root** A root at which the slope is nonzero (scalar functions). 125

**singular value decomposition** Factorization of a matrix into two unitary matrices and a diagonal matrix. 293

**sparse matrix** A matrix in which most of the entries are known to be exactly zero. 87, 312

**spectral convergence** Exponentially rapid decrease in error as the number of interpolation nodes increases, as observed in Chebyshev polynomial and trigonometric interpolation. 375

**stability region** Region of the complex plane describing when numerical solution of a linear initial-value problem is bounded as $t \to \infty$. 464

**stiff** Characteristic of a differential equation having widely varying time scales present simultaneously, typically favoring the use of an implicit time stepping scheme. 271, 471

**subtractive cancellation** The loss of accuracy resulting when floating point numbers are added or subtracted to give a result much smaller (in absolute value) than the operands. Also called loss of significance. 15

**superlinear convergence** Rate of convergence of a sequence in which the ratios of successive errors (differences from the limiting value) vanish in the limit. 146

**symmetric matrix** A square matrix $A$ is symmetric if $A^T = A$. In the complex case, a matrix is hermitian if $A^* = A$. 37

**symmetric positive definite** A matrix $A$ which is both symmetric and has $x^T A x > 0$ for all nonzero vectors $x$. 89

**tensor-product domain** A domain that can be parameterized using variables that lie in a logical rectangle or cuboid; i.e., each variable independently varies in an interval. 507

**trapezoid formula** Numerical integration formula derived by exact integration of a piecewise linear interpolant. 210

**triangular** A matrix $A$ is upper triangular if $A_{ij} = 0$ for all $i > j$ and lower triangular if $A_{ij} = 0$ for all $i < j$. The term is also used to refer to a linear system of equations whose matrix is triangular. 46

**tridiagonal matrix** A matrix whose bandwidth is three. 85

**trigonometric interpolation** Interpolation of a periodic function by a linear combination of real or complex trigonometric functions. 384

**truncation error** Error resulting from truncation of a Taylor series to a finite number of terms, typically arising when using discretization of differentiation or integration. 202

**unit upper/lower triangular** A triangular matrix whose diagonal entries are all ones. 55

**unit vector** A vector whose norm equals one. It represents a pure direction. Any nonzero vector can be divided by its norm to make a unit vector in the same direction. 75

**unitary** A complex-valued matrix whose hermitian is its inverse; in the real case the matrix is also called *orthogonal*. 286

**unstable algorithm** An algorithm that amplifies errors to an extent beyond that explained by a problem's condition number. 22

**upwind** Direction in which information from the past influenced the solution at a certain location in space-time. 490

**Vandermonde** A square matrix whose columns are evaluations of the monomials $1, t, t^2, \ldots, t^{n-1}$ at a fixed set of $n$ distinct points. 32

**vec** Operator that stacks the columns of a matrix to make an equivalent vector. 515

# List of MATLAB Commands

**any** Return logical value 1 (true) if any element of a vector is nonzero or a logical 1. 60

**\\** Solve linear or least squares system via direct solution methods. Detects whether the system is square, triangular, rectangular, or sparse. 33, 45

**ceil** Round up to the next integer. 401

**chol** Compute the Cholesky factorization of a matrix, checking whether the matrix is positive definite. 90

**clf** Clear the current figure window. 34

**cond** Compute the condition number of a matrix, optionally in a given norm. 81

**contourf** Make a filled contour plot for a function of two variables. Similar to **contour**. 508

**cos** Evaluate the cosine function. 42

**diag** Depending on context, extract a specified diagonal from a matrix, or create a matrix with specified diagonals. 85

**eigs** Compute some of the eigenvalues and eigenvectors of a sparse matrix. 316

**eig** Compute the eigenvalues or eigenvalue decomposition of a matrix. 287

**end** Specify the last entry in a vector, or of a dimension in a multidimensional array, without having to know its length. 38, 40

**eps** Return machine epsilon in double precision ($2^{-52}$). 12

**expm** Compute the matrix exponential (not the same as **exp** for a matrix). 243

**exp** Evaluate the exponential function. 42

**fft** Compute the fast Fourier transform for trigonometric interpolation. 388

**format** Control output appearance in the command window. 59

**fplot** Plot a function over a given interval. 176

**fzero** Find a root of a function of one variable. 122

**gallery** Create commonly used and specialized matrices. 60

**gmres** Apply GMRES to solve a large linear system of equations. 340

**hold** Keep current plot objects when making new ones (toggle). 34

**integral** Adaptive numerical integration. 223

**interp1** Interpolate data by a piecewise polynomial of one variable. 178

**inv** Compute the inverse of a square matrix. *Not* used to solve a linear system of equations; instead, use the backslash \. 45

**kron** Kronecker product of two matrices. 525

**legend** Create a legend for a plot. 34

**linspace** Create an evenly spaced vector of values between two given endpoints. 34

**loglog** Create a two-dimensional plot with a logarithmic scale on both axes. 99

**log** Evaluate the natural logarithm (base $e$). 42

**lu** Compute the row-pivoted LU factorization of a square matrix. 66

**max** Maximum value elementwise between two arguments, or maximum element of a vector and its location within the vector, or maximum along one dimension of a matrix. 322

**mesh** Make a three-dimensional plot of a two-dimensional array on a corresponding grid or mesh. 331

**ndgrid** Create coordinate matrices defining a tensor-product grid. 508

**norm** Compute the norm of a vector or matrix. 74

**ode45** Solve an initial-value problem using adaptive Runge–Kutta. 229

**ones** Create an array of specified size with each element having a value of one. 38

**pcolor** Make a two-dimensional plot for a function of two variables in which color represents function value. 508

**pinv** Compute the pseudoinverse of a rectangular matrix. 104

**plotyy** Create a two-dimensional plot with two differently labeled ordinates. 416

**plot** Create a two-dimensional plot. 33

**polyfit** Find the coefficients of a polynomial given appropriate data. 133

**polyval** Evaluate a polynomial. 21

**qr** Compute the full or thin QR factorization of a matrix. 110

**reshape** Reinterpret the dimensions of a matrix, e.g., to change between matrix and vector shapes for grid data. 515

**roots** Find the roots of a polynomial. 24

**round** Round a number to the nearest decimal or integer. 60

**semilogx** Create a two-dimensional plot with a logarithmic abscissa and linear ordinate. 67

**semilogy** Create a two-dimensional plot with a logarithmic ordinate and linear abscissa. 99

**sin** Evaluate the sine function. 42

**size** Return the size of a matrix or array. 38

**spdiags** Create a sparse matrix by specifying its diagonals. 314

**spy** Plot the locations of the nonzero elements of a matrix. 85

**sqrt** Compute the square root. 42

**subplot** Create an array of plots in a single figure window. 78

**surf** Make a three-dimensional surface plot for a function of two variables. 508

**svd** Compute the full or thin singular value decomposition of a matrix. 295

**tic** Start stopwatch timing of a computation. 62

**title** Create a title for a plot. 34

**toc** Stop stopwatch timing of a computation. 62

**tril** Create an lower triangular matrix from a given matrix, replacing all entries above a diagonal by zero. 48

**triu** Create an upper triangular matrix from a given matrix, replacing all entries below a diagonal by zero. 57

**waterfall** Make a three-dimensional plot for a function of time and space. 487

**xlabel** Create a label for the abscissa on a two-dimensional plot, or the $x$-axis on a three-dimensional plot. 34

**ylabel** Create a label for the abscissa on a two-dimensional plot, or the $y$-axis on a three-dimensional plot. 34

**ylim** Set limits on the $y$-axis. 234

**zeros** Create an array of specified size with each element having a value of zero. 38

# Prologue:
# Two imperfect ideas

> Luke, you're going to find that many of the truths we cling to depend greatly on our own point of view.
> —Obi-Wan Kenobi, *Return of the Jedi*

Over 2,000 years ago, the Greek mathematician Archimedes published the fact that

$$\frac{265}{153} < \sqrt{3} < \frac{1351}{780}.$$

Considering that the rational bounds are $1.73202614\ldots$ and $1.73205128\ldots$, this was quite an achievement. Archimedes said nothing in his text about how he went about finding this statement, but it's possible that he used an even more ancient algorithm for finding square roots that goes back to the Babylonians.

We know that $\sqrt{3}$ is the length of each side of a square whose area is 3. Suppose that $x$ is meant to be an estimate of $\sqrt{3}$. We know that the rectangle of length $x$ and width $3/x$ also has area 3. It's geometrically clear that if $x$ is too small, then $3/x$ is too large, and vice versa. So the side length of the square we are looking for is between $x$ and $3/x$, and it makes sense that the average of those two values will be a better estimate than $x$ is. This reasoning suggests the iteration

$$x_{n+1} = \frac{1}{2}\left(x_n + \frac{3}{x_n}\right), \quad n = 1, 2, 3, \ldots. \tag{P.1}$$

The iteration has to be seeded with an initial estimate $x_1$, which then defines the value of $x_2$ (with $n = 1$ in the formula), which then defines $x_3$ ($n = 2$), and so on. If we start with a rational number for $x_1$, then all of the $x_n$ that follow will also be rational numbers, so none of them can ever exactly equal $\sqrt{3}$, but we hope that the estimates keep improving. In fact one can prove that

$$\lim_{n \to \infty} x_n = \sqrt{3},$$

as long as $x_1 > 0$.

It's not hard to figure out that $\sqrt{3}$ is close to 1.7. A convenient nearby simple rational estimate is $x_1 = 5/3$. Using *Mathematica*, one can find the next few estimates easily.

x₁ = 5 / 3;

n = 1; $x_{n+1} = \frac{1}{2}\left(x_n + \frac{3}{x_n}\right)$

$$\frac{26}{15}$$

n = 2; $x_{n+1} = \frac{1}{2}\left(x_n + \frac{3}{x_n}\right)$

$$\frac{1351}{780}$$

After two iterations, we've arrived at Archimedes' upper bound (though with no proof that it's an upper bound). There's nothing to say that we can't keep going, however.

n = 3; $x_{n+1} = \frac{1}{2}\left(x_n + \frac{3}{x_n}\right)$

$$\frac{3\,650\,401}{2\,107\,560}$$

n = 4; $x_{n+1} = \frac{1}{2}\left(x_n + \frac{3}{x_n}\right)$

$$\frac{26\,650\,854\,921\,601}{15\,386\,878\,263\,120}$$

n = 5; $x_{n+1} = \frac{1}{2}\left(x_n + \frac{3}{x_n}\right)$

$$\frac{1\,420\,536\,136\,104\,448\,487\,712\,806\,401}{820\,146\,920\,573\,494\,197\,299\,310\,240}$$

Roughly speaking, the number of digits in the numerator and denominator doubles with each iteration. While these expressions are all small in an absolute sense, in principle it appears that each additional iteration will have to take at least twice as long as the previous one. However, the number of accurate decimal places in the approximations is also roughly doubling, as Table 1 shows. The price of doubling the expression length may therefore be considered worthwhile.

**Table 1.** *Errors in the Mathematica iterations of the square root algorithm.*

| $n$ | $x_n - \sqrt{3}$ | $n$ | $x_n - \sqrt{3}$ |
|---|---|---|---|
| 1 | −0.06538 | 4 | $6.499 \times 10^{-14}$ |
| 2 | $1.283 \times 10^{-3}$ | 5 | $1.219 \times 10^{-27}$ |
| 3 | $4.745 \times 10^{-7}$ | 6 | $4.292 \times 10^{-55}$ |

Now let's look at the square root algorithm in a different computing environment, MATLAB.

```
>> x = 5/3;
>> n = 1;   x(n+1) = ( x(n)+3/x(n) ) / 2;
>> n = 2;   x(n+1) = ( x(n)+3/x(n) ) / 2;
>> n = 3;   x(n+1) = ( x(n)+3/x(n) ) / 2;
>> n = 4;   x(n+1) = ( x(n)+3/x(n) ) / 2;
>> n = 5;   x(n+1) = ( x(n)+3/x(n) ) / 2;
>> format long,  x'

ans =

   1.666666666666667
   1.733333333333333
   1.732051282051282
   1.732050807568942
   1.732050807568877
   1.732050807568877
```

Aside from the obvious differences in syntax, there is a much more profound underlying difference. MATLAB represents every value just as it is shown to you, with about 16 significant decimal digits. That means that there is no growth in the length of the internal representations, and each iteration will take the same amount of time as the one before it. On the other hand, it also means that the accuracy will be limited to about $10^{-16}$:

```
>> x' - 1.73205080756887729352745

ans =

  -0.065384140902210
   0.001282525764456
   0.000000474482405
   0.000000000000065
                   0
                   0
```

Don't be fooled by the "zero" errors! Even our 24-digit input of $\sqrt{3}$ gets rounded off to 16 digits, so we're just seeing a report that the first inaccurate digit is somewhere beyond that. The other errors match up with those in Table 1.

So far you would be justified in wondering why anyone would use MATLAB's model of computing. Let's generalize the problem a bit. It turns out that the Babylonian square root iteration (P.1) is a special case of what we now call *Newton's iteration* for finding roots. If we want to find, say, the fifth root of 3, Newton's iteration becomes the formula

$$x_{n+1} = \frac{1}{5}\left(4x_n + \frac{3}{x_n^4}\right), \quad n = 1, 2, 3, \ldots. \tag{P.2}$$

Figure 1 shows what happens when we give *Mathematica* this iteration. Now the length of the representation grows much faster, essentially quintupling at each iteration! This

$x_1 = 5/3;$

$n = 1; \quad x_{n+1} = \frac{1}{5}\left(4x_n + \frac{3}{x_n^4}\right)$

$$\frac{13\,229}{9375}$$

$n = 2; \quad x_{n+1} = \frac{1}{5}\left(4x_n + \frac{3}{x_n^4}\right)$

$$\frac{1\,837\,930\,723\,586\,197\,456\,721}{1\,435\,652\,549\,600\,928\,796\,875}$$

$n = 3; \quad x_{n+1} = \frac{1}{5}\left(4x_n + \frac{3}{x_n^4}\right)$

102 185 626 549 503 452 093 625 963 764 838 087 672 399 460 421 572 161 799 051 892 976 454 163 647 502 ╲
719 007 746 845 396 681 742 986 531 029 /
81 909 806 796 833 340 717 385 554 095 262 985 361 987 071 184 330 385 818 719 020 963 863 097 846 957 ╲
804 224 852 407 472 404 863 318 984 375

$n = 4; \quad x_{n+1} = \frac{1}{5}\left(4x_n + \frac{3}{x_n^4}\right)$

55 627 715 978 772 395 939 904 241 051 451 108 941 865 651 904 407 406 904 177 977 947 212 228 494 664 709 ╲
229 149 952 912 126 017 608 118 958 276 102 334 827 486 938 502 293 477 387 264 240 656 240 719 743 021 ╲
236 916 246 309 469 881 402 346 774 338 589 355 527 922 758 549 498 269 264 360 050 599 883 381 375 936 ╲
608 219 739 018 440 129 798 032 325 906 681 581 055 505 966 362 825 009 374 932 141 376 765 093 467 408 ╲
535 529 428 242 750 475 591 277 899 542 925 225 585 172 028 854 787 981 681 160 209 174 532 136 130 370 ╲
349 173 671 015 419 616 773 318 828 978 612 508 415 099 486 980 060 415 843 479 752 999 817 421 271 872 ╲
856 018 889 748 227 875 431 185 395 056 116 263 830 740 017 430 690 029 085 075 717 721 /
44 654 492 141 730 529 408 414 095 177 874 377 094 478 282 618 031 681 644 604 817 132 472 859 323 929 ╲
026 158 568 176 850 888 526 110 128 146 372 894 347 779 118 261 507 051 446 752 580 029 336 445 752 860 ╲
852 720 158 917 992 215 323 085 947 587 642 431 368 945 457 468 357 993 847 933 140 731 999 048 772 533 ╲
024 710 240 203 183 634 831 998 174 726 474 325 221 934 777 702 223 145 928 132 463 518 045 029 670 128 ╲
814 097 756 389 382 860 577 655 630 611 416 013 940 051 455 998 405 845 919 779 661 494 507 421 379 054 ╲
495 788 122 962 758 152 069 663 727 425 926 981 233 024 845 853 777 022 547 771 401 527 245 066 628 774 ╲
468 430 610 084 769 033 021 285 618 513 659 800 382 751 883 603 407 135 628 849 501 171 875

**Figure 1.** Mathematica *implementation of the iteration* (P.2) *for* $3^{1/5}$.

observation should give us pause, especially considering that the results are not any more accurate as representations of $3^{1/5}$; the sequence of errors is 0.42, 0.17, 0.034, $1.8 \times 10^{-3}$, and $5.2 \times 10^{-6}$, consistent with doubling the number of accurate digits at each iteration. When you consider that $x_5$ stores 536 digits in both the numerator and the denominator in order to obtain just 6 digits of accuracy, MATLAB's use of a fixed representation length seems a lot wiser.[1]

The differences between the two approaches are the most fundamental ones in computing for mathematical problems. The approach of *Mathematica* is called **symbolic computing**, and MATLAB is an example of **numerical computing**.[2] The contrast goes much deeper than how they represent numbers. In symbolic computing one can manipulate expressions with variables that hold unknown values, as one does when applying algebra and calculus manually. In numerical computing, every named variable must have a numerical value at all times.

---

[1] We are indebted to Nick Trefethen for the inspiration behind this demonstration.
[2] Both environments offer both types of computing in practice. Nevertheless they are best known as specialists in the two domains.

The two approaches to computing are complementary. Symbolic computing can give formulas that yield much more insight than numbers alone. Numerical computing is able to solve a scope and scale of problems that symbolic computing can't begin to touch. The approaches also require very different mathematical tools to understand, apply, and analyze.

This book concentrates exclusively on numerical computing. There is more to learn about numerical computing than can fit into one course, or two, or six. Our aim is to introduce you to much of what could be called the most essential and universal mathematical aspects of numerical computing, or what we call "numerical mathematics" for short. To do so, we make omissions and simplifications that serve the overarching goal.

Numerical mathematics isn't like other branches of mathematics. For one thing, it's strongly problem driven. Beautiful theorems are celebrated, but they aren't the ultimate goal and can't be allowed to limit our investigations. For the same reason, the tools in play come from all over mathematics; in particular, both linear algebra and elementary analysis are indispensable. For that matter, computer science also has its say in the subject.

Another interesting but confounding aspect of numerical mathematics is that we collect a lot of different ways to solve a single type of mathematical problem. In elementary linear algebra, for instance, you learn that there is an algorithm, Gaussian elimination, that solves any consistent system of linear equations. In a purely mathematical sense, that problem is therefore "solved" once and for all. But when it comes to asking a computer to give you answers to real problems, having only Gaussian elimination at your disposal is a bit like playing a round of golf with nothing but a putter—possible, but far from optimal.

Finally, numerical mathematics doesn't play by all of the old rules. Inherent in the roundoff process are some surprising consequences. For example, suppose we round off all results of arithmetic to three significant digits. Then

$$(1.11 + 0.00411) + 0.00411 = 1.11\cancel{411} + 0.00411 = 1.11\cancel{411},$$

for a result of 1.11. But if we do operations in a different order, we get

$$1.11 + (0.00411 + 0.00411) = 1.11 + 0.00822 = 1.11\cancel{822},$$

for a rounded result of 1.12. This looks like a trivial difference here, but the repercussions of losing the associative law of arithmetic are far-reaching. Expressions that are mathematically equivalent can and do behave very differently from one another in numerical computing. In practice, we don't really disregard the familiar rules of mathematics. But you do have to learn how to use them with more care than you have become used to.

# Part I

*Much of your education in mathematics progressed in some order through understanding numbers, arithmetic, algebra, functions, calculus, and differential equations. In numerical computation, the rules for all of these items are changed. Even as simple a property as $(a + b) + c = a + (b + c)$ no longer holds! Foruntately, computation is used as a simulation of the mathematics you know, so we can understand its properties as perturbations of their mathematical analogs.*

*In the first six chapters we present the most fundamental problems and methods of numerical computing. Not only are they important in their own right, but they serve as a key introduction to many topics seen throughout computation as well. They also begin to address deep questions you may not have thought to ask in a while: Why do we solve problems? What does it even mean to "solve" a problem? Is having any formula or procedure for arriving at an answer sufficient? And how can one represent the solution of a mathematical problem using only numbers? As you work through the details of the techniques and analysis of the following chapters, bear in mind the big questions as well.*

# Chapter 1
# Numbers, problems, and algorithms

> You must unlearn what you have learned.
> — Yoda, *The Empire Strikes Back*

Our first step is to discretize the real numbers—specifically, to replace them with a finite surrogate set of numbers. This step keeps the time and storage requirements for operating with each number at constant levels, but (almost) every data set and arithmetic operation is perturbed slightly away from its idealized mathematical value. We can easily keep the individual roundoff errors very small, so small that simple random accumulation is unlikely to bother us. However, some problems are extremely sensitive to these perturbations, a trait we quantify using a *condition number*. Problems with large condition numbers are difficult to solve accurately. Furthermore, even when the condition number of a problem is not large, some algorithms for solving it allow errors to grow enormously. We call these algorithms *unstable*. In this chapter we discuss these ideas in simple settings before moving on to the more realistic problems in the rest of the book.

## 1.1 ▪ Floating point numbers

The real number set $\mathbb{R}$ is infinite in two ways: it is unbounded and continuous. In most practical computing, the second kind of infiniteness is much more consequential than the first kind, so we turn our attention there first. We replace $\mathbb{R}$ with the set $\mathbb{F}$ of **floating point numbers**, whose members are zero and all numbers of the form

$$\pm(1+f) \times 2^e, \qquad (1.1.1)$$

where $e$ is an integer called the **exponent**, and $1+f$ is the **mantissa**, in which

$$f = \sum_{i=1}^{d} b_i \, 2^{-i}, \qquad b_i \in \{0,1\}, \tag{1.1.2}$$

for a fixed integer $d$. Equation (1.1.2) represents the mantissa as a number in $[1,2)$ in base-2 form. Equivalently,

$$f = 2^{-d} \sum_{i=1}^{d} b_i \, 2^{d-i} = 2^{-d} z$$

for an integer $z$ in the set $\{0,1,\ldots,2^d-1\}$. Consequently, starting at $2^e$ and ending just before $2^{e+1}$ there are exactly $2^d$ evenly spaced numbers belonging to $\mathbb{F}$.

> **Example 1.1.1**
>
> Suppose $d=2$. Hence with $e=0$ in (1.1.1), we see that 1, 5/4, 3/2, and 7/4 are floating point numbers. They are the only members of $\mathbb{F}$ in the half-interval $[1,2)$. Taking $e=1$ gives the floating point numbers in $[2,4)$, specifically 2, 2.5, 3, and 3.5. Generally the spacing between the elements of $\mathbb{F}$ in $[2^e, 2^{e+1})$ is $2^{e-d}$.

Observe that *the smallest element of $\mathbb{F}$ that is greater than 1 is $1+2^{-d}$*. We define **machine epsilon** (or **machine precision**[3]) as $\varepsilon_{\text{mach}} = 2^{-d}$.

We also suppose the existence of a rounding function $\text{fl}(x)$ that maps every real $x$ to the nearest member of $\mathbb{F}$. If $x$ is positive, we know that it lies in some interval $[2^e, 2^{e+1})$, where the spacing between elements of $\mathbb{F}$ is $2^{e-d}$. Therefore we conclude that $|\text{fl}(x) - x| \leq \frac{1}{2}(2^{e-d})$, which leads to the bound

$$\frac{|\text{fl}(x) - x|}{|x|} \leq \frac{2^{e-d-1}}{2^e} \leq \tfrac{1}{2} \varepsilon_{\text{mach}}. \tag{1.1.3}$$

Equation (1.1.3) holds true for negative $x$ as well. A mathematically equivalent (see Exercise 1.1.2) and sometimes more convenient statement is that

$$\text{fl}(x) = x(1+\epsilon) \quad \text{for some } |\epsilon| \leq \tfrac{1}{2} \varepsilon_{\text{mach}}. \tag{1.1.4}$$

The value of $\epsilon$ depends on $x$, but this dependence is not usually shown explicitly.

## Precision and accuracy

The upshot of floating point representation, as stated in (1.1.3), is that every real number is represented with a uniformly bounded relative precision. Except for the use of

---

[3] The terms machine epsilon, machine precision, and unit roundoff aren't used consistently across references, but the differences are minor for our purposes.

## 1.1. Floating point numbers

base 2 rather than base 10, floating point representation is a form of scientific notation. For example, Planck's constant is $6.626068 \times 10^{-34}$ m²kg/s. If we alter just the last digit from 8 to 9, the relative change is

$$\frac{0.000001 \times 10^{-34}}{6.626068 \times 10^{-34}} \approx 1.51 \times 10^{-7}.$$

This observation justifies the statement that the constant is given with 7 **significant digits** of precision in base 10. That's in contrast to saying that the value is given to 40 decimal places. A major advantage of floating point is that the relative precision does not depend on the choice of physical units. For instance, when expressed in eV · sec, Planck's constant is $4.135668 \times 10^{-15}$, which still has 7 digits but only 21 decimal places.

It can be easy to confuse the terms *accuracy* and *precision*, especially when looking at the result of a calculation on the computer. The precision of a floating point number is always $d$ binary digits, but not all of those digits may accurately represent an intended value. Suppose $x$ is a number of interest and $\tilde{x}$ is an approximation to it. The **absolute accuracy** of $\tilde{x}$ is $|\tilde{x} - x|$, while the **relative accuracy** is $|\tilde{x} - x|/|x|$. Absolute accuracy has the same units as $x$, while relative accuracy is dimensionless. We can also express the relative accuracy as

$$\text{accurate digits} = -\log_{10}\left|\frac{\tilde{x} - x}{x}\right|. \tag{1.1.5}$$

We often round this down to an integer, but it does make sense to speak of "almost seven digits" or "ten and a half digits."

### Example 1.1.2

Recall the grade school rational approximation to the number $\pi$.

```
p = 22/7

p =
    3.1429
```

Note that not all of the digits displayed for $p$ are the same as for $\pi$. As an approximation, its absolute and relative accuracy are

```
abs_accuracy = abs(p-pi)
rel_accuracy = abs(p-pi)/pi
accurate_digits = -log10(rel_accuracy)

abs_accuracy =
    0.0013
rel_accuracy =
    4.0250e-04
accurate_digits =
    3.3952
```

## Double precision

Most numerical computing today is done in the **IEEE 754 double precision** standard. This standard uses $d = 52$ binary digits for $f$ above, so that

$$\varepsilon_{\text{mach}} = 2^{-52} \approx 2.2 \times 10^{-16}. \tag{1.1.6}$$

In MATLAB, this value is automatically available in the variable **eps**.[4] We often speak of double precision floating point numbers as having about 16 decimal digits. The 52-bit mantissa is paired with a sign bit and 11 binary bits to represent the exponent $e$ in (1.1.1), for a total of 64 binary bits per floating point number.

Our theoretical description of $\mathbb{F}$ did not place limits on the exponent, but in double precision its range is limited to $-1022 \leq e \leq 1023$. Thus, the largest number is just short of $2^{1024} \approx 2 \times 10^{308}$, which is more than enough in most applications. Results that should be larger are said to **overflow** and will actually result in the value Inf. Similarly, the smallest positive number is $2^{-1022} \approx 2 \times 10^{-308}$, and smaller values are said to **underflow** to zero.[5]

Note the crucial difference between $\varepsilon_{\text{mach}} = 2^{-52}$, which is the distance between 1 and the next larger double precision number, and $2^{-1022}$, which is the first double precision number larger than zero. The former has to do with relative precision, while the latter is about absolute precision. Getting close to zero always requires a shift in thinking to absolute precision, because any finite error is infinite relative to zero.

One more double precision value is worthy of note: NaN, which stands for **not a number**. It can be used to hold the place for a missing data value, and it is also the result of an undefined arithmetic operation such as 0/0.

## Floating point arithmetic

Computer arithmetic is performed on floating point numbers and returns floating point results. We assume the existence of machine-analog operations for real functions such as $+, -, \times, /, \sqrt{\phantom{x}}$, and so on. Without getting into the details, we will suppose that each elementary machine operation creates a floating point result whose relative error is bounded by $\varepsilon_{\text{mach}}$. For example, if $x$ and $y$ are in $\mathbb{F}$, then for machine addition $\oplus$ we have the bound

$$\frac{|(x \oplus y) - (x + y)|}{|x + y|} \leq \varepsilon_{\text{mach}}.$$

Hence the relative error in arithmetic is practically the same as for the floating point representation itself. However, even playing by these rules can lead to disturbing results.

---

[4] While MATLAB will blithely let you change the value assigned to this name, doing so has no effect on the precision of calculations.
[5] Actually, there is a special trick to define some still smaller *denormalized* numbers, but we won't use that level of detail.

### Example 1.1.3

There is no double precision number between 1 and $1+\varepsilon_{\text{mach}}$. Thus the following difference is zero despite its appearance.

```
( 1 + eps/2 ) - 1

ans =
     0
```

However, $-(1-\varepsilon_{\text{mach}}/2)$ is a double precision number, so it is represented exactly:

```
1 + ( eps/2 - 1 )

ans =
   1.1102e-16
```

This is now the "correct" result. But we have found a rather shocking breakdown of the associative law of addition!

There are two ways to look at Example 1.1.3. On one hand, its two versions of the result differ by less than $1.2 \times 10^{-16}$, which is very small—not just in everyday terms, but with respect to the operands, which are all close to 1 in absolute value. On the other hand, the difference is as large as the exact result itself! We formalize and generalize this observation in the next section. In the meantime, keep in mind that exactness cannot be taken for granted in floating point computation. Even ideally, we should not expect that two mathematically equivalent results will be equal, only that they be relatively close together.

*Exercises marked with ✎ are intended to be done by hand or with the aid of a simple calculator. Exercises marked with ⌨ are intended to be solved by using a computer.*

## Exercises

1.1.1. ✎ Consider a floating point set $\mathbb{F}$ defined by (1.1.1) and (1.1.2) with $d=4$.

   (a) How many elements of $\mathbb{F}$ are there in the real interval $[1/2, 4]$, including the endpoints?

   (b) What is the element of $\mathbb{F}$ closest to the real number $1/10$?

   (c) What is the smallest positive integer not in $\mathbb{F}$?

1.1.2. ✎ Prove that (1.1.3) is equivalent to (1.1.4). (This means showing first that (1.1.3) implies (1.1.4), and then separately that (1.1.4) implies (1.1.3).)

1.1.3. ⌨ There are much better rational approximations to $\pi$ than 22/7. For each one below, find its absolute and relative accuracy and (rounding down to an integer) the number of accurate digits. (In MATLAB, the variable `pi` is set to a 16-digit approximation to $\pi$ by default.)

(a) 355/113, (b) 103638/32989.

1.1.4. ⚠ IEEE 754 *single precision* specifies that 23 binary bits are used for the mantissa $f$ in (1.1.2). Because they can be accessed and operated on more quickly than double precision values, single precision values can be useful in time-sensitive computations not requiring a great deal of accuracy.

  (a) In base-10 terms, what is the first single precision number greater than 1 in this system?

  (b) What is the smallest positive integer that is not a single precision number?

1.1.5. ▣ In MATLAB, you can enter `format hex` to cause all subsequent results to be printed out in hexadecimal notation. Each group of four binary bits is equivalent to a decimal integer from 0 to 15, where the integers greater than 9 are represented as the letters a, b, ..., f. Because a double precision value has 64 bits, it is represented by 16 hexadecimal digits.

  (a) Find the hexadecimal representations of the values 1, 1+eps, 1+15*eps, and 1+16*eps. (Recall that `eps` in MATLAB is always set by default to be $\varepsilon_{\text{mach}}$ for double precision.)

  (b) Find a double precision number whose last four hexadecimal digits are the word "face."

1.1.6. ▣ It's reasonable to expect that floating point errors accumulate randomly during a long computation, creating what is known as a *random walk*. On average we expect as many errors to be negative as positive, so they tend to partially cancel out. Suppose we define a random sequence by $x_0 = 0$ and $x_n = x_{n-1} \pm 1$ for $n \geq 1$, with the signs chosen by tossing a fair coin for each $n$. Let $\alpha_n$ and $\beta_n$ be the average values of $x_n$ and $|x_n|$, respectively, over all such walks. Then a classic result of probability is that $\alpha_n \to 0$ and

$$\lim_{n \to \infty} \frac{\pi \beta_n^2}{2n} = 1.$$

  (a) In MATLAB the function `randn` simulates drawing numbers from the normal or Gaussian distribution (i.e., the bell curve) with mean zero and variance 1. Choose a unique positive integer seed value $s$ (for example, use the last 5 digits of your phone number) and enter `rng(s)` to initialize the random number generator. Then the following code generates one random walk for $n = 10^4$:

```
r = randn(10000,1);     % draw random numbers
x = cumsum(sign(r));    % cumulative summation
```

  We can plot every fifth number in the sequence using

```
plot(x(1:5:end)),  hold on
```

  Plot 50 such random walks all together on one graph.

  (b) Perform a million random walks, computing the average values of $x_{10000}$ and $|x_{10000}|$. Compare these to $\alpha_n \approx 0$ and $\beta_n \approx \sqrt{2n/\pi}$ at $n = 10000$.

## 1.2 • Problems and conditioning

Let's think a bit about what must be the easiest math problem you've dealt with in quite some time: adding 1 to a number. Formally, we describe this problem as a function $f(x) = x + 1$, where $x$ is any real number. On a computer, $x$ will be represented by its floating point counterpart, fl($x$). Given the property (1.1.3), we have fl($x$) = $x(1+\epsilon)$ for some $\epsilon$ satisfying $|\epsilon| < \varepsilon_{\text{mach}}/2$. There is no error in representing the value 1. Let's suppose that we are fortunate and that the addition proceeds exactly, with no additional errors. Then the machine result is just

$$y = x(1+\epsilon) + 1. \tag{1.2.1}$$

We can compute (in exact arithmetic) the relative error in this result:

$$\frac{|y - f(x)|}{|f(x)|} = \frac{|(x + \epsilon x + 1) - (x + 1)|}{|x + 1|} = \frac{|\epsilon x|}{|x + 1|}. \tag{1.2.2}$$

This error could be quite large if the denominator is small. In fact, we can make the relative error as large as we please by taking $x$ very close to $-1$. This is essentially what happened in Example 1.1.3.

You may have encountered this situation before when using significant digits for scientific calculations. Suppose we round all results to five decimal digits, and we add $-1.0012$ to $1.0000$. The result is $-0.0012$, or $-1.2 \times 10^{-3}$ in scientific notation. Notice that even though both operands are specified to five digits, it makes no sense to write more than two digits in the answer, because there is no information in the problem beyond their decimal places. This phenomenon is known as **subtractive cancellation**, or *loss of significance*. We may say that three digits were "lost" in the mapping from $-1.0012$ to $-0.0012$. There's no way the loss could be avoided, *regardless of the algorithm*, once we decided to round off everything to a fixed number of digits.

In double precision, all of the values are represented to about 16 significant decimal digits, but it's understood that subtractive cancellation may render some of those digits essentially meaningless. *Subtractive cancellation is one of the most common mechanisms introducing dramatic growth of errors in floating point computation.*

### Condition numbers

Now we consider problems more generally. As above, we represent a problem as a function $f$ that maps a real data value $x$ to a real result $f(x)$. We abbreviate this situation by the notation $f : \mathbb{R} \mapsto \mathbb{R}$, where $\mathbb{R}$ represents the real number set.

When the problem $f$ is approximated in floating point on a computer, the data $x$ is represented as a floating point value $\tilde{x} = \text{fl}(x)$. Ignoring all other sources of error, we define the quantitative measure

$$\frac{\dfrac{|f(x) - f(\tilde{x})|}{|f(x)|}}{\dfrac{|x - \tilde{x}|}{|x|}}, \tag{1.2.3}$$

which is the ratio of the relative changes in result and data. We make this expression more convenient if we recall that floating point arithmetic gives $\tilde{x} = x(1+\epsilon)$ for some value $|\epsilon| \le \varepsilon_{\text{mach}}/2$:

$$\frac{|f(x)-f(x(1+\epsilon))|}{|\epsilon f(x)|}. \tag{1.2.4}$$

Finally, we imagine what happens in the ideal case of a perfect computer by taking a limit as $\varepsilon_{\text{mach}} \to 0$:

$$\kappa_f(x) = \lim_{\epsilon \to 0} \frac{|f(x)-f(x(1+\epsilon))|}{|\epsilon f(x)|}. \tag{1.2.5}$$

This quantity, which we call *the relative **condition number** of the problem $f(x)$, is an idealized ratio of the relative error of the output to the relative error of the input.* It depends only on the problem and the data, not the computer or the algorithm.

Assuming that $f$ has at least one continuous derivative, we can simplify the expression (1.2.5) through some straightforward manipulations:

$$\begin{aligned}
\kappa_f(x) &= \lim_{\epsilon \to 0} \left| \frac{f(x+\epsilon x)-f(x)}{\epsilon f(x)} \right| \\
&= \lim_{\epsilon \to 0} \left| \frac{f(x+\epsilon x)-f(x)}{\epsilon x} \cdot \frac{x}{f(x)} \right| \\
&= \left| \frac{xf'(x)}{f(x)} \right|.
\end{aligned} \tag{1.2.6}$$

In retrospect, it should come as no surprise that the change in values of $f(x)$ due to small changes in $x$ involves the derivative of $f$. In fact, if we were making measurements of changes in absolute rather than relative terms, the condition number would simply be $|f'(x)|$.

---

**Example 1.2.1**

Let's return to our "add one" problem and generalize it slightly to $f(x) = x - c$ for constant $c$ (previously, we had $c = -1$). We easily compute, using (1.2.6),

$$\kappa_f(x) = \left| \frac{(x)(1)}{x-c} \right| = \left| \frac{x}{x-c} \right|. \tag{1.2.7}$$

The result is simply the relative change (1.2.2) normalized by the size of the perturbation $\epsilon$. The condition number is large when $|x| \gg |x-c|$. There is of course no meaningful difference between addition and subtraction over real numbers. Furthermore, the situation is symmetric in $x$ and $c$; that is, if we perturbed $c$ and not $x$, the result would be $|c|/|x-c|$. In words, *cancellation error occurs whenever the result of addition or subtraction is much smaller in magnitude than the operands.*

## 1.2. Problems and conditioning

**Example 1.2.2**

Another elementary operation is to multiply by a constant: $f(x) = cx$ for nonzero $c$. We compute

$$\kappa_f(x) = \left|\frac{xf'(x)}{f(x)}\right| = \left|\frac{(x)(c)}{cx}\right| = 1. \tag{1.2.8}$$

We conclude that multiplication by a real number leads to the same relative error in the result as in the data. In other words, multiplication does not have the potential for cancellation error that addition does.

Condition numbers of elementary functions are given in Table 1.1. As you are asked to show in Exercise 1.2.4, when two functions $f$ and $g$ are combined in a chain $h(x) = f(g(x))$, the composite condition number is

$$\kappa_h(x) = \kappa_f(g(x)) \cdot \kappa_g(x). \tag{1.2.9}$$

Roughly speaking, if $|\epsilon|$ is small, we expect (refer to (1.2.5))

$$\left|\frac{f(x+\epsilon x) - f(x)}{f(x)}\right| \approx \kappa_f(x)|\epsilon|.$$

That is, whenever the data $x$ is perturbed by a small amount, we expect the relative change to be magnified by a factor of $\kappa_f(x)$ in the result. *Large condition numbers signal when errors cannot be expected to remain comparable in size to the roundoff error.* We call a problem poorly conditioned or *ill-conditioned* when $\kappa_f(x)$ is large. If $\kappa_f \approx 10^d$, then we expect to "lose" up to $d$ decimal digits of accuracy in computing $f(x)$ from $x$. If $\kappa_f \approx 1/\varepsilon_{\text{mach}}$, then we can expect the result to be changed by as much as 100% simply by expressing $x$ in finite precision.

**Example 1.2.3**

Consider the problem $f(x) = \cos(x)$. Table 1.1 says that $\kappa_f(x) = |x \tan x|$. There are two different ways in which $\kappa$ might become large:

1. If $|x|$ is very large, then perturbations that are small relative to $x$ may still be large compared to 1. Because $|f(x)| \leq 1$ for all $x$, this implies that the perturbation will be large relative to the result, too.
2. The condition number grows without bound as $x$ approaches an odd integer multiple of $\pi/2$, where $f(x) = 0$. A perturbation that is small relative to a nonzero $x$ may not be small relative to $f(x)$ in such a case.

You may have noticed that for some functions, such as the square root, the condition number can be less than one. This means that relative changes get *smaller* in the passage from input to output. However, every result in floating point arithmetic is still subject to rounding error at the relative level of $\varepsilon_{\text{mach}}$, so $\kappa < 1$ is no better than $\kappa = 1$ in context.

**Table 1.1.** *Condition numbers of elementary functions. Note that here and throughout the book, "log" denotes the natural logarithm.*

| Function | Condition number |
|---|---|
| $f(x) = x + c$ | $\kappa_f(x) = \frac{|x|}{|x+c|}$ |
| $f(x) = cx$ | $\kappa_f(x) = 1$ |
| $f(x) = x^p$ | $\kappa_f(x) = |p|$ |
| $f(x) = e^x$ | $\kappa_f(x) = |x|$ |
| $f(x) = \sin(x)$ | $\kappa_f(x) = |x \cot(x)|$ |
| $f(x) = \cos(x)$ | $\kappa_f(x) = |x \tan(x)|$ |
| $f(x) = \log(x)$ | $\kappa_f(x) = 1/|\log(x)|$ |

## Multidimensional problems

Most problems have multiple input and output values. These introduce some complications into the formal definition of the condition number. Rather than worry over those details here, we can still look at variations in only one output and one input value at a time.

**Example 1.2.4**

Consider the problem of finding the roots of a quadratic polynomial, that is, the values of $t$ for which $at^2 + bt + c = 0$. Here the data are the coefficients $a$, $b$, and $c$ that define the polynomial, and the solution to the problem is the two (maybe complex-valued) roots $t_1$ and $t_2$. Formally, we might write $f([a,b,c]) = [t_1, t_2]$ using vector notation.

In order to simplify matters, we will pick one root called $r$, and first consider what happens as we vary just the leading coefficient $a$. This suggests a scalar function $f(a) = r$. We could use the quadratic formula to express $f$ explicitly, but it's a bit easier to start from $ar^2 + br + c = 0$ and use the technique of implicit differentiation to find $dr/da$, while $b$ and $c$ are held fixed. Taking $d/da$ of both sides and applying the chain rule, we get

$$r^2 + 2ar\left(\frac{dr}{da}\right) + b\frac{dr}{da} = 0.$$

Solving for the derivative, we obtain

$$\frac{dr}{da} = \frac{-r^2}{2ar+b} = \frac{-r^2}{\pm\sqrt{b^2-4ac}}, \qquad (1.2.10)$$

where in the last step we have applied the quadratic formula for the root $r$. Finally, the condition number for the problem $f(a) = r$ is

$$\kappa(a) = \left|\frac{a}{r} \cdot \frac{dr}{da}\right| = \left|\frac{ar}{\sqrt{b^2-4ac}}\right| = \left|\frac{r}{t_1 - t_2}\right|, \qquad (1.2.11)$$

where we used the fact that the two roots of the original polynomial satisfy $|t_1 - t_2| = |\sqrt{b^2-4ac}/a|$. Thus, we can expect poor conditioning in the rootfinding problem if and only if $|r| \gg |t_1 - t_2|$, i.e., if the two roots of the quadratic are much closer to each other than to the origin. Similar conclusions apply for variations with respect to the coefficients $b$ and $c$ while the others are held fixed.

Example 1.2.4 shows that the condition number of a root of a quadratic polynomial, as a function of its leading coefficient, can be arbitrarily large. In the extreme case of a double root, the condition number is formally infinite, which implies that the ratio of changes in the root to change in the coefficient $a$ cannot be bounded. While we usually think about these sensitivities in terms of numerical roundoff error, they apply to all sources of error, including measurement error or model uncertainty.

## Exercises

1.2.1. 🖋 Use (1.2.6) to derive the relative condition numbers of the following functions appearing in Table 1.1.
  (a) $f(x) = x^p$.  (b) $f(x) = \log(x)$.  (c) $f(x) = \cos(x)$.  (d) $f(x) = e^x$.

1.2.2. 🖋 Use the chain rule (1.2.9) to find the condition number of the given function. Then check your result by applying (1.2.6) directly.
  (a) $f(x) = \sqrt{x+5}$.  (b) $f(x) = \cos(2\pi x)$.  (c) $f(x) = e^{-x^2}$.

1.2.3. 🖋 Calculate the condition number of each function, and identify all values of $x$ at which $\kappa_f(x) \to \infty$ (including limits as $x \to \pm\infty$).
  (a) $f(x) = \tanh(x)$.  (b) $f(x) = \frac{e^x - 1}{x}$.  (c) $f(x) = \frac{1 - \cos(x)}{x}$.

1.2.4. 🖋 Suppose that $f$ and $g$ are real-valued functions that have condition numbers $\kappa_f$ and $\kappa_g$, respectively. Define a new function $h(x) = f(g(x))$. Show that for $x$ in the domain of $h$, the condition number of $h$ satisfies (1.2.9).

1.2.5. 🖋 Suppose that $f$ is a function with condition number $\kappa_f$, and that $f^{-1}$ is its inverse function. Show that the condition number of $f^{-1}$ satisfies

$$\kappa_{f^{-1}}(x) = \frac{1}{\kappa_f(f^{-1}(x))},$$

provided the denominator is nonzero.

1.2.6. 🖋 Referring to Example 1.2.4, derive an expression for the relative condition number of a root of $ax^2 + bx + c = 0$ due to perturbations in $b$ only.

1.2.7. The polynomial $x^2 - 2x + 1$ has a double root $r = 1$.
  (a) 🖋 Using a computer or calculator, make a table of the roots of $x^2 + (2 + \epsilon)x + 1$ for $\epsilon = 10^{-4}, 10^{-6}, \ldots, 10^{-12}$.
  (b) 🖋 What do the results of part (a) seem to imply about the condition number of the root?

1.2.8. 🖋 Generalize Example 1.2.4 to finding a root of the $n$th-degree polynomial $p(x) = a_n x^n + \cdots + a_1 x + a_0$, and show that the relative condition number of a root $r$ with respect to perturbations only in $a_k$ is

$$\kappa_r(a_k) = \left| \frac{a_k r^{k-1}}{p'(r)} \right|.$$

## 1.3 • Stability of algorithms

When we idealize a problem as the mathematical function $f(x)$, we are explicitly stating that each input (data) has exactly one correct output. When it comes to implementing a computational version of $f$, though, we usually have some choices to make. A complete set of instructions for mapping data to a result is called an **algorithm**. In most cases it is reasonable to represent each algorithm by another function, which for this section we denote $\tilde{f}(x)$.

Even simple problems can be associated with algorithms that have surprisingly different characteristics.

> **Example 1.3.1**
>
> Suppose we want to find an algorithm that maps a given $x$ to the value of the polynomial $f(x) = 5x^3 + 4x^2 + 3x + 2$. Representing $x^2$ as $(x)(x)$, we can find it with one multiplication. We can then find $x^3 = (x)(x^2)$ with one more multiplication. We can then apply all the coefficients (three more multiplications) and add all the terms (three additions), for a total of 8 arithmetic operations.
>
> There is a more efficient algorithm, however: organize the polynomial according to **Horner's rule**,
> $$f(x) = 2 + x\big(3 + x(4 + 5x)\big).$$
> In this form you can see that evaluation takes only 3 additions and 3 multiplications. The savings represent 25% of the original computational effort, which could be significant if repeated billions of times.

Descriptions of algorithms vary widely. Sometimes they are presented as a mixture of mathematics, words, and computer-style instructions called *pseudocode*, which varies in syntax, precision, and formality. In this book we present nontrivial algorithms as MATLAB functions.[6] As a result, we sometimes sacrifice a little readability and brevity, and in some cases we have to address aspects of the algorithm that are peripheral to our mathematical interests. We think these drawbacks are more than outweighed by the lack of ambiguity and the ability to execute the algorithms yourself—not to mention the opportunity to address some issues that are crucial even if not very mathematical.

Of all the desirable traits of code, we emphasize clarity the most. We do not represent our programs as being the shortest, fastest, most elegant, or most conforming to MATLAB style. Our primary goal is to illustrate and complement the mathematical underpinnings. Hopefully the language and codes are clear enough that if you would rather use a different computing environment to implement them, you will experience few difficulties. We do recommend an *interactive* environment, however, because being able to tweak inputs and pick apart the results is crucial to understanding.

---

[6]To get started with MATLAB as a first-time user, search for "matlab academy" online.

## 1.3. Stability of algorithms

**Function 1.3.1 (horner)** Evaluate a polynomial by Horner's rule.

```
function y = horner(c,x)
% HORNER    Evaluate a polynomial using Horner's rule.
% Input:
%   c       Coefficients of polynomial, in descending order (vector)
%   x       Evaluation point (scalar)
% Output:
%   y       Value of the polynomial at x (scalar)

n = length(c);
y = c(1);
for k = 2:n
    y = x*y + c(k);
end
```

As our first example, Function 1.3.1 implements an algorithm that applies Horner's rule on a general polynomial, through the identity

$$p(x) = c_1 x^{n-1} + c_2 x^{n-2} + \cdots + c_n$$
$$= \Big(\big((c_1 x + c_2)x + c_3\big)x + \cdots + c_{n-1}\Big)x + c_n. \quad (1.3.1)$$

In MATLAB, it is conventional to represent a polynomial by a vector of its coefficients in descending order of degree. As you can see from Function 1.3.1, MATLAB deals with vectors very naturally. You can query the vector to find out how many entries it has, and you can access the $k$th element of a vector $c$ by typing $c(k)$. The nested evaluation in (1.3.1) is accomplished through a **for** loop construction.

**Example 1.3.2**

Here we evaluate the polynomial $p(x) = (x-1)^3 = x^3 - 3x^2 + 3x - 1$. First, we define a vector of its coefficients in descending degree order.

```
c = [ 1 -3 3 -1];
```

Here is the evaluation of $p(1.6)$.

```
horner(c,1.6)
```

```
ans =
    0.2160
```

MATLAB has a built-in function **polyval** that does essentially the same thing as Function 1.3.1. This will often be the case for codes in this book, because the problems we study are classic and important. In a more practical setting you would take implementations of well-known methods for granted and build on top of them.

## Stability

If we solve a problem using a computer algorithm and see a large error in the result, we might suspect poor conditioning in the original mathematical problem. But algorithms can also be sources of errors. *When error in the result of an algorithm exceeds what conditioning can explain, we call the algorithm* **unstable**.

Recall from Example 1.2.4 that finding the roots of a quadratic polynomial $ax^2+bx+c$ is poorly conditioned if and only if the roots are close to each other relative to their size. Thus, for the polynomial

$$p(x) = (x-10^6)(x-10^{-6}) = x^2 - (10^6 + 10^{-6})x + 1, \qquad (1.3.2)$$

finding roots is a very well conditioned problem. An obvious algorithm for finding those roots is to directly apply the quadratic formula:

$$x_1 = \frac{-b+\sqrt{b^2-4ac}}{2a}, \qquad x_2 = \frac{-b-\sqrt{b^2-4ac}}{2a}. \qquad (1.3.3)$$

### Example 1.3.3

We apply the quadratic formula to find the roots of (1.3.2).

```
format long     % show all the digits
a = 1;  b = -(1e6+1e-6);  c = 1;

x1 = (-b + sqrt(b^2-4*a*c)) / (2*a)

x1 =
       1000000

x2 = (-b - sqrt(b^2-4*a*c)) / (2*a)

x2 =
     1.000007614493370e-06
```

The first value is correct to all stored digits, but the second has fewer than six accurate digits:

```
-log10( abs(1e-6-x2)/1e-6 )

ans =
   5.118358987126217
```

The instability is easily explained. Since $a = c = 1$, we treat them as exact numbers. First, we compute the condition numbers with respect to $b$ for each elementary step in finding the "good" root:

## 1.3. Stability of algorithms

| Calculation | Result | $\kappa$ |
|---|---|---|
| $u_1 = b^2$ | $1.000000000002000 \times 10^{12}$ | 2 |
| $u_2 = u_1 - 4$ | $9.999999999980000 \times 10^{11}$ | $\lvert u_1 \rvert / \lvert u_2 \rvert \approx 1.00$ |
| $u_3 = \sqrt{u_2}$ | $999999.9999990000$ | $1/2$ |
| $u_4 = u_3 - b$ | $2000000$ | $\lvert u_3 \rvert / \lvert u_4 \rvert \approx 0.500$ |
| $u_5 = u_4 / 2$ | $1000000$ | 1 |

Using (1.2.8), the chain rule for condition numbers, the conditioning of the entire chain is the product of the individual steps, so there is essentially no growth of relative error here. However, if we use the quadratic formula for the "bad" root, the next-to-last step becomes $u_4 = (-u_3) - b$, and now $\kappa = \lvert u_3 \rvert / \lvert u_4 \rvert \approx 5 \times 10^{11}$. So we can expect to lose 11 digits of accuracy, which is what we observed. The key issue is the subtractive cancellation in this one step.

Example 1.3.3 suggests that the venerable quadratic formula is an *unstable* means of computing roots in finite precision. The roots themselves were not sensitive to the data or arithmetic—it's the specific computational path we chose that caused the huge growth in errors. We can confirm this conclusion by finding a different path that avoids subtractive cancellation. A little algebra using (1.3.3) confirms the additional formula $x_1 x_2 = c/a$. So given one root $r$, we compute the other root using $c/(ar)$, which has only multiplication and division and therefore creates no numerical trouble.

### Example 1.3.4

We repeat the rootfinding experiment of Example 1.3.3 with an alternative algorithm.

```
format long    % show all the digits
a = 1;   b = -(1e6+1e-6);   c = 1;
```

First, we find the "good" root using the quadratic formula.

```
x1 = (-b + sqrt(b^2-4*a*c)) / (2*a);
```

Then we use the better formula for computing the other root.

```
x2 = c/(a*x1)

x2 =
      1.000000000000000e-06
```

Both algorithms for calculating roots are equivalent when using real numbers and exact arithmetic. When perturbations are introduced into each intermediate value, though, their effects may depend dramatically on the specific sequence of steps taken. The sensitivity of the target problem $f(x)$ is governed only by $\kappa_f$, but *the sensitivity of an algorithm depends on the condition numbers of all of its steps*. In this example, the direct application of the quadratic formula included a poorly conditioned step—subtractive cancellation—that can be avoided.

This situation may sound hopelessly complicated. But, as you can see in Table 1.1, the elementary operations we take for granted are quite well conditioned in most circumstances. The glaring exceptions occur when $|f(x)|$ is much smaller than $|x|$, which is not inherently hard to detect; however, not all such situations create sensitivity. A practical characterization of instability is that results are much less accurate than the conditioning of the problem suggests. Typically one should apply an algorithm to test problems whose answers are well known, or for which other programs are known to work well, in order to spot likely instabilities. In the rest of this book we will see some specific ways in which instability is manifested for different types of problems.

## Backward error

In the presence of poor conditioning, even a good algorithm $\tilde{f}$ for a problem $f$ may not have a small error $|\tilde{f}(x)-f(x)|/|f(x)|$. Just the act of rounding the data to floating point may introduce a large change in the result. There is another way to characterize the error that can be a useful alternative measurement, as illustrated in Figure 1.1.

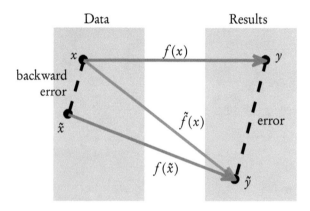

Figure 1.1. *Backward error.*

Let $\tilde{y}=\tilde{f}(x)$ be a computed result for the original data $x$. If there is a value $\tilde{x}$ such that

$$f(\tilde{x})=\tilde{y}=\tilde{f}(x), \qquad (1.3.4)$$

then we call $|\tilde{x}-x|/|x|$ the (relative) **backward error** of the result. Instead of asking, "How close to the true answer is your answer?", backward error asks, "How close to the true question is the question you answered?"

Let's have yet another look at the case of polynomial rootfinding. MATLAB has two functions related to polynomial roots: **roots**, which finds the roots of a polynomial given a vector of its coefficients (in decreasing degree order, as always), and **poly**, which does the opposite, returning the coefficients of a monic polynomial (i.e., having leading coefficient equal to one) given a vector of its roots.

## 1.3. Stability of algorithms

**Example 1.3.5**

We first construct a polynomial of degree six with known integer roots.

```
r = [-2 -1 1 1 3 6]';    % column vector of exact roots
p = poly(r)              % polynomial having those roots

p =
    1    -8     6    44   -43   -36    36
```

Now we use MATLAB's built-in function `roots` for finding them.

```
r_computed = sort( roots(p) )    % algorithmically computed
    roots

r_computed =
   -2.0000
   -1.0000
    1.0000
    1.0000
    3.0000
    6.0000
```

These are the relative errors in each computed root.

```
format short e
abs(r - r_computed) ./ r

ans =
  -6.6613e-16
  -9.9920e-16
   9.7221e-09
   9.7221e-09
   1.4803e-16
   1.3323e-15
```

It seems that the forward error is acceptably small in all cases except the double root at $x = 1$. This is not a surprise, though, given the poor conditioning at such roots.

Let's consider the backward error. The data in the rootfinding problem are the polynomial coefficients. We can apply `poly` to find the coefficients of the polynomial (that is, the data) whose roots were actually computed.

```
p_computed = poly(r_computed)'

p_computed =
   1.0000e+00
  -8.0000e+00
   6.0000e+00
   4.4000e+01
  -4.3000e+01
```

```
    -3.6000e+01
     3.6000e+01
```

We find that in a relative sense, these coefficients are very close to those of the original, exact polynomial:

```
(p_computed - p') ./ p'

ans =
            0
  -1.1102e-15
  -4.4409e-15
  -4.8446e-16
  -6.6097e-16
  -2.3685e-15
  -1.9737e-15
```

In summary, even though there are some computed roots relatively far from their correct values, they are nevertheless the roots of a polynomial that is very close to the original.

Small backward error is the best we can hope for in a poorly conditioned problem. Without getting into the formal details, know that *if an algorithm always produces small backward errors, then it is stable.* But the converse is not always true: some stable algorithms may produce a large backward error.

### Example 1.3.6

One stable algorithm that is not backward stable is floating point evaluation for our old standby, $f(x) = x + 1$. If $|x| < \varepsilon_{mach}/2$, then the computed result is $\tilde{f}(x) = 1$, since there are no floating point numbers between 1 and $1 + \varepsilon_{mach}$. Hence the only possible choice for a real number $\tilde{x}$ satisfying (1.3.4) is $\tilde{x} = 0$. But then $|\tilde{x} - x|/|x| = 1$, which indicates 100% backward error!

## Exercises

1.3.1. ▬ Write a MATLAB function

```
function r = polyadd(p,q)
```

that returns the coefficient vector $r$ for the sum of two polynomials $p$ and $q$, specified as vectors of coefficients in decreasing degree. You should *not* assume that $p$ and $q$ are given by vectors of the same length.

1.3.2. ▬ In statistics, one defines the variance of sample values $x_1, \ldots, x_n$ by

$$s^2 = \frac{1}{n-1}\sum_{i=1}^n (x_i - \overline{x})^2, \qquad \overline{x} = \frac{1}{n}\sum_{i=1}^n x_i. \qquad (1.3.5)$$

Write a MATLAB function

```
function s2 = samplevar(x)
```

that takes as input a vector x of any length and returns $s^2$ as calculated by the formula. You should test your function on some data and compare it to MATLAB's built-in var for some data.

1.3.3. ▣ Let x and y be vectors whose entries give the coordinates of the vertices of a polygon, given in counterclockwise order. Write a function

```
function A = polyarea(x,y)
```

that computes the area of the polygon, using this formula based on Green's theorem:

$$A = \frac{1}{2}\left|\sum_{k=1}^{n} x_k y_{k+1} - x_{k+1} y_k\right|.$$

Here $n$ is the number of polygon vertices, and by definition $x_{n+1} = x_1$ and $y_{n+1} = y_1$. Test your functions on a square and an equilateral triangle.

1.3.4. (a) ✎ Find the relative condition number for $f(x) = (1 - \cos x)/\sin x$.

(b) ✎ Explain carefully how many digits will be lost to cancellation when computing $f$ directly by the formula in (a) for $x = 10^{-6}$.

(c) ✎ Show that the mathematically identical formula

$$f(x) = \frac{2\sin^2(x/2)}{\sin(x)}$$

contains no poorly conditioned steps for $|x| < 1$.

(d) ▣ Compute and compare the formulas from (a) and (c) numerically at $x = 10^{-6}$.

1.3.5. Let $f(x) = (e^x - 1)/x$.

(a) ✎ Find the condition number $\kappa_f(x)$. What is the maximum of $\kappa_f(x)$ over $-1 \le x \le 1$?

(b) ▣ Use the "obvious" algorithm

```
y = (exp(x)-1) / x;
```

to compute $f(x)$ at 1000 evenly spaced points in the interval $[-1, 1]$.

(c) Use the first 18 terms of the Taylor series

$$f(x) = 1 + \frac{1}{2!}x + \frac{1}{3!}x^2 + \frac{1}{4!}x^3 + \cdots$$

to create a second algorithm, and evaluate it at the same set of points.

(d) ▣ Plot the relative difference between the two algorithms as a function of $x$. Which one do you believe is more accurate, and why?

1.3.6. ▣ The function

$$x = \cosh(t) = \frac{e^t + e^{-t}}{2}$$

can be inverted to yield a formula for acosh(x):

$$t = \log(x - \sqrt{x^2 - 1}). \qquad (1.3.6)$$

In MATLAB, let t=-4:-4:-16 and x=cosh(t).

(a) Find the condition number of the problem $f(x) = \operatorname{acosh}(x)$. (You can use (1.3.6)), or look up a formula for $f'$ in a calculus book.) Evaluate $\kappa_f$ at the entries of **x** in MATLAB. Would you consider the problem well-conditioned at these inputs?

(b) Use (1.3.6) on **x** to approximate **t**. Record the accuracy of the answers, and explain. (Warning: You should use `format long` to get the true picture.)

(c) An alternate formula is

$$t = -2\log\left(\sqrt{\frac{x+1}{2}} + \sqrt{\frac{x-1}{2}}\right). \qquad (1.3.7)$$

Apply (1.3.7) to **x** as before, and comment on the accuracy.

(d) Based on your experiments, which of the formulas (1.3.6) and (1.3.7) is unstable? What is the problem with that formula?[7]

1.3.7. ▣ (Continuation of problem Exercise 1.3.2. Adapted from [33].) One problem with the formula (1.3.5) for sample variance is that one computes a sum for $\bar{x}$, then another sum to find $s^2$. Some statistics textbooks quote a "one-pass" formula,

$$s^2 = \frac{1}{n-1}\left(u - \frac{1}{n}v^2\right)$$

$$u = \sum_{i=1}^{n} x_i^2$$

$$v = \sum_{i=1}^{n} x_i.$$

"One-pass" means that both $u$ and $v$ can be computed in a single loop.[8] Try this formula for the two data sets

```
x = [ 1e6, 1+1e6, 2+1e6 ],    x = [ 1e9, 1+1e9, 2+1e9 ],
```

compare this to using `var` in each case, and explain the results.

## Key ideas in this chapter

1. The smallest floating point number greater than 1 is $1 + \varepsilon_{\text{mach}}$, where $\varepsilon_{\text{mach}}$ is machine epsilon (page 10).
2. Cancellation error is one of the most common sources of loss of precision (page 15).
3. The condition number measures the ratio of error in the result to error in the data (page 16).
4. A large condition number implies that the error in a result may be much greater than the roundoff error used to compute it (page 17).
5. When an algorithm produces much more error than can be explained by the condition number, the algorithm is unstable (page 22).

---

[7] According to a MathWorks newsletter, for a long time MATLAB used the *unstable* formula.
[8] Loops can be avoided altogether in MATLAB by using the `sum` command.

6. An avoidable ill-conditioned step in an algorithm is a common source of instability (page 23).
7. Small backward error implies stability (page 26).

## Where to learn more

An accessible but more advanced discussion of machine arithmetic and roundoff error can be found in Higham [33].

Interesting and more advanced discussion of the numerical difficulties of finding the roots of polynomials can be found in the article by Wilkinson, "The Perfidious Polynomial" [77], and in the ripostes from Cohen, "Is the Polynomial So Perfidious?" [18] and from Trefethen, "Six myths of polynomial interpolation and quadrature," (Myth 6) [70].

# Chapter 2
# Square linear systems

> It's all a lot of simple tricks and nonsense.
> —Han Solo, *Star Wars: A New Hope*

One of the most frequently encountered tasks in scientific computation is the solution of the linear system of equations $Ax = b$ for a given square matrix $A$ and vector $b$. This problem can be solved in a finite number of steps, using an algorithm equivalent to Gaussian elimination. Describing the algorithm is mostly an exercise in organizing some linear algebra.

Analyzing the algorithm requires new tools. Because the computations will take place in floating point, we must first discuss a system for measuring the "size" of a perturbation to a vector or matrix data. Once that is understood, we find that the conditioning of the square linear system problem is quite straightforward to describe. Finally, we will see that the algorithm may change when certain things are known about the matrix $A$.

## 2.1 • Polynomial interpolation

The United States carries out a census of its population every 10 years. Suppose we want to know the population at times in between the census years, or to estimate future populations.[9] One technique is to use **interpolation**. Interpolation is the process of constructing a mathematical function of a continuous variable that passes through a given set of points. Once an interpolating function (interpolant) is found, it can be evaluated to estimate or predict values. By definition, the interpolant "predicts" the correct values at all of the known data points.

---

[9] We're quite certain that the U.S. Census Bureau uses more sophisticated modeling techniques than the one we present here!

Polynomials are one of the most popular and useful function types for interpolation. Suppose data $(t_i, y_i)$ are given for $i = 1, \ldots, n$. If we assume that $n$ values should be enough to uniquely determine $n$ unknown coefficients, then we can use a polynomial of degree $n-1$ to interpolate the data. So we assume interpolation by $y = f(t)$, where

$$f(t) = c_1 + c_2 t + c_3 t^2 + \cdots + c_n t^{n-1}. \tag{2.1.1}$$

We determine the coefficients $c_1 \ldots, c_n$ by setting $y_i = f(t_i)$ for all the values of $i$. Writing out these equations gives

$$\begin{aligned}
c_1 + c_2 t_1 + \cdots + c_{n-1} t_1^{n-2} + c_n t_1^{n-1} &= y_1, \\
c_1 + c_2 t_2 + \cdots + c_{n-1} t_2^{n-2} + c_n t_2^{n-1} &= y_2, \\
c_1 + c_2 t_3 + \cdots + c_{n-1} t_3^{n-2} + c_n t_3^{n-1} &= y_3, \\
&\vdots \\
c_1 + c_2 t_n + \cdots + c_{n-1} t_n^{n-2} + c_n t_n^{n-1} &= y_n.
\end{aligned}$$

These equations are not linear in the $t_i$. However, they do form a linear system for the coefficients $c_i$:

$$\begin{bmatrix} 1 & t_1 & \cdots & t_1^{n-2} & t_1^{n-1} \\ 1 & t_2 & \cdots & t_2^{n-2} & t_2^{n-1} \\ 1 & t_3 & \cdots & t_3^{n-2} & t_3^{n-1} \\ \vdots & \vdots & & \vdots & \vdots \\ 1 & t_n & \cdots & t_n^{n-2} & t_n^{n-1} \end{bmatrix} \begin{bmatrix} c_1 \\ c_2 \\ c_3 \\ \vdots \\ c_n \end{bmatrix} = \begin{bmatrix} y_1 \\ y_2 \\ y_3 \\ \vdots \\ y_n \end{bmatrix}, \tag{2.1.2}$$

or, more simply, $Vc = y$. The matrix $V$ is of a special type known as a **Vandermonde matrix**. *Polynomial interpolation can be posed as a linear system of equations with a Vandermonde matrix.*

---

**Example 2.1.1**

We create two vectors for data about the population of China. The first has the years of census data, the other has the numbers of millions of people.

```
year = (1980:10:2010)'
pop = [984.736; 1148.364; 1263.638; 1330.141];
```

```
year =
        1980
        1990
        2000
        2010
```

It's convenient to measure time in years since 1980.

```
t = year - 1980;
y = pop;
```

## 2.1. Polynomial interpolation

Now we have four data points $(t_1, y_1), \ldots, (t_4, y_4)$, so $n = 4$ and we seek an interpolating cubic polynomial. We construct the associated Vandermonde matrix:

```
V = zeros(4,4);
for i = 1:4
    V(i,:) = [1 t(i) t(i)^2 t(i)^3];
end
V
```

```
V =
    1      0      0       0
    1     10    100    1000
    1     20    400    8000
    1     30    900   27000
```

To solve for the vector of polynomial coefficients, we use a **backslash**:

```
c = V \ y
```

```
c =
   984.7360
    18.7666
    -0.2397
    -0.0001
```

The algorithms used by the backslash operator are the main topic of this chapter. For now, observe that the coefficients of the cubic polynomial vary over several orders of magnitude, which is typical in this context. By our definitions, these coefficients are given in ascending order of power in $t$. MATLAB always expects the decreasing-degree order, so we convert ours to this convention here:

```
c = c(end:-1:1);        % reverse the ordering
```

We can use the resulting polynomial to estimate the population of China in 2005:

```
polyval(c,2005-1980)    % apply the 1980 time shift
```

```
ans =
   1.3030e+03
```

The official figure is 1297.8, so our result is not bad.

We can visualize the interpolation process. First, we **plot** the data as points. We'll shift the $t$ variable back to actual years.

```
plot(1980+t,y,'.')
```

**34**  Chapter 2. Square linear systems

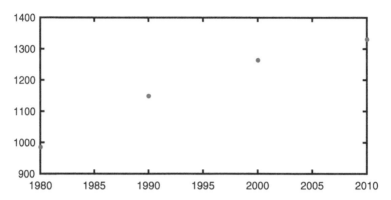

We want to superimpose a plot of the polynomial. In order to add to a plot, we must use the **hold** command:

```
hold on
```

To plot the interpolating polynomial, we create a vector with many points in the time interval using **linspace**.

```
tt = linspace(0,30,300)';    % 300 times from 1980 to 2010
yy = polyval(c,tt);          % evaluate the cubic
plot(1980+tt,yy)
```

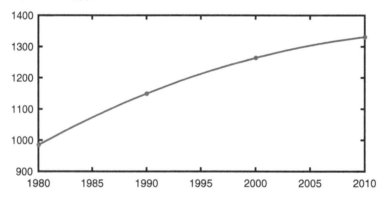

Let's clear the figure (**clf**) and redo it, this time continuing the curve outside of the original date range. We'll also annotate the graph (using **title**, **xlabel**, **ylabel**, and **legend**) to make its purpose clear.

```
clf    % clear figure
plot(1980+t,y,'.')
hold on
tt = linspace(-10,50,300)';
plot(1980+tt,polyval(c,tt))
title('Population of China')
xlabel('year'), ylabel('population (millions)')
legend('data','interpolant','location','northwest')
```

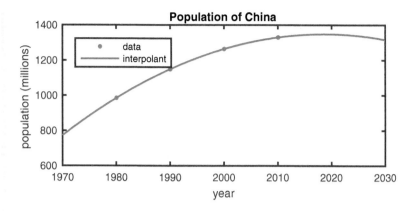

While the interpolation is plausible, the extrapolation to the future is highly questionable! As a rule, extrapolation more than a short distance beyond the original interval is not reliable.

Take a closer look at the construction of the matrix $V$ in Example 2.1.1:

```
for i = 1:4
    V(i,:) = [1 t(i) t(i)^2 t(i)^3];
end
```

Note that every iteration in $i$ is independent of all the other iterations. In other words, there are no references to $i-1$, $i-2$, etc., within the body of the loop. This observation usually suggests that we can replace the loop with a vector or matrix operation. In this case, we apply the componentwise-power operator to the vector t:

```
V = [ t.^0, t, t.^2, t.^3 ];
```

Because t is a column vector, the result of t.^3 is a column vector with entries $t_i^3$, and so on. The matrix $V$ is the horizontal concatenation of those columns into a single matrix.

This kind of code rewriting is referred to as **vectorization**. It can affect performance significantly in interpreted languages such as MATLAB. We don't often make much of a fuss about vectorization in this book unless there are nested loops. In this case, however, it emphasizes thinking about a matrix in terms of its columns, something that proves to be an important point of view conceptually.

## Exercises

2.1.1. Suppose you want to interpolate the points $(-1, 0)$, $(0, 1)$, $(2, 0)$, $(3, 1)$, and $(4, 2)$ by a polynomial of as low a degree as possible.

(a) ✍ What degree should you expect this polynomial to be? (The degree could be lower in special cases where some coefficients are exactly zero.)

(b) ✍ Write out a linear system of equations for the coefficients of the interpolating polynomial.

(c) 💻 Use MATLAB to solve the system in (b) numerically.

2.1.2. 💻 Here are population figures for three countries over the same 30-year period as in Example 2.1.1.

| Year | United States | China | Germany |
|---|---|---|---|
| 1980 | 227.225 | 984.736 | 78.298 |
| 1990 | 249.623 | 1,148.364 | 79.380 |
| 2000 | 282.172 | 1,263.638 | 82.184 |
| 2010 | 308.282 | 1,330.141 | 81.644 |

(a) Use cubic polynomial interpolation to estimate the population of China in 1992.

(b) Use cubic polynomial interpolation to estimate the population of the USA in 1984.

(c) Use cubic polynomial interpolation to make a plot of the German population from 1985 to 2000. Your plot should show a smooth curve and be well annotated.

2.1.3. 💻 The shifting of years in Example 2.1.1, so that $t=0$ means 1980, was more significant than it might seem.

(a) Using Example 2.1.1 as a model, find the ratio $a_4/a_1$ using the original code.

(b) Repeat the computation without shifting by 1980; that is, the entries of the time vector are the actual years. When you solve for the coefficients of the cubic polynomial, you will probably get a cryptic warning from MATLAB. Find $a_4/a_1$ again.

(c) Continuing part (b), again estimate the population of China in 2005. How different is the new answer? (The best way to compare two values is to take their difference.)

The phenomena observed in these experiments will be explained later in this chapter.

2.1.4. ✍ Say you want to find a cubic polynomial $p$ such that $p(0)=0$, $p'(0)=1$, $p(1)=2$, and $p'(1)=-1$. (This is known as a *Hermite interpolant.*) Write out a linear system of equations for the coefficients of $p$.

## 2.2 ▪ Computing with matrices

*We recommend that you review the linear algebra material in Appendix A before reading this section.*

We use capital letters in bold to refer to matrices, and lowercase bold letters for vectors. *All vectors in this book are column vectors*—in other words, matrices with just one column. The bold symbol **0** may refer to a vector of all zeros or to a zero matrix, depending on context; we use 0 as the scalar zero only.

## 2.2. Computing with matrices

To refer to a specific element of a matrix, we use the uppercase name of the matrix *without* boldface, as in $A_{24}$ to mean the $(2,4)$ element of $\mathbf{A}$.[10] To refer to an element of a vector, we use just one subscript, as in $x_3$. If you see a boldface character with one or more subscripts, then you know that it is a matrix or vector that belongs to a sequence or indexed collection.

We will have frequent need to refer to the individual columns of a matrix as vectors. Our convention is to use a lowercase bold version of the matrix name with a subscript to represent the column number. Thus, $\mathbf{a}_1, \mathbf{a}_2, \ldots, \mathbf{a}_n$ are the columns of the $m \times n$ matrix $\mathbf{A}$. Conversely, whenever we define a sequence of vectors $\mathbf{v}_1, \ldots, \mathbf{v}_p$, we can implicitly consider them to be columns of a matrix $\mathbf{V}$. Sometimes we might write $\mathbf{V} = \begin{bmatrix} \mathbf{v}_j \end{bmatrix}$ to emphasize the connection.

The notation $\mathbf{A}^T$ is used for the transpose of a matrix, whether it is real or complex. But in the case of complex matrices, it's almost always more desirable to use the **hermitian** $\mathbf{A}^*$, which is the transpose with the complex conjugate of each element.[11] If $\mathbf{A}$ is real, then $\mathbf{A}^* = \mathbf{A}^T$. A square matrix is **symmetric** if $\mathbf{A}^T = \mathbf{A}$ and **hermitian** if $\mathbf{A}^* = \mathbf{A}$. Nonsquare matrices cannot have these properties.

The **identity matrix** of size $n$ is denoted $\mathbf{I}$, or sometimes $\mathbf{I}_n$ if emphasizing the size is important in context. For columns of the identity we break with our usual naming convention and denote them by $\mathbf{e}_j$.

### Block matrix expressions

We will often find it useful to break a matrix into separately named pieces. For example, we might write

$$\mathbf{A} = \begin{bmatrix} \mathbf{A}_{11} & \mathbf{A}_{12} & \mathbf{A}_{13} \\ \mathbf{A}_{21} & \mathbf{A}_{22} & \mathbf{A}_{23} \end{bmatrix}, \qquad \mathbf{B} = \begin{bmatrix} \mathbf{B}_1 \\ \mathbf{B}_2 \\ \mathbf{B}_3 \end{bmatrix}.$$

It's understood that blocks that are on top of one another have the same number of columns, and blocks that are side by side have the same number of rows. As a rule, if the blocks all have compatible dimensions, then they can be multiplied as though the blocks were scalars. For instance, continuing with the definitions above, we say that $\mathbf{A}$ is block-$2 \times 3$ and $\mathbf{B}$ is block-$3 \times 1$, so we can write

$$\mathbf{AB} = \begin{bmatrix} \mathbf{A}_{11}\mathbf{B}_1 + \mathbf{A}_{12}\mathbf{B}_2 + \mathbf{A}_{13}\mathbf{B}_3 \\ \mathbf{A}_{21}\mathbf{B}_1 + \mathbf{A}_{22}\mathbf{B}_2 + \mathbf{A}_{23}\mathbf{B}_3 \end{bmatrix},$$

---

[10] This aspect of our notation is slightly unusual. More frequently one would see the lowercase $a_{24}$ in this context. We feel that our notation lends more consistency and clarity to expressions with mixed symbols, and it is more like how computer code is written.

[11] The conjugate of a complex number is found by replacing all references to the imaginary unit $i$ by $-i$. We will not see much of complex numbers until Chapter 7.

provided that the individual block products are well defined. For transposes we have, for example,

$$A^T = \begin{bmatrix} A_{11}^T & A_{21}^T \\ A_{12}^T & A_{22}^T \\ A_{13}^T & A_{23}^T \end{bmatrix}.$$

## Matrices in MATLAB

MATLAB originally stood for "matrix laboratory." It was designed from the beginning to treat matrices and vectors as the fundamental objects in scientific computing. A lot of how MATLAB deals with matrices is easy to remember once learned. There's a lot of this to learn, though, so we give just some of the basics here, and we will pick up more as we go from code used in our examples and functions. We begin with some handy functions for working with matrices (**ones, zeros, size, end**) and see how these work with familiar functions from algebra and calculus.

### Example 2.2.1

Square brackets are used to enclose elements of a matrix or vector. Use spaces or commas for horizontal concatenation, and semicolons or new lines to indicate vertical concatenation.

```
A = [ 1, 2, 3, 4, 5; 50 40 30 20 10
    pi, sqrt(2), exp(1), (1+sqrt(5))/2, log(3) ]

A =
    1.0000    2.0000    3.0000    4.0000    5.0000
   50.0000   40.0000   30.0000   20.0000   10.0000
    3.1416    1.4142    2.7183    1.6180    1.0986

[m,n] = size(A)

m =
    3
n =
    5
```

A vector is considered to be a matrix with one singleton dimension.

```
x = [ 3; 3; 0; 1; 0 ]
size(x)

x =
    3
    3
    0
    1
```

## 2.2. Computing with matrices

```
             0
ans =
     5       1
```

Concatenated elements within brackets may be matrices for a block representation, as long as all the block sizes are compatible.

```
AA = [ A; A ]
B = [ zeros(3,2), ones(3,1) ]

AA =
    1.0000    2.0000    3.0000    4.0000    5.0000
   50.0000   40.0000   30.0000   20.0000   10.0000
    3.1416    1.4142    2.7183    1.6180    1.0986
    1.0000    2.0000    3.0000    4.0000    5.0000
   50.0000   40.0000   30.0000   20.0000   10.0000
    3.1416    1.4142    2.7183    1.6180    1.0986
B =
     0     0     1
     0     0     1
     0     0     1
```

The dot-quote `.'` transposes a matrix. A single quote `'` on its own performs the hermitian (transpose and complex conjugation). For a real matrix, the two operations are the same.

```
A'

ans =
    1.0000   50.0000    3.1416
    2.0000   40.0000    1.4142
    3.0000   30.0000    2.7183
    4.0000   20.0000    1.6180
    5.0000   10.0000    1.0986

x'

ans =
     3     3     0     1     0
```

There are some convenient shorthand ways of building vectors and matrices other than entering all of their entries directly or in a loop. To get a row vector with evenly spaced entries between two endpoints, you have two options.

```
row = 1:4                    % start:stop
col = ( 0:3:12 )'            % start:step:stop

row =
     1     2     3     4
col =
     0
```

```
    3
    6
    9
   12
```

```
s = linspace(-1,1,5)'    % start,stop,number
```

```
s =
   -1.0000
   -0.5000
         0
    0.5000
    1.0000
```

Accessing an element is done by giving one (for a vector) or two index values in parentheses. The keyword **end** as an index refers to the last position in the corresponding dimension.

```
a = A(2,end-1)
```

```
a =
   20
```

```
x(2)
```

```
ans =
    3
```

The indices can be vectors, in which case a block of the matrix is accessed.

```
A(1:2,end-2:end)    % first two rows, last three columns
```

```
ans =
    3    4    5
   30   20   10
```

If a dimension has only the index : (a colon), then it refers to all the entries in that dimension of the matrix.

```
A(:,1:2:end)    % all of the odd columns
```

```
ans =
    1.0000    3.0000    5.0000
   50.0000   30.0000   10.0000
    3.1416    2.7183    1.0986
```

The matrix and vector senses of addition, subtraction, scalar multiplication, multiplication, and power are all handled by the usual symbols. If matrix sizes are such that the operation is not defined, an error message will result.

```
B = diag( [-1 0 -5] )    % create a diagonal matrix
```

## 2.2. Computing with matrices

```
B =
    -1     0     0
     0     0     0
     0     0    -5

BA = B*A      % matrix product

BA =
   -1.0000   -2.0000   -3.0000   -4.0000   -5.0000
         0         0         0         0         0
  -15.7080   -7.0711  -13.5914   -8.0902   -5.4931
```

A*B causes an error. (We trap it here using a special syntax.)

```
try A*B, catch lasterr, end

Error using *
Inner matrix dimensions must agree.
```

A square matrix raised to an integer power is the same as repeated matrix multiplication.

```
B^3     % same as B*B*B

ans =
    -1     0     0
     0     0     0
     0     0  -125
```

In many cases, one instead wants to treat a matrix or vector as a mere array and simply apply a single operation to each element of it. For multiplication, division, and power, the corresponding operators start with a dot.

```
C = -A;
```

A*C would be an error, but the elementwise product is defined.

```
elementwise = A.*C

elementwise =
   1.0e+03 *
   -0.0010   -0.0040   -0.0090   -0.0160   -0.0250
   -2.5000   -1.6000   -0.9000   -0.4000   -0.1000
   -0.0099   -0.0020   -0.0074   -0.0026   -0.0012
```

The two operands of a dot operator must have the same size—unless one is a scalar, in which case it is expanded or "broadcast" to be the same size as the other operand.

```
xtotwo = x.^2
```

```
xtotwo =
     9
     9
     0
     1
     0
```

```
twotox = 2.^x
```

```
twotox =
     8
     8
     1
     2
     1
```

Most of the mathematical functions, such as **cos**, **sin**, **log**, **exp** and **sqrt**, also operate elementwise on a matrix.

```
cos(pi*x')
```

```
ans =
    -1    -1     1    -1     1
```

## Row and column operations

A critical identity in matrix multiplication is

$$Ae_j = a_j.$$

In words, *multiplication on the right by column $j$ of the identity extracts the $j$th column.* Furthermore, the expression

$$A\begin{bmatrix} e_1 & e_3 & e_5 \end{bmatrix}$$

extracts three columns. An equivalent expression in MATLAB would be A(:,1:2:5). We can extend the same idea to rows by using the general identity $(RS)^T = S^T R^T$. Let $B = A^T$ have columns $\begin{bmatrix} b_j \end{bmatrix}$, and note

$$(b_j)^T = (Be_j)^T = e_j^T B^T = e_j^T A.$$

But $e_j^T$ is the $j$th *row* of $I$, and $b_j^T$ is the transpose of the $j$th column of $B$, which is the $j$th *row* of $A$ by $B = A^T$. Thus, *multiplication on the left by row $j$ of the identity extracts the $j$th row.* Extracting the single element $(i,j)$ from the matrix is, therefore, $e_i^T A e_j$.

Being able to extract specific rows and columns of a matrix makes it straightforward to do row- and column-oriented operations, such as linear combinations. For instance, adding twice the third column of $A$ to its first column is done by $A(e_1 + 2e_3)$. Or, if we wanted to do this operation "in place," meaning replacing the first column of $A$ with this value and leaving the other four columns of $A$ alone, we can say

$$A\begin{bmatrix} e_1+2e_3 & e_2 & e_3 & e_4 & e_5 \end{bmatrix}.$$

In MATLAB, the syntax would be

```
A(:,1) = A(:,1) + 2*A(:,3);
```

It's understood here that the right-hand side is evaluated using the current value of A, then the result is stuffed into the first column, thereby changing it.

## Exercises

2.2.1. Suppose $C = \begin{bmatrix} I & A \\ -I & B \end{bmatrix}$. Using block notation, find $C^2$ and $C^3$.

2.2.2. Let

$$A = \begin{bmatrix} 2 & 1 & 1 & 0 \\ 0 & -1 & 4 & 1 \\ 2 & 2 & 0 & -2 \\ 1 & 3 & -1 & 5 \end{bmatrix}, \quad B = \begin{bmatrix} 3 & -1 & 0 & 0 & 2 \\ 7 & 1 & 0 & 0 & 2 \end{bmatrix},$$

$$u = \begin{bmatrix} 2 \\ -1 \\ 3 \\ 1 \end{bmatrix}, \quad v = \begin{bmatrix} \pi \\ e \end{bmatrix}.$$

(Do not round off the values in $v$—find them using native MATLAB commands.) For each expression below, use MATLAB to find the result, or explain why the result does not exist.

(a) $AB$. (b) $B^T A$. (c) $v^T B$. (d) $Bu$. (e) $\begin{bmatrix} u & Au & A^2 u & A^3 u \end{bmatrix}$.

2.2.3. Let

$$u = \begin{bmatrix} 1 \\ 3 \\ 5 \\ 7 \\ 9 \\ 11 \end{bmatrix}, \quad v = \begin{bmatrix} -60 \\ -50 \\ -40 \\ -30 \\ -20 \\ -10 \end{bmatrix}.$$

Find the inner products $u^T v$ and $v^T u$, and the outer products $uv^T$ and $vu^T$.

2.2.4. In MATLAB, give a demonstration of the identity $(AB)^T = B^T A^T$ for some arbitrarily chosen nontrivial matrices $A$ and $B$.

2.2.5. Prove that if $A$ and $B$ are invertible, then $(AB)^{-1} = B^{-1}A^{-1}$ (and therefore $AB$ is invertible as well).

2.2.6. Suppose $B$ is an arbitrary $4 \times 3$ matrix. In each part below a matrix $A$ is described in terms of $B$. Express $A$ as a product of $B$ with one or more other matrices.

(a) $A \in \mathbb{R}^{4 \times 1}$ is the result of adding the first column of $B$ to $-2$ times the last column of $B$.

(b) The rows of $A \in \mathbb{R}^{4 \times 3}$ are the rows of $B$ in reverse order.

(c) The first column of $A \in \mathbb{R}^{4 \times 3}$ is 1 times the first column of $B$, the second column of $A$ is 2 times the second column of $B$, and the third column of $A$ is 3 times the third column of $B$.

(d) $A \in \mathbb{R}$ is the sum of all elements of $B$.

2.2.7. ✍ Prove true, or give a counterexample: The product of symmetric matrices is symmetric.

2.2.8. (a) ✍ Prove that for real vectors $v$ and $w$ of the same length, the inner products $v^T w$ and $w^T v$ are equal.

(b) ✍ Prove true, or give a counterexample for, the equivalent statement about outer products, $vw^T$ and $wv^T$.

## 2.3 ▪ Linear systems

We now attend to the central problem of this chapter: Given a square $n \times n$ matrix $A$ and an $n$-vector $b$, find an $n$-vector $x$ such that $Ax = b$. Writing out these equations, we obtain

$$a_{11}x_1 + a_{12}x_2 + \cdots + a_{1n}x_n = b_1,$$
$$a_{21}x_1 + a_{22}x_2 + \cdots + a_{2n}x_n = b_2,$$
$$\vdots$$
$$a_{n1}x_1 + a_{n2}x_2 + \cdots + a_{nn}x_n = b_n.$$

If $A$ is nonsingular, then the mathematical expression of the solution is $x = A^{-1}b$, because

$$A^{-1}b = A^{-1}(Ax) = (A^{-1}A)x = Ix = x.$$

When $A$ is singular, then $Ax = b$ may have no solution or infinitely many solutions.

---

**Example 2.3.1**

If we define

$$S = \begin{bmatrix} 0 & 1 \\ 0 & 0 \end{bmatrix},$$

then it is easy to check that for any real value of $\alpha$ we have

$$S \begin{bmatrix} \alpha \\ 1 \end{bmatrix} = \begin{bmatrix} 1 \\ 0 \end{bmatrix}.$$

Hence the linear system $Sx = b$ with $b = \begin{bmatrix} 1 \\ 0 \end{bmatrix}$ has infinitely many solutions. For some other choices of $b$ the system can be proven to have no solutions.

## 2.3. Linear systems

Matrix inverses are indispensable for mathematical discussion and derivations. However, as you may remember from a linear algebra course, they are not trivial to compute from the entries of the original matrix. While it can be done numerically by computer, it almost never is, because when the goal is to solve a linear system of equations, the inverse is not needed—and the process of finding it is slower than solving the original problem. MATLAB does have a command **inv** that finds the inverse of a matrix. But, as demonstrated in Section 2.1, in order to solve a linear system of equations, you should use **backslash** (the \ symbol, not to be confused with the slash /) instead.

### Example 2.3.2

For a square matrix **A**, the command A\b is mathematically equivalent to $\mathbf{A}^{-1}\mathbf{b}$.

```
A = magic(3)
b = [1;2;3];
x = A\b

A =
     8     1     6
     3     5     7
     4     9     2
x =
    0.0500
    0.3000
    0.0500
```

One way to check the answer is to compute a quantity known as the residual. It is (hopefully) close to machine precision, scaled by the size of the entries of the data.

```
residual = b - A*x

residual =
     0
     0
     0
```

If the matrix is singular, a warning is produced, but you get an answer anyway.

```
A = [0 1; 0 0];    % known to be singular
b = [1;2];
x = A\b

x =
  -Inf
   Inf
```

When you get a warning, it's important to check the result rather than blindly accepting it as correct.

## Triangular systems

The solution process is especially easy to demonstrate for systems with **triangular** matrices. For example, consider the lower triangular system

$$\begin{bmatrix} 4 & 0 & 0 & 0 \\ 3 & -1 & 0 & 0 \\ -1 & 0 & 3 & 0 \\ 1 & -1 & -1 & 2 \end{bmatrix} x = \begin{bmatrix} 8 \\ 5 \\ 0 \\ 1 \end{bmatrix}.$$

The first row of this system states simply that $4x_1 = 8$, which is easily solved as $x_1 = 8/4 = 2$. Now, the second row states that $3x_1 - x_2 = 5$. As $x_1$ is already known, it can be replaced to find that $x_2 = -(5 - 3 \cdot 2) = 1$. Similarly, the third row gives $x_3 = (0 + 1 \cdot 2)/3 = 2/3$, and the last row yields $x_4 = (1 - 1 \cdot 2 + 1 \cdot 1 + 1 \cdot 2/3)/2 = 1/3$. Hence the solution is

$$x = \begin{bmatrix} 2 \\ 1 \\ 2/3 \\ 1/3 \end{bmatrix}.$$

The process just described is called **forward substitution**. In the $4 \times 4$ lower triangular case of $Lx = b$ it leads to the formulas

$$\begin{aligned} x_1 &= \frac{b_1}{L_{11}}, \\ x_2 &= \frac{b_2 - L_{21} x_1}{L_{22}}, \\ x_3 &= \frac{b_3 - L_{31} x_1 - L_{32} x_2}{L_{33}}, \\ x_4 &= \frac{b_4 - L_{41} x_1 - L_{42} x_2 - L_{43} x_3}{L_{44}}. \end{aligned} \quad (2.3.1)$$

For upper triangular systems $Ux = b$ an analogous process of **backward substitution** begins by solving for the last component $x_n = b_n/U_{nn}$ and working backward. For the $4 \times 4$ case we have

$$\begin{bmatrix} U_{11} & U_{12} & U_{13} & U_{14} \\ 0 & U_{22} & U_{23} & U_{24} \\ 0 & 0 & U_{33} & U_{34} \\ 0 & 0 & 0 & U_{44} \end{bmatrix} x = \begin{bmatrix} b_1 \\ b_2 \\ b_3 \\ b_4 \end{bmatrix}.$$

Solving the system backward, starting with $x_4$ first and then proceeding in descending

## 2.3. Linear systems

order, gives

$$x_4 = \frac{b_4}{U_{44}},$$
$$x_3 = \frac{b_3 - U_{34}x_4}{U_{33}},$$
$$x_2 = \frac{b_2 - U_{23}x_3 - U_{24}x_4}{U_{22}},$$
$$x_1 = \frac{b_1 - U_{12}x_2 - U_{13}x_3 - U_{14}x_4}{U_{11}}.$$

It should be clear that forward or backward substitution fails if, and only if, one of the diagonal entries of the system matrix is zero. We have essentially proved the following theorem.

> **Theorem 2.3.1**
>
> A triangular matrix is singular if and only if at least one of its diagonal elements is zero.

### Implementation

Let's consider how to implement the sequential process implied by equation (2.3.1). It seems clear that we want to loop through the elements of $x$ in order. Within each iteration of that loop, we have an expression whose length depends on the iteration number. One way we could do this would be with a nested loop:

```
for i = 1:4
    x(i) = b(i);
    for j = 1:i-1
        x(i) = x(i) - L(i,j)*x(j);
    end
    x(i) = x(i)/L(i,i);
end
```

In the first pass though the outer loop, $i = 1$. This means that the vector 1:i-1 is actually 1:0, which is the empty vector. So the inner loop will be skipped on that pass. Thereafter, the inner loop only refers to elements of $x$ that were computed previously.

As mentioned in Section 2.1, we should avoid nested loops in MATLAB when possible, for performance reasons. To see how we can do this, consider the expression $b_4 - L_{41}x_1 - L_{42}x_2 - L_{43}x_3$, which we can rewrite as

$$b_4 - (L_{41}x_1 + L_{42}x_2 + L_{43}x_3) = b_4 - \begin{bmatrix} L_{41} & L_{42} & L_{43} \end{bmatrix} \begin{bmatrix} x_1 \\ x_2 \\ x_3 \end{bmatrix}.$$

**Function 2.3.1 (forwardsub)** Solve a lower triangular linear system.

```
function x = forwardsub(L,b)
% FORWARDSUB   Solve a lower triangular linear system.
% Input:
%   L    lower triangular square matrix (n by n)
%   b    right-hand side vector (n by 1)
% Output:
%   x    solution of Lx=b (n by 1 vector)

n = length(L);
x = zeros(n,1);
for i = 1:n
    x(i) = ( b(i) - L(i,1:i-1)*x(1:i-1) ) / L(i,i);
end
```

**Function 2.3.2 (backsub)** Solve an upper triangular linear system.

```
function x = backsub(U,b)
% BACKSUB   Solve an upper triangular linear system.
% Input:
%   U    upper triangular square matrix (n by n)
%   b    right-hand side vector (n by 1)
% Output:
%   x    solution of Ux=b (n by 1 vector)

n = length(U);
x = zeros(n,1);
for i = n:-1:1
    x(i) = ( b(i) - U(i,i+1:n)*x(i+1:n) ) / U(i,i);
end
```

The inner product above can be computed directly by matrix multiplication, without using a loop. The result of using this change is shown in Function 2.3.1. Line 10 of the function is used to make sure that the output vector x will be a column vector. The references in line 12 of Function 2.3.1 to L(i,1:i-1) and x(1:i-1) create the vectors in the inner product; when $i = 1$, the reference will give empty results and their inner product will be considered to be zero.

The implementation of backward substitution is much like forward substitution and is given in Function 2.3.2.

**Example 2.3.3**

It's easy to get just the lower triangular part of any matrix using the **tril** command.

```
A = magic(5)
L = tril(A)
```

```
A =
    17    24     1     8    15
    23     5     7    14    16
```

## 2.3. Linear systems

$$L = \begin{matrix} 4 & 6 & 13 & 20 & 22 \\ 10 & 12 & 19 & 21 & 3 \\ 11 & 18 & 25 & 2 & 9 \\ 17 & 0 & 0 & 0 & 0 \\ 23 & 5 & 0 & 0 & 0 \\ 4 & 6 & 13 & 0 & 0 \\ 10 & 12 & 19 & 21 & 0 \\ 11 & 18 & 25 & 2 & 9 \end{matrix}$$

We'll set up and solve a linear system with this matrix.

```
b = ones(5,1);
x = forwardsub(L,b)

x =
    0.0588
   -0.0706
    0.0914
   -0.0228
   -0.0684
```

It's not clear what the error in this answer is. However, the residual, while not zero, is virtually $\varepsilon_{\text{mach}}$ in size.

```
b - L*x

ans =
   1.0e-15 *
         0
         0
    0.1110
   -0.2220
    0.4441
```

Next we'll engineer a problem to which we know the exact answer. You should be able to convince yourself that for any $\alpha$ and $\beta$,

$$\begin{bmatrix} 1 & -1 & 0 & \alpha-\beta & \beta \\ 0 & 1 & -1 & 0 & 0 \\ 0 & 0 & 1 & -1 & 0 \\ 0 & 0 & 0 & 1 & -1 \\ 0 & 0 & 0 & 0 & 1 \end{bmatrix} \begin{bmatrix} 1 \\ 1 \\ 1 \\ 1 \\ 1 \end{bmatrix} = \begin{bmatrix} \alpha \\ 0 \\ 0 \\ 0 \\ 1 \end{bmatrix}.$$

```
alpha = 0.3;
beta = 2.2;
U = eye(5) + diag([-1 -1 -1 -1],1);
U(1,[4 5]) = [ alpha-beta, beta ];
x_exact = ones(5,1);
b = [alpha;0;0;0;1];
```

```
x = backsub(U,b);
err = x - x_exact

err =
    0
    0
    0
    0
    0
```

Everything seems OK here. But another example, with a different value for $\beta$, is more troubling.

```
alpha = 0.3;
beta = 1e12;
U = eye(5) + diag([-1 -1 -1 -1],1);
U(1,[4 5]) = [ alpha-beta, beta ];
b = [alpha;0;0;0;1];
x = backsub(U,b);
err = x - x_exact

err =
   1.0e-04 *
   -0.4883
         0
         0
         0
         0
```

It's not so good to get four digits of accuracy after starting with sixteen! But the source of the error is not hard to track down. Solving for $x_1$ performs $(\alpha - \beta) + \beta$ in the first row. Since $|\alpha|$ is so much smaller than $|\beta|$, this a recipe for losing digits to subtractive cancellation.

Example 2.3.3 is our first clue that linear system problems may have large condition numbers, making inaccurate solutions inevitable in floating point arithmetic. We will learn how to spot such problems in Section 2.8. Before reaching that point, however, we need to discuss how to solve general linear systems, not just triangular ones.

## Exercises

2.3.1. Find a right-hand side vector $b$ such that the system $Sx = b$, where $S$ is defined as in Example 2.3.1, has no solution.

2.3.2. Solve the following triangular systems by hand.

(a) $\begin{aligned} -2x_1 &= -4, \\ x_1 - x_2 &= 2, \\ 3x_1 + 2x_2 + x_3 &= 1. \end{aligned}$ (b) $\begin{bmatrix} 4 & 0 & 0 & 0 \\ 1 & -2 & 0 & 0 \\ -1 & 4 & 4 & 0 \\ 2 & -5 & 5 & 1 \end{bmatrix} x = \begin{bmatrix} -4 \\ 1 \\ -3 \\ 5 \end{bmatrix}.$

(c) $\begin{aligned} 3x_1 + 2x_2 + x_3 &= 1, \\ x_2 - x_3 &= 2, \\ 2x_3 &= -4. \end{aligned}$

2.3.3. 🖳 Use Function 2.3.1 to solve the systems from the previous problem. Verify that the solution is correct by computing $Lx$ and subtracting $b$.

2.3.4. 🖳 Use Function 2.3.2 to solve the following systems. Verify that the solution is correct by computing $Ux$ and subtracting $b$.

(a) $\begin{bmatrix} 3 & 1 & 0 \\ 0 & -1 & -2 \\ 0 & 0 & 3 \end{bmatrix} x = \begin{bmatrix} 1 \\ 1 \\ 6 \end{bmatrix}.$ (b) $\begin{bmatrix} 3 & 1 & 0 & 6 \\ 0 & -1 & -2 & 7 \\ 0 & 0 & 3 & 4 \\ 0 & 0 & 0 & 5 \end{bmatrix} x = \begin{bmatrix} 4 \\ 1 \\ 1 \\ 5 \end{bmatrix}.$

2.3.5. 🖳 If $B \in \mathbb{R}^{n \times p}$ has columns $b_1, \ldots, b_p$, then we can pose $p$ linear systems at once by writing $AX = B$, where $X$ is $n \times p$. Specifically, this equation implies $Ax_j = b_j$ for $j = 1, \ldots, p$.

(a). Modify Functions 2.3.1 and 2.3.2 so that they solve the case where the second input is $n \times p$ for $p \geq 1$.

(b). If $AX = I$, then $X = A^{-1}$. Use this fact to write a function ltinverse that uses your modified forwardsub to compute the inverse of a lower triangular matrix. Test your function on at least two nontrivial matrices. (We remind you here that this is just an exercise; matrix inverses are rarely a good idea in numerical practice!)

2.3.6. 🖳 Example 2.3.3 showed solutions of $Ax = b$, where

$$A = \begin{bmatrix} 1 & -1 & 0 & \alpha - \beta & \beta \\ 0 & 1 & -1 & 0 & 0 \\ 0 & 0 & 1 & -1 & 0 \\ 0 & 0 & 0 & 1 & -1 \\ 0 & 0 & 0 & 0 & 1 \end{bmatrix}, \quad b = \begin{bmatrix} \alpha \\ 0 \\ 0 \\ 0 \\ 1 \end{bmatrix}.$$

Solve with $\alpha = 0.1$ and $\beta = 10, 100, \ldots, 10^{12}$, making a table of the values of $\beta$ and $|x_1 - 1|$.

## 2.4 • LU factorization

Every first linear algebra course introduces **Gaussian elimination** for a general square system $Ax = b$. In Gaussian elimination one uses row operations on an augmented matrix $[A \; b]$ to reduce it to an equivalent triangular system (usually upper triangular). Rather than writing out the process in full generality, we use an example to refresh your memory, and we get arithmetic support from MATLAB.

### Example 2.4.1

We create a 4 × 4 linear system with the matrix

```
A = [
    2    0    4    3
   -4    5   -7  -10
    1   15    2   -4.5
   -2    0    2  -13
    ];
```

and with the right-hand side

```
b = [ 4; 9; 29; 40 ];
```

We define an *augmented matrix* by tacking **b** on the end as a new column.

```
S = [A, b]
```

```
S =
    2.0000         0    4.0000    3.0000    4.0000
   -4.0000    5.0000   -7.0000  -10.0000    9.0000
    1.0000   15.0000    2.0000   -4.5000   29.0000
   -2.0000         0    2.0000  -13.0000   40.0000
```

The goal is to introduce zeros into the lower triangle of this matrix. By using only elementary row operations, we ensure that the matrix **S** always represents a linear system that is equivalent to the original. We proceed from left to right and top to bottom. The first step is to put a zero in the (2,1) location using a multiple of row 1:

```
mult21 = S(2,1)/S(1,1)
S(2,:) = S(2,:) - mult21*S(1,:)
```

```
mult21 =
    -2
S =
    2.0000         0    4.0000    3.0000    4.0000
         0    5.0000    1.0000   -4.0000   17.0000
    1.0000   15.0000    2.0000   -4.5000   29.0000
   -2.0000         0    2.0000  -13.0000   40.0000
```

We repeat the process for the (3,1) and (4,1) entries.

```
mult31 = S(3,1)/S(1,1)
S(3,:) = S(3,:) - mult31*S(1,:);
mult41 = S(4,1)/S(1,1)
S(4,:) = S(4,:) - mult41*S(1,:);
S
```

```
mult31 =
    0.5000
mult41 =
```

## 2.4. LU factorization

```
        -1
S =
         2         0         4         3         4
         0         5         1        -4        17
         0        15         0        -6        27
         0         0         6       -10        44
```

The first column has the zero structure we want. To avoid interfering with that, we no longer add multiples of row 1 to anything. Instead, to handle column 2, we use multiples of row 2. We'll also exploit the highly repetitive nature of the operations to write them as a loop.

```
for i = 3:4
    mult = S(i,2)/S(2,2);
    S(i,:) = S(i,:) - mult*S(2,:);
end
S

S =
         2         0         4         3         4
         0         5         1        -4        17
         0         0        -3         6       -24
         0         0         6       -10        44
```

We finish out the triangularization with a zero in the (4,3) place. It's a little silly to use a loop for just one iteration, but the point is to establish a pattern.

```
for i = 4
    mult = S(i,3)/S(3,3);
    S(i,:) = S(i,:) - mult*S(3,:);
end
S

S =
         2         0         4         3         4
         0         5         1        -4        17
         0         0        -3         6       -24
         0         0         0         2        -4
```

Recall that $S$ is an augmented matrix: it represents the system $Ux = z$, where

```
U = S(:,1:4)
z = S(:,5)

U =
         2         0         4         3
         0         5         1        -4
         0         0        -3         6
         0         0         0         2
z =
         4
        17
```

```
            -24
             -4
```

The solutions to this system are identical to those of the original system, but this one can be solved by backward substitution.

```
x = backsub(U,z)

x =
    -3
     1
     4
    -2

b - A*x

ans =
     0
     0
     0
     0
```

## The algebra of Gaussian elimination

In Section 2.2 we observed that row and column operations can be expressed as linear algebra using columns from the identity matrix. This connection allows us to express Gaussian elimination using matrices. We will ignore the augmentation step, set aside $b$ for now, and consider only the square system matrix $A$.

Reconsider Example 2.4.1. As the first step, we get the multiplier $A_{21}/A_{11} = -2$. The first row of $A$ is extracted by $e_1^T A$. After $-2$ times this row is subtracted from row 2, with the other rows being left alone, we arrive at the matrix

$$\begin{bmatrix} e_1^T \\ e_2^T + 2e_1^T \\ e_3^T \\ e_4^T \end{bmatrix} A.$$

This expression can be manipulated into

$$A = \left( I + \begin{bmatrix} 0e_1^T \\ 2e_1^T \\ 0e_1^T \\ 0e_1^T \end{bmatrix} \right) A = \left( I + 2 \begin{bmatrix} 0 \\ 1 \\ 0 \\ 0 \end{bmatrix} e_1^T \right) A = (I + 2e_2 e_1^T) A.$$

## 2.4. LU factorization

In general, adding $\alpha$ times row $j$ of $\boldsymbol{A}$ to row $i$ in place is done via the expression

$$(\boldsymbol{I} + \alpha \boldsymbol{e}_i \boldsymbol{e}_j^T)\boldsymbol{A}. \tag{2.4.1}$$

Following many introductory texts on linear algebra, we refer to the matrix in parentheses above as an *elementary matrix*.

### Example 2.4.2

We revisit the previous example using algebra to express the row operations on $\boldsymbol{A}$.

```
A = [2 0 4 3 ; -4 5 -7 -10 ; 1 15 2 -4.5 ; -2 0 2 -13];
```

We use the identity and its columns heavily.

```
I = eye(4);
```

The first step is to put a zero in the (2,1) location using a multiple of row 1:

```
mult21 = A(2,1)/A(1,1);
L21 = I - mult21*I(:,2)*I(:,1)';
A = L21*A
```

```
A =
    2.0000         0    4.0000    3.0000
         0    5.0000    1.0000   -4.0000
    1.0000   15.0000    2.0000   -4.5000
   -2.0000         0    2.0000  -13.0000
```

We repeat the process for the (3,1) and (4,1) entries.

```
mult31 = A(3,1)/A(1,1);
L31 = I - mult31*I(:,3)*I(:,1)';
A = L31*A;
mult41 = A(4,1)/A(1,1);
L41 = I - mult41*I(:,4)*I(:,1)';
A = L41*A
```

```
A =
    2     0     4     3
    0     5     1    -4
    0    15     0    -6
    0     0     6   -10
```

And so on, following the pattern as before.

The elementary matrix factors found in Example 2.4.2, each in the form (2.4.1), have some important properties. First, in addition to being triangular, each has all ones on the diagonal, a property we call **unit triangular**. Theorem 2.3.1 implies that all unit triangular matrices are invertible, which is about to become important.

Let's review. The Gaussian elimination procedure in Example 2.4.1 did six row operations in order to introduce six zeros into the lower triangle of $A$. Each row operation can be expressed using multiplication by an elementary matrix $L_{ij}$. At the end we get an upper triangular matrix, $U$:

$$U = L_{43}L_{42}L_{32}L_{41}L_{31}L_{21}A. \tag{2.4.2}$$

Now we multiply both sides on the left by $L_{43}^{-1}$. On the right-hand side, it can be grouped together with $L_{43}$ to form an identity matrix. Then we multiply both sides on the left by $L_{42}^{-1}$, which knocks out the next term on the right side, etc. Eventually we get

$$L_{21}^{-1}L_{31}^{-1}L_{41}^{-1}L_{32}^{-1}L_{42}^{-1}L_{43}^{-1}U = A. \tag{2.4.3}$$

We come next to an interesting property of these elementary matrices. If $i \neq j$, then for any scalar $\alpha$ we can calculate that

$$\left(I + \alpha e_i e_j^T\right)\left(I - \alpha e_i e_j^T\right) = I + \alpha e_i e_j^T - \alpha e_i e_j^T - \alpha^2 e_i e_j^T e_i e_j^T$$
$$= I - \alpha^2 e_i \left(e_j^T e_i\right) e_j^T = I,$$

since the inner product between any two different columns of $I$ is zero. We have shown that

$$\left(I + \alpha e_i e_j^T\right)^{-1} = I - \alpha e_i e_j^T.$$

All that is needed to invert $I + \alpha e_i e_j^T$ is to flip the sign of $\alpha$, the lone element in the lower triangle.

We need one more remarkably convenient property of the elementary matrices. Looking, for example, at the first two matrices on the left in (2.4.3), we calculate that

$$\left(I + \alpha e_2 e_1^T\right)\left(I + \beta e_3 e_1^T\right) = I + \alpha e_2 e_1^T + \beta e_3 e_1^T + \alpha\beta e_2 e_1^T e_3 e_1^T.$$

We can use associativity in the last term to group together $e_1^T e_3$, which is another inner product that equals zero thanks to the structure of the identity matrix. In summary: the product of these elementary factors just combines the nonzero terms that each one has below the diagonal.

That reasoning carries across each of the new terms in the product on the left side of (2.4.3). The conclusion is that *Gaussian elimination finds a unit lower triangular matrix L and an upper triangular matrix U such that*

$$LU = A. \tag{2.4.4}$$

Furthermore, the lower triangular entries of $L$ are the row multipliers we found as in Example 2.4.1, and the entries of $U$ are those found at the end of the elimination process. Equation (2.4.4) is called an **LU factorization** of the matrix $A$.

## An algorithm—for now

LU factorization reduces any linear system to two triangular ones. From this, solving $Ax = b$ follows immediately:

## 2.4. LU factorization

**Function 2.4.1** (lufact) LU factorization for a square matrix.

```
function [L,U] = lufact(A)
% LUFACT   LU factorization (demo only--not stable!).
% Input:
%   A    square matrix
% Output:
%   L,U  unit lower triangular and upper triangular such that LU=A

n = length(A);
L = eye(n);    % ones on diagonal

% Gaussian elimination
for j = 1:n-1
  for i = j+1:n
    L(i,j) = A(i,j) / A(j,j);   % row multiplier
    A(i,j:n) = A(i,j:n) - L(i,j)*A(j,j:n);
  end
end

U = triu(A);
```

1. Factor $LU = A$ using Gaussian elimination.

2. Solve $Lz = b$ for $z$ using forward substitution.

3. Solve $Ux = z$ for $x$ using backward substitution.

One of the important aspects of this algorithm is that the factorization step depends only on the matrix $A$; the right-hand side $b$ is not involved. Thus if one has to solve multiple systems with a single matrix $A$, the factorization needs to be performed only once for all systems. As we show in Section 2.5, the factorization is by far the most computationally expensive step, so this note is of more than academic interest.

Based on the examples and discussion above, a code for LU factorization is given in Function 2.4.1.[12] The multipliers are stored in the lower triangle of $L$ as they are found. When operations are done to put zeros in column $j$, they are carried out only in lower rows to create an upper triangular matrix. (Only columns $j$ through $n$ are accessed, since the other entries should already be zero.) At the end of the process the matrix $A$ should be upper triangular, but since roundoff errors could create some small nonzeros the **triu** command is used to make them exactly zero.

### Example 2.4.3

We find the factors of the matrix from Example 2.4.1.

```
A = [2 0 4 3; -4 5 -7 -10; 1 15 2 -4.5; -2 0 2 -13];
[L,U] = lufact(A)

L =
```

---

[12] We did not follow the usual advice to avoid nested loops in MATLAB. There is a one-loop version (see Exercise 2.4.7), but this one is a little clearer for beginners.

```
           1.0000          0          0          0
          -2.0000     1.0000          0          0
           0.5000     3.0000     1.0000          0
          -1.0000          0    -2.0000     1.0000
U =
           2          0          4          3
           0          5          1         -4
           0          0         -3          6
           0          0          0          2
```

```
LtimesU = L*U
```

```
LtimesU =
           2.0000          0     4.0000     3.0000
          -4.0000     5.0000    -7.0000   -10.0000
           1.0000    15.0000     2.0000    -4.5000
          -2.0000          0     2.0000   -13.0000
```

Because MATLAB doesn't show all the digits by default, it's best to compare two quantities by taking their difference.

```
A - LtimesU
```

```
ans =
     0     0     0     0
     0     0     0     0
     0     0     0     0
     0     0     0     0
```

(Usually we can expect "zero" only up to machine precision. However, all the exact numbers in this example are also floating point numbers.)

To solve a linear system, we no longer need the matrix $A$.

```
b = [4;9;29;40];
z = forwardsub(L,b);
x = backsub(U,z)
```

```
x =
    -3
     1
     4
    -2
```

```
b - A*x
```

```
ans =
     0
     0
     0
     0
```

Observe from Function 2.4.1 that the factorization can fail if $A_{jj} = 0$ when it is put in the denominator in line 14. This does *not* necessarily mean there is a zero in the diagonal of the original $A$, because $A$ is changed during the computation. Moreover, there are perfectly good nonsingular matrices for which this type of failure occurs, such as

$$\begin{bmatrix} 0 & 1 \\ 1 & 0 \end{bmatrix}.$$

Fortunately, this defect can be repaired for all nonsingular matrices at minor cost, as we will see in Section 2.6.

## Exercises

2.4.1. ✍ For each matrix, perform Gaussian elimination by hand to produce an LU factorization. Write out the $L$ matrix using outer products of standard basis vectors.

(a) $\begin{bmatrix} 2 & 3 & 4 \\ 4 & 5 & 10 \\ 4 & 8 & 2 \end{bmatrix}.$
(b) $\begin{bmatrix} 6 & -2 & -4 & 4 \\ 3 & -3 & -6 & 1 \\ -12 & 8 & 21 & -8 \\ -6 & 0 & -10 & 7 \end{bmatrix}.$

2.4.2. 🖥 The matrices

$$T(x,y) = \begin{bmatrix} 1 & 0 & 0 \\ 0 & 1 & 0 \\ x & y & 1 \end{bmatrix}, \quad R(\theta) = \begin{bmatrix} \cos\theta & \sin\theta & 0 \\ -\sin\theta & \cos\theta & 0 \\ 0 & 0 & 1 \end{bmatrix}$$

are used to represent translations and rotations of plane points in computer graphics. For the following, let

$$A = T(3,-1)R(\pi/5)T(-3,1), \quad z = \begin{bmatrix} 2 \\ 2 \\ 1 \end{bmatrix}.$$

(a) Find $b = Az$.
(b) Find the LU factorization of $A$.
(c) Use the factors with triangular substitutions in order to solve $Ax = b$, and find $x - z$.

2.4.3. 🖥 In MATLAB, define

$$A = \begin{bmatrix} 1 & 0 & 0 & 0 & 10^{12} \\ 1 & 1 & 0 & 0 & 0 \\ 0 & 1 & 1 & 0 & 0 \\ 0 & 0 & 1 & 1 & 0 \\ 0 & 0 & 0 & 1 & 0 \end{bmatrix}, \quad \hat{x} = \begin{bmatrix} 0 \\ 1/3 \\ 2/3 \\ 1 \\ 4/3 \end{bmatrix}, \quad b = A\hat{x}.$$

(a) The **format** command can adjust output appearance in MATLAB; enter format long in MATLAB. Using Function 2.4.1 and triangular substitutions, solve the linear system $Ax = b$ and let MATLAB print out the result. About how many (to the nearest integer) accurate digits are in the result? (The answer is much less than the default 16 of double precision.)

(b) Repeat part (a) with $10^{20}$ as the corner element. (The result is even less accurate. We will study the causes of such low accuracy later in the chapter.)

2.4.4. ▣ Let $D_n$ be the matrix created using MATLAB's **gallery** function using

`gallery('dramadah',n)`

where $n$ is a positive integer. It has interesting properties: the entries of $D_n$ are all 0 or 1, and the entries of $D_n^{-1}$ are all integers. Run an experiment that verifies that if $D_n = LU$ is an LU factorization, then the entries of $L, U, L^{-1}$, and $U^{-1}$ are all integers for $n = 2, 3, \ldots, 50$. You will have to do something more clever than visual inspection of the matrix entries to determine that they are integers; the **round** and **any** commands may be helpful.

2.4.5. ▣ Function 2.4.1 factors $A = LU$ in such a way that $L$ is a *unit* lower triangular matrix—that is, has all ones on the diagonal. It is also possible to define the factorization so that $U$ is a unit upper triangular matrix instead. Write a function [L,U]=lufact2(A) that uses Function 2.4.1 *without modification* to produce this version of the factorization. (Hint: Begin with the standard LU factorization of $A^T$.) Demonstrate on a 4 × 4 example in MATLAB.

2.4.6. When computing the determinant of a matrix by hand, it's common to use cofactor expansion and apply the definition recursively. But this is terribly inefficient as a function of the matrix size.

(a) ✍ Explain why, if $A = LU$ is an LU factorization,

$$\det(A) = U_{11} U_{22} \cdots U_{nn} = \prod_{i=1}^{n} U_{ii}.$$

(b) ▣ Using the result of part (a), write a function determinant(A) that computes the determinant using Function 2.4.1. Use your function and the built-in det on the matrices magic(n) for $n = 3, 4, \ldots, 7$, and make a table showing $n$, the value from your function, and the relative error when compared to det.

(c) ▣ Show that determinant fails for magic(8) but is fine for magic(9). Speculate on what property of these two matrices makes the results so different.

2.4.7. ▣ Consider the portion of Function 2.4.1 in the innermost loop. Because the different iterations in $i$ are all independent, it is possible to rewrite this group of operations without a loop. In fact, the necessary changes are to delete the keyword **for** in the inner loop, and delete the following **end** line. (You should also put a semicolon at the end of i = j+1:n to suppress extra output.)

(a) Make the changes as directed and verify that the function works properly on the matrix from Example 2.4.3.

(b) Write out symbolically (i.e., using ordinary elementwise vector and matrix notation) what the new version of the function does in the case $n = 5$ for the iteration with $j = 3$.

## 2.5 ▪ Efficiency of matrix computations

Predicting how long an algorithm will take to solve a particular problem, on a particular computer, as written in a particular way in a particular programming language, is an enormously difficult undertaking. It's a bit more practical to predict how the required time will scale as a function of the size of the problem. In the case of a linear system of equations, the problem size is $n$, the number of equations/variables in the system. Because expressions of computational time are necessarily approximate, it's traditional to suppress all but the term that is dominant as $n \to \infty$. We first need to build some notation and definitions for these expressions.

For positive functions $f(n)$ and $g(n)$, we say $f(n) = O(g(n))$ ("$f$ is **big-O** of $g$") as $n \to \infty$ if $f(n)/g(n)$ is bounded above for all sufficiently large $n$. We say $f(n) \sim g(n)$ ("$f$ is **asymptotic** to $g$") as $n \to \infty$ if $f(x)/g(x) \to 1$ as $n \to \infty$. One immediate result is that $f \sim g$ implies $f = O(g)$.[13]

### Example 2.5.1

Consider the functions $f(n) = a_1 n^3 + b_1 n^2 + c_1 n$ and $g(n) = a_2 n^3$ in the limit $n \to \infty$. Then
$$\lim_{n \to \infty} \frac{f(n)}{g(n)} = \lim_{n \to \infty} \frac{a_1 + b_1 n^{-1} + c_1 n^{-2}}{a_2} = \frac{a_1}{a_2}.$$
Since $a_1/a_2$ is a constant, $f(n) = O(g(n))$; if $a_1 = a_2$, then $f \sim g$.

### Example 2.5.2

Consider $f(n) = \sin(1/n)$, $g(n) = 1/n$ and $h(n) = 1/n^2$. For large $n$, Taylor's theorem with remainder implies that
$$f(n) = \frac{1}{n} - \cos(1/\xi) \frac{1}{6n^3},$$
where $n < \xi < \infty$. But
$$\lim_{n \to \infty} \frac{f}{g} = \lim_{n \to \infty} 1 - \cos(1/\xi) \frac{1}{6n^2} = 1,$$
and so $f \sim g$. On the other hand, comparing $f$ and $h$, we find
$$\lim_{n \to \infty} \frac{f}{h} = \lim_{n \to \infty} n - \cos(1/\xi) \frac{1}{6n} = \infty,$$
so we cannot say that $f = O(h)$. A consideration of $h/f$ will show that $h = O(f)$, however.

It's conventional to use asymptotic notation that is as specific as possible. For instance, while it is true that $n^2 + n = O(n^{10})$, it's more informative, and usually expected, to say

---

[13] More precisely, $O(g)$ and $\sim g$ are *sets* of functions, and $\sim g$ is a subset of $O(g)$. That we write $f = O(g)$ rather than $f \in O(g)$ is a quirk of convention.

$n^2 + n = O(n^2)$. There is additional notation that enforces this requirement strictly, but we will just stick to the informal understanding.

## Flop counting

Traditionally, in numerical linear algebra we count "floating point operations," or **flops** for short. In our interpretation each scalar addition, subtraction, multiplication, division, and square root counts as one flop. Given any algorithm, we can simply add up the number of scalar flops and ignore everything else.

**Example 2.5.3**

Here is a simple algorithm implementing the multiplication of an $n \times n$ matrix and an $n \times 1$ vector. We use only scalar operations here for maximum transparency.

```
n = 6;
A = magic(n);
x = ones(n,1);
y = zeros(n,1);
for i = 1:n
    for j = 1:n
        y(i) = y(i) + A(i,j)*x(j);     % 2 flops
    end
end
```

Each of the loops implies a summation of flops. The total flop count for this algorithm is

$$\sum_{i=1}^{n}\sum_{j=1}^{n} 2 = \sum_{i=1}^{n} 2n = 2n^2.$$

Since the matrix $A$ has $n^2$ elements, all of which have to be involved in the product, it seems unlikely that we could get a flop count that is smaller than $O(n^2)$.

Let's run an experiment with the built-in matrix-vector multiplication. We use **tic** and **toc** to start and end timing of the computation.

```
n_ = (400:400:4000)';
t_ = 0*n_;
for i = 1:length(n_)
    n = n_(i);
    A = randn(n,n);   x = randn(n,1);
    tic   % start a timer
    for j = 1:10      % repeat ten times
        A*x;
    end
    t = toc;          % read the timer
    t_(i) = t/10;     % seconds per instance
end
```

## 2.5. Efficiency of matrix computations

The reason for doing multiple repetitions at each value of $n$ is to avoid having times so short that the resolution of the timer is a factor.

```
table(n_,t_,'variablenames',{'size','time'})
```

```
ans =
  10x2 table
    size           time
    ____        _____

     400        5.6961e-05
     800        5.4235e-05
    1200        0.00046657
    1600        0.00092544
    2000        0.0015169
    2400        0.0022416
    2800        0.0031205
    3200        0.0041771
    3600        0.005121
    4000        0.00613
```

It's clear from the example that the time increases at a function of $n$—but at what rate? Suppose that the time obeys a function that is not just $O(n^p)$, but actually equal to $Cn^p$ for some constants $C$ and $p$. For large enough $n$, this should be a good approximation. Then

$$t = Cn^p \quad \Longrightarrow \quad \log t = p(\log n) + (\log C). \tag{2.5.1}$$

Hence, a graph of $\log n$ as a function of $\log t$ will be a straight line of slope $p$.

### Example 2.5.4

Let's repeat the experiment of the previous figure for more, and larger, values of $n$.

```
n_ = (400:200:6000)';
t_ = 0*n_;
for i = 1:length(n_)
    n = n_(i);
    A = randn(n,n);  x = randn(n,1);
    tic   % start a timer
    for j = 1:10  % repeat ten times
        A*x;
    end
    t = toc;  % read the timer
    t_(i) = t/10;
end
```

Plotting the time as a function of $n$ on log-log scales is equivalent to plotting the logs of the variables, but is formatted more neatly.

```
loglog(n_,t_,'.-')
xlabel('size of matrix'), ylabel('time (sec)')
```

```
title('Timing of matrix-vector multiplications')
```

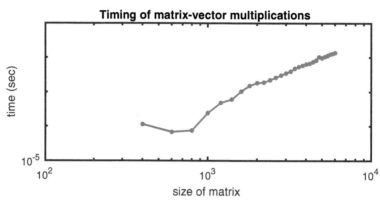

You can see that the graph is trending to a straight line of positive slope. For comparison, we can plot a line that represents $O(n^2)$ growth exactly. (All such lines have slope equal to 2.)

```
hold on, loglog(n_,t_(1)*(n_/n_(1)).^2,'--')
axis tight
legend('data','O(n^2)','location','southeast')
```

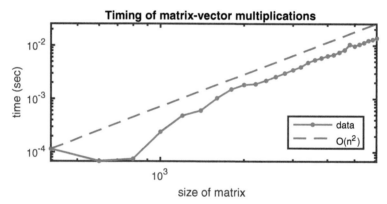

The full story of the execution times is complicated, but the asymptotic approach to $O(n^2)$ is unmistakable.

## Flops for solving linear equations

Recall how we have proposed to solve the system $Ax = b$:

1. Factor $LU = A$ using Gaussian elimination.

2. Solve $Lz = b$ for $z$ using forward substitution.

3. Solve $Ux = z$ for $x$ using backward substitution.

## 2.5. Efficiency of matrix computations

The second and third steps are solved by `forwardsub` (Function 2.3.1) and `backsub` (Function 2.3.2). Only one line in each of these functions performs any arithmetic. Take `forwardsub`, for instance. First is the multiplication

```
L(i,1:i-1)*x(1:i-1)
```

This is the product of a $1 \times (i-1)$ matrix and an $(i-1) \times 1$ vector, i.e., an inner product. It requires $(i-1)$ multiplications and $(i-2)$ additions, for a total of $2i-3$ flops, resulting in a scalar. This scalar is subtracted from `b(i)` and the result is divided by `L(i,i)`, which adds two more flops. Finally, the loop is performed as $i$ ranges from 1 to $n$, so the total count is

$$\sum_{i=1}^{n}(2i-1). \tag{2.5.2}$$

It is not hard to find an exact formula for this sum, but we will use asymptotics to simplify it. Using calculus, it can be proved that

$$\sum_{k=1}^{n} k \sim \frac{n^2}{2} = O(n^2) \quad \text{as } n \to \infty, \tag{2.5.3a}$$

$$\sum_{k=1}^{n} k^2 \sim \frac{n^3}{3} = O(n^3) \quad \text{as } n \to \infty, \tag{2.5.3b}$$

$$\vdots$$

$$\sum_{k=1}^{n} k^p \sim \frac{n^{p+1}}{p+1} = O(n^{p+1}) \quad \text{as } n \to \infty, \tag{2.5.3c}$$

which holds for any positive integer $p$. This formula has memorable similarity to the antiderivative "power rule" for $\int x^p \, dx$. Applying it to (2.5.2) leads us to conclude that *solving a triangular linear system of size $n \times n$ takes $\sim n^2$ flops.* An analysis of backward substitution yields the same result.

Now let's count flops for LU factorization from the listing of Function 2.4.1. Line 14 requires one division, while line 15 requires $n-j+1$ scalar multiplications and the same number of subtractions. Summing over the two loops, the exact flop count is

$$\sum_{j=1}^{n-1} \sum_{i=j+1}^{n} [2(n-j)+3] = \sum_{j=1}^{n-1} (n-j)[2(n-j)+3]. \tag{2.5.4}$$

If we transform the index using $k = n - j$, this becomes

$$\sum_{k=1}^{n-1} k(2k+3) = 2\sum_{k=1}^{n-1} k^2 + 3\sum_{k=1}^{n-1} k$$

$$\sim \frac{2}{3}(n-1)^3 + \frac{3}{2}(n-1)^2$$

$$\sim \frac{2}{3}n^3.$$

In conclusion, *LU factorization takes* $\sim \frac{2}{3}n^3$ *flops as* $n \to \infty$. This dwarfs the $O(n^2)$ count of the triangular system solves.

---

**Example 2.5.5**

We'll test the conclusion of $O(n^3)$ flops experimentally, using the built-in **lu** function instead of the purely instructive `lufact`.

```
n_ = (200:100:2400)';
t_ = 0*n_;
for i = 1:length(n_)
    n = n_(i);
    A = randn(n,n);
    tic    % start a timer
    for j = 1:6,  [L,U] = lu(A);   end
    t = toc;    % read the timer
    t_(i) = t/6;
end
```

We plot the performance on a log–log graph and compare it to $O(n^3)$. The result could vary significantly from machine to machine.

```
loglog(n_,t_,'.-')
hold on, loglog(n_,t_(end)*(n_/n_(end)).^3,'--')
axis tight
xlabel('size of matrix'), ylabel('time (sec)')
title('Timing of LU factorization')
legend('lu','O(n^3)','location','southeast')
```

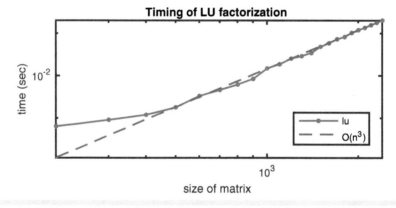

---

In practice, of course, flops are not the only aspect of an implementation that occupies significant time. Our position is that counting flops as a measure of performance is a useful oversimplification. We will assume that LU factorization (and, as a result, the solution of a linear system of $n$ equations) requires a real-world time that is roughly $O(n^3)$. This growth rate is a great deal more tolerable than, say, $O(2^n)$, but it does mean that for (at this writing) $n$ greater than 10,000 or so, something other than general LU factorization will have to be used.

## Exercises

2.5.1. ✍ The following are asymptotic assertions about the limit $n \to \infty$. In each case, prove the statement true or false.

(a) $n^2 = O(\log n)$.

(b) $n^a = O(n^b)$ if $a \le b$.

(c) $e^n \sim e^{2n}$.

(d) $n + \sqrt{n} \sim n + 2\sqrt{n}$.

2.5.2. ✍ The following are asymptotic assertions about the limit $h \to 0$. In each case, prove the statement true or false.

(a) $h^2 \log(h) = O(h^3)$.

(b) $h^a = O(h^b)$ if $a < b$.

(c) $\sin(h) \sim h$.

(d) $(e^{2h} - 1) \sim h$.

2.5.3. ✍ Show that the inner product of two $n$-vectors takes exactly $2n - 1$ flops.

2.5.4. ✍ Show that the multiplication of two $n \times n$ matrices takes $\sim 2n^3$ flops.

2.5.5. ✍ This problem is about evaluation of a polynomial $c_1 x^{n-1} + c_2 x^{n-2} + \cdots + c_{n-1} x + c_n$.

(a) Here is a little code to do the evaluation.

```
y = c(n);
xpow = 1;
for i = 1:n-1
    xpow = xpow * x;
    y = y + c(n-i)*xpow;
end
```

Assuming that **x** is a scalar, how many flops does this function take, as a function of $n$?

(b) Compare the count from (a) to the flop count for Horner's rule, Function 1.3.1.

2.5.6. The exact sums for $p = 1, 2$ in (2.5.3) are as follows:

$$\sum_{k=1}^{n} k = \frac{n(n+1)}{2}, \quad \sum_{k=1}^{n} k^2 = \frac{n(n+1)(2n+1)}{6}.$$

(a) ✍ Use these to find the exact result for (2.5.4).

(b) 💻 On one log–log graph, plot the exact expression from (a) together with the asymptotic result $2n^3/3$ for n=logspace(1,4). (We are assuming that we can extend $n$ from the integers to real numbers.)

(c) 💻 Plot the ratio of the two expressions as a function of $n$ on a **semilogx** plot.

2.5.7. ✍ Show that for any nonnegative constant integer $m$,

$$\sum_{k=0}^{n-m} k^p \sim \frac{n^{p+1}}{p+1}.$$

## 2.6 ▪ Row pivoting

As mentioned in Section 2.4, the $A = LU$ factorization does not work for every nonsingular $A$. A simple rearrangement of Example 2.4.1 demonstrates this fact.

---

**Example 2.6.1**

Here is the previously solved system.

```
A = [2 0 4 3 ; -4 5 -7 -10 ; 1 15 2 -4.5 ; -2 0 2 -13]
b = [ 4; 9; 29; 40 ]
```

```
A =
    2.0000         0    4.0000    3.0000
   -4.0000    5.0000   -7.0000  -10.0000
    1.0000   15.0000    2.0000   -4.5000
   -2.0000         0    2.0000  -13.0000
b =
     4
     9
    29
    40
```

It has a perfectly good solution, obtainable through LU factorization.

```
[L,U] = lufact(A);
x = backsub( U, forwardsub(L,b) )
```

```
x =
    -3
     1
     4
    -2
```

If we swap the second and fourth equations, nothing essential is changed, and MATLAB still finds the solution.

```
A([2 4],:) = A([4 2],:);
b([2 4]) = b([4 2]);
x = A\b
```

```
x =
   -3.0000
    1.0000
    4.0000
   -2.0000
```

However, LU factorization fails.

## 2.6. Row pivoting

```
[L,U] = lufact(A);
L

L =
    1.0000         0         0         0
   -1.0000    1.0000         0         0
    0.5000       Inf    1.0000         0
   -2.0000       Inf       NaN    1.0000
```

The breakdown in Example 2.6.1 is easily explained. After elimination is performed successfully in the first column, we have the matrix

$$\begin{bmatrix} 2 & 0 & 4 & 3 \\ 0 & 0 & 6 & -10 \\ 0 & 15 & 0 & -6 \\ 0 & 5 & 1 & -4 \end{bmatrix}.$$

At this point there is no way to proceed, because trying to find the next row multiplier means division by the zero now sitting in the (2,2) position. In this particular case, we know that we can remove the problem by swapping two of the rows. Fortunately, that capability is all that is needed for *any* invertible matrix.

A row swap is necessary when the algorithm would otherwise require a division by zero. The only context in which this can occur is if the $(j,j)$ element is zero just before the eliminations for column $j$. This number is called the **pivot element**. If it is zero, we need to swap row $j$ with another row in order to create a nonzero pivot. But we don't want to swap with a row *above* row $j$, because that could destroy the upper triangular structure we have partially built. So we search only in rows $j+1$ through $n$ of column $j$ to find a nonzero element to serve as pivot. This process is called **row pivoting** or *partial pivoting*.

What if no nonzero pivot can be found? The following theorem ties up this loose end.

### Theorem 2.6.1

If a pivot element and all the elements below it are zero, then the original matrix $A$ is singular. In other words, if $A$ is nonsingular, then Gaussian elimination with row pivoting will run to completion.

*Proof.* The proof of the first statement is considered in Exercise 2.6.5. The second statement follows from it, because the only way that the Gaussian elimination algorithm can have an undefined step is when all of the choices for a pivot element are zero. □

It's important to keep in mind that Theorem 2.6.1 does not refer to elements of the *original* matrix $A$. Only the numbers that appear during the elimination process are relevant.

## The algebra of row pivoting

Recall that the algebraic expression of LU factorization culminated in (2.4.2) and (2.4.3). Pivoting introduces a new type of elementary matrix called a **permutation matrix**, which is an identity matrix with its rows (or, depending on your point of view, its columns) reordered. The structure of the elimination process implies

$$U = L_{43}P_3(L_{42}L_{32})P_2(L_{41}L_{31}L_{21})P_1 A, \tag{2.6.1}$$

for permutation matrices $P_j$ each featuring a single row swap. With a little more algebra, which we won't give here, this equation can be rearranged to

$$PA = LU, \tag{2.6.2}$$

where $P$ is the permutation matrix resulting from all of the row swaps in the factorization, and $L$ is unit lower triangular. (The construction of $L$ also has to account for the row swaps; we do not give the details.) We emphasize that $P$ cannot be computed all at once from inspection of the original elements of $A$. Rather, the elimination process has to be run in full in order to deduce the row swaps.

Permutation matrices have the property $P^T = P^{-1}$ (see Exercise 2.6.8). As a result, we can rewrite equation (2.6.2) as $A = P^T LU$, and we have a modified factorization of $A$ (which we call a **PLU factorization**). Either way, we could rewrite the linear system $Ax = b$ as $LUx = Pb$, which implies that we can use backward and forward substitution following a permutation of the elements of $b$. However, MATLAB offers an alternative shortcut for this situation.

**Example 2.6.2**

Here is the system that "broke" LU factorization for us.

```
A = [ 2 0 4 3; -2 0 2 -13 ; 1 15 2 -4.5 ; -4 5 -7 -10 ];
b = [ 4; 40; 29; 9 ];
```

When we use the built-in lu function with three outputs, we get the elements of the PLU factorization.

```
[L,U,P] = lu(A)
```

```
L =
    1.0000         0         0         0
   -0.2500    1.0000         0         0
    0.5000   -0.1538    1.0000         0
   -0.5000    0.1538    0.0833    1.0000
U =
   -4.0000    5.0000   -7.0000  -10.0000
         0   16.2500    0.2500   -7.0000
         0         0    5.5385   -9.0769
         0         0         0   -0.1667
P =
         0         0         0         1
         0         0         1         0
```

## 2.6. Row pivoting

```
    0      1      0      0
    1      0      0      0
```

We can solve this as before by incorporating the permutation.

```
x = backsub( U, forwardsub(L,P*b) )

x =
   -3.0000
    1.0000
    4.0000
   -2.0000
```

However, if we use just two outputs with lu, we get $\mathbf{P}^T\mathbf{L}$ as the first result.

```
[PtL,U] = lu(A)

PtL =
   -0.5000    0.1538    0.0833    1.0000
    0.5000   -0.1538    1.0000         0
   -0.2500    1.0000         0         0
    1.0000         0         0         0
U =
   -4.0000    5.0000   -7.0000  -10.0000
         0   16.2500    0.2500   -7.0000
         0         0    5.5385   -9.0769
         0         0         0   -0.1667
```

MATLAB has engineered the backslash so that systems with triangular *or permuted triangular* structure are solved with the appropriate style of triangular substitution.

```
x = U \ (PtL\b)

x =
   -3.0000
    1.0000
    4.0000
   -2.0000
```

The pivoted factorization and triangular substitutions are done silently and automatically when backslash is called on the original matrix.

```
x = A\b

x =
   -3.0000
    1.0000
    4.0000
   -2.0000
```

Row pivoting does not add any flops to the factorization process. It does introduce $O(n^2)$ numerical comparisons, which should be inconsequential compared to the $O(n^3)$ flop requirement.

## Stability and pivoting

If you're very attentive, you may have noticed something curious in Example 2.6.2. The matrix we introduced had a single swap between rows 2 and 4, and therefore the elimination process could have been fixed by making just that swap again. But this is not what is shown by the $P$ computed by MATLAB. The reason is that there is an additional rule about row pivoting that is essential to the success of the algorithm in the presence of rounding errors.

A simple example illustrates the key issue. Let

$$A = \begin{bmatrix} -\epsilon & 1 \\ 1 & -1 \end{bmatrix}, \quad b = \begin{bmatrix} 1-\epsilon \\ 0 \end{bmatrix},$$

where $\epsilon$ is a small positive number. Elimination can be performed without any pivoting:

$$\begin{bmatrix} -\epsilon & 1 & 1-\epsilon \\ 0 & -1+\epsilon^{-1} & \epsilon^{-1}-1 \end{bmatrix} \quad \Rightarrow \quad \begin{aligned} x_2 &= 1, \\ x_1 &= \frac{(1-\epsilon)-1}{-\epsilon}. \end{aligned}$$

In exact arithmetic, this produces the correct solution $x_1 = x_2 = 1$. But look at how $x_1$ is computed. It involves the subtraction of nearby numbers, which is sure to lead to subtractive cancellation in floating point arithmetic. As a result, the solution will lose an arbitrary amount of accuracy, depending on the value of $\epsilon$. In the language of Chapter 1, the computation has a poorly conditioned step, introducing instability into the complete algorithm.

Now suppose we swapped the rows of the matrix before elimination, even though it isn't actually required:

$$\begin{bmatrix} 1 & -1 & 0 \\ 0 & 1-\epsilon & 1-\epsilon \end{bmatrix} \quad \Rightarrow \quad \begin{aligned} x_2 &= 1, \\ x_1 &= \frac{0-(-1)}{1}. \end{aligned}$$

Each of the arithmetic steps is well-conditioned now, and the solution is computed stably.

In general, a small (in absolute value) pivot element means weak dependence on the variable that is about to be eliminated. Even though only a zero pivot makes elimination technically impossible, in floating point using a pivot close to zero can cause instability. As a result, the pivoting rule chosen in practice is the following: *When performing elimination in column $j$, swap row $j$ with the row below it whose entry in column $j$ is the largest (in absolute value).* One consequence of this rule is that all the below-diagonal elements in the unit lower triangular matrix $L$ are bounded above by 1 in absolute value.

## Exercises

2.6.1. ✍ Suppose that $A$ is a square matrix and $b$ is a column vector of compatible length. On the left is correct MATLAB code to solve $Ax = b$; on the right is similar but incorrect code. Explain using mathematical notation exactly what vector is found by the right-hand version.

```
[L,U] = lu(A);              [L,U] = lu(A);
x = U \ (L\b);              x = U \ L \ b;
```

2.6.2. ⌨ Rework Example 2.4.1 with row pivoting, using MATLAB to do the elimination step by step and using visual inspection of the values as you go to determine the correct pivot element for each column. Start with $P = I$ and perform all the same row swaps in $P$ as you do in the process of creating $U$. (There is no need to compute $L$.)

2.6.3. Suppose a string is stretched with tension $\tau$ horizontally between two anchors at $x = 0$ and $x = 1$. At each of the $n-1$ equally spaced positions $x_k = k/n$, $k = 1, \ldots, n-1$, we attach a little mass $m_i$ and allow the string to come to equilibrium. This causes vertical displacement of the string. Let $q_k$ be the amount of displacement at $x_k$. If the displacements are not too large, then an approximate force balance equation is

$$n\tau(q_k - q_{k-1}) + n\tau(q_k - q_{k+1}) = m_k g, \qquad k = 1, \ldots, n-1,$$

where $g = -9.8$ m/s$^2$ is the acceleration due to gravity, and we naturally define $q_0 = 0$ and $q_n = 0$ due to the anchors.

(a) ✍ Show that the force balance equations can be written as a linear system $Aq = f$, where $q$ is a vector of displacements and $A$ is a tridiagonal matrix (that is, $A_{ij} = 0$ if $|i-j| > 1$) of size $(n-1) \times (n-1)$.

(b) ⌨ Let $\tau = 10$ N, and let $m_k = (1/10n)$ kg for every $k$. Find the displacements in MATLAB when $n = 4$ and $n = 40$, and superimpose plots of $q$ over $0 \le x \le 1$ for the two cases. (Be sure to include the zero values at $x = 0$ and $x = 1$ in your plots of the string.)

(c) ⌨ Repeat (b) for the case $m_k = (k/5n^2)$ kg.

2.6.4. ⌨ Repeat the experiment of Exercise 2.4.4 using $PA = LU$ in place of $A = LU$. Does the hypothesis on the entries of $L$, $U$, and their inverses remain true?

2.6.5. ✍ Suppose that $A$ is an $n \times n$ matrix such that its first $k$ columns are upper triangular and $A_{kk} = 0$. Show that $A$ is singular. (This can be used to complete the proof of Theorem 2.6.1, because such a matrix occurs when there is no possible choice of a nonzero pivot.)

2.6.6. ✍ How many unique $6 \times 6$ permutation matrices are there?

2.6.7. ✍ Suppose that A is a $4 \times 6$ matrix in MATLAB and you let

```
B = A(end:-1:1,end:-1:1);
```

Show that $B = PAQ$ for certain matrices $P$ and $Q$.

2.6.8. ✍ Suppose an $n \times n$ permutation matrix $P$ has the effect of moving rows $1, 2, \ldots, n$ to new positions $i_1, i_2, \ldots, i_n$. Then $P$ can be expressed as

$$P = e_{i_1} e_1^T + e_{i_2} e_2^T + \cdots + e_{i_n} e_n^T.$$

(a) For the case $n = 4$ and $i_1 = 3$, $i_2 = 2$, $i_3 = 4$, $i_4 = 1$, write out separately, as matrices, all four of the terms in the sum. Then add them together to find $P$.

(b) Use the formula in the general case to show that $P^{-1} = P^T$.

## 2.7 • Vector and matrix norms

The manipulations on matrices and vectors so far in this chapter have been algebraic, much like those in an introductory linear algebra course. In order to progress to the analysis of the algorithms we have introduced, we need a way to measure the "size" of vectors and matrices—not the number of rows and columns, but a notion of distance from the origin. For vectors we use a **norm** $\|\cdot\|$, which is a function from $\mathbb{C}^n$ to $\mathbb{R}$ with the following properties:

$$\|x\| \geq 0 \quad \text{for all } x \in \mathbb{R}^n, \tag{2.7.1a}$$
$$\|x\| = 0 \quad \text{if and only if } x = 0, \tag{2.7.1b}$$
$$\|\alpha x\| = |\alpha|\|x\| \quad \text{for all } x, \alpha, \tag{2.7.1c}$$
$$\|x + y\| \leq \|x\| + \|y\| \quad \text{for all } x, y. \tag{2.7.1d}$$

Property (2.7.1d) is known as the **triangle inequality**. Just as $\|x\| = \|x - 0\|$ is the distance from $x$ to the origin, $\|x - y\|$ is the distance from $x$ to $y$. The three most important vector norms are

$$\|x\|_2 = \left(\sum_{i=1}^n |x_i|^2\right)^{\frac{1}{2}} = \sqrt{x^*x} \qquad \text{2-norm,} \tag{2.7.2}$$

$$\|x\|_\infty = \max_{i=1,\ldots,n} |x_i| \qquad \text{$\infty$-norm, or max-norm,} \tag{2.7.3}$$

$$\|x\|_1 = \sum_{i=1}^n |x_i| \qquad \text{1-norm.} \tag{2.7.4}$$

The 2-norm corresponds to ordinary Euclidean distance. Most of the time, when just $\|x\|$ is written, the 2-norm is implied. However, in this section we use it to mean a generic, unspecified vector norm.

**Example 2.7.1**

Given the vector $x = \begin{bmatrix} 2 & -3 & 1 & -1 \end{bmatrix}^T$, we have

$$\|x\|_2 = \sqrt{4 + 9 + 1 + 1} = \sqrt{15}, \qquad \|x\|_\infty = \max\{2, 3, 1, 1\} = 3,$$
$$\|x\|_1 = 2 + 3 + 1 + 1 = 7.$$

In MATLAB one uses the **norm** command.

```
x = [2;-3;1;-1];
twonorm = norm(x)             % or norm(x,2)
```

## 2.7. Vector and matrix norms

```
infnorm = norm(x,inf)
onenorm = norm(x,1)

twonorm =
    3.8730
infnorm =
    3
onenorm =
    7
```

In any norm, we refer to a vector $x$ satisfying $\|x\| = 1$ as a **unit vector**. For any nonzero vector $v$ we can find a unit vector through the normalization $x = v/\|v\|$, thanks to property (2.7.1c) of the norm. Thus, we can interpret

$$v = \|v\| \frac{v}{\|v\|} \qquad (2.7.5)$$

as writing $v$ in magnitude–direction form. We say that a sequence of vectors $x_n$ **converges** to $x$ if

$$\lim_{n \to \infty} \|x_n - x\| = 0. \qquad (2.7.6)$$

By definition, a sequence is convergent in the infinity norm if and only if it converges componentwise. The same is true for a convergent sequence in *any* norm.

> **Theorem 2.7.1**
>
> In a finite-dimensional space, convergence in any norm implies convergence in all norms.

### Induced matrix norms

Most useful matrix norms are defined in terms of vector norms. Using a vector norm $\|\cdot\|_a$, we define for any $m \times n$ matrix $A$

$$\|A\|_a = \max_{\|x\|_a = 1} \|Ax\|_a = \max_{x \neq 0} \frac{\|Ax\|_a}{\|x\|_a}. \qquad (2.7.7)$$

The last equality follows from linearity (see Exercise 2.7.5). Thus in the 2-norm, for instance,

$$\|A\|_2 = \max_{\|x\|_2 = 1} \|Ax\|_2.$$

For the rest of this section we will continue to omit subscripts when we want to refer to an unspecified norm; after this section, an unsubscripted norm is understood to be the 2-norm.

The definition of an induced matrix norm may seem complicated. However, there are some key properties that follow directly from the definition.

> **Theorem 2.7.2**
>
> Let $\|\cdot\|$ designate a matrix norm and the vector norm that induced it. Then for all matrices and vectors of compatible sizes,
>
> $$\|Ax\| \leq \|A\| \cdot \|x\|, \qquad (2.7.8a)$$
> $$\|AB\| \leq \|A\| \cdot \|B\|, \qquad (2.7.8b)$$
> $$\|A^k\| \leq \|A\|^k \text{ for any integer } k \geq 0 \text{ if } A \text{ is square.} \qquad (2.7.8c)$$

*Proof.* The first result is trivial if $x = 0$; otherwise,

$$\frac{\|Ax\|}{\|x\|} \leq \max_{x \neq 0} \frac{\|Ax\|}{\|x\|} = \|A\|.$$

Inequality (2.7.8b) then follows because

$$\|ABx\| = \|A(Bx)\| \leq \|A\| \cdot \|Bx\| \leq \|A\| \cdot \|B\| \cdot \|x\|,$$

and then

$$\|AB\| = \max_{x \neq 0} \frac{\|ABx\|}{\|x\|} \leq \max_{x \neq 0} \|A\| \cdot \|B\| = \|A\| \cdot \|B\|.$$

Finally, (2.7.8c) results from repeated application of (2.7.8b). $\qquad \square$

One can interpret the definition of an induced norm geometrically. Each vector $x$ on the unit "sphere" (as defined by the chosen vector norm) is mapped to its image $Ax$, and the norm of $A$ is the radius of the smallest "sphere" that encloses all such images. In addition, two of the vector norms we have encountered lead to equivalent formulas that are easy to compute from the matrix elements:

$$\|A\|_\infty = \max_{1 \leq i \leq n} \sum_{j=1}^n |A_{ij}|, \qquad (2.7.9)$$

$$\|A\|_1 = \max_{1 \leq j \leq n} \sum_{i=1}^n |A_{ij}|. \qquad (2.7.10)$$

In words, the infinity norm is the maximum *row* sum, and the 1-norm is the maximum *column* sum. (One way to keep this straight: the horizontal orientation of the $\infty$ symbol suggests the row sum for the $\infty$-norm, while the vertical orientation of the 1 suggests the column sum of the 1-norm.) The 2-norm of a matrix, however, cannot be summarized by a comparably simple formula.

## 2.7. Vector and matrix norms

**Example 2.7.2**

```
A = [ 2   0;  1   -1 ]

A =
    2    0
    1   -1
```

The default norm returned by the **norm** command is the 2-norm.

```
twonorm = norm(A)

twonorm =
    2.2882
```

You can get the 1-norm as well.

```
onenorm = norm(A,1)

onenorm =
    3
```

The 1-norm is equivalent to

```
max( sum(abs(A),1) )    % sum along rows (1st matrix
    dimension)

ans =
    3
```

Similarly, we can get the ∞-norm and check our formula for it.

```
infnorm = norm(A,inf)
max( sum(abs(A),2) )    % sum along columns (2nd matrix
    dimension)

infnorm =
    2
ans =
    2
```

Here we illustrate the geometric interpretation of the 2-norm. First, we will sample a lot of vectors on the unit circle in $\mathbf{R}^2$.

```
theta = linspace(0,2*pi,601);
x = [ cos(theta); sin(theta) ];    % 601 unit columns
subplot(1,2,1), plot(x(1,:),x(2,:)), axis equal
title('Unit circle in 2-norm')
xlabel('x_1'),  ylabel('x_2')
```

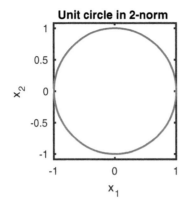

We can apply A to every column of x simply by using

```
Ax = A*x;
```

We superimpose the image of the unit circle with the circle whose radius is $\|\mathbf{A}\|_2$, and display multiple plots with the **subplot** command.

```
subplot(1,2,2), plot(Ax(1,:),Ax(2,:)), axis equal
hold on, plot(twonorm*x(1,:),twonorm*x(2,:),'--')
title('Image of Ax, with ||A||')
xlabel('x_1'), ylabel('x_2')
```

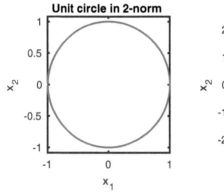
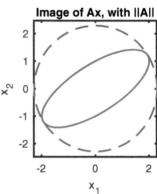

The geometric interpretation of the matrix 2-norm shown in Example 2.7.2, as the radius of the smallest circle (or hypersphere in higher dimensions) containing the images of all unit vectors, is not a practical means of computing the norm. The computation of the matrix 2-norm is discussed further in Section 7.3.

## Exercises

2.7.1. ✍ Why is the vector 1-norm also called the "taxicab norm"?

2.7.2. (a) ✍ Draw the unit "circle" in the $\infty$-norm; that is, the set of all vectors $x \in \mathbb{R}^2$ such that $\|x\|_\infty = 1$.

(b) ✍ Draw the unit "circle" in the 1-norm.

2.7.3. ✍ Show that for all vectors $x \in \mathbb{R}^n$,

(a) $\|x\|_\infty \leq \|x\|_2$.  (b) $\|x\|_2 \leq \|x\|_1$.

2.7.4. ✍ Show that for any vectors $x, y$ in $\mathbb{R}^n$, $|x^T y| \leq \|x\|_1 \|y\|_\infty$.

2.7.5. ✍ Prove using the definition that for any induced matrix norm, matrix $A$, and scalar $c$, $\|cA\| = |c| \cdot \|A\|$.

2.7.6. ✍ Let $A = \begin{bmatrix} -1 & 1 \\ 2 & 2 \end{bmatrix}$.

(a) Find all vectors satisfying $\|x\|_\infty = 1$ and $\|Ax\|_\infty = \|A\|_\infty$.

(b) Find a vector satisfying $\|x\|_1 = 1$ and $\|Ax\|_1 = \|A\|_1$.

(c) Find a vector satisfying $\|x\|_2 = 1$ such that $\|Ax\|_2 = \|A\|_2$. (Hint: A unit two-dimensional vector is a function only of its angle. Use the definition of $\|A\|_2$ as the maximum of $\|Ax\|_2$, which is also a function of the angle.)

2.7.7. ✍ Prove the equivalence of the two formulas for a matrix norm in (2.7.7).

2.7.8. ✍ Explain why, for any permutation matrix $P$, $\|P\|_2 = 1$.

2.7.9. ✍ Show that for any induced matrix norm and nonsingular matrix $A$, $\|A^{-1}\| \geq (\|A\|)^{-1}$. (Hint: Apply Theorem 2.7.2.)

2.7.10. (a) ✍ Show that for any $v \in \mathbb{R}^n$,

$$\|v\|_p \geq \max_{i=1,\ldots,n} |v_i|,$$

where $p = 1, 2$, or $\infty$.

(b) ✍ Show that for any $A \in \mathbb{R}^{n \times n}$,

$$\|A\|_p \geq \max_{i,j=1,\ldots,n} |A_{ij}|,$$

where $p = 1, 2$, or $\infty$. (Hint: For $p = 2$, rearrange (2.7.8a) for a well-chosen particular value of $x$.)

2.7.11. ✍ Show that if $D$ is a diagonal matrix, then $\|D\|_2 = \max_i |D_{ii}|$. You may assume the matrix is real and square, but that does not affect the result or the proof in any significant way. (Hint: Let $M = \max_i |D_{ii}|$. Proceed in two stages, showing that $\|D\|_2 \geq M$ and separately that $\|D\|_2 \leq M$.)

2.7.12. ✍ Suppose that $A$ is $n \times n$ and that $\|A\| < 1$ in some induced matrix norm.

(a) Show that $(I - A)$ is nonsingular. (Hint: Show that $(I - A)x = 0$ for nonzero $x$ implies that $\|A\| \geq 1$, using the definition of an induced matrix norm.)

(b) Show that $\lim_{m \to \infty} A^m = 0$. (For matrices as with vectors, we say $B_m \to L$ if $\|B_m - L\| \to 0$.)

(c) Use (a) and (b) to show that we may obtain the geometric series

$$(I - A)^{-1} = \sum_{k=0}^{\infty} A^k.$$

(Hint: Start with $\left(\sum_{k=0}^{m} A^k\right)(I - A)$ and take the limit.)

## 2.8 • Conditioning of linear systems

We are ready to consider the conditioning of solving the square linear system $Ax = b$. Recall that the condition number is the relative change in the solution divided by a relative change in the data. In this case the data are $A$ and $b$, and the solution is $x$.

For simplicity we start by allowing perturbations to $b$ only while $A$ remains unchanged. Let $Ax = b$ be perturbed to

$$A(x+h) = b+d.$$

We seek to bound $\|h\|$ in terms of $\|d\|$:

$$Ax + Ah = b + d,$$
$$Ah = d,$$
$$h = A^{-1}d,$$
$$\|h\| \leq \|A^{-1}\|\|d\|,$$

where we have used $Ax = b$ and (2.7.8a). Since furthermore $b = Ax$ and therefore $\|b\| \leq \|A\|\|x\|$, we derive

$$\frac{\frac{\|h\|}{\|x\|}}{\frac{\|d\|}{\|b\|}} = \frac{\|h\|\,\|b\|}{\|d\|\,\|x\|} \leq \frac{\left(\|A^{-1}\|\,\|d\|\right)\left(\|A\|\,\|x\|\right)}{\|d\|\,\|x\|} = \|A^{-1}\|\,\|A\|.$$

It is possible to show that this bound is "tight," in the sense that the inequalities are in fact equalities for some choices of $b$ and $d$. Motivated by the definition of the condition number as the ratio of relative changes in solution and data, we define the **matrix condition number**

$$\kappa(A) = \|A^{-1}\|\,\|A\|. \tag{2.8.1}$$

Note that $\kappa(A)$ depends on the choice of norm; a subscript on $\kappa$ such as 1, 2, or $\infty$ is used if clarification is needed. *The matrix condition number is equal to the condition number of solving a linear system of equations.* Although we derived this only for perturbations of $b$, a similar statement holds when $A$ is perturbed. Using a traditional $\Delta$ notation for the perturbation in a quantity, we can write

$$\frac{\|\Delta x\|}{\|x\|} \leq \kappa(A)\frac{\|\Delta b\|}{\|b\|}, \tag{2.8.2a}$$

$$\frac{\|\Delta x\|}{\|x\|} \leq \kappa(A)\frac{\|\Delta A\|}{\|A\|}. \tag{2.8.2b}$$

Equation (2.8.2a) is true without qualification; equation (2.8.2b) is strictly true only in the limit of infinitesimal perturbations $\Delta A$.

Observe that for any induced matrix norm,

$$1 = \|I\| = \|AA^{-1}\| \leq \|A\|\,\|A^{-1}\| = \kappa(A).$$

A condition number of 1 is the best we can hope for—the relative perturbation of the solution has the same size as that of the data. A condition number of size $10^t$ indicates that in floating point arithmetic, roughly $t$ digits are "lost" (i.e., become incorrect) in computing the solution $x$.

## 2.8. Conditioning of linear systems

### Example 2.8.1

MATLAB has a function **cond** to compute $\kappa_2(A)$. The family of *Hilbert matrices* is famously badly conditioned. Here is the $7 \times 7$ case.

```
A = hilb(7);
kappa = cond(A)

kappa =
   4.7537e+08
```

Next we engineer a linear system problem to which we know the exact answer.

```
x_exact = (1:7)';
b = A*x_exact;
```

Now we perturb the data randomly but with norm $10^{-12}$.

```
randn('state',333);          % reproducible results
dA = randn(size(A));   dA = 1e-12*(dA/norm(dA));
db = randn(size(b));   db = 1e-12*(db/norm(db));
```

We solve the perturbed problem using built-in pivoted LU and see how the solution was changed.

```
x = (A+dA) \ (b+db);
dx = x - x_exact;
```

Here is the relative error in the solution.

```
rel_error = norm(dx) / norm(x_exact)

rel_error =
   1.7603e-05
```

And here are upper bounds predicted using the condition number of the original matrix.

```
b_bound = kappa * 1e-12/norm(b)
A_bound = kappa * 1e-12/norm(A)

b_bound =
   4.0852e-05
A_bound =
   2.8621e-04
```

Even if we don't make any manual perturbations to the data, machine epsilon does when we solve the linear system numerically.

```
x = A\b;
rel_error = norm(x - x_exact) / norm(x_exact)
rounding_bound = kappa*eps
```

```
rel_error =
   6.2203e-09
rounding_bound =
   1.0555e-07
```

Because $\kappa \approx 10^8$, it's possible to lose 8 digits of accuracy in the process of passing from $A$ and $b$ to $x$. That's independent of the algorithm; it's inevitable once the data are expressed in double precision.

Now we choose an even more poorly conditioned matrix from this family.

```
A = hilb(14);
kappa = cond(A)
```

```
kappa =
   2.5515e+17
```

Before we compute the solution, note that $\kappa$ exceeds 1/eps. In principle we might end up with an answer that is completely wrong.

```
rounding_bound = kappa*eps
```

```
rounding_bound =
   56.6547
```

MATLAB will notice the large condition number and warn us not to expect much from the result.

```
x_exact = (1:14)';
b = A*x_exact;   x = A\b;
```

In fact the error does exceed 100%.

```
relative_error = norm(x_exact - x) / norm(x_exact)
```

```
relative_error =
   5.7285
```

If $\kappa(A) > \varepsilon_{mach}^{-1}$, then for computational purposes the matrix is singular. If $A$ is exactly singular, it is customary to say that $\kappa(A) = \infty$.

## Residual and backward error

Suppose that $Ax = b$ and $\tilde{x}$ is a computed estimate of the solution $x$. The most natural quantity to study is the error, $x - \tilde{x}$. Normally we can't compute it because we don't know the exact solution. However, we can certainly compute the **residual**, defined as

$$r = b - A\tilde{x}. \tag{2.8.3}$$

Obviously a zero residual means that $\tilde{x} = x$ and we get the exact solution. What does a "small" residual mean? Note that $A\tilde{x} = b - r$, so $\tilde{x}$ solves the linear system problem for a right-hand side that is changed by $-r$. This is precisely what is meant by backward error: the perturbation from the original problem to the one that is solved exactly.

But does a small residual mean that the error is also small? We can reconnect with (2.8.2) by the definition $h = \tilde{x} - x$, in which case $d = A(x + h) - b = Ah = -r$. Hence (2.8.2) is equivalent to

$$\frac{\|x - \tilde{x}\|}{\|x\|} \leq \kappa(A) \frac{\|r\|}{\|b\|}. \tag{2.8.4}$$

Equation (2.8.4) says that the relative error can be much larger than the relative residual when the matrix condition number is large. To put it another way: *When solving a linear system, all that can be expected is that the backward error, not the error, be small.*

## Exercises

2.8.1. A *Hilbert matrix* is a square matrix whose $(i, j)$ entry is $1/(i + j - 1)$. The $n \times n$ version $H_n$ can be generated using the MATLAB command `hilb(n)`. The condition number of a Hilbert matrix grows very rapidly as a function of $n$, showing that even simple, small linear systems can be badly conditioned. Using MATLAB, make a table of the values of $\kappa(H_n)$ in the 2-norm for $n = 2, 3, \ldots, 16$. Why does the growth of $\kappa$ appear to considerably slow down at $n = 13$?

2.8.2. The purpose of this problem is to verify, like in Example 2.8.1, the error bound
$$\frac{\|x - \tilde{x}\|}{\|x\|} \leq \kappa(A) \frac{\|d\|}{\|b\|}.$$
Here $\tilde{x}$ is a numerical approximation to the exact solution $x$, and $d$ is an unknown perturbation caused by machine roundoff. We will assume that $\|d\|/\|b\|$ is of size roughly `eps` in MATLAB.
For each $n = 10, 20, \ldots, 70$ let

```
A = gallery('prolate',n,0.4);
```

The exact solution $x$ should have components $x_k = k/n$ for $k = 1, \ldots, n$, and you should define $b$ as $Ax$. Then $\tilde{x}$ is the solution produced numerically by the backslash.
Make a table of the results, including columns for $n$, the relative error in $\tilde{x}$, the condition number of $A$, and the right-hand side of the inequality above. Does the inequality hold in every case?

2.8.3. Make a table of the $\infty$-norm condition numbers of the matrices in Exercise 2.3.6. (These explain the errors observed in that problem.)

2.8.4. The condition number of the formulation of polynomial interpolation suggested in Section 2.1 is terrible as the degree of the polynomial increases.

Let $A_n$ denote the $(n+1) \times (n+1)$ version of the Vandermonde matrix in equation (2.1.2) based on the equally spaced interpolation nodes $t_i = i/n$ for $i = 0, \ldots, n$.

(a) Using the 1-norm, graph $\kappa(A_n)$ as a function of $n$ for $n = 4, 5, 6, \ldots, 20$, using a log scale on the $y$-axis. (The graph is nearly a straight line.)

(b) Show that if $\log f(n)$ is a linear function of $n$ with positive slope, then $f(n) = C\alpha^n$ for constants $C > 0$ and $\alpha > 1$.

2.8.5. ✍ Define $A_n$ as the $n \times n$ matrix

$$\begin{bmatrix} 1 & -2 & & & \\ & 1 & -2 & & \\ & & \ddots & \ddots & \\ & & & 1 & -2 \\ & & & & 1 \end{bmatrix}.$$

(a) Write out $A_2^{-1}$ and $A_3^{-1}$.

(b) Write out $A_n^{-1}$ in the general case $n > 1$. (If necessary, look at a few more cases in MATLAB until you are certain of the pattern.) Make a clear argument why it is correct.

(c) Using the $\infty$-norm, find $\kappa(A_n)$.

2.8.6. (a) ✍ Prove that for $n \times n$ nonsingular matrices $A$ and $B$, $\kappa(AB) \leq \kappa(A)\kappa(B)$.

(b) ✍ Show by means of an example that the result of part (a) cannot be an equality in general.

2.8.7. ✍ Let $D$ be a diagonal $n \times n$ matrix, not necessarily invertible. Prove that in the 2-norm,

$$\kappa(D) = \frac{\max_i |D_{ii}|}{\min_i |D_{ii}|}.$$

(Hint: See Exercise 2.7.11.)

## 2.9 • Exploiting matrix structure

A common situation in computation is that a problem has certain properties or structure that can be used to get a faster or more accurate solution. There are many properties of a matrix that can affect LU factorization. For example, an $n \times n$ matrix $A$ is **diagonally dominant** if

$$|A_{ii}| > \sum_{\substack{j=1 \\ j \neq i}}^{n} |A_{ij}| \quad \text{for each } i = 1, \ldots, n. \tag{2.9.1}$$

It turns out that a diagonally dominant matrix is guaranteed to be nonsingular, and row pivoting is not required for stability; i.e., $A = LU$ is just as good as $PA = LU$.

We next consider three important types of matrices that cause the LU factorization to be specialized in some important way.

## 2.9. Exploiting matrix structure

### Banded matrices

A matrix $A$ has **upper bandwidth** $b_u$ if $j - i > b_u$ implies $A_{ij} = 0$, and **lower bandwidth** $b_\ell$ if $i - j > b_\ell$ implies $A_{ij} = 0$. We say the total **bandwidth** is $b_u + b_\ell + 1$. When $b_u = b_\ell = 1$, we have the important case of a **tridiagonal matrix**. The **spy** command is handy for visualizing the location of nonzero elements.

#### Example 2.9.1

Here is a matrix with both lower and upper bandwidth equal to one. Such a matrix is called tridiagonal.

```
A = [ 2 -1  0  0  0  0
      4  2 -1  0  0  0
      0  3  0 -1  0  0
      0  0  2  2 -1  0
      0  0  0  1  1 -1
      0  0  0  0  0  2 ]

A =
    2   -1    0    0    0    0
    4    2   -1    0    0    0
    0    3    0   -1    0    0
    0    0    2    2   -1    0
    0    0    0    1    1   -1
    0    0    0    0    0    2
```

We can extract the elements on any diagonal using the `diag` command. The "main" or central diagonal is numbered zero, above and to the right of that is positive, and below and to the left is negative.

```
diag_main = diag(A,0)'

diag_main =
    2    2    0    2    1    2

diag_plusone = diag(A,1)'

diag_plusone =
   -1   -1   -1   -1   -1

diag_minusone = diag(A,-1)'

diag_minusone =
    4    3    2    1    0
```

We can also put whatever numbers we like onto any diagonal with the `diag` command.

```
A = A + diag([5 8 6 7],2)
```

```
A =
    2   -1    5    0    0    0
    4    2   -1    8    0    0
    0    3    0   -1    6    0
    0    0    2    2   -1    7
    0    0    0    1    1   -1
    0    0    0    0    0    2
```

Here is what happens when we factor this matrix without pivoting.

```
[L,U] = lufact(A);
subplot(1,2,1), spy(L), title('L')
subplot(1,2,2), spy(U), title('U')
```

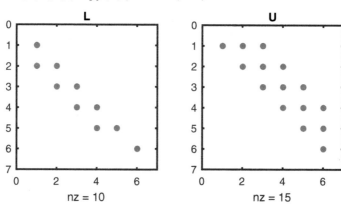

Observe that the factors preserve the lower and upper bandwidth of the original matrix. However, if we introduce row pivoting, this structure may be destroyed.

```
[L,U,P] = lu(A);
subplot(1,2,1), spy(L), title('L')
subplot(1,2,2), spy(U), title('U')
```

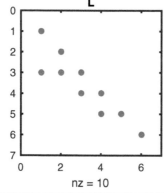
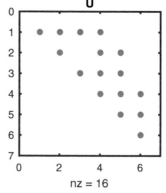

If no row pivoting is used, the $L$ and $U$ factors preserve the lower and upper bandwidths of $A$. This is a consequence of the row elimination process—fewer zeros need to be introduced into each column (equivalently, most of the row multipliers are zero), and adding rows downward cannot introduce new nonzeros into the upper triangle.

## 2.9. Exploiting matrix structure

This observation implies computational savings in both the factorization and the triangular substitutions, because the zeros appear predictably and we can skip multiplication and addition with them. In fact, *the number of flops needed by LU factorization is $O(b_u b_\ell n)$ when the upper and lower bandwidths are $b_u$ and $b_\ell$*.

In order to take advantage of the savings, we would need minor modifications to Function 2.4.1 and the triangular substitution routines. Alternatively, we can get MATLAB to take advantage of the structure automatically by converting the matrix into a special type called **sparse**. Sparse matrices are covered in more detail in Chapter 8.

### Example 2.9.2

We'll use a large banded matrix to observe the speedup possible in LU factorization.

```
n = 8000;
A = diag(1:n) + diag(n-1:-1:1,1) + diag(ones(n-1,1),-1);
```

If we factor the matrix as is, MATLAB has no idea that it could be exploiting the fact that it is tridiagonal.

```
tic, [L,U] = lu(A); toc
```

```
Elapsed time is 2.850122 seconds.
```

But if we convert the matrix to a sparse one, the time required gets a lot smaller.

```
tic, [L,U] = lu(sparse(A)); toc
```

```
Elapsed time is 0.149473 seconds.
```

## Symmetric matrices

If $A^T = A$, then $A$ is symmetric. Symmetric matrices arise frequently because so many types of interactions are symmetric: gravitation, social-network befriending, etc. Symmetry in linear algebra simplifies many properties and algorithms. As a rule of thumb, if your matrix has symmetry, you want to exploit and preserve it.

Now, if we create an LU factorization $A = LU$ of a symmetric $A$, at first glance it seems that because $A^T = U^T L^T$, it might be that $U$ and $L$ are transposes of one another. But that's not possible in general, because we required only $L$ to have ones on the diagonal, and that broke the symmetry. However, it's straightforward to restore the symmetry, as can be demonstrated with a numerical example.

### Example 2.9.3

We begin with a symmetric $A$.

```
A = [  2    4    4     2
       4    5    8    -5
       4    8    6     2
       2   -5    2   -26 ];
```

Carrying out our usual elimination in the first column leads us to

```
L1 = eye(4);  L1(2:4,1) = [-2;-2;-1];
A1 = L1*A
```

```
A1 =
    2    4    4     2
    0   -3    0    -9
    0    0   -2    -2
    0   -9   -2   -28
```

But now let's note that if we transpose this result, we have the same first column as before! So we could apply $L_1$ again and then transpose back.

```
A2 = (L1*A1')'
```

```
A2 =
    2    0    0     0
    0   -3    0    -9
    0    0   -2    -2
    0   -9   -2   -28
```

Using transpose identities, this is just

```
A2 = A1*L1'
```

```
A2 =
    2    0    0     0
    0   -3    0    -9
    0    0   -2    -2
    0   -9   -2   -28
```

Now you can see how we proceed down and to the right, eliminating in a column and then symmetrically in the corresponding row.

```
L2 = eye(4);  L2(3:4,2) = [0;-3];
A3 = L2*A2*L2'
```

```
A3 =
    2    0    0     0
    0   -3    0     0
    0    0   -2    -2
    0    0   -2    -1
```

## 2.9. Exploiting matrix structure

Finally, we arrive at a diagonal matrix.

```
L3 = eye(4); L3(4,3) = -1;
D = L3*A3*L3'

D =
     2     0     0     0
     0    -3     0     0
     0     0    -2     0
     0     0     0     1
```

If no pivoting is done, then the symmetric Gaussian elimination process yields $A = LDL^T$, where $L$ is unit lower triangular and $D$ is diagonal. In practice we don't actually have to carry out any arithmetic in the upper triangle as we work, since the operations are the mirror image of those in the lower triangle. As a result, it can be shown that $LDL^T$ factorization takes about half as much work as the standard LU, or $\sim \frac{1}{3}n^3$ flops.

In the general case we know that row pivoting is necessary to stabilize LU factorization. Pivoting is also needed to keep $LDL^T$ stable, but it has to be done symmetrically. We won't go into the details, as the resulting factorization isn't used very often. Instead, we'll explore the case when the matrix also possesses another important property.

### Symmetric positive definite matrices

Suppose that $A$ is $n \times n$ and $x \in \mathbb{R}^n$. Observe that $x^T A x$ is the product of $1 \times n$, $n \times n$, and $n \times 1$ matrices, so it is a scalar, sometimes referred to as a *quadratic form*. In componentwise terms it becomes

$$x^T A x = \sum_{i=1}^{n} \sum_{j=1}^{n} A_{ij} x_i x_j. \tag{2.9.2}$$

A real matrix $A$ is called **symmetric positive definite** (or SPD) if it is symmetric and

$$x^T A x > 0 \quad \text{for all nonzero } x \in \mathbb{R}^n. \tag{2.9.3}$$

The definiteness property is usually difficult to check directly from the definition. There are equivalent conditions, though; for instance, a symmetric matrix is positive definite if and only if its eigenvalues are all real positive numbers. SPD matrices have important properties and appear in applications in which the definiteness is known for theoretical reasons.

Let us consider what definiteness means to the $LDL^T$ factorization (itself the adaptation of LU factorization to symmetric matrices). We compute

$$0 < x^T A x = x^T L D L^T x = z^T D z,$$

where $z = L^T x$. Note that since $L$ is unit lower triangular, it is nonsingular, so $x = L^{-T} z$. By taking $z = e_k$ for $k = 1,\ldots,n$, we can read the equalities from right to left and conclude that $D_{kk} > 0$ for all $k$. That permits us to write a "square root" formula[14]

$$D = \begin{bmatrix} D_{11} & & & \\ & D_{22} & & \\ & & \ddots & \\ & & & D_{nn} \end{bmatrix} = \begin{bmatrix} \sqrt{D_{11}} & & & \\ & \sqrt{D_{22}} & & \\ & & \ddots & \\ & & & \sqrt{D_{nn}} \end{bmatrix}^2 = \left(D^{1/2}\right)^2. \quad (2.9.4)$$

Now we have $A = LD^{1/2}D^{1/2}L^T = R^T R$, where $R = D^{1/2}L^T$ is an upper triangular matrix whose diagonal entries are positive.

Recall that the unpivoted $LDL^T$ (like unpivoted LU) factorization is not stable and not even always possible. However, in the SPD case one can prove that pivoting is not necessary for the existence nor the stability of the factorization $A = R^T R$, which is known as the **Cholesky factorization**. The elimination process is readily adapted into an algorithm for Cholesky factorization. Like $LDL^T$, *the Cholesky factorization requires* $\sim \frac{1}{3}n^3$ *flops asymptotically in the $n \times n$ case*, half as many as standard LU factorization.

The speed and stability of the Cholesky factorization make it the top choice for solving linear systems with SPD matrices. As a side benefit, the Cholesky algorithm fails, in the sense of trying to take a negative square root or divide by zero, if and only if the matrix $A$ is indefinite (i.e., not positive definite). In practice this is a good way to test the definiteness of a symmetric matrix about which nothing else is known.

**Example 2.9.4**

Here is a simple trick for turning any square matrix into a symmetric one.

```
A = magic(4) + eye(4);
B = A+A'
```

```
B =
     34      7     12     17
      7     24     17     22
     12     17     14     27
     17     22     27      4
```

Picking a symmetric matrix at random, there is little chance that it will be positive definite. Fortunately, the built-in Cholesky factorization **chol** always detects this property. The following would cause an error if run:

```
R = chol(B)
```

There is a different trick for making an SPD matrix from (almost) any other matrix.

---

[14] Except for the diagonal, positive definite case, it's not trivial to define the square root of a matrix, so don't misuse the notation.

```
B = A'*A

B =
   411   213   224   377
   213   393   385   234
   224   385   383   233
   377   234   233   381

R = chol(B)

R =
   20.2731   10.5065   11.0491   18.5960
         0   16.8111   15.9961    2.2973
         0         0    2.2453   -4.1053
         0         0         0    3.6133

norm( R'*R - B )

ans =
   7.0520e-14
```

A word of caution: chol does not check symmetry; in fact, it doesn't even look at the lower triangle of the input matrix.

```
chol( triu(B) )

ans =
   20.2731   10.5065   11.0491   18.5960
         0   16.8111   15.9961    2.2973
         0         0    2.2453   -4.1053
         0         0         0    3.6133
```

## Exercises

2.9.1. For each matrix, use (2.9.1) to determine whether it is diagonally dominant.

$$A = \begin{bmatrix} 3 & 1 & 0 & 1 \\ 0 & -2 & 0 & 1 \\ -1 & 0 & 4 & -1 \\ 0 & 0 & 0 & 6 \end{bmatrix}, \quad B = \begin{bmatrix} 1 & 0 & 0 & 0 & 0 \\ 0 & 1 & 0 & 0 & 0 \\ 0 & 0 & 1 & 0 & 0 \\ 0 & 0 & 0 & 1 & 0 \\ 0 & 0 & 0 & 0 & 0 \end{bmatrix},$$

$$C = \begin{bmatrix} 2 & -1 & 0 & 0 \\ -1 & 2 & -1 & 0 \\ 0 & -1 & 2 & -1 \\ 0 & 0 & -1 & 2 \end{bmatrix}.$$

2.9.2. 🖮 For each matrix, use inspection or chol in MATLAB to determine whether it is SPD.

$$A = \begin{bmatrix} 1 & 0 & -1 \\ 0 & 4 & 5 \\ -1 & 5 & 10 \end{bmatrix}, \quad B = \begin{bmatrix} 1 & 0 & 1 \\ 0 & 4 & 5 \\ -1 & 5 & 10 \end{bmatrix}, \quad C = \begin{bmatrix} 1 & 0 & 1 \\ 0 & 4 & 5 \\ 1 & 5 & 1 \end{bmatrix}.$$

2.9.3. ✍ Show that the diagonal entries of a positive definite matrix are positive numbers. (Hint: Use special cases of (2.9.3).)

2.9.4. 🖮 Using Function 2.4.1 as a guide, write a function

```
function [L,U] = luband(A,upper,lower)
```

that accepts upper and lower bandwidth values and computes LU factors in a way that avoids doing arithmetic using the locations that are known to stay zero.

2.9.5. 🖮 In this problem you will explore the backslash and how it handles banded matrices. To do this you will generate tridiagonal matrices using the following code:

```
A = triu( tril(rand(n),1), -1);
A(1:n+1:end) = a;
```

The result is $n \times n$, with each entry on the sub- and superdiagonals chosen randomly from $(0, 1)$ and each diagonal entry equaling a.

(a) Write a script that solves 200 linear systems whose matrices are generated as above, with $n = 1000$ and $a = 2$. (Use randomly generated right-hand side vectors.) Record the total time used by the solution process A\b only, using the built-in cputime.

(b) Repeat the experiment of part (a), but add the command A=sparse(A); right after the two lines above. How does the timing change?

(c) Repeat parts (a) and (b) with $a = 1$. In which case is there a major change?

(d) Based on these observations, state a hypothesis on how backslash solves tridiagonal linear systems given in standard dense form and in sparse form. (Hint: What is the mathematical significance of $a = 2$ versus $a = 1$?)

2.9.6. ✍ A matrix $A$ is called *skew-symmetric* if $A^T = -A$. Explain why unpivoted LU factorization of a skew-symmetric matrix is impossible.

2.9.7. ✍ Prove that if $A$ is any real nonsingular square matrix, then $A^T A$ is SPD.

**Key ideas in this chapter**

1. Polynomial interpolation leads to a linear system of equations with a Vandermonde matrix (page 32).
2. Gaussian elimination is equivalent to LU factorization (page 56).
3. Solving a triangular linear system of size $n \times n$ takes $\sim n^2$ flops (page 65).
4. LU factorization takes $\sim \frac{2}{3}n^3$ flops as $n \to \infty$ (page 66).
5. Row pivoting is required for numerical stability (page 72).
6. The matrix condition number is equal to the condition number of solving a linear system of equations (page 80).

7. If $\kappa(A) > \varepsilon_{mach}^{-1}$, then for computational purposes the matrix is singular (page 82).
8. When the linear system matrix is ill-conditioned, one can expect only a small residual, not a small error (page 83).
9. LU factorization for a $(p,q)$-banded $n \times n$ matrix takes $O(pqn)$ flops (page 87).
10. An SPD matrix has a Cholesky factorization, which can be found in $\sim n^3/3$ flops (page 90).

## Where to learn more

The underlying theory for linear systems is covered in numerous linear algebra textbooks. Some popular choices include those by Strang [62], Lay [38], and Leon [39], but there are many other good choices.

More advanced texts specifically on numerical linear algebra include the classic texts by Trefethen and Bau [72] and Golub and Van Loan [27]. Numerical analysis of the fundamental algorithms is emphasized in Higham [33]; there are many interesting quotations there as well. Here is a sample:

> Many years ago we made out of half a dozen transformers a simple and rather inaccurate machine for solving simultaneous equations—the solutions being represented as flux in the cores of the transformers. During the course of our experiments we set the machine to solve the equations—
>
> $$X + Y + Z = 1$$
> $$X + Y + Z = 2$$
> $$X + Y + Z = 3$$
>
> The machine reacted sharply—it blew the main fuse and put all the lights out.
> —B. V. Bowden, *The Organization of a Typical Machine* (1953)

The reader may find historical information on numerical linear algebra of interest at histories.siam.org. The materials include presentations, oral histories, articles, and links on a wide range of topics in numerical analysis and scientific computing, including numerical linear algebra.

# Chapter 3
# Overdetermined linear systems

> I must have hit her pretty close to the mark to get her all riled up like that, huh, kid?
> —Han Solo, *The Empire Strikes Back*

So far we have considered $Ax = b$ only when $A$ is a square matrix. In this chapter we consider how to interpret and solve the problem for an $m \times n$ matrix where $m > n$—and in practice, $m$ is often *much* larger than $n$. This is called an **overdetermined** linear system because, in general, the system has more equations to satisfy than the variables allow. The complementary *underdetermined* case $m < n$ turns up less frequently and will not be considered in this book.

Since we cannot solve all of the system's equations, we need to define what the "best possible" answer is. There are multiple useful options, but the most important version of the overdetermined problem occurs using the *least squares* criterion—*the sum of the squares of the equation residuals is minimized*. This is far from an arbitrary criterion. Mathematically, we recognize the sum-of-squares as a vector 2-norm and therefore tied to inner products; physically, the 2-norm may coincide with energy, which is often minimized by natural systems, and statistically, least squares leads to the estimates of maximum likelihood for certain models. Furthermore the solution of the least squares problem requires only linear algebra and is about as easily to compute as in the square case.

The linear least squares problem serves as our introduction to the vast field of *optimization*. It is one of the simplest problems of this type. We will see an extension to a nonlinear version in Chapter 4.

# 3.1 • Fitting functions to data

In Section 2.1 we showed how a polynomial can be used to interpolate data—that is, derive a continuous function that evaluates to give a set of prescribed values. But interpolation may not be appropriate in many applications.

### Example 3.1.1

Here are 5-year averages of the worldwide temperature anomaly as compared to the 1951–1980 average (source: NASA).

```
t = (1955:5:2000)';
y = [ -0.0480; -0.0180; -0.0360; -0.0120; -0.0040;
    0.1180; 0.2100; 0.3320; 0.3340; 0.4560 ];
plot(t,y,'.')
```

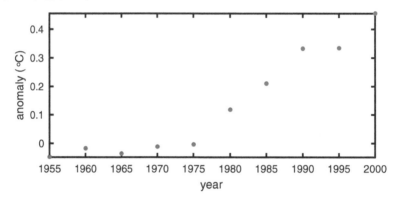

A polynomial interpolant can be used to fit the data. Here we build one using a Vandermonde matrix. First, though, we express time as decades since 1950, as it improves the condition number of the matrix.

```
t = (t-1950)/10;
V = t.^0;        % vector of ones
n = length(t);
for j = 1:n-1
    V(:,j+1) = t.*V(:,j);
end
c = V\y;
p = @(x) polyval(flipud(c),(x-1950)/10);
hold on, fplot(p,[1955 2000])
```

## 3.1. Fitting functions to data

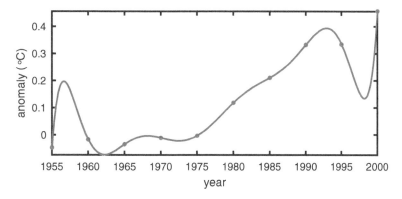

As you can see, the interpolant does represent the data, in a sense. However, it's a crazy-looking curve for the application. Trying too hard to reproduce all the data exactly is known as *overfitting*.

In many cases we can get better results by relaxing the interpolation requirement. In the polynomial case this allows us to lower the degree of the polynomial, which limits the number of local max and min points. Let $(t_i, y_i)$ for $i = 1, \ldots, m$ be the given points. We will represent the data by the polynomial

$$y \approx f(t) = c_1 + c_2 t + \cdots + c_{n-1} t^{n-2} + c_n t^{n-1}, \qquad (3.1.1)$$

with $n < m$. Just as in (2.1.2), we can express a vector of $f$ values by a matrix-vector multiplication:

$$\begin{bmatrix} f(t_1) \\ f(t_2) \\ f(t_3) \\ \vdots \\ f(t_m) \end{bmatrix} = \begin{bmatrix} 1 & t_1 & \cdots & t_1^{n-1} \\ 1 & t_2 & \cdots & t_2^{n-1} \\ 1 & t_3 & \cdots & t_3^{n-1} \\ \vdots & \vdots & & \vdots \\ 1 & t_m & \cdots & t_m^{n-1} \end{bmatrix} \begin{bmatrix} c_1 \\ c_2 \\ \vdots \\ c_m \end{bmatrix}. \qquad (3.1.2)$$

Note that $V$ has the same structure as the Vandermonde matrix in (2.1.2) but is $m \times n$, thus taller than it is wide. It's impossible in general to satisfy $m$ conditions with $n < m$ variables, so rather than solving a linear system $Vc = y$, we have to find an approximation $Vc \approx y$. Below we specify precisely what is meant by this, but first we note that MATLAB conveniently uses the same backslash notation to solve the problem for both square and rectangular matrices.

### Example 3.1.2

Here are the 5-year temperature averages again.

```
year = (1955:5:2000)';
y = [ -0.0480; -0.0180; -0.0360; -0.0120; -0.0040;
      0.1180;  0.2100;  0.3320;  0.3340;  0.4560 ];
```

The standard best-fit line results from using a linear polynomial that meets the least squares criterion.

```
t = year - 1955;            % better matrix conditioning
     later
V = [ t.^0 t ];     % Vandermonde-ish matrix
c = V\y;
f = @(x) polyval(c(end:-1:1),x-1955);
fplot(f,[1955 2000])
hold on, plot(year,y,'.')
```

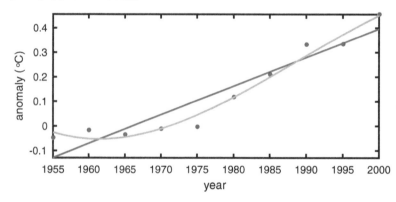

If we use a global cubic polynomial, the points are fit more closely.

```
V = [ t.^0 t t.^2 t.^3];        % Vandermonde-ish matrix
c = V\y;   f = @(x) polyval(c(end:-1:1),x-1955);
fplot(f,[1955 2000])
```

If we were to continue increasing the degree of the polynomial, the residual at the data points would get smaller, but overfitting would increase.

## The least squares formulation

In the most general terms, our fitting functions take the form

$$f(t) = c_1 f_1(t) + \cdots + c_n f_n(t), \tag{3.1.3}$$

where $f_1, \ldots, f_n$ are all known functions with no undetermined parameters. This leaves only $c_1, \ldots, c_n$ to be determined. *The essential feature of a linear least squares problem is that the fit depends only* linearly *on the unknown parameters.* For instance, a function of the form $f(t) = c_1 + c_2 e^{c_3 t}$ is not of this type.

## 3.1. Fitting functions to data

At each observation $(t_i, y_i)$, we define a residual, $y_i - f(t_i)$. A sensible formulation of the fitting criterion is to minimize

$$R(c_1, \ldots, c_n) = \sum_{i=1}^{m} [y_i - f(t_i)]^2$$

over all possible choices of parameters $c_1, \ldots, c_n$. We can apply linear algebra to write the problem in the form $R = r^T r$, where

$$r = \begin{bmatrix} y_1 \\ y_2 \\ \vdots \\ y_{m-1} \\ y_m \end{bmatrix} - \begin{bmatrix} f_1(t_1) & f_2(t_1) & \cdots & f_n(t_1) \\ f_1(t_2) & f_2(t_2) & \cdots & f_n(t_2) \\ & & \vdots & \\ f_1(t_{m-1}) & f_2(t_{m-1}) & \cdots & f_n(t_{m-1}) \\ f_1(t_m) & f_2(t_m) & \cdots & f_n(t_m) \end{bmatrix} \begin{bmatrix} c_1 \\ c_2 \\ \vdots \\ c_n \end{bmatrix}.$$

Recalling that $r^T r = \|r\|_2^2$, and renaming the variables to standardize the statement, we arrive at the general **linear least squares problem**: Given $A \in \mathbb{R}^{m \times n}$ and $b \in \mathbb{R}^m$, with $m > n$, find

$$\operatorname{argmin}_{x \in \mathbb{R}^n} \|b - Ax\|_2^2. \tag{3.1.4}$$

The notation "argmin" means to find an $x$ that produces the minimum value. While we could choose to minimize in any other vector norm, the 2-norm is the most common and convenient choice. *For the rest of this chapter we exclusively use the 2-norm.* An $x$ that achieves the minimum residual is considered to be a solution of the problem. In the edge case $m = n$ for a nonsingular $A$, the definitions of the linear least squares and linear systems problems coincide: the solution of $Ax = b$ implies $r = 0$, which is a global minimum of $\|r\|_2^2 \geq 0$.

### Change of variables

The most familiar and common case of a polynomial least squares fit is the straight line, $f(t) = c_1 + c_2 t$. Certain other fit functions can be transformed into this situation. For example, suppose we want to fit data using $g(t) = a_1 e^{a_2 t}$. Then

$$\log y \approx \log g(t) = (\log a_1) + a_2 t = c_1 + c_2 t. \tag{3.1.5}$$

While the fit of the $y_i$ to $f(t)$ is nonlinearly dependent on fitting parameters, the fit of $\log y_i$ to a straight line is a linear problem. Similarly, the power-law relationship $y \approx f(t) = a_1 t^{a_2}$ is equivalent to

$$\log y \approx (\log a_1) + a_2 (\log t). \tag{3.1.6}$$

Thus the variable $z = \log y$ can be fit linearly in terms of the variable $s = \log t$. In MATLAB these two cases—exponential fit and power law—are easily discerned by using **semilogy** or **loglog** plots, respectively.

### Example 3.1.3

Finding numerical approximations to $\pi$ has fascinated people for millennia. One famous formula is

$$\frac{\pi^2}{6} = 1 + \frac{1}{2^2} + \frac{1}{3^2} + \cdots.$$

Say $s_k$ is the sum of the first $k$ terms of the series above, and $p_k = \sqrt{6 s_k}$. Here is a fancy way to compute these sequences in a compact code.

```
k = (1:100)';
s = cumsum( 1./k.^2 );   % cumulative summation
p = sqrt(6*s);
plot(k,p,'.-')
xlabel('k'), ylabel('p_k')
title('Sequence converging to \pi')
```

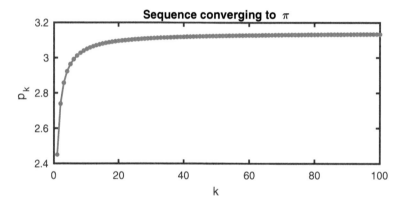

This graph suggests that $p_k \to \pi$ but doesn't give much information about the rate of convergence. Let $\epsilon_k = |\pi - p_k|$ be the sequence of errors. By plotting the error sequence on a log–log scale, we can see a nearly linear relationship.

```
ep = abs(pi-p);    % error sequence
loglog(k,ep,'.'), title('log-log convergence')
xlabel('k'), ylabel('error'),   axis tight
```

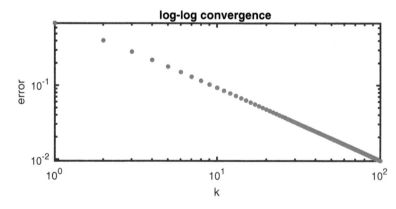

This suggests a power-law relationship where $\epsilon_k \approx a k^b$, or $\log \epsilon_k \approx b(\log k) + \log a$.

```
V = [ k.^0, log(k) ];        % fitting matrix
c = V \ log(ep)              % coefficients of linear fit

c =
    -0.1824
    -0.9674
```

In terms of the parameters $a$ and $b$ used above, we have

```
a = exp(c(1)),    b = c(2),

a =
    0.8333
b =
    -0.9674
```

It's tempting to conjecture that $b = -1$ asymptotically. Here is how the numerical fit compares to the original convergence curve.

```
hold on, loglog(k,a*k.^b,'r'), title('power-law fit')
```

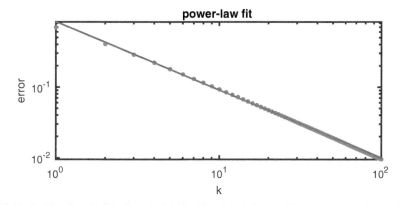

## Exercises

3.1.1. ✎ Suppose $f(x)$ is a differentiable nonnegative real function. Use calculus to prove that $f(x)$ and $[f(x)]^2$ have local minima at the same values of $x$.

3.1.2. ▦ MATLAB comes with data from the U.S. census, collected every ten years from 1790 to 1990. To work with this data, enter

```
load census
t = cdate - 1790;   % years since 1790
y = pop;            % population in millions
```

(a) Find a best-fitting cubic polynomial for the data. Plot the data as points superimposed on a (smooth) graph of the cubic over the full range of time. Label the axes. What is the predicted population for the year 2000? 2010?

(b) Repeat (a) but use a fitting function of the form $y(t) \approx ae^{bt}$.

(c) Look up the actual U.S. population in 2000 and 2010 and compare to the predictions of parts (a) and (b).

3.1.3. In this problem you are trying to find an approximation to the periodic function $f(t) = e^{\sin(t-1)}$ over one period, $0 \le t \le 2\pi$. In MATLAB, let `t=linspace(0,2*pi,200)'` and let b be a column vector of evaluations of $f$ at those points.

(a) Find the coefficients of the least squares fit
$$f(t) \approx c_1 + c_2 t + \cdots + c_7 t^7.$$

(b) Find the coefficients of the least squares fit
$$f(t) \approx d_1 + d_2 \cos(t) + d_3 \sin(t) + d_4 \cos(2t) + d_5 \sin(2t).$$

(c) Plot the original function $f(t)$ and the two approximations from (a) and (b) together on a well-labeled graph.

3.1.4. Define the following data in MATLAB:

`t = (0:.5:10)';   y = tanh(t);`

(a) Fit the data to a cubic polynomial and plot the data together with the polynomial fit.

(b) Fit the data to the function $c_1 + c_2 z + c_3 z^2 + c_4 z^3$, where $z = t^2/(1+t^2)$. Plot the data together with the fit. What feature of $z$ makes this fit much better than the original cubic?

3.1.5. One series for finding $\pi$ is
$$\frac{\pi}{2} = 1 + \frac{1}{3} + \frac{1 \cdot 2}{3 \cdot 5} + \frac{1 \cdot 2 \cdot 3}{3 \cdot 5 \cdot 7} + \cdots.$$

Define $s_k$ to be the sum of the first $k$ terms (that is, up to $j = k-1$), and let $e_k = |\pi - 2s_k|$. Do a numerical experiment in MATLAB to determine which is a better fit, $e_k \approx ab^k$ or $e_k \approx ak^b$. Then determine the parameters in that fit.

3.1.6. Kepler found that the orbital period $\tau$ of a planet depends on its mean distance $R$ from the sun according to $\tau = cR^\alpha$ for a simple rational number $\alpha$. Validate Kepler's result in MATLAB by using a linear least squares fit and the following table.

| Planet  | Distance from sun ($R$, in km $\times 10^6$) | Orbital period ($\tau$, in days) |
|---------|------|---------|
| Mercury | 57.59 | 87.99 |
| Venus   | 108.11 | 224.7 |
| Earth   | 149.57 | 365.26 |
| Mars    | 227.84 | 686.98 |
| Jupiter | 778.14 | 4332.4 |
| Saturn  | 1427 | 10759 |
| Uranus  | 2870.3 | 30684 |
| Neptune | 4499.9 | 60188 |
| Pluto   | 5909 | 90710 |

3.1.7. Show that finding a best fit of the form
$$y(t) \approx \frac{a}{t+b}$$

can be transformed into a linear fitting problem (with different undetermined coefficients) by rewriting the equation.

3.1.8. ✍ Show how to find the constants $a$ and $b$ in a data fitting problem of the form $y(t) \approx t/(at+b)$ by transforming it into a linear least squares fitting problem.

## 3.2 ▪ The normal equations

We seek to solve the general linear least squares problem: Given $A \in \mathbb{R}^{m \times n}$ and $b \in \mathbb{R}^m$, with $m > n$, find $x \in \mathbb{R}^n$ such that $\|b - Ax\|_2$ is minimized. There is a concise explicit solution to the problem. In the following proof we make use of the elementary algebraic fact that for two vectors $u$ and $v$,

$$(u+v)^T(u+v) = u^T u + u^T v + v^T u + v^T v = u^T u + 2 v^T u + v^T v.$$

> **Theorem 3.2.1**
>
> If $x$ satisfies $A^T(Ax - b) = 0$, then $x$ solves the linear least squares problem, i.e., $x$ minimizes $\|b - Ax\|_2$.

*Proof.* Let $y \in \mathbb{R}^n$ be any vector. Then $A(x+y) - b = Ax - b + Ay$, and

$$\begin{aligned}
\|A(x+y) - b\|_2^2 &= [(Ax - b) + (Ay)]^T [(Ax - b) + (Ay)] \\
&= (Ax - b)^T (Ax - b) + 2(Ay)^T (Ax - b) + (Ay)^T (Ay) \\
&= \|Ax - b\|_2^2 + 2 y^T A^T (Ax - b) + \|Ay\|_2^2 \\
&= \|Ax - b\|_2^2 + \|Ay\|_2^2 \\
&\geq \|Ax - b\|_2^2.
\end{aligned}$$
□

The condition $A^T(Ax - b) = 0$ is often written as

$$A^T A x = A^T b, \tag{3.2.1}$$

called the **normal equations**. They have a straightforward geometric interpretation, as shown in Figure 3.1. The vector in the range (column space) of $A$ that lies closest to $b$ makes the vector difference $Ax - b$ perpendicular to the range. Thus for any $z$, we must have $(Az)^T (Ax - b) = 0$, which is satisfied if $A^T(Ax - b) = 0$.

If we group the left-hand side of the normal equations as $(A^T A)x$, we recognize (3.2.1) as a *square* linear system to solve for $x$. *The normal equations express an $m \times n$ linear least squares problem as an $n \times n$ linear system of equations.*

### Pseudoinverse and definiteness

The $n \times m$ matrix

$$A^+ = (A^T A)^{-1} A^T \tag{3.2.2}$$

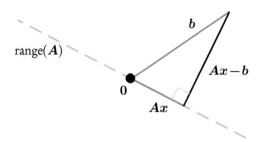

**Figure 3.1.** *Geometry of the normal equations.*

is called the **pseudoinverse** of $A$. Mathematically, the overdetermined least squares problem $Ax \approx b$ has the solution $x = A^+ b$. Hence we can generalize our earlier observation on the backslash: *In MATLAB, backslash is equivalent mathematically to left-multiplication by the inverse (square case) or pseudoinverse (rectangular case) of a matrix.* One may also compute the pseudoinverse directly using **pinv**, but as with matrix inverses, this is rarely necessary in practice.

The matrix $A^T A$ appearing in the pseudoinverse has some important properties.

> **Theorem 3.2.2**
>
> For any real $m \times n$ matrix $A$ with $m \geq n$, the following are true:
>
> 1. $A^T A$ is symmetric.
>
> 2. $A^T A$ is singular if and only if the columns of $A$ are linearly dependent. (Equivalently, the rank of $A$ is less than $n$.)
>
> 3. If $A^T A$ is nonsingular, then it is positive definite.

*Proof.* The first part is left as an exercise (Exercise 3.2.3). For the second part, suppose that $A^T A z = 0$. Note that $A^T A$ is singular if and only if $z$ may be nonzero. Left-multiplying by $z^T$, we find that

$$0 = z^T A^T A z = (Az)^T (Az) = \|Az\|_2^2,$$

which is equivalent to $Az = 0$. Then $z$ may be nonzero if and only if the columns of $A$ are linearly dependent.

Finally, we can repeat the manipulations above to show that for any nonzero $n$-vector $v$, $v^T (A^T A) v = \|Av\|_2^2 \geq 0$, and equality is not possible thanks to the second part of the theorem. □

The definition of the pseudoinverse involves taking the inverse of a matrix and is therefore not advisable to use computationally. Instead, we simply use the definition of the normal equations to set up a linear system, which we already know how to solve.

## 3.2. The normal equations

**Function 3.2.1** (lsnormal) Solve linear least squares by normal equations.

```
function x = lsnormal(A,b)
% LSNORMAL   Solve linear least squares by normal equations.
% Input:
%   A        coefficient matrix (m by n, m>n)
%   b        right-hand side (m by 1)
% Output:
%   x        minimizer of || b-Ax ||

N = A'*A;   z = A'*b;
R = chol(N);
w = forwardsub(R',z);                  % solve R'z=c
x = backsub(R,w);                      % solve Rx=z
```

In summary, the steps for solving the linear least squares problem $Ax \approx b$ are as follows:

1. Compute $N = A^T A$.
2. Compute $z = A^T b$.
3. Solve the $n \times n$ linear system $Nx = z$ for $x$.

In the last step we can exploit the fact, proved in Theorem 3.2.2, that $N$ is symmetric and positive definite, and use Cholesky factorization as in Section 2.9. (The backslash in MATLAB does this automatically.)

### Conditioning and stability

We have already used A\b as the native way to solve the linear least squares problem $Ax \approx b$ in MATLAB. The algorithm employed by the backslash does *not* proceed through the normal equations, and the reason is stability.

The conditioning of the linear least squares problem relates changes in the solution $x$ to those in the data, $A$ and $b$. A full accounting of the condition number is too messy to present here, but we can generalize from the linear system problem $Ax = b$, where $m = n$. Recall that the condition number of solving $Ax = b$ is $\kappa(A) = \|A\| \cdot \|A^{-1}\|$. Provided that the residual norm $\|b - Ax\|$ at the least squares solution is relatively small, the conditioning of the linear least squares problem is similar. The condition number of the problem generalizes via the pseudoinverse:

$$\kappa(A) = \|A\| \cdot \|A^+\|. \tag{3.2.3}$$

These are rectangular matrices, but the induced matrix norm is defined by (2.7.7) just as in the square case. If $A$ has rank less than $n$, then $\kappa(A) = \infty$. The MATLAB function **cond** computes condition numbers of rectangular matrices in the 2-norm.

As an algorithm, the normal equations begin by computing the $n \times n$ system $(A^T A)x = A^T b$. *When these equations are solved, perturbations to the data can be amplified by a factor* $\kappa(A^T A)$. In Exercise 7.3.8 you are asked to prove that

$$\kappa(A^T A) = \kappa(A)^2 \tag{3.2.4}$$

in the 2-norm. If $\kappa(A)$ is large, the squaring of it can destabilize the normal equations: while the solution of the least squares problem is sensitive, finding it via the normal equations makes it doubly so.

### Example 3.2.1

Because the functions $\sin^2(t)$, $\cos^2(t)$, and 1 are linearly dependent, we should find that the following matrix is somewhat ill-conditioned.

```
t = linspace(0,3,400)';
A = [ sin(t).^2, cos((1+1e-7)*t).^2, t.^0 ];
kappa = cond(A)
```

```
kappa =
    1.8253e+07
```

Now we set up an artificial linear least squares problem with a known exact solution that actually makes the residual zero.

```
x = [1;2;1];
b = A*x;
```

Using backslash to find the solution, we get a relative error that is about $\kappa$ times machine epsilon.

```
x_BS = A\b;
observed_err = norm(x_BS-x)/norm(x)
max_err = kappa*eps
```

```
observed_err =
    1.3116e-10
max_err =
    4.0530e-09
```

If we formulate and solve via the normal equations, we get a much larger relative error. With $\kappa^2 \approx 10^{14}$, we may not be left with more than about 2 accurate digits.

```
N = A'*A;
x_NE = N\(A'*b);
observed_err = norm(x_NE-x)/norm(x)
digits = -log10(observed_err)
```

```
observed_err =
    0.0150
digits =
    1.8226
```

## Exercises

3.2.1. Work out the least squares solution when
$$A = \begin{bmatrix} 2 & -1 \\ 0 & 1 \\ -2 & 2 \end{bmatrix}, \quad b = \begin{bmatrix} 1 \\ -5 \\ 6 \end{bmatrix}.$$

3.2.2. Find the pseudoinverse $A^+$ of the matrix $A = \begin{bmatrix} 1 & -2 & 3 \end{bmatrix}^T$.

3.2.3. Prove the first statement of Theorem 3.2.2: $A^T A$ is symmetric for any $m \times n$ matrix $A$ with $m \geq n$.

3.2.4. Prove that if $A$ is a nonsingular square matrix, then $A^+ = A^{-1}$.

3.2.5. (a) Show that for any $m \times n$ $A$ ($m > n$) for which $A^T A$ is nonsingular, $A^+ A$ is the $n \times n$ identity.

(b) Show using an example in MATLAB that $A A^+$ is not an identity matrix. (This matrix has rank no greater than $n$, so it can't be an $m \times m$ identity.)

3.2.6. Prove that the vector $AA^+ b$ is the vector in the column space (i.e., range) of $A$ that is closest to $b$, in the sense of the 2-norm.

3.2.7. Show that the flop count for Function 3.2.1 is asymptotically $\sim 2mn^2 + \frac{1}{3}n^3$. (In finding the asymptotic count you can ignore terms like $mn$ whose total degree is less than three.)

3.2.8. Let $t_1, \ldots, t_m$ be $m+1$ equally spaced points in $[0, 2\pi]$.

(a) Let $A_\beta$ be the matrix in (3.1.2) that corresponds to fitting data with the function $c_1 + c_2 \sin(t) + c_3 \cos(\beta t)$. Using the identity (3.2.4), make a table of the condition numbers of $A_\beta$ for $\beta = 2, 1.1, 1.01, \ldots, 1 + 10^{-8}$.

(b) Repeat part (a) using the fitting function $c_1 + c_2 \sin^2(t) + c_3 \cos^2(\beta t)$.

(c) Why does it make sense that $\kappa(A_\beta) \to \infty$ as $\beta \to 1$ in part (b) but not in part (a)?

3.2.9. When $A$ is $m \times n$ with rank less than $n$, the pseudoinverse is still defined and can be computed in MATLAB using `pinv`. However, the behavior in this case is not always intuitive. Let
$$A_s = \begin{bmatrix} 1 & 1 \\ 0 & 0 \\ 0 & s \end{bmatrix}.$$

Then $A_0$ has rank equal to one. Demonstrate experimentally that $A_0^+ \neq \lim_{s \to 0} A_s^+$.

## 3.3 • The QR factorization

We say that two vectors $u$ and $v$ in $\mathbb{R}^n$ are **orthogonal** if $u^T v = 0$. For $n = 2$ or $n = 3$ this means the vectors are perpendicular. We say that a collection of vectors $q_1, \ldots, q_k$ is orthogonal if
$$q_i^T q_j = 0 \quad \text{whenever } i \neq j. \tag{3.3.1}$$

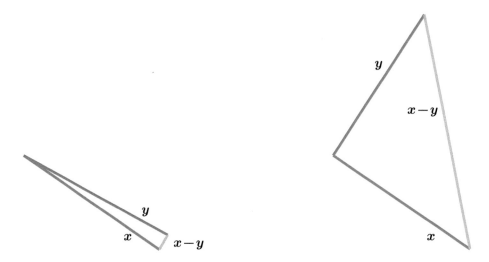

**Figure 3.2.** *Cancellation and orthogonality.*

If also $q_i^T q_i = 1$ for all $i = 1, \ldots, n$, we say the vectors are **orthonormal**.

Orthogonal vectors are convenient theoretically and computationally. In theoretical calculations they make many terms of inner products vanish. For example, if $q_1$ and $q_2$ are orthogonal, then (in the 2-norm)

$$\|q_1 - q_2\|^2 = (q_1 - q_2)^T (q_1 - q_2) = q_1^T q_1 - 2 q_1^T q_2 + q_2^T q_2 = \|q_1\|^2 + \|q_2\|^2. \quad (3.3.2)$$

This calculation is also the key to the computational attractiveness of orthogonality. Figure 3.2 shows how nonorthogonal vectors can allow a multidimensional version of subtractive cancellation, in which $\|x - y\|$ is much smaller than $\|x\|$ and $\|y\|$. As the figure and (3.3.2) show, orthogonal vectors do not allow this phenomenon. In other words, *the addition and subtraction of vectors is guaranteed to be well-conditioned when the vectors are orthogonal.*

## Orthogonal and ONC matrices

Statements about orthogonal vectors are often made more easily in matrix form. Let $Q$ be an $n \times k$ matrix whose columns $q_1, \ldots, q_k$ are orthogonal vectors. The orthogonality conditions (3.3.1) become simply that $Q^T Q$ is a diagonal matrix, since

$$Q^T Q = \begin{bmatrix} q_1^T \\ q_2^T \\ \vdots \\ q_k^T \end{bmatrix} \begin{bmatrix} q_1 & q_2 & \cdots & q_k \end{bmatrix} = \begin{bmatrix} q_1^T q_1 & q_1^T q_2 & \cdots & q_1^T q_k \\ q_2^T q_1 & q_2^T q_2 & \cdots & q_2^T q_k \\ \vdots & \vdots & & \vdots \\ q_k^T q_1 & q_k^T q_2 & \cdots & q_k^T q_k \end{bmatrix}.$$

If the columns of $Q$ are orthonormal, then $Q^T Q$ is the $k \times k$ identity matrix. This is such an important property that we will break with common practice here and give this type of matrix a name: an **ONC matrix** is one whose columns are an orthonormal set of vectors. We summarize their important properties here.

## 3.3. The QR factorization

> **Theorem 3.3.1**
>
> Suppose $Q$ is a real $n \times k$ ONC matrix (matrix with orthonormal columns). Then
>
> 1. $Q^T Q = I$ ($k \times k$ identity);
> 2. $\|Qx\|_2 = \|x\|_2$ for all $k$-vectors $x$;
> 3. $\|Q\|_2 = 1$.

*Proof.* The first part is derived above. The second part follows a pattern that has become well established by now:

$$\|Qx\|_2^2 = (Qx)^T(Qx) = x^T Q^T Q x = x^T I x = \|x\|_2^2.$$

The last part of the theorem is left to the exercises. □

Of particular interest is a *square* ONC matrix, for which $Q^T Q = I$, where all three matrices are $n \times n$. Hence $Q^{-1} = Q^T$. Such a matrix is called an **orthogonal matrix**.[15] These matrices have properties beyond the general ONC type. The proofs of these are left to the exercises.

> **Theorem 3.3.2**
>
> Suppose $Q$ is an $n \times n$ real orthogonal matrix. Then
>
> 1. $Q^T$ is also an orthogonal matrix;
> 2. $\kappa(Q) = 1$ in the 2-norm;
> 3. for any other $n \times n$ matrix $A$, $\|AQ\|_2 = \|A\|_2$;
> 4. if $U$ is another $n \times n$ orthogonal matrix, then $QU$ is also orthogonal.

### The QR factorization

Now we come to another important way to factor a matrix, the **QR factorization**. As we will show below, *the QR factorization plays a role in linear least squares analogous to the role of LU factorization in linear systems.*

> **Theorem 3.3.3**
>
> Every real $m \times n$ matrix $A$ ($m \geq n$) can be written as $A = QR$, where $Q$ is an $m \times m$ orthogonal matrix and $R$ is an $m \times n$ upper triangular matrix.

---

[15]Confusingly, a square matrix whose columns are orthogonal is not necessarily an orthogonal matrix; the columns must be orthonormal, which is a stricter condition.

In most introductory books on linear algebra, the QR factorization is derived through a process known as **Gram–Schmidt orthogonalization**. However, while it is an important tool for theoretical work, the Gram–Schmidt process is numerically unstable. We will consider an alternative construction in the next section.

When $m$ is much larger than $n$, which is often the case, there is a compressed form of the factorization that is more efficient. In the product

$$A = \begin{bmatrix} q_1 & q_2 & \cdots & q_m \end{bmatrix} \begin{bmatrix} r_{11} & r_{12} & \cdots & r_{1n} \\ 0 & r_{22} & \cdots & r_{2n} \\ \vdots & & \ddots & \vdots \\ 0 & 0 & \cdots & r_{nn} \\ 0 & 0 & \cdots & 0 \\ \vdots & \vdots & & \vdots \\ 0 & 0 & \cdots & 0 \end{bmatrix},$$

the vectors $q_{n+1}, \ldots, q_m$ always get multiplied by zero. Nothing about $A$ is lost if we delete them and reduce the factorization to the equivalent form

$$A = \begin{bmatrix} q_1 & q_2 & \cdots & q_n \end{bmatrix} \begin{bmatrix} r_{11} & r_{12} & \cdots & r_{1n} \\ 0 & r_{22} & \cdots & r_{2n} \\ \vdots & & \ddots & \vdots \\ 0 & 0 & \cdots & r_{nn} \end{bmatrix} = \widehat{Q}\widehat{R}, \qquad (3.3.3)$$

in which $\widehat{Q}$ is an $m \times n$ ONC matrix and $\widehat{R}$ is $n \times n$ and upper triangular. We refer to this as a **thin QR factorization**, as the number of columns in $\widehat{Q}$ is $n$ rather than $m$. MATLAB returns either type of QR factorization from the **qr** command.

---

**Example 3.3.1**

MATLAB gives direct access to both the full and thin forms of the QR factorization.

```
A = magic(5);
A = A(:,1:4);
[m,n] = size(A)
```

```
m =
     5
n =
     4
```

Here is the full form:

```
[Q,R] = qr(A);
szQ = size(Q)
szR = size(R)
```

```
szQ =
     5     5
```

## 3.3. The QR factorization

```
szR =
     5     4
```

We can test that **Q** is orthogonal.

```
QTQ = Q'*Q
norm(QTQ - eye(m))

QTQ =
    1.0000    0.0000    0.0000    0.0000   -0.0000
    0.0000    1.0000    0.0000   -0.0000    0.0000
    0.0000    0.0000    1.0000   -0.0000   -0.0000
    0.0000   -0.0000   -0.0000    1.0000   -0.0000
   -0.0000    0.0000   -0.0000   -0.0000    1.0000
ans =
   3.7889e-16
```

With a second input argument given, the thin form is returned.

```
[Q,R] = qr(A,0);
szQ = size(Q)
szR = size(R)

szQ =
     5     4
szR =
     4     4
```

Now **Q** cannot be an orthogonal matrix, because it is not square, but it is still ONC.

```
Q'*Q - eye(4)

ans =
   1.0e-15 *
   -0.2220    0.0527    0.0751    0.0270
    0.0527    0.2220    0.0400   -0.0337
    0.0751    0.0400   -0.2220   -0.0153
    0.0270   -0.0337   -0.0153   -0.2220
```

### Least squares and QR

If we substitute the thin factorization (3.3.3) into the normal equations (3.2.1), we can simplify expressions a great deal.

$$A^T A x = A^T b,$$
$$\widehat{R}^T \widehat{Q}^T \widehat{Q} \widehat{R} x = \widehat{R}^T \widehat{Q}^T b,$$
$$\widehat{R}^T \widehat{R} x = \widehat{R}^T \widehat{Q}^T b.$$

**Function 3.3.1** (`lsqrfact`) Solve linear least squares by QR factorization.

```
function x = lsqrfact(A,b)
% LSQRFACT   Solve linear least squares by QR factorization.
% Input:
%   A        coefficient matrix (m by n, m>n)
%   b        right-hand side (m by 1)
% Output:
%   x        minimizer of || b-Ax ||

[Q,R] = qr(A,0);                       % compressed factorization
c = Q'*b;
x = backsub(R,c);
```

In order to have the normal equations be well posed, we require that $A$ is not rank-deficient (Theorem 3.2.2). This is enough to guarantee that $\widehat{R}$ is nonsingular (see Exercise 3.3.4). Therefore, its transpose is nonsingular as well, and we arrive at

$$\widehat{R}x = \widehat{Q}^T b. \tag{3.3.4}$$

This is a triangular $n \times n$ linear system that is easily solved by backward substitution, as demonstrated in Function 3.3.1. The function itself is superfluous, however, as this is essentially the algorithm used internally by MATLAB when A\b is called. Most importantly, even though we derived (3.3.4) from the normal equations, *the solution of least squares problems via QR factorization does not suffer from the instability seen when the normal equations are solved directly using Cholesky factorization.*

## Exercises

3.3.1. ✍ Prove part 3 of Theorem 3.3.1.

3.3.2. ✍ Prove Theorem 3.3.2. For the third part, use the definition of the 2-norm as an induced matrix norm, then apply some of our other results as needed.

3.3.3. ⌨ Let $t_0, \ldots, t_m$ be $m+1$ equally spaced points in $[-1, 1]$. Let $A$ be the matrix in (3.1.2) for $m = 400$ and fitting by polynomials of degree less than 5. Find the thin QR factorization of $A$, and, on a single graph, plot every column of $\widehat{Q}$ as a function of the vector $t$.

3.3.4. ✍ Prove that if the $m \times n$ ($m \geq n$) matrix $A$ is not rank-deficient, then the factor $\widehat{R}$ of the thin QR factorization is nonsingular. (Hint: Suppose on the contrary that $\widehat{R}$ is singular. Show using the factored form of $A$ that this would imply that $A$ is rank-deficient.)

3.3.5. ✍ Let $A$ be $m \times n$ with $m > n$. Show that if $A = QR$ is a QR factorization and $R$ has rank $n$, then $A^+ = R^+ Q^T$.

3.3.6. ✍ Let $A$ be $m \times n$ with $m > n$. Show that if $A = \widehat{Q}\widehat{R}$ is a thin QR factorization and $\widehat{R}$ is nonsingular, then $A^+ = \widehat{R}^{-1}\widehat{Q}^T$.

3.3.7. ⌨ Repeat Exercise 3.1.2, but use thin QR factorization rather than the backslash to solve the least squares problem.

3.3.8. ✍ The matrix $P = \widehat{Q}\widehat{Q}^T$ derived from the thin QR factorization has some interesting and important properties.

(a) Show that $P = AA^+$.

(b) Prove that $P^2 = P$. (This property defines a *projection matrix*.)

(c) Clearly, any vector $x$ may be written as $x = u+v$, where $u = Px$ and $v = (I-P)x$. Prove that for $P = \widehat{Q}\widehat{Q}^T$, $u$ and $v$ are orthogonal. (Together with part (b), this means that $P$ is an *orthogonal projector*.)

## 3.4 ▪ Computing QR factorizations

To compute an LU factorization, we follow elimination rules to introduce zeros into the lower triangle of the matrix, leaving only the $U$ factor. The row operations are themselves triangular and can be combined into the $L$ factor. For the QR factorization, the game is again to introduce zeros into the lower triangle, but the rules have changed; now the row operations must be done orthogonally. Thanks to Theorem 3.3.2, the product of orthogonal operations will also be orthogonal, providing us with the final $Q$ of the QR.

### Householder reflections

A **Householder reflection** is a particular type of orthogonal matrix $P$. The reflection is customized for a particular given vector $z$ so that $Pz$ is nonzero only in the first element. Since orthogonal matrices preserve the 2-norm, we must have

$$Pz = \begin{bmatrix} \pm\|z\| \\ 0 \\ \vdots \\ 0 \end{bmatrix} = \pm\|z\|e_1. \tag{3.4.1}$$

(Recall that $e_k$ is the $k$th column of the identity matrix.) We choose the positive sign for our discussion, but see Function 3.4.1 and Exercise 3.3.4 for important computational details.

Given $z$, let

$$v = \|z\|e_1 - z. \tag{3.4.2}$$

Then the reflector $P$ is defined by

$$P = I - 2\frac{vv^T}{v^Tv} \tag{3.4.3}$$

if $v \neq 0$, or $P = I$ if $v = 0$. Note that $v^Tv$ is a scalar and can appear in a denominator, while the outer product $vv^T$ is $n \times n$. It is straightforward to show that $P$ has the following key properties.

> **Theorem 3.4.1**
>
> Let $v = \|z\|e_1 - z$ and let $P$ be given by (3.4.3). Then $P$ is symmetric and orthogonal, and $Pz = \|z\|e_1$.

*Proof.* The case $v = 0$ is obvious. For $v \neq 0$, the proofs of symmetry and orthogonality are left to the exercises. As for the last fact, we simply compute

$$Pz = z - 2\frac{vv^T z}{v^T v} = z - 2\frac{v^T z}{v^T v}v, \tag{3.4.4}$$

and, since $e_1^T z = z_1$,

$$v^T v = \|z\|^2 - 2\|z\|z_1 + z^T z = 2\|z\|(\|z\| - z_1),$$
$$v^T z = \|z\|z_1 - z^T z = -\|z\|(\|z\| - z_1),$$

leading finally to

$$Pz = z - 2 \cdot \frac{-\|z\|(\|z\| - z_1)}{2\|z\|(\|z\| - z_1)}v = z + v = \|z\|e_1. \qquad \square$$

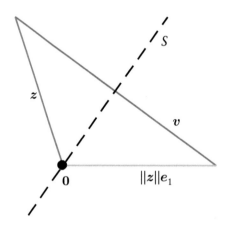

**Figure 3.3.** *Action of a Householder reflector.*

The reason $P$ is called a reflector is sketched in Figure 3.3. The vector $v$ defines an $n-1$-dimensional subspace $S$ perpendicular to it. Elementary vector analysis shows that $v(v^T z)/(v^T v)$ is the vector projection of $z$ along $v$. By subtracting this projection from $z$, we end up in $S$, but by subtracting twice the projection we get a reflection through $S$. This reflection occurs when $P$ is applied to any vector, but when it is applied to $z$, the result ends up on the $x_1$-axis.

## Factorization algorithm

*The QR factorization is computed by using successive Householder reflections to introduce zeros in one column at a time.* We first show the process for a small numerical example.

## 3.4. Computing QR factorizations

**Example 3.4.1**

We will use Householder reflections to produce a QR factorization of the matrix

```
A = magic(6);
A = A(:,1:4);
[m,n] = size(A);
```

Our first step is to introduce zeros below the diagonal in column 1. Define the vector

```
z = A(:,1);
```

Applying the Householder definitions gives us

```
v = z - norm(z)*eye(m,1);
P = eye(m) - 2/(v'*v)*(v*v');    % reflector
```

By design we can use the reflector to get the zero structure we seek:

```
P*z

ans =
   56.3471
    0.0000
    0.0000
    0.0000
    0.0000
    0.0000
```

Now we let

```
A = P*A

A =
   56.3471   16.4693   30.0459   39.0969
    0.0000   29.8260    3.6207   19.1594
    0.0000  -13.4643  -32.9191    2.9808
         0   22.2028   23.9886   12.0918
    0.0000  -16.7397    0.2073   -6.4056
         0   33.1014   24.4943   10.5459
```

We are set to put zeros into column 2. We must not use row 1 in any way, lest it destroy the zeros we just introduced. To put it another way, we can repeat the process we just did on the smaller submatrix

```
A(2:m,2:n)

ans =
   29.8260    3.6207   19.1594
  -13.4643  -32.9191    2.9808
   22.2028   23.9886   12.0918
  -16.7397    0.2073   -6.4056
   33.1014   24.4943   10.5459
```

```
z = A(2:m,2);
v = z - norm(z)*eye(m-1,1);
P = eye(m-1) - 2/(v'*v)*(v*v');    % reflector
```

We now apply the reflector to the submatrix.

```
A(2:m,2:n) = P*A(2:m,2:n)
```

```
A =
    56.3471   16.4693   30.0459   39.0969
     0.0000   54.2196   34.8797   23.1669
     0.0000   -0.0000  -15.6653    5.1928
          0   -0.0000   -4.4630    8.4443
     0.0000    0.0000   21.6583   -3.6556
          0   -0.0000  -17.9233    5.1079
```

We need two more iterations of this process.

```
for j = 3:n
    z = A(j:m,j);
    k = m-j+1;
    v = z - norm(z)*eye(k,1);
    P = eye(k) - 2/(v'*v)*(v*v');
    A(j:m,j:n) = P*A(j:m,j:n);
end
```

We have now reduced the original $A$ to an upper triangular matrix using four orthogonal Householder reflections:

```
R = A
```

```
R =
    56.3471   16.4693   30.0459   39.0969
     0.0000   54.2196   34.8797   23.1669
     0.0000   -0.0000   32.4907   -8.9182
          0   -0.0000   -0.0000    7.6283
     0.0000    0.0000    0.0000   -0.0000
          0   -0.0000   -0.0000    0.0000
```

You may be wondering what happened to the $Q$ in Example 3.4.1. Each Householder reflector is orthogonal but not full-size. We have to pad it out to represent algebraically the fact that a block of the first rows are left alone. Given a reflector $P_k$ that is of square size $m-k+1$, we define

$$Q_k = \begin{bmatrix} I_{k-1} & 0 \\ 0 & P_k \end{bmatrix}.$$

It is trivial to show that $Q_k$ is also orthogonal. Then

$$Q_n Q_{n-1} \cdots Q_1 A = R.$$

But $Q_n Q_{n-1} \cdots Q_1$ is orthogonal too, and we multiply on the left by its transpose to get $A = QR$, where $Q = (Q_n Q_{n-1} \cdots Q_1)^T$. We don't even need to form these

**Function 3.4.1** (`qrfact`) QR factorization by Householder reflections.

```
function [Q,R] = qrfact(A)
% QRFACT   QR factorization by Householder reflections.
% (demo only--not efficient)
% Input:
%   A      m-by-n matrix
% Output:
%   Q,R    A=QR, Q m-by-m orthogonal, R m-by-n upper triangular

[m,n] = size(A);
Q = eye(m);
for k = 1:n
  z = A(k:m,k);
  v = [ -sign(z(1))*norm(z) - z(1); -z(2:end) ];
  nrmv = norm(v);
  if nrmv < eps, continue, end    % nothing is done in this iteration
  v = v / nrmv;                   % removes v'*v in other formulas
  % Apply the reflection to each relevant column of A and Q
  for j = 1:n
    A(k:m,j) = A(k:m,j) - v*( 2*(v'*A(k:m,j)) );
  end
  for j = 1:m
    Q(k:m,j) = Q(k:m,j) - v*( 2*(v'*Q(k:m,j)) );
  end
end

Q = Q';
R = triu(A);                      % enforce exact triangularity
```

matrices explicitly. Writing

$$Q^T = Q_n Q_{n-1} \cdots Q_1 = Q_n\big(Q_{n-1}(\cdots(Q_1 I)\cdots)\big),$$

we can build $Q^T$ iteratively by starting with the identity and doing the same row operations as on $A$. Creating $Q^T$ with row operations on $I$ uses much less memory than building the $Q_k$ matrices explicitly.

The algorithm we have described is encapsulated in Function 3.4.1. There is one more refinement in it, though. As indicated by (3.4.4), the application of a reflector $P$ to a vector does not require the formation of the matrix $P$ itself. Its effect can be computed directly from the vector $v$, as is shown in Function 3.4.1.

The MATLAB command `qr` works similarly to, but more efficiently than, Function 3.4.1. It finds the factorization in $\sim (2mn^2 - n^3/3)$ flops asymptotically.

## Exercises

3.4.1. Find a Householder reflector $P$ such that

$$P \begin{bmatrix} -6 \\ 2 \\ 9 \end{bmatrix} = \begin{bmatrix} 11 \\ 0 \\ 0 \end{bmatrix}.$$

3.4.2. ✍ Prove the unfinished items in Theorem 3.4.1, namely, that a Householder reflector $P$ is symmetric and orthogonal.

3.4.3. ✍ Let $P$ be a Householder reflector as in (3.4.3).

   (a) Find a vector $u$ such that $Pu = -u$. (Figure 3.3 may be of help.)

   (b) What algebraic condition is necessary and sufficient for a vector $w$ to satisfy $Pw = w$? In $n$ dimensions, how many such linearly independent vectors are there?

3.4.4. ✍ Under certain circumstances, computing the vector $v$ in (3.4.2) could lead to subtractive cancellation, which is why line 13 of Function 3.4.1 reads as it does. Devise an example that causes subtractive cancellation if (3.4.2) is used.

3.4.5. ✍ Suppose QR factorization is used to compute the solution of a *square* linear system, $Ax = b$; i.e., let $m = n$.

   (a) Find an asymptotic flop count for this procedure, and compare to the LU factorization algorithm.

   (b) Show that $\kappa_2(A) = \kappa_2(R)$.

3.4.6. ✍ Show that $\kappa_2(A) = \kappa_2(R)$ when $A$ is not square. (Hint: You can't take an inverse of $A$ or $R$.)

3.4.7. ⌨ Modify Function 3.4.1 so that the loops in lines 18–23 are removed. In other words, update the matrices A and Q with a single statement for each. Test your code against the original code on an example to verify it.

3.4.8. Another algorithmic technique for orthogonally introducing zeros into a matrix is the *Givens rotation*. Given a 2-vector $[\alpha\ \beta]^T$, it defines an angle $\theta$ such that
$$\begin{bmatrix} \cos(\theta) & \sin(\theta) \\ -\sin(\theta) & \cos(\theta) \end{bmatrix} \begin{bmatrix} \alpha \\ \beta \end{bmatrix} = \begin{bmatrix} \sqrt{\alpha^2 + \beta^2} \\ 0 \end{bmatrix}.$$

   (a) ✍ Given $\alpha$ and $\beta$, show how to compute $\theta$.

   (b) ⌨ Given the vector $z = [1\ 2\ 3\ 4\ 5]^T$, use MATLAB to find a sequence of Givens rotations that transforms $z$ into the vector $\|z\|e_1$. (Hint: You can operate only on pairs of elements at a time, introducing a zero at the lower of the two positions.)

## Key ideas in this chapter

Each item below links back to its introduction in the text.

1. The most common solution to overdetermined systems is obtained by least squares (page 95).
2. Least squares is used to find fitting functions that depend linearly on the unknown parameters (page 98).
3. The backslash in MATLAB is theoretically equivalent to left-multiplying by a pseudoinverse or inverse (page 104).
4. The condition number of the normal equations is the square of the condition number of the original matrix (page 105).

5. Orthogonal sets of vectors are preferable to nonorthogonal ones in computing (page 108).
6. Any matrix $A$ can be factored into the product of an orthogonal matrix $Q$ and an upper triangular matrix $R$. This factorization plays a role in linear least squares similar to that of LU factorization for linear systems (page 109).
7. The QR approach to solving the least squares problem is more stable than the normal equations approach (page 112).
8. The QR factorization is computed by using successive Householder reflections to introduce zeros in one column at a time (page 114).

## Where to learn more

The least squares problem has been widely studied and used, and only seems to become more important in this era of ever-increasing amounts of data. A good reference for numerical methods is the monograph by Björck [10]. Some theoretical results can be found in Higham [33]; a brief and advanced discussion can be found in Golub and Van Loan [27].

Note that a vast literature can also be found in statistics for what is referred to as "data regression," or simply "regression." Nonlinear methods for least squares fitting of data will be discussed in the following chapter.

In modern applications one may have to deal with so-called online fitting, in which new data must continually be incorporated with old. More recent sources address related issues, e.g., in Hansen, Pereyra, and Scherer [31] and in Teunissen [66]. The problem of geodesy and GPS positioning are discussed in some detail in Strang and Borre [63]; for these applications, they describe how the updating of least squares leads to Kalman filtering.

# Chapter 4

# Roots of nonlinear equations

> He says, "I found her" and keeps repeating, "She's here."
>
> —C3PO, *Star Wars: A New Hope*

In this chapter we extend from linear algebra to deal with *nonlinear* algebraic problems. This kind of problem arises when there is a parameter or variable that can be changed in order to satisfy a constraint or achieve some goal. We start with scalar functions of a single variable, then generalize to $n$ variables and $n$ nonlinear equations. Finally, we generalize the problem of linear least squares to situations with more nonlinear constraints to satisfy than there are variables. In every case the strategy used is one of the cornerstones of numerical computing: *replace a problem you can't solve with an approximate one that you can.* In the context of nonlinear algebraic problems, the particular tactic is to set up and solve a sequence of linear problems of the types covered in the two previous chapters.

## 4.1 • The rootfinding problem

For the time being we will focus on single equations in one variable.

> **Rootfinding problem**: Given a continuous scalar function $f$ of a scalar variable, find a real number $r$ such that $f(r) = 0$.

We call $r$ a **root** of the function $f$. The formulation $f(x) = 0$ is general enough to solve any equation, for if we are given an equation $g(x) = h(x)$, we can define $f = g - h$ and find a root of $f$.

Unlike the linear problems of the earlier chapters, the usual situation here is that the root cannot be produced in a finite number of operations, even in exact arithmetic. Instead, we seek a sequence of approximations that formally converge to the root, stopping when some member of the sequence seems to be "good enough." In MATLAB, the general-purpose rootfinding tool is called **fzero**.

### Example 4.1.1

In the theory of vibrations of a circular drum, the displacement of the drumhead can be expressed in terms of pure harmonic modes,

$$J_m(\omega_{k,m} r)\cos(m\theta)\cos(c\omega_{k,m} t),$$

where $(r,\theta)$ are polar coordinates, $0 \le r \le 1$, $t$ is time, $m$ is a positive integer, $c$ is a material parameter, and $J_m$ is a *Bessel function of the first kind*. The quantity $\omega_{k,m}$ is a resonant frequency and is a positive root of the equation

$$J_m(\omega_{k,m}) = 0,$$

which states that the drumhead is clamped around the rim. Tabulating approximations to the zeros of Bessel functions has occupied countless mathematician-hours throughout the centuries.

Too bad they didn't have MATLAB back in the day. Bessel functions are built in.

```
J3 = @(x) besselj(3,x);
fplot(J3,[0 20]), grid on
```

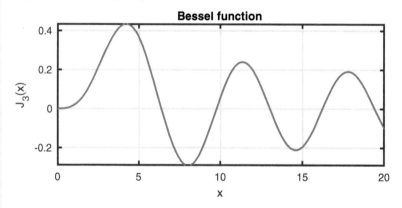

From the graph we see roots near 6, 10, 13, 16, and 19. We use `fzero` to find these roots accurately.

```
omega = [];
for guess = [6,10,13,16,19]
    omega = [omega;fzero(J3,guess)];
end
omega
```

```
omega =
    6.3802
```

```
   9.7610
  13.0152
  16.2235
  19.4094
```

```
hold on, plot(omega,J3(omega),'.')
```

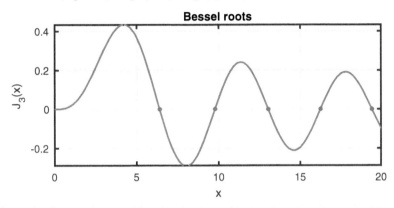

## Conditioning, error, and residual

In the rootfinding problem the data is a continuous function $f$ whose root we seek. Let's assume $f$ has at least one continuous derivative near a particular root $r$. Say that $f$ is perturbed to $\tilde{f}(x) = f(x) + \epsilon$. As a result, the root will be perturbed to $\tilde{r} = r + \delta$ satisfying, by definition, $\tilde{f}(\tilde{r}) = 0$. We now compute an absolute condition number $\kappa_r$, which is the ratio $|\delta/\epsilon|$ as $\epsilon \to 0$.

Using Taylor series expansions,

$$0 = f(r+\delta) + \epsilon = f(r) + f'(r)\delta + \epsilon + O(\delta^2).$$

Given that $f(r) = 0$, this implies

$$\kappa_r = |f'(r)|^{-1}. \tag{4.1.1}$$

Recall that the absolute condition number for the evaluation of $f$ at $x$ is simply $|f'(x)|$. *The condition number of the rootfinding problem is equivalent to that of evaluating the inverse function.* The influence of $|f'(r)|$ on rootfinding is easily illustrated.

**Example 4.1.2**

Consider first the function

```
f = @(x) (x-1).*(x-2);
```

At the root $r = 1$, we have $f'(r) = -1$. If the values of $f$ were perturbed at any point by noise of size, say, 0.05, we can imagine finding the root of the function as though drawn with a thick line, whose edges we show here.

```
interval = [0.8 1.2];
fplot(f,interval), grid on, hold on
fplot(@(x) f(x)+0.02,interval,'k')
fplot(@(x) f(x)-0.02,interval,'k')
```

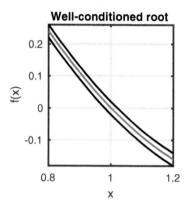

The possible values for a perturbed root all lie within the interval where the black lines intersect the x-axis. The width of that zone is about the same as the vertical distance between the lines.

By contrast, consider the function

```
f = @(x) (x-1).*(x-1.01);
```

Now $f'(1) = -0.01$, and the graph of $f$ will be much shallower near $x = 1$. Look at the effect this has on our thick rendering:

```
axis(axis), cla
fplot(f,interval)
fplot(@(x) f(x)+0.02,interval,'k')
fplot(@(x) f(x)-0.02,interval,'k')
```

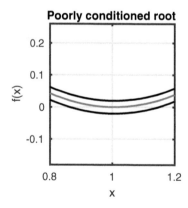

The vertical displacements in this picture are exactly the same as before. But the potential *horizontal* displacement of the root is much wider. In fact, if we perturb the function upward by the amount drawn here, the root disappears entirely!

We must accept that when $|f'|$ is small at the root, it may not be possible to get a small error in a computed root estimate. As always, the error is not a quantity we can

## 4.1. The rootfinding problem

compute without knowing the exact answer. But we can compute the **residual** of a root estimate, which is the value of $f$ there. Since $f(r) = 0$, it stands to reason that a small residual might be associated with a small error.

Suppose that we find an approximation $\tilde{r}$ to the actual root $r$. Define the new function $g(x) = f(x) - f(\tilde{r})$. Trivially, $g(\tilde{r}) = 0$, meaning that the root estimate is a true root of $g$. Since the difference between $g$ and the original $f$ is the residual value $f(\tilde{r})$, the residual is the distance to a rootfinding problem that our root estimate solves exactly. That is, the residual is the backward error of the estimate.

To summarize: *In general, it is not always realistic to expect a small error in a root approximation. However, the backward error is the same as the residual of the estimate.*

## Multiple roots

The condition number (4.1.1) naturally leads to the question of what happens if $f'(r) = 0$ at a root $r$. Suppose first that $f$ is a polynomial of degree $n > 0$, so that

$$f(x) = (x - r)q(x)$$

for a polynomial $q$ of degree $n-1$. If $r$ is a simple root of $f$—that is, it appears just once in the list of the $n$ roots—then it follows that $q(r) \neq 0$. Conversely, if $q(r) = 0$, then $r$ appears among the roots of $q$ and is a multiple root of $f$. However, we don't need to know the quotient polynomial $q$ explicitly in order to make the determination. Consider that

$$f'(x) = (x - r)q'(x) + q(x),$$

so that $f'(r) = q(r)$. Hence $r$ is simple if and only if $f'(r) \neq 0$.

This conclusion extends to nonpolynomial differentiable functions $f$. If $r$ is a root of $f$, define $q(x) = f(x)/(x - r)$. By L'Hôpital's rule, $q$ is well defined at $x = r$ as long as $f'$ is. Now we can again write $f(x) = (x - r)q(x)$ for $x \neq r$, and by continuity it works at $x = r$ as well. So the reasoning we applied to polynomials can be repeated: *$r$ is a simple root of $f$ if and only if $f'(r) \neq 0$.*

Now suppose that $f'(r) = q(r) = 0$, so that $r$ is not simple. If $q$ is differentiable, we may apply the same logic to it that we did to $f$. Hence $r$ is not simple for $q$ if and only if $q'(r) = 0$. Now observe that

$$f''(x) = (x - r)q''(x) + 2q'(x),$$

and thus $f''(r) = 2q'(r)$, so $f''(r) = 0$ if and only if $r$ is a multiple root of $q$. In general we define $r$ as a **root of multiplicity** $m$ if $f(r) = f'(r) = \cdots = f^{(m-1)}(r) = 0$, but $f^{(m)}(r) \neq 0$. If $m = 1$, we say $r$ is a **simple root**.

It's useful to think through the consequences of these definitions for the Taylor series at the point $r$,

$$f(x) = a_0 + a_1(x - r) + a_2(x - r)^2 + \cdots,$$

where $a_n = f^{(n)}(r)/n!$. The fact that $r$ is a root implies $f(r) = a_0 = 0$. If $r$ is a simple root, then $a_1 \neq 0$, and conversely. If $r$ is a double root, then $a_2 \neq 0$, and so on. Simply put, if $r$ is a root of order $m$, then the series expansion begins with $(x - r)^m$.

When $r$ is a multiple root, the condition number (4.1.1) is apparently infinite.[16] However, even if $r$ is technically simple, we should expect difficulty if the condition number is very large. This occurs when $|f'(r)|$ is very small, which means that quotient $q$ satisfies $q(r) \approx 0$ and another root of $f$ is very close to $r$. The situation is reminiscent of the linear system problem: the degenerate case (singular matrix/multiple root) is mathematically isolated when considered exactly, but the effect on fixed precision computation is just as drastic in a neighborhood of the singularity.

## Exercises

For each equation and given interval, do the following steps:

(a) Rewrite the equation into the standard form for rootfinding, $f(x) = 0$. Make a plot of $f$ over the given interval and determine how many roots lie in the interval.

(b) Use `fzero` to find each root.

(c) Compute the condition number of each root found in part (b).

4.1.1. $x^2 = e^{-x}$, over $[-2, 2]$.

4.1.2. $2x = \tan x$, over $[-0.2, 1.4]$.

4.1.3. $e^{x+1} = 2 + x$, over $[-2, 2]$.

4.1.4. A basic safe type of investment is an annuity: One makes monthly deposits of size $P$ for $n$ months at a fixed annual interest rate $r$, and at maturity collects the amount

$$\frac{12P}{r}\left(\left(1+\frac{r}{12}\right)^n - 1\right).$$

Say you want to create an annuity for a term of 300 months and final value of $1,000,000. Using `fzero`, make a table of the interest rate you will need to get for each of the different contribution values $P = 500, 550, \ldots, 1000$.

4.1.5. The most easily observed properties of the orbit of a celestial body around the sun are the period $\tau$ and the elliptical eccentricity $\epsilon$. (A circle has $\epsilon = 0$.) From these it is possible to find at any time $t$ the angle $\theta(t)$ made between the body's position and the major axis of the ellipse. This is done through

$$\tan \frac{\theta}{2} = \sqrt{\frac{1+\epsilon}{1-\epsilon}} \tan \frac{\psi}{2}, \qquad (4.1.2)$$

where the eccentric anomaly $\psi$ satisfies Kepler's equation:

$$\psi - \epsilon \sin \psi - \frac{2\pi t}{\tau} = 0. \qquad (4.1.3)$$

---

[16] Based on our definitions, this means that the relative change to the root when $f$ is changed by a perturbation of size $\epsilon$ is not $O(\epsilon)$ as $\epsilon \to 0$.

Equation (4.1.3) must be solved numerically to find $\psi(t)$, and then (4.1.2) can be solved analytically to find $\theta(t)$.

The asteroid Eros has $\tau = 1.7610$ years and $\epsilon = 0.2230$. Using `fzero` for (4.1.3), make a plot of $\theta(t)$ for 100 values of $t$ between 0 and $\tau$ (one full period).

4.1.6. Lambert's $W$ function is defined as the inverse of $xe^x$. That is, $y = W(x)$ if and only if $x = ye^y$. Write a function `y=lambertW(x)` that computes $W$ using `fzero`. Make a plot of $W(x)$ for $0 \le x \le 4$.

4.1.7. For each function, find the multiplicity of the given root. If it is a simple root, find its absolute condition number.

  (a) $f(x) = x^3 - 2x^2 + x - 2$, root $r = 2$.
  
  (b) $f(x) = (\cos x + 1)^2$, root $r = \pi$.
  
  (c) $f(x) = \frac{\sin^2 x}{x}$, root $r = 0$ (define $f(0) = 0$).
  
  (d) $f(x) = (x-1)\log(x)$, root $r = 1$.

4.1.8. For any $\epsilon > 0$, let $f_\epsilon(x) = \sin[(x-1+\epsilon)(x-1)]$. This function has roots at $1-\epsilon$ and $1$.

  (a) Find $|f'_\epsilon(1)|$. According to (4.1.1), the condition number of the root $r=1$ is inversely proportional to this quantity.
  
  (b) Define a perturbation function by $g(x) = \cos[10x + \sin(20x)]$. Verify that $f_\epsilon(x) + 10^{-10} g(x)$ has a root in the interval $[1-\epsilon, 1+10\epsilon]$ for $\epsilon = 10^{-3}$.
  
  (c) For $\epsilon = 10^{-3}, 10^{-4}, 10^{-5}, 10^{-6}$, use `fzero` to find the root in the interval given in part (b). Make a table of $\epsilon$, $1/|f'_\epsilon(1)|$, $|r-1|$, and $|(r-1)f'_\epsilon(1)|$. The last value should be approximately constant.

4.1.9. (continuation) The condition number theory (4.1.1) suggests that if $f$ has simple roots at $r_1$ and $r_2$ that are close to each other, then the condition number is large. But then the function $\tilde{f}(x) = f(x)/(x-r_2)$ no longer has a root at $r_2$ and should have a much better condition number if there are no other nearby roots. This trick (called *deflation*) can work even if the division factor does not use the exact root $r_2$. For the steps below, use the same definitions as in Exercise 4.1.8.

  (a) Define $\tilde{f}_\epsilon(x) = f_\epsilon(x)/(x+1.01)$. Find $|\tilde{f}'_\epsilon(1)|$. (You may want to use computer algebra for this.)
  
  (b) Repeat part (c) of Exercise 4.1.8 but using the interval $[1-\epsilon/200, 1+\epsilon/100]$ each time. By what factor are the errors improved over using $f_\epsilon$? Do you still find that $|(x-1)\tilde{f}'_\epsilon(1)|$ is roughly constant?

## 4.2 • Fixed point iteration

The rootfinding problem $f(x) = 0$ can always be transformed into another form, $g(x) = x$, known as the **fixed point problem**. Given $f$, one such transformation is to define $g(x) = x - f(x)$. Then the fixed point equation is true at, and only at, a root of $f$. The next example shows that evaluations of the function $g$ can be used to try to locate a fixed point.

## Example 4.2.1

Let's convert the roots of a quadratic polynomial $f(x)$ to a fixed point problem.

```
f = @(x) x.^2 - 4*x + 3.5;
r = roots([1 -4 3.5])
```

```
r =
    2.7071
    1.2929
```

We'll define $g(x) = x - f(x)$. Intersections of its graph with the line $y = x$ are fixed points of $g$ and thus roots of $f$. (Only one is shown in the chosen plot range.)

```
g = @(x) x - f(x);
fplot(g,[2 3])
hold on, plot([2 3],[2 3],'k')
```

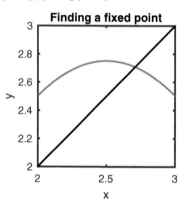

If we evaluate $g(2.1)$, we get a value of almost 2.6.

```
x = 2.1;   y = g(x)
```

```
y =
    2.5900
```

So $g(x)$ is considerably closer to a fixed point than $x$ was. The value $y = g(x)$ ought to become our new $x$ value! Changing the $x$ coordinate in this way is the same as following a horizontal line over to the graph of $y = x$.

```
plot([x y],[y y],'-')
x = y;
```

## 4.2. Fixed point iteration

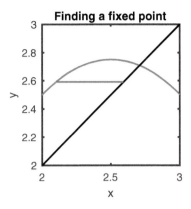

Now we can compute a new value for $y$. We leave $x$ alone here, so we travel along a vertical line to the graph of $g$.

```
y = g(x)
plot([x x],[x y],'-')
```

```
y =
    2.7419
```

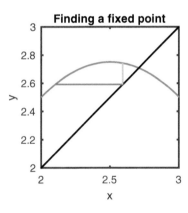

You see that we are in a position to repeat these steps as often as we like. Let's apply them a few times and see the result.

```
for k = 1:5
    plot([x y],[y y],'-'),  x = y;      % y --> new x
    y = g(x);  plot([x x],[x y],'-')    % g(x) --> new y
end
```

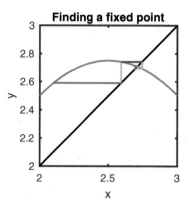

**Finding a fixed point**

The process spirals in beautifully toward the fixed point we seek. Our last estimate has almost 4 accurate digits.

```
abs(y-r(1))/r(1)
```

```
ans =
    1.6531e-04
```

Now let's try to find the other fixed point in the same way. We'll use 1.3 as a starting approximation.

```
cla
fplot(g,[1 2])
hold on, plot([1 2],[1 2],'k'), ylim([1 2])
x = 1.3; y = g(x);
for k = 1:5
    plot([x y],[y y],'-'),   x = y;      % y --> new x
    y = g(x);   plot([x x],[x y],'-')    % g(x) --> new y
end
```

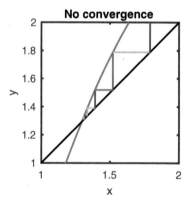

**No convergence**

We started near the true fixed point, which is at about 1.293. But this time, the iteration is pushing us *away* from the correct answer.

This is our first example of an **iteration**. The idea is to generate not a single answer but a sequence of values that one hopes will converge to the correct result. Often the iteration is constructed by defining a formula to map one member of the sequence to

## 4.2. Fixed point iteration

the next one. In this case we have

$$x_{k+1} = g(x_k), \qquad k = 1, 2, \ldots, \qquad (4.2.1)$$

which is known as the **fixed point iteration**. In order to fully define the process, we must also provide a starting value $x_1$. Then (4.2.1) defines the rest of the sequence $x_2, x_3, \ldots$ for as many terms as we care to generate.

### Series analysis

In Example 4.2.1, the two computed iterations differ only in the choice of $x_1$. In the first case we evidently generated a sequence that converged to one of the fixed points. In the second case, however, the generated sequence diverged.[17] The easiest way to uncover the essential difference between the two cases is to use Taylor series expansions.

Suppose a fixed point $r$ is the desired limit of an iteration $x_1, x_2, \ldots$. It's often easier to express quantities in terms of the error sequence $\epsilon_1, \epsilon_2, \ldots$, where $\epsilon_k = x_k - r$. Starting from (4.2.1), we have

$$\epsilon_{k+1} + r = g(\epsilon_k + r) = g(r) + g'(r)\epsilon_k + \frac{1}{2}g''(r)\epsilon_k^2 + \cdots,$$

assuming that $g$ has at least two continuous derivatives. But by definition, $g(r) = r$, so

$$\epsilon_{k+1} = g'(r)\epsilon_k + O(\epsilon_k^2). \qquad (4.2.2)$$

If the iteration is to converge to $r$, the errors must approach zero. In this case we can neglect the second-order term and conclude that $\epsilon_{k+1} \approx g'(r)\epsilon_k$. This is satisfied if $\epsilon_k \approx C[g'(r)]^k$ for a constant $C$ and all sufficiently large $k$. Hence if $|g'(r)| > 1$, we are led to the contradictory conclusion that the errors must *grow*, not vanish. More precisely, if a term in the sequence really does get close to the fixed point $r$, the iteration will start producing new values that are farther away from it. However, if $|g'(r)| < 1$, we do see errors that converge to zero.

#### Example 4.2.2

The role of $g'(r)$ is clear in Example 4.2.1. We have $g(x) = -x^2 + 5x - 3.5$ and $g'(x) = -2x + 5$. For the first fixed point, near 2.71, we get $g'(r) \approx -0.42$, indicating convergence. For the second fixed point, near 1.29, we get $g'(r) \approx 2.42$, which is consistent with the observed divergence.

### Linear convergence

In numerical computation we want to know not just whether an iteration converges but also the *rate* at which convergence occurs, i.e., how quickly the errors approach

---

[17] We can only ever generate a finite sample from an infinite sequence, which in principle does not guarantee anything whatsoever about the limit or divergence of that sequence. However, in practical computing one usually assumes that well-established trends in the sequence will continue, and we complement the observed experiences with rigorous theory where possible.

zero. Other things being equal, faster convergence is preferred to slower convergence, as it usually implies that the computation will take less time to achieve a desired accuracy.

The prediction of the series analysis above is that if the fixed point iteration converges, the errors satisfy

$$|\epsilon_k| = |x_k - r| \approx C\sigma^k, \quad \sigma = |g'(r)| < 1. \tag{4.2.3}$$

Taking logs, we get

$$\log|\epsilon_k| \approx k(\log \sigma) + (\log C).$$

This is in the form $\log|\epsilon_k| \approx \alpha k + \beta$, which is a linear relationship, and we refer to this situation as **linear convergence**. The formal definition of linear convergence for the sequence $x_1, x_2, \ldots$ to the number $r$ is that there exists a number $0 < \sigma < 1$ such that

$$\lim_{k \to \infty} \frac{|x_{k+1} - r|}{|x_k - r|} = \sigma. \tag{4.2.4}$$

Practically speaking, linear convergence is identified by two different observations about the errors $\epsilon_k$:

1. The errors lie on a straight line on a log–linear graph, as implied by $\log|\epsilon_k| \approx \alpha k + \beta$.

2. The error is reduced by a constant factor in each step, as implied by $|\epsilon_k| \approx C\sigma^k$.

Both statements are approximate and only apply for sufficiently large values of $k$, so a certain amount of judgment has to be applied.

### Example 4.2.3

We revisit Example 4.2.1 and investigate the observed convergence more closely. Recall that in Example 4.2.2 we calculated $g'(r) \approx -0.42$ at the convergent fixed point.

```
f = @(x) x.^2 - 4*x + 3.5;
r = roots([1 -4 3.5]);
```

Here is the fixed point iteration. This time we keep track of the whole sequence of approximations.

```
g = @(x) x - f(x);
x = 2.1;
for k = 1:12
    x(k+1) = g(x(k));
end
```

It's easiest to construct and plot the sequence of errors.

```
err = abs(x-r(1));
semilogy(err,'.-'), axis tight
xlabel('iteration'), ylabel('error')
```

## 4.2. Fixed point iteration

```
title('Convergence of fixed point iteration')
```

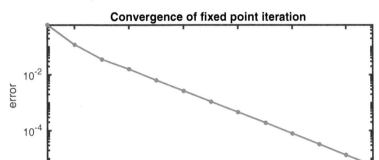

It's quite clear that the convergence quickly settles into a linear rate. We could estimate this rate by doing a least squares fit to a straight line using MATLAB's built-in **polyfit**. Keep in mind that the values for small $k$ should be left out of the computation, as they don't represent the linear trend.

```
p = polyfit(5:12,log(err(5:12)),1)

p =
    -0.8807    -0.6681
```

The first value is the slope, which must be negative. We can exponentiate it to get the convergence constant $\sigma$.

```
sigma = exp(p(1))

sigma =
    0.4145
```

The numerical values of the error should decrease by a factor of $\sigma$ at each iteration. We can check this easily with an elementwise division.

```
err(9:12) ./ err(8:11)

ans =
    0.4138    0.4144    0.4141    0.4142
```

The methods for finding $\sigma$ agree well.

### Contraction maps

The convergence condition $\sigma = |g'(r)| < 1$ derived by series expansion is a special case of a more general condition.

**Definition 4.2.1.** A function $g$ is said to satisfy a **Lipschitz condition** with constant $L$ on the interval $S \subset \mathbb{R}$ if

$$|g(s) - g(t)| \leq L|s - t| \quad \text{for all } s, t \in S. \tag{4.2.5}$$

It can be shown that a function satisfying (4.2.5) is continuous in $S$. If $L < 1$, we call $g$ a **contraction map**, because distances between points decrease after an application of $g$.

From the Fundamental Theorem of Calculus, which asserts $g(s) - g(t) = \int_s^t g'(x)\, dx$, it's not difficult to conclude that an upper bound of $|g'(x)| \leq L$ for all $x$ results in (4.2.5). But the weaker Lipschitz condition alone is enough to guarantee the success of fixed point iteration.

> **Theorem 4.2.1: Contraction Mapping Theorem**
>
> Suppose that $g$ satisfies (4.2.5) with $L < 1$ on an interval $S$. Then $S$ contains exactly one fixed point $r$ of $g$. If $x_1, x_2, \ldots$ are generated by the fixed point iteration (4.2.1), and $x_1, x_2, \ldots$ all lie in $S$, then $|x_k - r| \leq L^{k-1}|x_1 - r|$ for all $k > 1$.

*Partial proof.* First we show there is at most one fixed point in $S$. Suppose $f(r) = r$ and $f(s) = s$ in $S$. Then by (4.2.5), $|r - s| = |g(r) - g(s)| \leq L|r - s|$, which for $L < 1$ is possible only if $|r - s| = 0$, so $r = s$.

Now suppose that for some $r \in S$, $g(r) = r$. By the definition of the fixed point iteration and the Lipschitz condition,

$$|x_{k+1} - r| = |g(x_k) - g(r)| \leq L|x_k - r|,$$

which shows that $x_k \to r$ as $k \to \infty$. To show that $r$ must exist and complete the proof, one needs to apply the Cauchy theory of convergence of a sequence, which is beyond the scope of this book. □

There are stronger and more general statements of Theorem 4.2.1. For instance, it's possible to show that all initial $x_1$ that are "sufficiently close" to the fixed point will lead to convergence of the iteration. Algorithmically the main virtue of the fixed point iteration is that it is incredibly easy to apply. However, as we are about to discover, it's not the fastest option.

## Exercises

4.2.1. ✎ In each case, show that the given $g(x)$ has a fixed point at the given $r$ and use (4.2.2) to show that fixed point iteration can converge to it.

(a) $g(x) = \frac{1}{2}\left(x + \frac{9}{x}\right)$, $r = 3$.

(b) $g(x) = \pi + \frac{1}{4}\sin(x)$, $r = \pi$.

(c) $g(x) = x + 1 - \tan(x/4)$, $r = \pi$.

**4.2.2.** 🖳 For each case in Exercise 4.2.1, apply fixed point iteration in MATLAB and use a log–linear graph of the error to verify linear convergence. Then use numerical values of the error to determine an approximate value for $\sigma$ in (4.2.3).

**4.2.3.** ✍ In each case, show that the given $g(x)$ has a fixed point of the given $r$. Then determine analytically whether the fixed point iteration could converge to that point given a close enough starting value.

(a) $g(x) = 3 + x - x^2$, $r = \sqrt{3}$.

(b) $g(x) = \sqrt{1+x}$, $r = (1+\sqrt{5})/2$.

(c) $g(x) = \sqrt{1+x}$, $r = (1-\sqrt{5})/2$.

(d) $g(x) = x + 1 - \tan(\pi x)$, $r = 1/4$.

**4.2.4.** In Example 4.2.1 we used $g(x) = x - f(x)$ to find a fixed point of the polynomial $f(x) = x^2 - 4x + 3.5$.

(a) ✍ Why does the iteration "spiral in" to the fixed point? (Refer to the series analysis.)

(b) ✍ Show that if $\hat{g}(x) = (x^2 + 3.5)/4$, then any fixed point of $g$ is a root of $f$.

(c) 🖳 Use fixed point iteration on $\hat{g}$ to try to find both roots of $f$, and note which case(s), if either, converge.

(d) ✍ Use (4.2.2) to explain the observed behavior in part (c).

**4.2.5.** ✍ The $m$th root of a positive real number $a$ is a fixed point of the function

$$g(x) = \frac{a}{x^{m-1}}.$$

For what positive integer values of $m$ will the fixed point iteration for $g$ converge (for close enough initial guesses)?

**4.2.6.** (a) ✍ Show that $r = 1/3$ is a fixed point of $g(x) = 2x - 3x^2$.

(b) ✍ Find $g'(1/3)$. How does this affect (4.2.2)?

(c) 🖳 Do an experiment with fixed point iteration on $g$ to converge to $r = 1/3$. Is the convergence linear?

**4.2.7.** ✍ Consider the iteration

$$x_{k+1} = x_k - \frac{f(x_k)}{c}, \quad k = 0, 1, \ldots.$$

Suppose $f(r) = 0$ and $f'(r)$ exist. Find one or more conditions on $c$ such that the iteration converges to $r$.

## 4.3 • Newton's method in one variable

Newton's method is the cornerstone of rootfinding. We introduce the key idea with an example.

### Example 4.3.1

Suppose we want to find a root of the function

```
f = @(x) x.*exp(x) - 2;
fplot(f,[0 1.5])
```

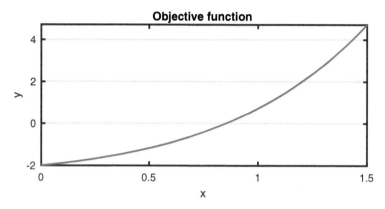

From the graph, it is clear that there is a root near $x = 1$. So we call that our initial guess, $x_1$.

```
x1 = 1;
f1 = f(x1)
hold on, plot(x1,f1,'.')
```

f1 =
    0.7183

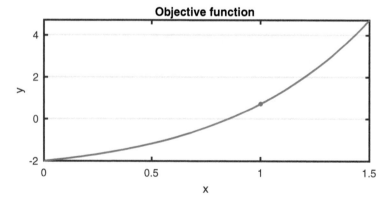

Next, we can compute the tangent line at the point $(x_1, f(x_1))$, using the derivative.

```
dfdx = @(x) exp(x).*(x+1);
slope1 = dfdx(x1);
tangent1 = @(x) f1 + slope1*(x-x1);
fplot(tangent1,[0 1.5],'--')
```

## 4.3. Newton's method in one variable

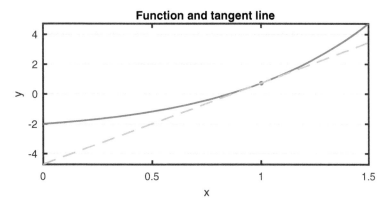

**Function and tangent line**

In lieu of finding the root of $f$ itself, we settle for finding the root of the tangent line approximation, which is trivial. Call this $x_2$, our next approximation to the root.

```
x2 = x1 - f1/slope1
plot(x2,0,'.')
f2 = f(x2)
```

```
x2 =
    0.8679
f2 =
    0.0672
```

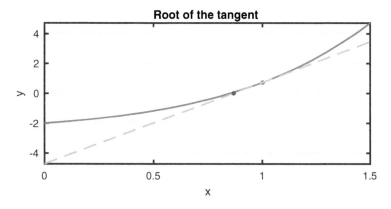

**Root of the tangent**

The residual (value of $f$) is smaller than before, but not zero. So we repeat the process with a new tangent line based on the latest point on the curve.

```
cla, fplot(f,[0.8 0.9])
plot(x2,f2,'.')
slope2 = dfdx(x2);
tangent2 = @(x) f2 + slope2*(x-x2);
fplot(tangent2,[0.8 0.9],'--')
x3 = x2 - f2/slope2;
plot(x3,0,'.')
f3 = f(x3)
```

```
f3 =
   7.7309e-04
```

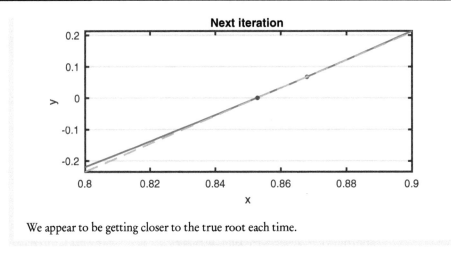

We appear to be getting closer to the true root each time.

Using general notation, if we have a root approximation $x_k$, we can construct a **linear model** of $f(x)$ using the classic formula for the tangent line of a differentiable function,

$$q(x) = f(x_k) + f'(x_k)(x - x_k). \qquad (4.3.1)$$

Finding the root of $q(x) = 0$ is trivial. We define the next approximation by the condition $q(x_{k+1}) = 0$, or

$$x_{k+1} = x_k - \frac{f(x_k)}{f'(x_k)}. \qquad (4.3.2)$$

Starting with an initial estimate $x_1$, this formula defines a sequence of estimates $x_2, x_3, \ldots$. The iteration so defined is what we call **Newton's method**.

## Quadratic convergence

The graphs of Example 4.3.1 suggest why the Newton iteration may converge to a root: Any differentiable function looks more and more like its tangent line as we zoom in to the point of tangency. Yet it is far from clear that it *must* converge, or at what rate it will do so. The matter of the convergence rate is fairly straightforward to resolve. Define the error sequence

$$\epsilon_k = r - x_k, \quad k = 1, 2, \ldots, \qquad (4.3.3)$$

where $r$ is the limit of the sequence and $f(r) = 0$. Exchanging $x$ values for $\epsilon$ values in (4.3.2) gives

$$\epsilon_{k+1} = \epsilon_k + \frac{f(r - \epsilon_k)}{f'(r - \epsilon_k)}.$$

We assume that $|\epsilon_k| \to 0$; eventually, the errors remain as small as we please forever. Then a Taylor expansion of $f$ about $x = r$ gives

$$\epsilon_{k+1} = \epsilon_k + \frac{f(r) - \epsilon_k f'(r) + \frac{1}{2}\epsilon_k^2 f''(r) + O(\epsilon_k^3)}{f'(r) - \epsilon_k f''(r) + O(\epsilon_k^2)}.$$

## 4.3. Newton's method in one variable

We use the fact that $f(r)=0$ and additionally assume now that $f'(r)\neq 0$. Then

$$\epsilon_{k+1} = \epsilon_k - \epsilon_k\left[1 - \frac{1}{2}\frac{f''(r)}{f'(r)}\epsilon_k + O(\epsilon_k^2)\right]\left[1 - \frac{f''(r)}{f'(r)}\epsilon_k + O(\epsilon_k^2)\right]^{-1}.$$

The series in the denominator is of the form $(1+z)^{-1}$. Provided $|z| < 1$, this is the limit of the geometric series $1-z+z^2-z^3+\cdots$. Keeping only the lowest-order terms, we derive

$$\begin{aligned}\epsilon_{k+1} &= \epsilon_k - \epsilon_k\left[1 - \frac{1}{2}\frac{f''(r)}{f'(r)}\epsilon_k + O(\epsilon_k^2)\right]\left[1 + \frac{f''(r)}{f'(r)}\epsilon_k + O(\epsilon_k^2)\right] \\ &= -\frac{1}{2}\frac{f''(r)}{f'(r)}\epsilon_k^2 + O(\epsilon_k^3).\end{aligned} \quad (4.3.4)$$

Equation (4.3.4) suggests that, *eventually, each iteration of Newton's method roughly squares the error*. This behavior is called **quadratic convergence**. The formal definition of quadratic convergence is that there exists a number $\alpha > 0$ such that

$$\lim_{k\to\infty}\frac{|x_{k+1}-r|}{|x_k-r|^2} = \alpha. \quad (4.3.5)$$

Recall that linear convergence is identifiable by trending toward a straight line on a log-linear plot of the error. When the convergence is quadratic, no such straight line exists—the convergence keeps getting steeper. Alternatively, note that (neglecting high-order terms)

$$\log(|\epsilon_{k+1}|) \approx 2\log(|\epsilon_k|) + \text{constant},$$

which is equivalent to saying that the number of accurate digits approximately doubles at each iteration, once the errors become small enough.

### Example 4.3.2

We again look at finding a solution of $xe^x = 2$ near $x = 1$. To apply Newton's method, we need to calculate values of both the residual function $f$ and its derivative.

```
f = @(x) x.*exp(x) - 2;
dfdx = @(x) exp(x).*(x+1);
```

We don't know the exact root, so we use the built-in `fzero` to determine the "true" value.

```
format long,  r = fzero(f,1)

r =
    0.852605502013726
```

We use $x_1 = 1$ as a starting guess and apply the iteration in a loop, storing the sequence of iterates in a vector.

```
x = 1;
for k = 1:6
```

```
    x(k+1) = x(k) - f(x(k)) / dfdx(x(k));
end
```

Here is the sequence of errors (leaving out the last value, now serving as our exact root).

```
format short e
err = x' - r

err =
   1.4739e-01
   1.5274e-02
   1.7787e-04
   2.4355e-08
   4.4409e-16
  -1.1102e-16
  -1.1102e-16
```

Glancing at the exponents of the errors, they roughly form a neat doubling sequence until the error is comparable to machine precision. We can see this more precisely by taking logs.

```
format short
logerr = log(abs(err))

logerr =
   -1.9146
   -4.1816
   -8.6344
  -17.5305
  -35.3505
  -36.7368
  -36.7368
```

Quadratic convergence isn't as graphically distinctive as linear convergence.

```
semilogy(abs(err),'.-')
```

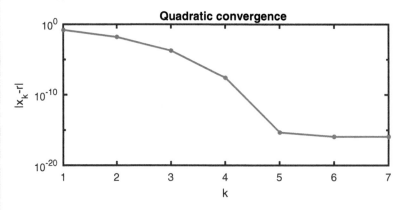

This looks faster than linear convergence, but it's not easy to say more. If we could use infinite precision, the curve would continue to steepen forever.

## 4.3. Newton's method in one variable

**Function 4.3.1 (newton)** Newton's method for a scalar rootfinding problem.

```
function x = newton(f,dfdx,x1)
% NEWTON    Newton's method for a scalar equation.
% Input:
%   f         objective function
%   dfdx      derivative function
%   x1        initial root approximation
% Output:
%   x         vector of root approximations (last one is best)

% Operating parameters.
funtol = 100*eps;  xtol = 100*eps;  maxiter = 40;

x = x1;
y = f(x1);
dx = Inf;    % for initial pass below
k = 1;

while (abs(dx) > xtol) && (abs(y) > funtol) && (k < maxiter)
    dydx = dfdx(x(k));
    dx = -y/dydx;           % Newton step
    x(k+1) = x(k) + dx;

    k = k+1;
    y = f(x(k));
end

if k==maxiter, warning('Maximum number of iterations reached.'), end
```

Let's revisit the assumptions made to derive quadratic convergence as given by (4.3.4):

1. The residual function $f$ has to have enough continuous derivatives to make the Taylor series expansion valid. Often this is stated as $f$ being "smooth enough." This is usually not a problem, but see Exercise 4.3.6.

2. We required $f'(r) \neq 0$—that is, $r$ must be a *simple* root. See Exercise 4.3.7 to investigate what happens at a multiple root.

3. We assumed that the sequence converged, which is not easy to guarantee in any particular case. In fact, finding a starting guess from which the Newton iteration converges is often the most challenging part of a rootfinding problem. We will try to deal with this issue in Section 4.6.

### Implementation

Our implementation of Newton's iteration is given in Function 4.3.1. It accepts mathematical functions $f$ and $f'$ and the starting guess $x_1$ as input arguments. Beginning programmers are tempted to embed the functions directly into the code, but there are two good reasons not to do so. First, you would need a new copy of the whole code for each new instance of the problem, even though very little code may need to change. Second, you may want to try more than one rootfinding implementation for a particular problem, and keeping the definition of the problem separate from the algorithm for its solution makes this task much easier. As a practical issue in MATLAB, you can define short functions inline as in Example 4.3.2. For longer functions, you can write separate function files and pass them as arguments to Function 4.3.1 by affixing an at-sign @ to the front of the name of the function file.

> **Example 4.3.3**
>
> Suppose we want to solve $e^x = x + c$ for multiple values of $c$. We can create functions for $f$ and $f'$ in each case.
>
> ```
> for c = [2 4 7.5 11]
>     f = @(x) exp(x) - x - c;
>     dfdx = @(x) exp(x) - 1;
>     x = newton(f,dfdx,1);   r = x(end);
>     fprintf('root with c = %4.1f is %.14f\n',c,r)
> end
> ```
>
> ```
> root with c =  2.0 is 1.14619322062058
> root with c =  4.0 is 1.74903138601270
> root with c =  7.5 is 2.28037814882306
> root with c = 11.0 is 2.61086863814988
> ```
>
> The definition of f locks in whatever value is defined for c at the moment of definition. Even if we later change the value assigned to c, the function is unaffected.
>
> ```
> f(r)
> c = 100;  f(r)
> ```
>
> ```
> ans =
>    1.7764e-15
> ans =
>    1.7764e-15
> ```
>
> This can get a little tricky, because the function is not executed or checked at definition time. You may discover an error using such a definition only later in the code.
>
> ```
> clear c
> f = @(x) exp(x) - x - c;   % executes OK
> c = 1;                     % does not change f
> ```
>
> A call such as f(1) would create an error, since the assignment of c did not come until after f was defined.

Function 4.3.1 also deals with a thorny practical issue: how to stop the iteration. It adopts a three-part criterion. First, it monitors the difference between successive root estimates, $|x_k - x_{k-1}|$, which is used as a stand-in for the unknown error $|r - x_k|$. In addition, it monitors the residual $|f(x_k)|$, which is equivalent to the backward error and more realistic to control in badly conditioned problems (see Section 4.1). If either of these quantities is considered to be sufficiently small, the iteration ends. Finally, we need to protect against the possibility of a nonconvergent iteration, so the procedure terminates with a warning if a maximum number of iterations is exceeded.[18]

---

[18] In more practical codes, the thresholds used to make these decisions are controllable through additional user inputs to the procedure.

## Exercises

For each equation and given interval, do the following steps:

(a) ✍ Rewrite the equation into the standard form for rootfinding, $f(x) = 0$, and compute $f'(x)$.

(b) 💻 Make a plot of $f$ over the given interval and determine how many roots lie in the interval.

(c) 💻 Use `fzero` to find an "exact" value for each root.

(d) 💻 Use Function 4.3.1 to find each root.

(e) 💻 For one of the roots, define **e** as a vector of the errors in the Newton sequence. Determine numerically whether the convergence is roughly quadratic.

4.3.1. $x^2 = e^{-x}$, over $[-2, 2]$.

4.3.2. $2x = \tan x$, over $[-0.2, 1.4]$.

4.3.3. $e^{x+1} = 2 + x$, over $[-2, 2]$.

---

4.3.4. 💻 Consider the equation $f(x) = x^{-2} - \sin x = 0$ on the interval $x \in [0.1, 4\pi]$. Use a plot to approximately locate the roots of $f$. To which roots do the following initial guesses converge when using Function 4.3.1? Is the root obtained the one that is closest to that guess?

(a) $x_0 = 1.5$, (b) $x_0 = 2$, (c) $x_0 = 3.2$, (d) $x_0 = 4$, (e) $x_0 = 5$, (f) $x_0 = 2\pi$.

4.3.5. ✍ Show that if $f(x) = x^{-1} - b$ for nonzero $b$, then Newton's iteration converging to the root $r = 1/b$ can be implemented without performing any divisions.

4.3.6. ✍ Discuss what happens when Newton's method is applied to find a root of $f(x) = \text{sign}(x)\sqrt{|x|}$, starting at $x_0 \neq 0$.

4.3.7. ✍ In the case of a multiple root, where $f(r) = f'(r) = 0$, the derivation of the quadratic error convergence in (4.3.4) is invalid. Redo the derivation to show that in this circumstance and with $f''(r) \neq 0$ the error converges only linearly.

4.3.8. ✍ In Function 4.3.1 and elsewhere, the actual error is not available, so we use $|x_k - x_{k-1}|$ as an approximate indicator of error to determine when to stop the iteration. Find an example that foils this indicator, that is, a sequence $\{x_k\}$ such that

$$\lim_{k \to \infty} (x_k - x_{k-1}) = 0,$$

but $\{x_k\}$ diverges. (Note: You have seen such sequences in calculus.) Hence the need for residual tolerances and escape hatches in the code!

## 4.4 • Interpolation-based methods

From a practical standpoint, one of the biggest drawbacks of Newton's method is the requirement to supply $f'$ in Function 4.3.1. It is both a programming inconvenience and a step that requires computational time. We can avoid using $f'$, however, by making a simple but easily overlooked observation: *when a step produces an approximate result, you are free to carry it out approximately.* Let's call this the "principle of approximate approximation."

In the Newton context, the principle of approximate approximation begins with the observation that the use of $f'$ is linked to the construction of a linear approximation $q(x)$ equivalent to a tangent line. The root of $q(x)$ is used to define the next iterate in the sequence. We can avoid calculating the value of $f'$ by choosing a different linear approximation.

### Example 4.4.1

We return to finding a root of the equation $xe^x = 2$.

```
f = @(x) x.*exp(x) - 2;
fplot(f,[0.25 1.25])
```

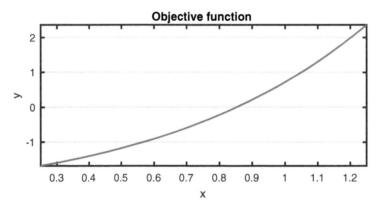

From the graph, it's clear that there is a root near $x = 1$. To be more precise, there is a root in the interval $[0.5, 1]$. So let us take the endpoints of that interval as *two* initial approximations.

```
x1 = 1;     f1 = f(x1);
x2 = 0.5;   f2 = f(x2);
hold on, plot([x1 x2],[f1 f2],'.')
```

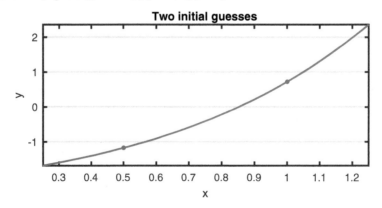

## 4.4. Interpolation-based methods

Instead of constructing the tangent line by evaluating the derivative, we can construct a linear model function by drawing the line between the two points $(x_1, f(x_1))$ and $(x_2, f(x_2))$. This is called a *secant line*.

```
slope2 = (f2-f1) / (x2-x1);
secant2 = @(x) f2 + slope2*(x-x2);
```

As before, the next value in the iteration is the root of this linear model.

```
fplot(secant2,[0.25 1.25],'k--')
x3 = x2 - f2/slope2;
f3 = f(x3)
plot(x3,0,'.')
```

```
f3 =
    -0.1777
```

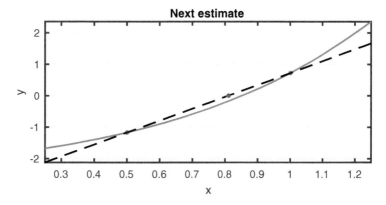

For the next linear model, we use the line through the two most recent points. The next iterate is the root of that secant line, and so on.

```
slope2 = (f3-f2) / (x3-x2);
x4 = x3 - f3/slope2;
f4 = f(x4)
```

```
f4 =
    0.0572
```

Example 4.4.1 demonstrates the **secant method**. *In the secant method, one finds the root of the linear approximation through the two most recent root estimates.* That is, given previous approximations $x_1, \ldots, x_k$, define the linear model function as the line through $(x_{k-1}, f(x_{k-1}))$ and $(x_k, f(x_k))$:

$$q(x) = f(x_k) + \frac{f(x_k) - f(x_{k-1})}{x_k - x_{k-1}}(x - x_k). \tag{4.4.1}$$

Solving $q(x_{k+1}) = 0$ for $x_{k+1}$ gives the formula

$$x_{k+1} = x_k - \frac{f(x_k)(x_k - x_{k-1})}{f(x_k) - f(x_{k-1})}, \quad n = 1, 2, \ldots. \tag{4.4.2}$$

**Function 4.4.1 (secant)** Secant method for scalar rootfinding.

```
1   function x = secant(f,x1,x2)
2   % SECANT   Secant method for a scalar equation.
3   % Input:
4   %   f          objective function
5   %   x1,x2      initial root approximations
6   % Output
7   %   x          vector of root approximations (last is best)
8
9   % Operating parameters.
10  funtol = 100*eps;  xtol = 100*eps;  maxiter = 40;
11
12  x = [x1 x2];
13  dx = Inf;  y1 = f(x1);
14  k = 2;  y2 = 100;
15
16  while (abs(dx) > xtol) && (abs(y2) > funtol) && (k < maxiter)
17      y2 = f(x(k));
18      dx = -y2 * (x(k)-x(k-1)) / (y2-y1);    % secant step
19      x(k+1) = x(k) + dx;
20
21      k = k+1;
22      y1 = y2;    % current f-value becomes the old one next time
23  end
24
25  if k==maxiter, warning('Maximum number of iterations reached.'), end
```

Our implementation of the method based on this formula is given in Function 4.4.1.

The secant method uses a different linear model in each iteration than the one Newton's method would use. Is it as good? As before, let $\epsilon_k = r - x_k$ be the errors in the successive root approximations. If the initial errors are small, then a tedious but straightforward Taylor expansion shows that, to lowest order,

$$\epsilon_{k+1} \approx -\frac{1}{2}\frac{f''(r)}{f'(r)}\epsilon_k \epsilon_{k-1}. \tag{4.4.3}$$

If we make an educated guess that

$$\epsilon_{k+1} = c(\epsilon_k)^\alpha, \qquad \epsilon_k = c(\epsilon_{k-1})^\alpha, \qquad \alpha > 0,$$

then (4.4.3) becomes

$$\left[\epsilon_{k-1}^\alpha\right]^\alpha \approx C\epsilon_{k-1}^{\alpha+1} \tag{4.4.4}$$

for an unknown constant $C$. Treating the implied approximation as an equality, this becomes solvable if and only if the exponents match, i.e., $\alpha^2 = \alpha + 1$. The only positive root of this equation is the golden ratio,

$$\alpha = \frac{1+\sqrt{5}}{2} \approx 1.618.$$

Hence the errors in the secant method converge like $\epsilon_{k+1} = c(\epsilon_k)^\alpha$ for $1 < \alpha < 2$, a situation called **superlinear convergence**.

## 4.4. Interpolation-based methods

**Example 4.4.2**

We check the convergence of the secant method from the previous example.

```
f = @(x) x.*exp(x) - 2;
x = secant(f,1,0.5);
```

We don't know the exact root, so we use the built-in `fzero` to get a substitute.

```
r = fzero(f,1);
```

Here is the sequence of errors.

```
format short e
err = r - x(1:end-1)'

err =
  -1.4739e-01
   3.5261e-01
   4.2234e-02
  -1.3026e-02
   4.2748e-04
   4.2699e-06
  -1.4055e-09
   4.6629e-15
```

It's not so easy to see the convergence rate by looking at these numbers. But we can check the ratios of the log of successive errors.

```
logerr = log(abs(err));
ratios = logerr(2:end) ./ logerr(1:end-1)

ratios =
   5.4444e-01
   3.0358e+00
   1.3717e+00
   1.7871e+00
   1.5938e+00
   1.6486e+00
   1.6190e+00
```

It seems to be heading toward a constant ratio of about 1.6 by the time it quits.

In terms of error as a function of the iteration number $k$, the secant method converges at a rate between linear and quadratic, which is slower than Newton's method. In that sense we must conclude that the secant line is inferior to the tangent line for approximating $f$ near a root. But error versus iteration count may not be the best means of comparison. Often we analyze rootfinding methods by assuming that the bulk of computing time is spent evaluating the user-defined functions $f$ and $f'$. (Our simple examples and exercises mostly don't support this assumption, but many practical applications do.) In this light we see that Newton's method requires two evaluations,

$f(x_k)$ and $f'(x_k)$, for each iteration. The secant method, on the other hand, while it *uses* the two function values $f(x_k)$ and $f(x_{k-1})$ at each iteration, only needs to *compute* a single new one. Note that Function 4.4.1 keeps track of one previous function value rather than recomputing it.

Now suppose that $|\epsilon_k| = \epsilon$. Roughly speaking, two units of work (i.e., function evaluations) in Newton's method brings us to an error of $\epsilon^2$. If one spreads out the improvement in the error evenly across the two geometric steps, using

$$\epsilon^2 = \left(\epsilon^{\sqrt{2}}\right)^{\sqrt{2}},$$

it seems reasonable to say that the rate of convergence *per function evaluation* is $\sqrt{2} \approx$ 1.41. This is actually less than the comparable rate of about 1.62 for the secant method. *Not only is the secant method easier to apply than Newton's method in practice, it's also more efficient when counting function evaluations*—a rare double victory!

## Inverse interpolation

At each iteration, the secant method constructs a linear model function that interpolates the two most recently found points on the graph of $f$. Two points determine a straight line, so this seems like a sensible choice. But as the iteration progresses, why use only the *two* most recent points? What would it mean to use more of them?

If we interpolate through three points by a polynomial, we get a unique quadratic function. Unfortunately, a parabola may have zero, one, or two crossings of the $x$-axis, leaving some doubt as to how to define the next root estimate. On the other hand, if we turn a parabola on its side, we get a graph that intersects the $x$-axis exactly once, which is ideal for defining the next root estimate.

This leads to the idea of defining $q(y)$ as the quadratic interpolant to the points $(y_{k-2}, x_{k-2}), (y_{k-1}, x_{k-1})$, and $(y_k, x_k)$, where $y_i = f(x_i)$ for all $i$, and setting $x_{k+1} = q(0)$. The process defined in this way (given three initial estimates) is called **inverse quadratic interpolation**. Rather than deriving lengthy formulas for it here, we demonstrate how to perform inverse quadratic interpolation using `polyfit` to perform the interpolation step.

### Example 4.4.3

Here we look for a root of $x + \cos(10x)$ that is close to 1.

```
f = @(x) x + cos(10*x);
interval = [0.5,1.5];
fplot(f,interval)
r = fzero(f,1)

r =
    0.9679
```

## 4.4. Interpolation-based methods

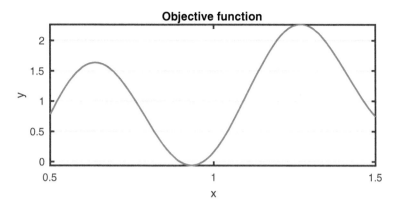

**Objective function**

We choose three values to get the iteration started.

```
x = [0.8 1.2 1]';
y = f(x);
hold on, plot(x,y,'.')
```

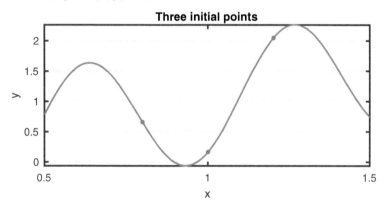

**Three initial points**

If we were using "forward" interpolation, we would ask for the polynomial interpolant of $y$ as a function of $x$. But that parabola has no real roots.

```
c = polyfit(x,y,2);    % coefficients of interpolant
q = @(x) polyval(c,x);
fplot(q,interval,'--')
```

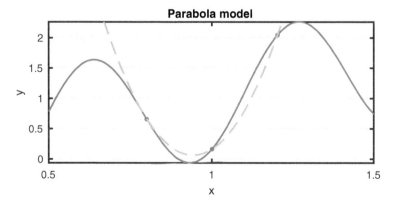

**Parabola model**

To do inverse interpolation, we swap the roles of $x$ and $y$ in the interpolation.

```
c = polyfit(y,x,2);     % coefficients of interpolating
    polynomial
q = @(y) polyval(c,y);
fplot(q,@(y) y,ylim,'--')     % plot x=q(y), y=y
```

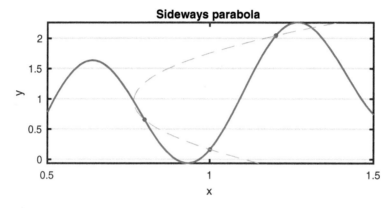

We seek the value of $x$ that makes $y$ zero. This means evaluating $q$ at zero.

```
x = [x; q(0)];
y = [y; f(x(end))]
```

```
y =
    0.6545
    2.0439
    0.1609
    1.1482
```

We repeat the process a few more times.

```
for k = 4:8
    c = polyfit(y(k-2:k),x(k-2:k),2);
    x(k+1) = polyval(c,0);
    y(k+1) = f(x(k+1));
end
```

Here is the sequence of errors.

```
format short e
err = x - r
```

```
err =
   -1.6789e-01
    2.3211e-01
    3.2112e-02
    1.3609e-01
    1.5347e-02
    3.2683e-03
    4.6174e-04
```

```
    6.2958e-06
    3.4390e-09
```

The error seems to be superlinear, but subquadratic.

```
logerr = log(abs(err));
ratios = logerr(2:end) ./ logerr(1:end-1)

ratios =
    8.1848e-01
    2.3543e+00
    5.8002e-01
    2.0943e+00
    1.3703e+00
    1.3419e+00
    1.5592e+00
    1.6273e+00
```

## Bracketing

Like Newton's method, the secant and inverse quadratic interpolation methods cannot guarantee convergence. One final new idea is needed to make a foolproof algorithm.

If $f$ is continuous on the interval $[a, b]$ and $f(a)f(b) < 0$—that is, $f$ changes sign on the interval—then $f$ must have (at least) one root in the interval, due to the Intermediate Value Theorem from calculus. If we come up with a new root estimate $c \in (a, b)$, then whatever sign $f(c)$ is, it is different from the sign at one of the endpoints. (Of course, if $f(c)$ is zero, we are done!) So either $[a, c]$ or $[c, b]$ is guaranteed to have a root too, and in this way we can maintain not just individual estimates but an interval that always contains a root.

The best algorithms blend the use of fast-converging methods with the guarantee provided by a bracket. For example, say that an iteration begins with a bracketing interval. Make a list of the inverse quadratic estimate, the secant estimate, and the midpoint of the current interval and pick the first member of the list that lies within the current interval. Replace the interval with the bracketing subinterval, and start a new iteration. The method employed by `fzero` in MATLAB works along these lines, but we do not give the details.

## Exercises

For each equation and given interval, do the following steps:

(a) Rewrite the equation into the standard form for rootfinding, $f(x) = 0$.

(b) Make a plot of $f$ over the given interval and determine how many roots lie in the interval.

(c) 🖳 Use `fzero` to find an "exact" value for each root.

(d) 🖳 Determine a bracketing interval for each root. Then use Function 4.4.1, starting with the endpoints of the interval, to find each root.

(e) 🖳 For one of the roots, define e as a vector of the errors in the secant sequence. Determine numerically whether the convergence is superlinear.

4.4.1. $x^2 = e^{-x}$, over $[-2, 2]$.

4.4.2. $2x = \tan x$, over $[-0.2, 1.4]$.

4.4.3. $e^{x+1} = 2 + x$, over $[-2, 2]$.

---

4.4.4. 🖳 Use a plot to approximately locate all the roots of $f(x) = x^{-2} - \sin(x)$ in the interval $[0.5, 4\pi]$. Then find a pair of initial points for each root such that Function 4.4.1 converges to that root.

4.4.5. ✍ Show analytically that the secant method converges in one step for a linear function, regardless of the initialization.

4.4.6. ✍ In general, the secant method formula (4.4.2) cannot be applied if $x_k = x_{k-1}$. However, suppose that $f(x) = ax^2 + bx + c$ for constants $a$, $b$, and $c$. Show that in this case the formula can be simplified to one that is well defined when $x_k = x_{k-1}$. Then show that the resulting $x_{k+1}$ is the same as the result of one step of Newton's method applied to $f$ at $x_k$.

4.4.7. ✍ Let $f(x) = x^2$. Show that if $(1/x_1)$ and $(1/x_2)$ are positive integers, and the secant iteration is applied, then the sequence $1/x_1, 1/x_2, 1/x_3, \ldots$ is a Fibonacci sequence.

4.4.8. ✍ Provide the details that show how to derive (4.4.3) from (4.4.2).

4.4.9. 🖳 Write a function `iqi(f,x1,x2,x3)` that performs inverse quadratic interpolation for finding a root of $f$, given three initial estimates. To find the quadratic polynomial $q(y)$ passing through the three most recent points, use the built-in `polyfit` command. Test your function on the first problem from this section.

## 4.5 • Newton for nonlinear systems

The rootfinding problem becomes much more difficult when multiple variables and equations are involved. We now let $\boldsymbol{f}$ be a vector function of a vector argument: $\boldsymbol{f}$ maps from $\mathbb{R}^n$ to $\mathbb{R}^n$. By $\boldsymbol{f}(\boldsymbol{x}) = \boldsymbol{0}$ we mean the simultaneous system of $n$ scalar equations,

$$\begin{aligned} f_1(x_1, \ldots, x_n) &= 0, \\ f_2(x_1, \ldots, x_n) &= 0, \\ &\vdots \\ f_n(x_1, \ldots, x_n) &= 0. \end{aligned}$$

When discussing a specific problem, it is often necessary to write out the equations componentwise, but in discussing methods for the general problem it's more conve-

## 4.5. Newton for nonlinear systems

nient to use the vector form. Proving the existence and uniqueness of a solution for any particular $f$ is typically quite difficult.

### Linear model

To extend rootfinding methods to systems, we will keep to the basic philosophy of constructing easily managed models of the exact function. As usual, the starting point is a linear model. We base it on the multidimensional Taylor series,

$$f(x+h) = f(x) + J(x)h + O(\|h\|^2), \tag{4.5.1}$$

where $J$ is called the **Jacobian matrix** of $f$ and is defined by

$$J(x) = \begin{bmatrix} \frac{\partial f_1}{\partial x_1} & \frac{\partial f_1}{\partial x_2} & \cdots & \frac{\partial f_1}{\partial x_n} \\ \frac{\partial f_2}{\partial x_1} & \frac{\partial f_2}{\partial x_2} & \cdots & \frac{\partial f_2}{\partial x_n} \\ \vdots & \vdots & & \vdots \\ \frac{\partial f_n}{\partial x_1} & \frac{\partial f_n}{\partial x_2} & \cdots & \frac{\partial f_n}{\partial x_n} \end{bmatrix} = \left[ \frac{\partial f_i}{\partial x_j} \right]_{i,j=1,\ldots,n}. \tag{4.5.2}$$

Because of the Jacobian's role in (4.5.1), we may write $J(x)$ as $f'(x)$. Like any derivative, it is a function of the independent variable $x$.

**Example 4.5.1**

Let

$$f_1(x_1, x_2, x_3) = -x_1 \cos(x_2) - 1,$$
$$f_2(x_1, x_2, x_3) = x_1 x_2 + x_3,$$
$$f_3(x_1, x_2, x_3) = e^{-x_3} \sin(x_1 + x_2) + x_1^2 - x_2^2.$$

Then

$$J(x) = \begin{bmatrix} -\cos(x_2) & x_1 \sin(x_2) & 0 \\ x_2 & x_1 & 1 \\ e^{-x_3} \cos(x_1 + x_2) + 2x_1 & e^{-x_3} \cos(x_1 + x_2) - 2x_2 & -e^{-x_3} \sin(x_1 + x_2) \end{bmatrix}.$$

If we were to start writing out the terms in (4.5.1), we would begin with

$$f_1(x_1 + h_1, x_2 + h_2, x_3 + h_3) = -x_1 \cos(x_2) - 1 - \cos(x_2)h_1 + x_1 \sin(x_2)h_2 + O(\|h\|^2),$$
$$f_2(x_1 + h_1, x_2 + h_2, x_3 + h_3) = x_1 x_2 + x_3 + x_2 h_1 + x_1 h_2 + h_3 + O(\|h\|^2),$$

and so on.

The terms $f(x) + J(x)h$ in (4.5.1) represent the "linear part" of $f$ near $x$. If $f$ is actually linear, i.e., $f(x) = Ax - b$, then the Jacobian matrix is the constant matrix $A$ and the higher-order terms in (4.5.1) disappear.

## The multidimensional Newton iteration

With a method in hand for constructing a linear model for the vector system $f(x)$, we can generalize Newton's method. Specifically, at a root estimate $x_k$, we set $h = x - x_k$ in (4.5.1) and get

$$f(x) \approx q(x) = f(x_k) + J(x_k)(x - x_k).$$

We define the next iteration value $x_{k+1}$ by requiring $q(x_{k+1}) = 0$,

$$0 = f(x_k) + J(x_k)(x_{k+1} - x_k),$$

which can be rearranged into

$$x_{k+1} = x_k - [J(x_k)]^{-1} f(x_k). \tag{4.5.3}$$

Note that $J^{-1}f$ now plays the role that $f/f'$ had in the scalar case; in fact the two are the same in one dimension. In computational practice, however, we don't compute matrix inverses. Instead, define the *Newton step* $s_k$ by the linear $n \times n$ system

$$J(x_k) s_k = -f(x_k), \tag{4.5.4}$$

so that $x_{k+1} = x_k + s_k$. *Computing the Newton step is equivalent to solving a linear system using the Jacobian matrix and the function value.*

An extension of our series analysis of the scalar Newton method shows that the vector version is also quadratically convergent in any vector norm, under suitable circumstances.

**Example 4.5.2**

Let us use Newton's method on the system defined by the function

```
function f = nlvalue(x)
    f = zeros(3,1);     % ensure a column vector output
    f(1) = exp(x(2)-x(1)) - 2;
    f(2) = x(1)*x(2) + x(3);
    f(3) = x(2)*x(3) + x(1)^2 - x(2);
end
```

Here is a function that computes the Jacobian matrix.

```
function J = nljac(x)
    J(1,:) = [-exp(x(2)-x(1)),exp(x(2)-x(1)), 0];
    J(2,:) = [x(2), x(1), 1];
    J(3,:) = [2*x(1), x(3)-1, x(2)];
end
```

(These functions could be written as separate files, or embedded within another function as we have done here.) Our initial guess at a root is the origin.

```
x = [0;0;0];    % column vector!
```

We need the value (residual) of the nonlinear function, and its Jacobian, at this value for x.

## 4.5. Newton for nonlinear systems

```
f = nlvalue(x);
J = nljac(x);
```

We solve for the Newton step and compute the new estimate.

```
s = -(J\f);
x(:,2) = x(:,1) + s;
```

Here is the new residual.

```
format short e
f(:,2) = nlvalue(x(:,2))

f =
  -1.0000e+00    7.1828e-01
            0             0
            0    1.0000e+00
```

We don't seem to be especially close to a root. Let's iterate a few more times.

```
for n = 2:7
    f(:,n) = nlvalue(x(:,n));
    s = -( nljac(x(:,n)) \ f(:,n) );
    x(:,n+1) = x(:,n) + s;
end
```

We find the infinity norm of the residuals.

```
residual_norm = max(abs(f),[],1)'    % max in dimension 1

residual_norm =
    1.0000e+00
    1.0000e+00
    2.0229e-01
    1.0252e-02
    2.1556e-05
    1.9900e-10
    1.3878e-17
```

We don't know an exact answer, so we will take the last computed value as its surrogate.

```
r = x(:,end);
x = x(:,1:end-1);
```

The following will subtract $r$ from every column of $x$.

```
z = size(x);
e = x - r*ones(1,z(2));
```

Now we take infinity norms and check for quadratic convergence.

```
logerrs = log(abs( max(e,[],1)' ));
ratios = logerrs(2:end) ./ logerrs(1:end-1)
```

```
ratios =
    2.8541e+00
    1.3490e+00
    1.6162e+00
    2.1967e+00
    2.3094e+00
           Inf
```

For a brief time, we see ratios around 2, but then the limitation of double precision makes it impossible for the doubling to continue.

## Implementation

A MATLAB implementation of Newton's method for systems is given in Function 4.5.1. It is a remarkable effect of the vector-friendliness of MATLAB that this program is hardly different from the scalar version in Section 4.3:

- The root estimates are stored as columns in an array.

- The Newton step is calculated using a backslash.

- The function **norm** is used for the magnitude of a vector, instead of **abs** for the magnitude of a scalar.

**Function 4.5.1 (newtonsys)** Newton's method for a system of equations.

```
1  function x = newtonsys(f,x1)
2  % NEWTONSYS    Newton's method for a system of equations.
3  % Input:
4  %   f          function that computes residual and Jacobian matrix
5  %   x1         initial root approximation (n-vector)
6  % Output
7  %   x          array of approximations (one per column, last is best)
8
9  % Operating parameters.
10 funtol = 1000*eps;  xtol = 1000*eps;  maxiter = 40;
11
12 x = x1(:);
13 [y,J] = f(x1);
14 dx = Inf;
15 k = 1;
16
17 while (norm(dx) > xtol) && (norm(y) > funtol) && (k < maxiter)
18     dx = -(J\y);    % Newton step
19     x(:,k+1) = x(:,k) + dx;
20
21     k = k+1;
22     [y,J] = f(x(:,k));
23 end
24
25 if k==maxiter, warning('Maximum number of iterations reached.'), end
```

## 4.5. Newton for nonlinear systems

Indeed, Function 4.5.1 is a proper generalization—it can be used on scalar problems as well as on systems.

**Example 4.5.3**

We repeat Example 4.5.2. The system is again defined by its residual and Jacobian, but this time we implement them as a single function, which in practice could go into a separate file.

```
function [f,J] = nlsystem(x)
    f = zeros(3,1);    % ensure a column vector output
    f(1) = exp(x(2)-x(1)) - 2;
    f(2) = x(1)*x(2) + x(3);
    f(3) = x(2)*x(3) + x(1)^2 - x(2);
    J(1,:) = [-exp(x(2)-x(1)),exp(x(2)-x(1)), 0];
    J(2,:) = [x(2), x(1), 1];
    J(3,:) = [2*x(1), x(3)-1, x(2)];
end
```

Our initial guess is the origin. The output has one column per iteration.

```
x1 = [0;0;0];    % column vector!
x = newtonsys(@nlsystem,x1);
[~,num_iter] = size(x)

num_iter =
    7
```

The last column contains the final Newton estimate. We'll compute the residual there in order to check the quality of the result.

```
r = x(:,end)
back_err = norm(nlsystem(r))

r =
   -0.4580
    0.2351
    0.1077
back_err =
   1.3878e-17
```

Let's use the convergence to the first component of the root as a proxy for the convergence of the vectors.

```
log10( abs(x(1,1:end-1)-r(1)) )'

ans =
   -0.3391
   -0.2660
   -0.9188
   -2.2920
```

```
-5.1929
-9.9685
```

The exponents approximately double, as is expected of quadratic convergence.

## Exercises

4.5.1. ✍ Suppose that
$$f(x) = \begin{bmatrix} x_1 x_2 + x_2^2 - 1 \\ x_1 x_2^3 + x_1^2 x_2^2 + 1 \end{bmatrix}.$$
Let $x_1 = [-2, 1]^T$. Use Newton's method to find $x_2$.

4.5.2. ✍ Suppose that $f(x) = Ax - b$ for a constant $n \times n$ matrix $A$ and constant $n \times 1$ vector $b$. Show that Newton's method converges to the exact root in one iteration.

4.5.3. Two curves in the $(u, v)$ plane are defined implicitly by the equations $u \log u + v \log v = -0.3$ and $u^4 + v^2 = 1$.

   (a) ✍ Write the intersection of these curves in the form $f(x) = 0$ for two-dimensional $f$ and $x$.

   (b) ✍ Find the Jacobian matrix of $f$.

   (c) 💻 Use Function 4.5.1 to find an intersection point near $u = 1$, $v = 0.1$.

   (d) 💻 Use Function 4.5.1 to find an intersection point near $u = 0.1$, $v = 1$.

4.5.4. Two elliptical orbits $(x_1(s), y_1(s))$ and $(x_2(t), y_2(t))$ are described by the equations
$$\begin{bmatrix} x_1(t) \\ y_1(t) \end{bmatrix} = \begin{bmatrix} -5 + 10\cos(t) \\ 6\sin(t) \end{bmatrix}, \quad \begin{bmatrix} x_2(t) \\ y_2(t) \end{bmatrix} = \begin{bmatrix} 8\cos(t) \\ 1 + 12\sin(t) \end{bmatrix},$$
where $t$ represents time.

   (a) 💻 Make a plot of the two orbits in the following code:
```
x1 = @(t) -5+10*cos(t);   y1 = @(t) 6*sin(t);
clf, fplot(x1,y1,[0 2*pi])
x2 = @(t) 8*cos(t);       y2 = @(t) 1+12*sin(t);
hold on, fplot(x2,y2,[0 2*pi])
```

   (b) ✍ Write out a 2 × 2 nonlinear system of equations that describes an intersection of these orbits. (Note: An intersection is not the same as a collision—they don't have to occupy the same point at the same time.)

   (c) ✍ Write out the Jacobian matrix of this nonlinear system.

   (d) 💻 Use Function 4.5.1 to find all of the unique intersections.

4.5.5. 💻 Suppose one wants to find the points on the ellipsoid $x^2/25 + y^2/16 + z^2/9 = 1$ that are closest to and farthest from the point $(5, 4, 3)$. The method of Lagrange multipliers implies that any such point satisfies

$$x - 5 = \frac{\lambda x}{25},$$
$$y - 4 = \frac{\lambda y}{16},$$
$$z - 3 = \frac{\lambda z}{9},$$
$$1 = \frac{1}{25}x^2 + \frac{1}{16}y^2 + \frac{1}{9}z^2$$

for an unknown value of $\lambda$.

(a) Write out this system in the form $f(u) = 0$. (Note that the system has four variables to go with the four equations.)

(b) Write out the Jacobian matrix of this system.

(c) Use Function 4.5.1 with different initial guesses to find the two roots of this system. Which is the closest point to $(5, 4, 3)$ and which is the farthest?

4.5.6. In this problem you are to fit a function of the form

$$P(t) = a_1 + a_2 e^{a_3 t}$$

to a subset of U.S. census data for the twentieth century:

| Year | 1910 | 1930 | 1950 | 1970 | 1990 |
|---|---|---|---|---|---|
| Population (millions) | 92.0 | 122.8 | 150.7 | 205.0 | 248.7 |

(a) Determine the unknown parameters $a_1$, $a_2$, $a_3$ in $P$ by requiring that $P$ exactly reproduce the data in the years 1910, 1950, and 1990. This creates three nonlinear equations for $a_1$, $a_2$, and $a_3$ that may be solved using Function 4.5.1.

(b) To obtain convergence, rescale the data using the time variable $t = \text{(year}-1900)/100$ and divide the population numbers above by 100. Using your model $P(t)$, predict the result of the 2000 census, and compare it to the true figure of 284.1 million.

## 4.6 • Quasi-Newton methods

Newton's method is a foundation for algorithms to solve equations and minimize quantities. But it is not ideal in its plain or "pure" form. Instead *there are different quasi-Newton methods that attempt to overcome two serious issues: the programming nuisance and computational expense of evaluating the Jacobian matrix, and the tendency of the iteration to diverge for many starting points.*

### Jacobian by finite differences

In the scalar case, we found an easy alternative to a direct evaluation of the derivative. Specifically, we may interpret the secant formula (4.4.2) as the Newton formula (4.3.2)

**Function 4.6.1** (fdjac) Finite difference approximation of a Jacobian.

```
function J = fdjac(f,x0,y0)
% FDJAC    Finite-difference approximation of a Jacobian.
% Input:
%    f          function to be differentiated
%    x0         evaluation point (n-vector)
%    y0         value of f at x0 (m-vector)
% Output
%    J          approximate Jacobian (m-by-n)

delta = sqrt(eps);    % FD step size
m = length(y0);  n = length(x0);
J = zeros(m,n);
I = eye(n);
for j = 1:n
    J(:,j) = ( f(x0+delta*I(:,j)) - y0) / delta;
end

end
```

with $f'(x_k)$ replaced by the quotient

$$\frac{f(x_k)-f(x_{k-1})}{x_k-x_{k-1}}. \tag{4.6.1}$$

If the sequence of $x_k$ values converges to a root $r$, then this quotient converges to $f'(r)$.

In the system case, replacing the Jacobian evaluation is more complicated: derivatives are needed with respect to $n$ variables, not just one. From (4.5.2), we note that the $j$th column of the Jacobian is

$$J(x)e_j = \begin{bmatrix} \frac{\partial f_1}{\partial x_j} \\ \frac{\partial f_2}{\partial x_j} \\ \vdots \\ \frac{\partial f_n}{\partial x_j} \end{bmatrix}.$$

(As always, $e_j$ represents the $j$th column of the identity matrix, here in $n$ dimensions.) Inspired by (4.6.1), we can replace the differentiation with a quotient involving a change in only $x_j$, while the other variables remain fixed:

$$J(x)e_j \approx \frac{f(x+\delta e_j)-f(x)}{\delta}, \qquad j=1,\ldots,n. \tag{4.6.2}$$

For reasons explained in Section 5.5, $\delta$ is usually chosen close to $\sqrt{\epsilon}$, where $\epsilon$ represents the expected noise level in evaluation of $f$. If the only source of noise is floating point roundoff, then $\delta = \sqrt{\epsilon_{\text{mach}}}$.

The finite difference formula (4.6.2) is implemented by the short code Function 4.6.1. (The code is written to accept the case where $f$ maps $n$ variables to $m$ values with $m \neq n$, in anticipation of Section 4.7.)

## Broyden's update

The finite-difference Jacobian is easy to conceive and use. But, as you can see from (4.6.2), it requires $n$ additional evaluations of the system function at each iteration, which can be unacceptably slow in some applications. Conceptually these function evaluations seem especially wasteful given that the root estimates, and thus presumably the Jacobian matrix, are supposed to change little as the iteration converges. This is a good time to step in with the principle of approximate approximation, which suggests looking for a shortcut in the form of a cheap-but-good-enough way to update the Jacobian from one iteration to the next.

Recall that the Newton iteration is derived by solving the linear model implied by equation (4.5.1):

$$0 \approx f(x_{k+1}) \approx f(x_k) + J(x_k)(x_{k+1} - x_k).$$

Let $s_k = x_{k+1} - x_k$ be the Newton step. We will make the notation simpler via $f_k = f(x_k)$, and now we replace $J(x_k)$ by a matrix $A_k$ that is meant to approximate the Jacobian. Hence the Newton step is considered to be defined by

$$A_k s_k = -f_k. \tag{4.6.3}$$

This equation gets us from $x_k$ to $x_{k+1}$. To continue the iteration, we want to update the approximate Jacobian to $A_{k+1}$. If we think one-dimensionally for a moment, the secant method would assume that $A_{k+1} = (f_{k+1} - f_k)/(x_{k+1} - x_k)$. It's not easy to generalize a fraction to vectors, but we can do it if we instead write it as

$$f_{k+1} - f_k = A_{k+1}(x_{k+1} - x_k) = A_{k+1} s_k.$$

This is used to justify the following requirement:

$$A_{k+1} s_k = f_{k+1} - f_k. \tag{4.6.4}$$

This isn't enough to uniquely determine $A_{k+1}$. However, if we also require that $A_{k+1} - A_k$ is a matrix of rank one, then one arrives at the **Broyden update formula**

$$A_{k+1} = A_k + \frac{1}{s_k^T s_k}(f_{k+1} - f_k - A_k s_k) s_k^T. \tag{4.6.5}$$

Observe that $A_{k+1} - A_k$, being proportional to the outer product of two vectors, is indeed a rank-1 matrix, and that computing it requires no extra evaluations of $f$. Remarkably, under reasonable assumptions the sequence of $x_k$ so defined converges superlinearly, even though the matrices $A_k$ do not necessarily converge to the Jacobian of $f$. In practice one typically uses finite differences to initialize the Jacobian at iteration $k = 1$. If the step computed by the update formula improves the solution, it is accepted and the iteration continues, but if the update formula fails to give a good result, the matrix $A_k$ is reinitialized by finite differences and the step is recalculated.

## Levenberg's method

The most difficult part of many rootfinding problems is finding a starting point that will lead to convergence. The linear model implicitly constructed during a Newton iteration—whether we use an exact, finite difference, or iteratively updated Jacobian matrix—becomes increasingly inaccurate as one ventures further from the most recent root estimate, eventually failing to resemble the exact function much at all. Although one could imagine trying to do a detailed accuracy analysis of each linear model as we go, in practice simple strategies are valuable here. Suppose, after computing the step suggested by the linear model, we ask a binary question: Would taking that step improve our situation? Since we are trying to find a root of $f$, we have a quantitative way to pose this question: Does the backward error $\|f\|$ decrease? If not, we should reject the step and find an alternative.

There are several ways to find alternatives to the standard step, but we will consider just one of them. Let $A_k$ be the (exact or approximate) Jacobian matrix for iteration number $k$. **Levenberg's method** introduces a positive parameter $\lambda$ into the calculation of the next step: define

$$(A_k^T A_k + \lambda I) s_k = -A_k^T f_k, \qquad (4.6.6)$$

where $x_{k+1} = x_k + s_k$. Some justification of (4.6.6) comes from considering extreme cases for $\lambda$. If $\lambda = 0$, then

$$A_k^T A_k s_k = -A_k^T f_k,$$

which is equivalent to the definition of the usual linear model (i.e., Newton or quasi-Newton) step (4.6.3). On the other hand, as $\lambda \to \infty$, equation (4.6.6) is increasingly close to

$$\lambda s_k = -A_k^T f_k. \qquad (4.6.7)$$

To interpret this equation, define the scalar residual function $\phi(x) = f(x)^T f(x) = \|f(x)\|^2$. Finding a root of $f$ is equivalent to minimizing $\phi$. A calculation shows that the gradient of $\phi$ is

$$\nabla \phi(x) = 2 J(x)^T f(x). \qquad (4.6.8)$$

Hence if $A_k = J(x_k)$, then $s_k$ from (4.6.7) is in the opposite direction from the gradient vector. In vector calculus you learn that this direction is the one of most rapid decrease; for this reason (4.6.7) is called the **steepest descent direction**. A small enough step in this direction is guaranteed (in all but pathological cases) to decrease $\phi$, which is exactly what we want from a backup plan.

In summary, the $\lambda$ parameter in (4.6.6) allows a smooth transition between the pure Newton step, for which convergence is very rapid near a root, and a small step in the descent direction, which guarantees some progress for the iteration when we are far from a root. To make the Levenberg step computation into an algorithm, we will combine it with an accept/reject strategy as described above.

## Implementation

To a large extent the incorporation of finite differences, Jacobian updates, and Levenberg step are independent. Function 4.6.2 shows how they might be combined. This function is one of the most logically complex we have encountered so far.

## 4.6. Quasi-Newton methods

**Function 4.6.2** (`levenberg`) Quasi-Newton method for nonlinear systems.

```
function x = levenberg(f,x1,tol)
% LEVENBERG    Quasi-Newton method for nonlinear systems.
% Input:
%   f          objective function
%   x1         initial root approximation
%   tol        stopping tolerance (default is 1e-12)
% Output
%   x          array of approximations (one per column)

% Operating parameters.
if nargin < 3, tol = 1e-12; end
ftol = tol;   xtol = tol;   maxiter = 40;

x = x1(:);      fk = f(x1);
k = 1;  s = Inf;
Ak = fdjac(f,x(:,1),fk);   % start with FD Jacobian
jac_is_new = true;
I = eye(length(x));

lambda = 10;
while (norm(s) > xtol) && (norm(fk) > ftol) && (k < maxiter)
    % Compute the proposed step.
    B = Ak'*Ak + lambda*I;
    z = Ak'*fk;
    s = -(B\z);

    xnew = x(:,k) + s;   fnew = f(xnew);

    % Do we accept the result?
    if norm(fnew) < norm(fk)     % accept
        y = fnew - fk;
        x(:,k+1) = xnew;   fk = fnew;
        k = k+1;

        lambda = lambda/10;   % get closer to Newton
        % Broyden update of the Jacobian.
        Ak = Ak + (y-Ak*s)*(s'/(s'*s));
        jac_is_new = false;
    else                          % don't accept
        % Get closer to steepest descent.
        lambda = lambda*4;
        % Re-initialize the Jacobian if it's out of date.
        if ~jac_is_new
            Ak = fdjac(f,x(:,k),fk);
            jac_is_new = true;
        end
    end
end

if (norm(fk) > 1e-3)
    warning('Iteration did not find a root.')
end

end
```

First observe that we have introduced a MATLAB convenience feature: *optional input parameters*. The keyword `nargin` inside a function evaluates to the number of input arguments that were provided by the caller. In this case we supply a default value for the third argument, `tol`, if none was specified by the caller. Most modern computing languages have analogous mechanisms for accepting a variable number of input parameters. A common use case is what we have done here, allowing the optional override of a default setting.

Each pass through the loop starts by using (4.6.6) to propose a step $s_k$. The algorithm then asks whether using this step would decrease the value of $\|f\|$ from its present value. If so, $x_k + s_k$ is the new root estimate; since the iteration is going well, we decrease $\lambda$ (i.e., get more Newton-like) and apply the Broyden formula to get a fast update of the Jacobian. If the proposed step is not successful, we increase $\lambda$ (more descent-like) and, if the current Jacobian was the result of a cheap update, use finite differences to reevaluate it. Whether or not the step was accepted, everything is set up for the next loop iteration. Note that the loop iterations no longer correspond to the (quasi-)Newton iteration counter $k$.

Finally, we draw attention to lines 50–52. Rather than issuing a warning if the number of iterations is too large, we do it when the final residual is fairly large. This is done to avoid silently returning a value that is easily seen to be incorrect.

### Example 4.6.1

We repeat Example 4.5.3. We need to code only the function defining the system, and not its Jacobian.

```
function f = nlsystem(x)
    f = zeros(3,1);    % ensure a column vector output
    f(1) = exp(x(2)-x(1)) - 2;
    f(2) = x(1)*x(2) + x(3);
    f(3) = x(2)*x(3) + x(1)^2 - x(2);
end
```

In all other respects usage is the same as for the `newtonsys` function.

```
x1 = [0;0;0];
x = levenberg(@nlsystem,x1);
```

It's always a good idea to check the accuracy of the root by measuring the residual (backward error).

```
r = x(:,end)
backward_err = norm(nlsystem(r))
```

```
r =
   -0.4580
    0.2351
    0.1077
backward_err =
   1.2708e-13
```

Looking at the convergence of the first component, we find a subquadratic convergence rate, just as with the secant method.

```
log10( abs(x(1,1:end-1)-r(1)) )'

ans =
   -0.3391
   -0.4271
   -1.4439
   -1.5517
   -2.7571
   -2.6208
   -3.4403
   -4.9947
   -7.3448
   -8.7366
  -10.0913
```

In some cases our simple logic in Function 4.6.2 can make $\lambda$ oscillate between small and large values; several better but more complicated strategies for controlling $\lambda$ are known. In addition, the linear system (4.6.6) is usually modified to get the well-known *Levenberg–Marquardt* algorithm, which does a superior job in some problems as $\lambda \to \infty$.

## Exercises

4.6.1.  Repeat Exercise 4.5.3, but without finding the Jacobian and using Function 4.6.2 instead of Function 4.5.1.

4.6.2.  Repeat Exercise 4.5.4, but without finding the Jacobian and using Function 4.6.2 instead of Function 4.5.1.

4.6.3.  Repeat Exercise 4.5.5, but without finding the Jacobian and using Function 4.6.2 instead of Function 4.5.1.

4.6.4.  The Broyden update formula (4.6.5) is just one instance of so-called rank-one updating. Verify the *Sherman–Morrison formula*,

$$(A + uv^T)^{-1} = A^{-1} - A^{-1}\frac{uv^T}{1 + u^T v}A^{-1},$$

valid whenever $A$ is invertible and the denominator above is nonzero. (Hint: Show that $A + uv^T$ times the matrix above simplifies to the identity matrix.)

4.6.5.  Derive equation (4.6.8).

4.6.6.  (see also Exercise 4.5.1) Suppose that

$$f(x) = \begin{bmatrix} x_1 x_2 + x_2^2 - 1 \\ x_1 x_2^3 + x_1^2 x_2^2 + 1 \end{bmatrix}.$$

Let $x_1 = [-2, 1]^T$ and let $A_1 = J(x_1)$ be the exact Jacobian.

(a) Solve (4.6.6) for $s_1$ with $\lambda = 0$; this is the "pure" Newton step. Show numerically that $\|f(x_1 + s_1)\| > \|f(x_1)\|$. (Thus, the Newton step made us go to a point seemingly farther from a root than where we started.)

(b) Now repeat part (a) with $\lambda = 0.01j$ for $j = 1, 2, 3, \ldots$. What is the smallest value of $j$ such that $\|f(x_1 + s_1)\| < \|f(x_1)\|$?

4.6.7. ⚠ Show that the Levenberg equation (4.6.6) is equivalent to the linear least squares problem

$$\min_v \left( \|J_k v + f_k\|_2^2 + \lambda^2 \|v\|_2^2 \right).$$

Thus another interpretation of Levenberg's method is that it is the Newton step plus a penalty, whose strength is determined by $\lambda$, for taking large steps (where the linear model may not be trustworthy). (Hint: Express the minimized quantity using block matrix notation.)

## 4.7 • Nonlinear least squares

After the solution of square linear systems, we generalized to the case of having more constraints to satisfy than available variables. Our next step is to do the same for nonlinear equations, thus filling out this table:

|                | linear            | nonlinear        |
| -------------- | ----------------- | ---------------- |
| square         | $Ax = b$          | $f(x) = 0$       |
| overdetermined | $\min \|Ax - b\|_2$ | $\min \|f(x)\|_2$ |

Given a function $f(x)$ mapping from $\mathbb{R}^n$ to $\mathbb{R}^m$ we define the **nonlinear least squares** problem

$$\text{Find } x \in \mathbb{R}^n \text{ such that } \|f(x)\|_2 \text{ is minimized.} \qquad (4.7.1)$$

As in the linear case, we consider only overdetermined problems, where $m > n$. Minimizing a positive quantity is equivalent to minimizing its square, so we could also define the result as minimizing $\phi(x) = f(x)^T f(x)$.

You should not be surprised to learn that we can formulate an algorithm by substituting a linear model function for $f$. At a current estimate $x_k$ we define

$$q(x) = f(x_k) + A_k(x - x_k),$$

where $A_k$ might be the exact $m \times n$ Jacobian matrix, $J(x_k)$, or a finite difference or Broyden approximation of it as described in Section 4.6.

In the square case we solved $q = 0$ to define the new value for $x$, leading to the condition $A_k s_k = -f_k$, where $s_k = x_{k+1} - x_k$. Now, with $m > n$, we cannot expect to solve $q = 0$, so instead we define $x_{k+1}$ as the value that minimizes $\|q\|_2$:

$$x_{k+1} = x_k + s_k, \text{ where } \|A_k s_k + f(x_k)\|_2 \text{ is minimized.} \qquad (4.7.2)$$

## 4.7. Nonlinear least squares

We have just described the **Gauss–Newton method.** *Gauss–Newton solves a series of linear least squares problems in order to solve a nonlinear least squares problem.*

Here we reap an amazing benefit from MATLAB: The functions `newtonsys` (Function 4.5.1) and `levenberg` (Function 4.6.2), which were introduced for the case of $m = n$ nonlinear equations, work *without modification* as the Gauss–Newton method for the overdetermined case! The reason is that the backslash operator applies equally well to the linear system and linear least squares problems, and nothing else in the function was written with explicit reference to $n$.

### Convergence

In the multidimensional Newton method for a nonlinear system, we expect quadratic convergence to a solution in the typical case. For the Gauss–Newton method, the picture is more complicated. As always in least squares problems, the residual $f(x)$ will not necessarily be zero when $\|f\|$ is minimized. Suppose that the minimum value of $\|f\|$ is $R > 0$. In general, we might observe quadratic-like convergence until the iterate $\|x_k\|$ is within distance $R$ of a true minimizer, and linear convergence thereafter. When $R$ is not sufficiently small, the convergence can be quite slow.

### Nonlinear data fitting

In Section 3.1 we saw how to fit functions to data values, provided that the set of candidate fitting functions depends linearly on the undetermined coefficients. We now have the tools to generalize that process to fitting functions that depend nonlinearly on unknown parameters. Suppose that $(t_i, y_i)$ for $i = 1, \ldots, m$ are given points. We wish to model the data by a function $g(t;c)$ that depends on unknown parameters $c_1, \ldots, c_n$ in an arbitrary way. A standard approach is to minimize the discrepancy between the model and the observations, in a least squares sense:

$$\min_{c \in \mathbb{R}^n} \sum_{i=1}^{m} \left[ g(t_i;c) - y_i \right]^2 = \min_{c \in \mathbb{R}^n} \|f(c)\|^2,$$

where $f(c)$ is the vector of values $g(t_i;c) - y_i$. We call $f$ a **misfit** function: the smaller the norm of the misfit, the better the fit.

The form of $g$ is up to the modeler. There may be compelling theoretical choices, or you may just be looking for enough algebraic power to express the data well. Naturally, in the special case where the dependence on $c$ is linear, i.e.,

$$g(t;c) = c_1 g_1(t) + c_2 g_2(t) + \cdots + c_m g_m(t),$$

then the misfit function is also linear in $c$ and the fitting problem reduces to linear least squares.

## Example 4.7.1

Inhibited enzyme reactions often follow what are known as *Michaelis–Menten* kinetics, in which a reaction rate $v$ follows a law of the form

$$v(x) = \frac{Vx}{K_m + x},$$

where $x$ is the concentration of a substrate. The real values $V$ and $K_m$ are parameters that are free to fit to data. For this example we cook up some artificial data with $V = 2$ and $K_m = 1/2$.

```
m = 25;
x = linspace(0.05,6,m)';
y = 2*x./(0.5+x);                      % exactly on the curve
y = y + 0.15*cos(2*exp(x/16).*x);      % noise added
plot(x,y,'.')
```

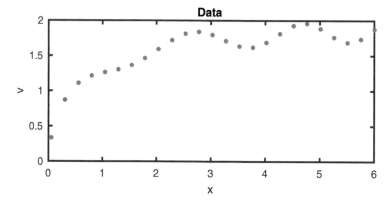

The idea is to pretend that we know nothing of the origins of this data and use nonlinear least squares on the misfit to find the parameters in the theoretical model function $v(x)$. Note in the Jacobian that the derivatives are *not* with respect to $x$, but with respect to the two parameters, which are contained in the vector c.

```
function [f,J] = misfit(c)
    V = c(1);    Km = c(2);
    f = V*x./(Km+x) - y;
    J(:,1) = x./(Km+x);                % d/d(V)
    J(:,2) = -V*x./(Km+x).^2;          % d/d(Km)
end

c1 = [1; 0.75];
c = newtonsys(@misfit,c1);
V = c(1,end),   Km = c(2,end)    % final values
model = @(x) V*x./(Km+x);
```

```
V =
    1.9687
Km =
    0.4693
```

## 4.7. Nonlinear least squares

The final values are close to the noise-free values of $V = 2$, $K_m = 0.5$ that we used to generate the data. We can calculate the amount of misfit at the end, although it's not completely clear what a "good" value would be. Graphically, the model looks reasonable.

```
final_misfit_norm = norm(model(x)-y)
hold on
fplot( model, [0 6] )

final_misfit_norm =
    0.5234
```

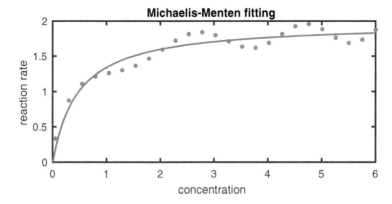

For this model, we also have the option of linearizing the fit process. Rewrite the model as $1/v = (a/x) + b$ for the new parameters $\alpha = K_m/V$ and $\beta = 1/V$. This corresponds to the misfit function whose entries are

$$f_i(\alpha, \beta) = \alpha \cdot \frac{1}{x_i} + \beta - \frac{1}{y_i}$$

for $i = 1, \ldots, m$. Although the misfit is nonlinear in $x$ and $y$, it's linear in the unknown parameters $\alpha$ and $\beta$, and so can be posed and solved as a linear least squares problem.

```
A = [ x.^(-1), x.^0 ];   u = 1./y;
z =   A\u;
alpha = z(1);   beta = z(2);
```

The two fits are different, because they do not optimize the same quantities.

```
linmodel = @(x) 1 ./ (beta + alpha./x);
final_misfit_linearized = norm(linmodel(x)-y)
fplot(linmodel, [0 6] )

final_misfit_linearized =
    0.7487
```

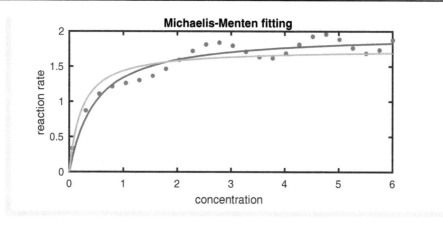

## Exercises

4.7.1. ⚠ Define $f(x) = [x-8,\ x^2-4]^T$.

    (a) Write out the linear model of $f$ at $x=2$.

    (b) Find the estimate produced by one step of the Gauss–Newton method, starting at $x=2$.

4.7.2. ⚠ (Continuation of Exercise 4.7.1.) The Gauss–Newton method replaces $f(x)$ by a linear model and minimizes the residual norm of it. An alternative is to replace $\|f(x)\|_2^2$ by a scalar *quadratic* model $q(x)$, and minimize that.

    (a) Using $f$ defined in Exercise 4.7.1, let $q(x)$ be defined by the first three terms in the Taylor series for $\|f(x)\|_2^2$ at $x=2$.

    (b) Find the unique $x$ that minimizes $q(x)$. Is the result the same as the estimate produced by Gauss–Newton?

4.7.3. 🖥 A famous result by Kermack and McKendrick in 1927 [37] suggests that in epidemics that kill only a small fraction of a susceptible population, the death rate as a function of time is well modeled by

$$w'(t) = A\operatorname{sech}^2[B(t-C)]$$

for constant values of the parameters $A, B, C$. Since the maximum of sech is $\operatorname{sech}(0) = 1$, $A$ is the maximum death rate and $C$ is the time of peak deaths. You will use this model to fit the deaths per week from plague recorded in Bombay (now Mumbai) in a period during 1906:

5, 10, 17, 22, 30, 50, 51, 90, 120, 180, 292, 395, 445, 775, 780, ...
700, 698, 880, 925, 800, 578, 400, 350, 202, 105, 65, 55, 40, 30, 20

    (a) Use Function 4.6.2 to find the best least squares fit to the data using the $\operatorname{sech}^2$ model. Make a plot of the model fit superimposed on the data. What are $A$ and $C$?

    (b) In practice, one would like a model to predict the full course of the epidemic before it has reached its peak and subsided. Redo the fitting from part (a) using only the first 12 data values. Add this model to your plot and report $A$ and $C$. Is this model a useful predictor of the value and timing of the maximum death rate?

(c) Repeat part (b) using the first 13 data values.

4.7.4. The "Rosenbrock banana function" is defined as $\|f(x)\|_2^2$, where

$$f(x) = \begin{bmatrix} 10(x_2 - x_1^2) \\ 1 - x_1 \end{bmatrix}.$$

Use Function 4.5.1 to find a minimizer of the banana function starting from $(-1.4, 5.1)$. (If you're curious about its name, make a contour plot of the residual over $-2 \le x_1 \le 3$, $-1 \le x_2 \le 4$.) Show all the digits of the final result.

4.7.5. In this problem you are to fit a function of the form

$$P(t) = c_1 + c_2 e^{c_3 t}$$

to U.S. census data for the twentieth century. Starting in 1900, the population in millions every ten years was:

76.0, 92.0, 105.7, 122.8, 131.7, 150.7, 179.0, 205.0, 226.5, 248.7.

(This is a subset of the data from the built-in file census.mat.) Use nonlinear least squares to determine the unknown parameters $c_1$, $c_2$, $c_3$ in $P$. To aid convergence, rescale the data using the time variable $t = (\text{year} - 1900)/100$ and divide the population numbers above by 100. Using your model $P(t)$, predict the result of the 2000 census, and compare it to the exact figure (which can be found easily on the Internet).

4.7.6. The position of the upper lid during an eye blink can be measured from high speed video [78], and it may be possible to classify blinks based in part on fits to the lid position [13]. The lid position functions proposed to fit blinks is a product of a monomial or polynomial multiplying a decaying exponential [9]. In this problem, you will generate representative data, add a small amount of noise to it, and then perform nonlinear least squares fits to the data.

(a) Consider the function $y(\mathbf{a}) = a_1 \exp(-a_2 t^{a_3})$, using the vector of coefficients $\mathbf{a} = [a_1\ a_2\ a_3]$, and create eyelid position data as follows:

```
N = 20;                                     % number of time values
t = linspace(1,N,N)'/N;                     % equally spaced to t=1
a = [10, 10, 2]';                           % baseline values
y = a(1)*t.^2.*exp(-a(2)*t.^a(3));          % ideal data
ym = y;                                     % vector for data
ir = 1:N-1;                                 % range to add noise
rng(13);                                    % set seed for rand
noise = 0.03;                               % amplitude of noise
ym(ir) = y(ir) + noise*rand(size(y(ir)));   % add noise
```

(b) Using the data (t, ym), find the nonlinear least squares fit using both the fminsearch function (with MaxFunEvals = 2000 and MaxIter = 1000) and the function levenberg using the default values from the text. How close are the answers for the coefficients?

(c) Plot the fits using np = 100 points over t=(1:np)'/np together with symbols for the N measured data points ym.

(d) Increase the noise to 5% and 10%. You may have to increase the number of measured points ($N$) and/or the maximum number of iterations. How close are the coefficients? Plot the data and the resulting fit for each case.

(e) Which method do you think is more robust? Justify your answer.

4.7.7. ▧ Repeat the previous problem using the fitting function $y(\mathbf{a}) = (a_1 + a_2 t + a_3 t^2)t^2 \exp(-a_4 t^{a_5})$, using the vector of coefficients $\mathbf{a} = [a_1\ a_2\ a_3\ a_4\ a_5]$. (This was the choice used in Brosch et al. [13].) Use a = [20, -10, -8, 7, 2]' to create the data and as an initial guess for the coefficients for the fit to the noisy data.

4.7.8. ▧ The following problems ask you to generate nonlinear fits using the specified functions with the same levels of noise and numbers of points as in the previous two problems; however, use randn rather than rand. For these functions, estimate to two digits the amplitude of the noise where the SSE becomes larger than unity. For all computations, use initial guess $\mathbf{a} = [2,\ 2\pi,\ 2.5]$.

(a) $y(\mathbf{a}) = a_1 t \sin(a_2 t^{a_3})$.

(b) $y(\mathbf{a}) = a_1(1-t)\cos(a_2 t^{a_3})$.

## Key ideas in this chapter

1. The conditioning of the rootfinding problem is the same as for evaluating the inverse function (page 123).
2. Small error is not always possible, but the backward error equals the residual (page 125).
3. A root $r$ of $f$ is simple if and only if $f'(r) \neq 0$ (page 125).
4. For a simple root, Newton's method exhibits quadratic convergence, in which the error is roughly squared at each iteration (page 139).
5. The principle of approximate approximation states that approximations may be solved only approximately if there is an advantage to gain by doing so (page 144).
6. In the secant method, one finds the root of the linear approximation through the two most recent root estimates (page 145).
7. The secant method is easier to implement than Newton's method and is faster when measured in terms of the number of function evaluations (page 148).
8. Newton's method for a system of nonlinear equations solves a sequence of linearized systems involving the Jacobian matrix (page 154).
9. Quasi-Newton methods attempt to avoid the need for an exact Jacobian and the tendency of the iteration to diverge for many starting points (page 159).
10. The Gauss–Newton approach uses a sequence of linear least squares problems to solve the nonlinear least squares problem (page 167).

## Where to learn more

The fixed point iteration has been used to prove results about convergence; see Quarteroni, Sacco, and Saleri [55] for more details.

Solving nonlinear systems is a complex task. Some classic texts on the subject include Ortega and Rheinboldt [52], Kelley [36], and many others.

Turning to nonlinear least squares, the widely used Levenberg–Marquardt algorithm has a connection to the authors' home institution. Donald Marquardt completed an MS in mathematics and statistics at the University of Delaware in 1956 titled "Application of Digital Computers to Statistics." He developed the method while working at DuPont Corporation's laboratories; he worked there from 1953 to 1991. He needed a better fitting method for experimental data generated at DuPont, and created just that. The method was published in 1963 in the *Journal of the Society for Industrial and Applied Mathematics* [44]. In a note at the end, he thanks a referee for pointing out that Levenberg had published related ideas elsewhere [40] and then cites that work. He also made a Fortran program available that contained an implementation of the program. There's a lesson there for budding mathematical software designers: If you want people to use your algorithm or method, give away software to do it! An interesting discussion of the evolution of computing during his career at DuPont can be found in this interview [29].

Optimization is closely linked to solving nonlinear equations as well. More info can be found, among many other places, in Strang [61], Quarteroni, Sacco, and Saleri [55], and Conn, Scheinberg, and Vicente [19].

Finally, we close with a remark about polynomials. Polynomials are common and finding their roots is a frequently occurring task. As we saw in Chapter 1, however, the condition number of polynomial rootfinding can be large, and the possibility of complex roots is another challenge. It turns out that while general-purpose rootfinding can work for polynomials, it's faster and more robust to use something tailored to the task. The MATLAB command `roots` is designed to find all roots of a polynomial given its coefficients by converting the problem to one of finding eigenvalues of a related companion matrix. A recent paper on the subject won an outstanding paper award from the Society for Industrial and Applied Mathematics [6]; it used a clever combination of the companion matrix together with a specialized QR factorization to develop a fast and backward stable method. That paper also contains a very good comparison among modern methods for polynomial rootfinding.

# Chapter 5
# Piecewise interpolation and calculus

> You must feel the Force around you. Here, between you... me... the tree... the rock... everywhere!
> —Yoda, *The Empire Strikes Back*

In many scientific problems the solution is a function. Accordingly, our next task is to represent functions numerically. This task is more difficult and complicated than the one we faced in representing real numbers. With numbers it's intuitively clear how one real value can stand for a small interval around it. But designating representatives for sets of functions is less straightforward—in fact, it's one of the core topics in computing. The process of converting functions into numerical representations of finite length is known as **discretization**.

Once we have selected a method of discretization, we can define numerical analogs of our two favorite operations on functions: differentiation and integration. These are linear operations, so the most natural numerical analogs are linear operations too. As we will see in many of the chapters following this one, a lot of numerical computing boils down to converting calculus to algebra, with discretization as the link between them.

## 5.1 • The interpolation problem

Formally, we now want to solve the following.

**Interpolation problem:** Given $n+1$ distinct points $(t_0, y_0)$, $(t_1, y_1)$,..., $(t_n, y_n)$, with $t_0 < t_1 < \cdots < t_n$, find a function $p(x)$, called the **interpolant**, such that $p(t_k) = y_k$ for $k = 0, \ldots, n$.

The values $t_0, \ldots, t_n$ are called the **nodes** of the interpolant. In this chapter, we use $t_k$ for the nodes and $x$ to denote the continuous independent variable. Take note that the nodes are numbered from zero to $n$. This is convenient for many of our mathematical statements, but less so in a language such as MATLAB in which all vectors must be indexed starting with one. Our approach is that *indices in a computer code have the same meaning as those identically named in the mathematical formulas*, and therefore must be increased by one whenever used in an indexing context.

## Polynomials

Polynomials are the obvious first candidate to serve as interpolating functions. They are easy to work with, and in Section 2.1 we saw that a linear system of equations can be used to determine the coefficients of a polynomial that passes through every member of a set of given points in the plane. However, *it's not hard to find examples for which polynomial interpolation leads to unusable results.*

### Example 5.1.1

Here are some points that we could consider to be observations of an unknown function on $[-1,1]$.

```
n = 5;
t = linspace(-1,1,n+1)';    y = t.^2 + t + 0.05*sin(20*t);
plot(t,y,'.')
```

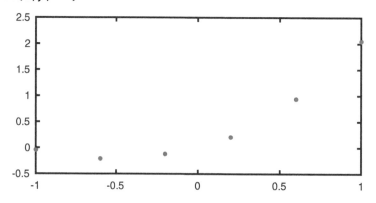

The polynomial interpolant, as computed using **polyfit**, looks very sensible. It's the kind of function you'd take home to meet your parents. The interpolant plot is created using **fplot**.

```
c = polyfit(t,y,n);         % polynomial coefficients
p = @(x) polyval(c,x);
hold on, fplot(p,[-1 1])
```

## 5.1. The interpolation problem

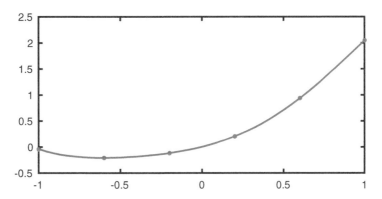

But now consider a different set of points generated in almost exactly the same way.

```
n = 18;
t = linspace(-1,1,n+1)';   y = t.^2 + t + 0.05*sin(20*t);
clf, plot(t,y,'.')
```

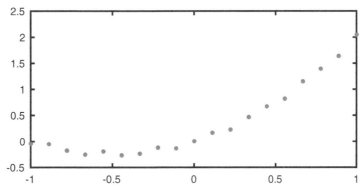

The points themselves are unremarkable. But take a look at what happens to the polynomial interpolant.

```
c = polyfit(t,y,n);      % polynomial coefficients
p = @(x) polyval(c,x);
hold on, fplot(p,[-1 1])
```

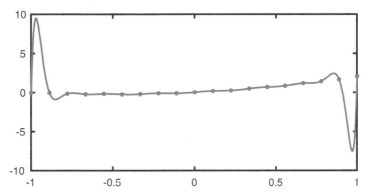

Surely there must be functions that are more intuitively representative of those points!

In Chapter 9 we explore the large oscillations in the last figure of Example 5.1.1; it turns out that one must abandon either equally spaced nodes or $n \to \infty$ for polynomials. In the rest of this chapter we will keep $n$ fairly small and let the nodes be unrestricted.

## Piecewise polynomials

In order to keep polynomial degrees small while interpolating large data sets, we will choose interpolants from the **piecewise polynomials**. Specifically, the interpolant $p$ must be a polynomial on each subinterval $[t_{k-1}, t_k]$ for $k = 1, \ldots, n$.

### Example 5.1.2

Some examples of piecewise polynomials for the nodes $t_0 = -2$, $t_1 = 0$, $t_2 = 1$, and $t_3 = 4$ are $p_1(x) = x + 1$, $p_2(x) = \text{sign}(x)$, $p_3(x) = |x - 1|^3$, and $p_4(x) = (\max\{0, x\})^4$. Note that $p_1$, $p_2$, and $p_4$ would also be piecewise polynomial on the node set $\{t_0, t_1, t_3\}$, but $p_3$ would not.

Usually we designate in advance a maximum degree $d$ for each polynomial piece of $p(x)$. An important property of the piecewise polynomials of degree $d$ is that they form a vector space: that is, any linear combination of piecewise polynomials of degree $d$ is another piecewise polynomial of degree $d$. If $p$ and $q$ share the same node set, then the combination is piecewise polynomial on that node set.

### Example 5.1.3

Let us repeat the data from Example 5.1.1 but use piecewise polynomials constructed using the **interp1** command.

```
n = 18;
t = linspace(-1,1,n+1)';   y = t.^2 + t + 0.05*sin(20*t);
clf, plot(t,y,'.')
```

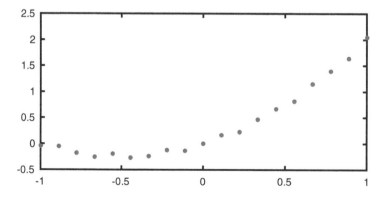

By default interp1 gives us an interpolating function that is linear between each pair of consecutive nodes.

## 5.1. The interpolation problem

```
x = linspace(-1,1,400)';
hold on, plot(x,interp1(t,y,x))
```

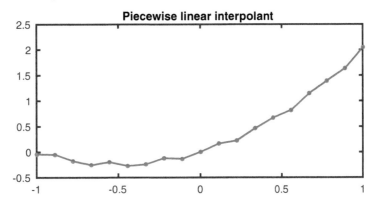

We may instead request a smoother interpolant that is piecewise cubic.

```
cla
plot(t,y,'.')
plot(x,interp1(t,y,x,'spline'))
```

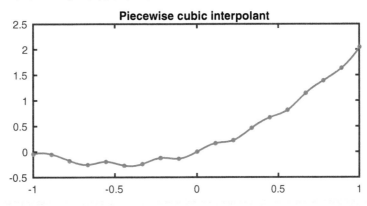

We will consider piecewise linear interpolation in more detail in Section 5.2, and we look at this type of piecewise cubic interpolation in Section 5.3.

## Conditioning of interpolation

In the interpolation problem we are given the values $(t_k, y_k)$ for $k = 0, \ldots, n$. Let us consider the nodes $t_k$ of the problem to be fixed, and let $a = t_0$, $b = t_n$. Then the data for the interpolation problem consist of a vector $y$, and the result of the problem is a function on $[a, b]$.

Let $\mathscr{I}$ be a prescription for producing the interpolant from a data vector. That is, $\mathscr{I}(y) = p$, where $p(t_k) = y_k$ for all $k$. The interpolation methods we will consider are all *linear* in the sense that

$$\mathscr{I}(\alpha y + \beta z) = \alpha \mathscr{I}(y) + \beta \mathscr{I}(z) \tag{5.1.1}$$

for all vectors $y, z$ and scalars $\alpha, \beta$.

Linearity greatly simplifies the analysis of the conditioning of interpolation. If the data are changed from $y$ to $y + \Delta y$, then

$$\delta \mathscr{I} = \mathscr{I}(y + \Delta y) - \mathscr{I}(y) = \mathscr{I}(\Delta y) = \sum_{k=0}^{n} (\Delta y)_k \mathscr{I}(e_k), \qquad (5.1.2)$$

where as always $e_k$ is a column of an identity matrix. We use $\|\Delta y\|_\infty$ to measure the size of the perturbation to the data, and for $\delta \mathscr{I}$, which is a function, we use the functional infinity norm or max norm defined by

$$\|f\|_\infty = \max_{x \in [a,b]} |f(x)|. \qquad (5.1.3)$$

The absolute condition number $\kappa(y)$ relates $\|\delta \mathscr{I}\|_\infty$ to $\|\Delta y\|_\infty$.

> **Theorem 5.1.1: Conditioning of interpolation**
>
> Suppose that $\mathscr{I}$ is a linear interpolation method. Then the absolute condition number of $\mathscr{I}$ satisfies
>
> $$\max_{0 \leq k \leq n} \|\mathscr{I}(e_k)\|_\infty \leq \kappa(y) \leq \sum_{k=0}^{n} \|\mathscr{I}(e_k)\|_\infty \qquad (5.1.4)$$
>
> if vectors and functions are measured in the infinity norm.

*Proof.* Because of (5.1.2), we have

$$\frac{\|\delta \mathscr{I}\|_\infty}{\|\Delta y\|_\infty} = \left\| \sum_{k=0}^{n} \frac{(\Delta y)_k}{\|\Delta y\|_\infty} \mathscr{I}(e_k) \right\|_\infty.$$

The absolute condition number maximizes this quantity over all $\Delta y$. Suppose $j$ is such that $\|\mathscr{I}(e_j)\|$ is maximal. Then let $\Delta y = e_j$, and the first inequality in (5.1.4) follows. The other inequality follows from the triangle inequality and the definition of $\|\Delta y\|_\infty$. □

Theorem 5.1.1 says that *assessing the condition number of interpolation for any data can be simplified to measuring the effect of interpolation with each node "switched on" one at a time.* The results of such interpolations are known as **cardinal functions**.

> **Example 5.1.4**
>
> In Examples 5.1.1 and 5.1.3 we showed a big difference between polynomial interpolation and piecewise polynomial interpolation of some arbitrarily chosen data. The same effects can be seen clearly in the cardinal functions.
>
> ```
> n = 18;
> t = linspace(-1,1,n+1)';
> y = [zeros(9,1);1;zeros(n-9,1)];    % 10th cardinal function
> plot(t,y,'.'), hold on
> ```

```
x = linspace(-1,1,400)';
plot(x,interp1(t,y,x,'spline'))
```

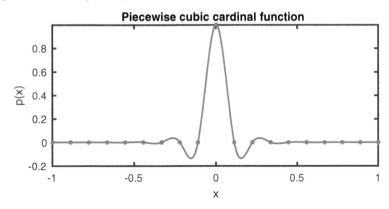

The piecewise cubic cardinal function is nowhere greater than one in absolute value. This happens to be true for all the cardinal functions, ensuring a good condition number for the interpolation. But the story for global polynomials is very different.

```
c = polyfit(t,y,n);
hold on, plot(x,polyval(c,x))
```

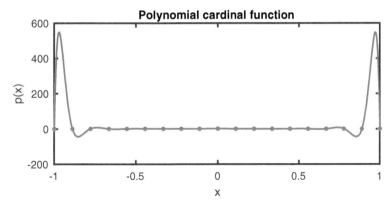

From the figure we can see that the condition number for polynomial interpolation on these nodes is at least 500.

## Exercises

5.1.1. ▦ Create data by entering

```
t = (-2:4)';   y = tanh(t);
```

    (a) Use `polyfit` to construct and plot the polynomial interpolant of the data.

    (b) Use `interp1` to construct and plot a piecewise cubic interpolant of the data.

**5.1.2.** 🖥 The following table gives the life expectancy in the U.S. by year of birth:

| 1980 | 1985 | 1990 | 1995 | 2000 | 2005 | 2010 |
|------|------|------|------|------|------|------|
| 73.7 | 74.7 | 75.4 | 75.8 | 77.0 | 77.8 | 78.7 |

(a) Defining "year since 1980" as the independent variable, use `polyfit` to construct and plot the polynomial interpolant of the data.

(b) Use `interp1` to construct and plot a piecewise cubic interpolant of the data.

(c) Use both methods to estimate the life expectancy for a person born in 2007. Which value is more believable?

**5.1.3.** 🖥 The following two point sets define the top and bottom of a flying saucer shape:

Top:

| $x$ | 0 | 0.51 | 0.96 | 1.06 | 1.29 | 1.55 | 1.73 | 2.13 | 2.61 |
|-----|---|------|------|------|------|------|------|------|------|
| $y$ | 0 | 0.16 | 0.16 | 0.43 | 0.62 | 0.48 | 0.19 | 0.18 | 0 |

Bottom:

| $x$ | 0 | 0.58 | 1.04 | 1.25 | 1.56 | 1.76 | 2.19 | 2.61 |
|-----|---|------|------|------|------|------|------|------|
| $y$ | 0 | -0.16 | -0.15 | -0.30 | -0.29 | -0.12 | -0.12 | 0 |

Use piecewise cubic interpolation to make a picture of the flying saucer.

**5.1.4.** ✎ Define

$$q(x) = a\frac{x(x-1)}{2} - b(x-1)(x+1) + c\frac{x(x+1)}{2}.$$

(a) Show that $q$ is a polynomial interpolant of the points $(-1, a), (0, b), (1, c)$.

(b) Use a change of variable to find a quadratic polynomial interpolant for the points $(x_0 - h, a), (x_0, b), (x_0 + h, c)$.

**5.1.5.** ✎ Use the formula of Exercise 5.1.4 and Theorem 5.1.1 to derive bounds on the condition number of quadratic polynomial interpolation at the nodes $-1, 0, 1$.

## 5.2 ▪ Piecewise linear interpolation

Piecewise linear interpolation is simply a game of "connect the dots." Let us assume the nodes are given in order, so that $t_0 < t_1 < \cdots < t_n$. Between each pair of adjacent nodes, we use a straight line segment. The resulting interpolant $p(x)$ is given by

$$p(x) = y_k + \frac{y_{k+1} - y_k}{t_{k+1} - t_k}(x - t_k) \quad \text{for } x \in [t_k, t_{k+1}]. \tag{5.2.1}$$

It should be clear that on each interval $[t_k, t_{k+1}]$, $p(x)$ is a linear function, and you can easily verify from the formula that it passes through both $(t_k, y_k)$ and $(t_{k+1}, y_{k+1})$.

### Hat functions

Rather than basing an implementation on equation (5.2.1), we return to the idea used in Section 2.1 of choosing the interpolant from among the linear combinations of a

## 5.2. Piecewise linear interpolation

preselected finite set of functions. In the present context we use

$$H_k(x) = \begin{cases} \dfrac{x - t_{k-1}}{t_k - t_{k-1}} & \text{if } x \in [t_{k-1}, t_k], \\ \dfrac{t_{k+1} - x}{t_{k+1} - t_k} & \text{if } x \in [t_k, t_{k+1}], \\ 0 & \text{otherwise,} \end{cases} \qquad k = 0, \ldots, n. \qquad (5.2.2)$$

The functions $H_0, \ldots, H_n$ are called **hat functions**. They depend on the node vector $t$, but this dependence is not usually indicated explicitly.

Each hat function is globally continuous and is linear inside every interval $[t_k, t_{k+1}]$. Consequently, any linear combination of them will have the same property. Furthermore, *any* such function is expressible as a unique linear combination of hat functions, i.e.,

$$\sum_{k=0}^{n} c_k H_k(x) \qquad (5.2.3)$$

for some choice of the coefficients $c_0, \ldots, c_n$. No smaller set of functions can have the same properties. We summarize these facts by calling the hat functions a **basis** of the set of functions that are continuous and piecewise linear relative to $t$. Another point of view, familiar from abstract linear algebra, is that a basis sets up a one-to-one correspondence between the spanned function space and the more familiar space $\mathbb{R}^{n+1}$, with each function being represented by its coefficients $c_0, \ldots, c_n$.

Note that the definitions of $H_0$ for $x < t_0$ and $H_n$ for $x > t_n$ are irrelevant—a fact exploited by the implementation given in Function 5.2.1 through the introduction of two fictitious nodes lying on either side of the interval $[t_0, t_n]$. This trick allows the use of an identical formula for all of the hat functions. Otherwise, we would need to take special action for the two edge cases $H_0$ and $H_n$.

Another trick in Function 5.2.1 is to exploit the observation that at each $x$, $H_k(x)$ is the larger of two options: zero, or the smaller of two linear functions (those appearing in (5.2.2)). This code is not the most efficient one possible, but it is more concise than detecting which particular subinterval each $x$ lies within.

### Example 5.2.1

Let's define a set of 6 nodes (i.e., $n = 5$ in our formulas).

```
t = [0 0.075 0.25 0.55 0.7 1]';
```

We plot the hat functions $H_0, \ldots, H_5$.

```
for k = 0:5
    fplot(@(x) hatfun(x,t,k),[0 1])
    hold on
end
xlabel('x'), ylabel('H_k(x)'), title('Hat functions')
```

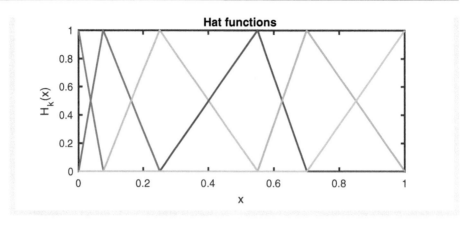

---

**Function 5.2.1 (hatfun)** Hat function/piecewise linear basis function.

```
function H = hatfun(x,t,k)
% HATFUN    Hat function/piecewise linear basis function.
% Input:
%   x       evaluation points (vector)
%   t       interpolation nodes (vector, length n+1)
%   k       node index (integer, in 0,...,n)
% Output:
%   H       values of the kth hat function

n = length(t)-1;
k = k+1;    % adjust for starting with index=1

% Fictitious nodes to deal with first, last funcs.
t = [ 2*t(1)-t(2); t(:); 2*t(n+1)-t(n) ];
k = k+1;    % adjust index for the fictitious first node

H1 = (x-t(k-1))/(t(k)-t(k-1));      % upward slope
H2 = (t(k+1)-x)/(t(k+1)-t(k));      % downward slope

H = min(H1,H2);
H = max(0,H);
```

---

## Cardinality conditions

For the purposes of interpolation, the most salient property of the hat functions is that they are cardinal functions for piecewise linear interpolation; that is, they satisfy the **cardinality conditions**

$$H_k(t_i) = \begin{cases} 1 & \text{if } i = k, \\ 0 & \text{otherwise.} \end{cases} \qquad (5.2.4)$$

The appeal of a cardinal basis is that it makes the expression of the interpolant trivial. All candidate piecewise linear (PL) functions can be expressed as a linear combination such as (5.2.3) for some coefficients $c_0, \ldots, c_n$. But because of the cardinality conditions and the necessity for $p(x)$ to interpolate the data values in $y$, *expressing the interpolant*

## 5.2. Piecewise linear interpolation

*using the hat functions is trivial:*

$$p(x) = \sum_{k=0}^{n} y_k H_k(x). \tag{5.2.5}$$

The resulting algorithmic simplicity is reflected in Function 5.2.2. Take note that the output of Function 5.2.2 is itself a function, meant to be called with a single (possibly vector) argument representing values of $x$. Our mathematical viewpoint is that the result of an interpolation process is a function, and our codes reflect this.[19]

### Example 5.2.2

We generate a piecewise linear interpolant of $f(x) = e^{\sin 7x}$.

```
f = @(x) exp(sin(7*x));
fplot(f,[0 1])
```

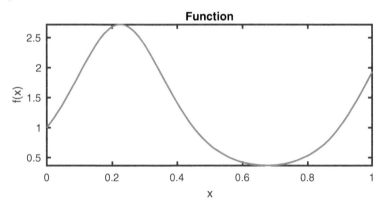

First we sample the function to create the data.

```
t = [0 0.075 0.25 0.55 0.7 1];    % nodes
y = f(t);
hold on, plot(t,y,'.')
```

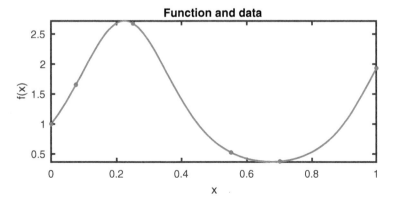

---

[19] By contrast, MATLAB's built-in interpolation function `interp1` returns values, not a function; values for $x$ must be supplied in the same call as $t$ and $y$.

Now we create a callable function that will evaluate the piecewise linear interpolant at any $x$.

```
p = plinterp(t,y);
fplot(p,[0 1])
```

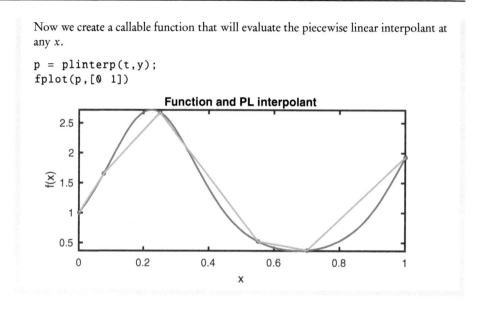

A final appealing characteristic of the hat function basis is that it depends only on the node locations, while the expansion coefficients in (5.2.3) depend only on the data values. This clean separation would be useful if we wanted to construct many interpolants on the same node set, and it has deeper theoretical uses as well.

**Function 5.2.2** (`plinterp`) Piecewise linear interpolation.

```
1  function p = plinterp(t,y)
2  % PLINTERP   Piecewise linear interpolation.
3  % Input:
4  %   t      interpolation nodes (vector, length n+1)
5  %   y      interpolation values (vector, length n+1)
6  % Output:
7  %   p      piecewise linear interpolant (function)
8
9  n = length(t)-1;
10 p = @evaluate;
11
12    % This function evaluates p when called.
13    function f = evaluate(x)
14        f = 0;
15        for k = 0:n
16            f = f + y(k+1)*hatfun(x,t,k);
17        end
18    end
19
20 end
```

## Conditioning and convergence

The condition number bounds from Theorem 5.1.1 are very simple for piecewise linear interpolation, because the interpolant of the data $e_k$ is just the hat function $H_k$. Hence $1 \leq \kappa \leq n+1$. However, there is an even simpler result.

## 5.2. Piecewise linear interpolation

> **Theorem 5.2.1**
>
> The absolute condition number of piecewise linear interpolation in the infinity norm equals one. More specifically, if $\mathscr{I}$ is the piecewise linear interpolation operator, then
> $$\|\mathscr{I}(y+z)-\mathscr{I}(y)\|_\infty = \|z\|_\infty. \tag{5.2.6}$$
> (The norm on the left side is on functions, while the norm on the right side is on vectors.)

*Proof.* By linearity,
$$\mathscr{I}(y+z)-\mathscr{I}(y) = \mathscr{I}(z) = \sum_{k=0}^n z_k H_k(x).$$

Call this piecewise linear function $p(x)$. Consider a maximum element of $z$, i.e., choose $i$ such that $|z_i| = \|z\|_\infty$. Then $|p(t_i)| = \|z\|_\infty$. Hence $\|p\|_\infty \geq \|z\|_\infty$. Now consider
$$|p(x)| = \left|\sum_{k=0}^n z_k H_k(x)\right| \leq \sum_{k=0}^n |z_k| H_k(x) \leq \|z\|_\infty \sum_{k=0}^n H_k(x) = \|z\|_\infty.$$

You are asked to prove the final step above in Exercise 5.2.4. We conclude that $\|p\|_\infty \leq \|z\|_\infty$, so that $\|p\|_\infty = \|z\|_\infty$, which completes the proof. □

Now suppose that $f$ is a "nice" function on an interval $[a,b]$ containing all of the nodes. We can play a game of sampling values of $f$ to get data, i.e., $y_k = f(t_k)$ for all $k$, then perform piecewise linear interpolation of the data to get a different function, the interpolant $p$. How close is $p$ to the original $f$? To make a simple statement, we will consider only the case of equally spaced nodes covering the interval.

> **Theorem 5.2.2**
>
> Suppose that $f(x)$ has a continuous second derivative in $[a,b]$ (often expressed as $f \in C^2[a,b]$). Let $p_n(x)$ be the piecewise linear interpolant of $(t_i, f(t_i))$ for $i = 0, \ldots, n$, where $t_i = a + ih$ and $h = (b-a)/n$. Then
> $$\|f - p_n\|_\infty = \max_{x \in [a,b]} |f(x) - p(x)| \leq M h^2, \tag{5.2.7}$$
> where $M = \|f''\|_\infty$.

*Proof.* See Exercise 5.2.5 for an outline. □

We normally don't have access to $f''$, so the importance of Theorem 5.2.2 is that the error in the interpolant is $O(h^2)$ as $h \to 0$. The exponent of 2 is indicated by saying that *piecewise linear interpolation is second-order accurate*. For instance, if we double the number of equally spaced nodes used to sample a function, the piecewise linear interpolant becomes about four times more accurate.

### Example 5.2.3

We measure the convergence rate for piecewise linear interpolation of $e^{\sin 7x}$.

```
f = @(x) exp(sin(7*x));
x = linspace(0,1,10001)';   % sample the difference at many
   points
n_ = 2.^(3:10)';
err_ = 0*n_;
for i = 1:length(n_)
    n = n_(i);
    t = linspace(0,1,n+1)';   % interpolation nodes
    p = plinterp(t,f(t));
    err = max(abs( f(x) - p(x) ));
    err_(i) = err;
end
```

Since we expect convergence that is $O(h^2) = O(n^{-2})$, we use a log-log graph of error and expect a straight line of slope $-2$.

```
loglog( n_, err_, '.-' )
hold on, loglog( n_, 0.1*(n_/n_(1)).^(-2), '--' )
```

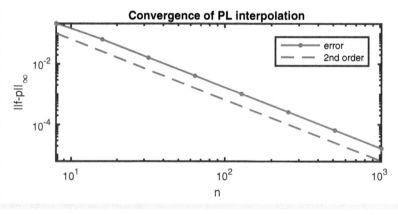

## Exercises

5.2.1. ▨ For each of the functions and intervals given, perform piecewise linear interpolation using Function 5.2.2 for equispaced nodes with $n = 10, 20, 40, 80, 160$. For each $n$, estimate the error

$$E(n) = \|f - p\|_\infty = \max_x |f(x) - p(x)|$$

by evaluating the function and interpolant at 1600 points in the interval. Make a log-log plot of $E$ as a function of $n$ and add the line $E = Cn^{-2}$ for a constant $C$ of your choosing.

(a) $\cos(\pi^2 x^2)$ on $[0, 1]$.

(b) $\log(x)$ on $[1,3]$.

(c) $\sin(x^2)$ on $[0, 2.5]$.

5.2.2. ✍ For this problem, let $H(x)$ be the hat function that passes through the three points $(-1, 0)$, $(0, 1)$, and $(1, 0)$.

(a) Write out a piecewise definition of $H$ in the style of (5.2.2).

(b) Define the function $Q$ by $Q(x) = \int_{x-1}^{x} H(t)\,dt$. Find a piecewise formula for $Q(x)$. (Hint: Perform the integration separately for the cases $-1 \le x \le 0$, $0 \le x \le 1$, etc.)

(c) Make a sketch of $Q(x)$ for $-2 \le x \le 2$.

(d) Show that $Q$ is continuous. Are $Q'$ and $Q''$?

5.2.3. ✍ Before electronic calculators, the function $\ln(x)$ was often computed using piecewise linear interpolation with a table of values. If you were using such a table at the nodes $1.01, 1.02, \ldots, 1.99, 2$, what is an upper bound on the error in the result?

5.2.4. ✍ Show that for any node distribution and any $x \in [t_0, t_n]$,

$$\sum_{k=0}^{n} H_k(x) = 1. \qquad (5.2.8)$$

(Hint: The simplest way is to apply (5.2.5).) This is called the *partition of unity* property.

5.2.5. ✍ Here we consider a proof of Theorem 5.2.2 using the mean value theorems from elementary calculus: If $f$ is continuously differentiable in $(a, b)$, then there exist points $s$ and $t$ in $(a, b)$ such that

$$\int_a^b f(z)\,dz = (b-a)f(s) \quad \text{and} \quad f'(t) = \frac{f(b) - f(a)}{b-a}.$$

For the following, suppose $x \in (t_k, t_{k+1})$.

(a) Show that for some $s \in (t_k, t_{k+1})$,

$$f(x) = y_k + (x - t_k)f'(s).$$

(b) Show that for some other values $u$ and $v$ in $(t_k, t_{k+1})$,

$$f'(s) - \frac{y_{k+1} - y_k}{t_{k+1} - t_k} = (s - u)f''(v).$$

(c) Use (5.2.1) to finish the proof of the theorem.

## 5.3 ▪ Cubic splines

A piecewise linear interpolant is continuous but has discontinuities in its derivative. We often desire a smoother interpolant, i.e., one that has one or more continuous derivatives. *A cubic spline is a piecewise cubic function that has two continuous derivatives everywhere.*[20] To respect the terminology we use $S(x)$ to denote the spline interpolant.

As before, suppose that distinct nodes $t_0 < t_1 < \cdots < t_n$ (not necessarily equally spaced) and data $y_0, \ldots, y_n$ are given. For any $k = 1, \ldots, n$, the spline $S(x)$ on the interval $[t_{k-1}, t_k]$ is by definition a cubic polynomial $S_k(x)$, which we express as

$$S_k(x) = a_k + b_k(x - t_{k-1}) + c_k(x - t_{k-1})^2 + d_k(x - t_{k-1})^3, \quad k = 1, \ldots, n, \quad (5.3.1)$$

where $a_k, b_k, c_k, d_k$ are values to be determined. Overall there are $4n$ such undetermined coefficients.

## Smoothness conditions

We are able to ensure that $S$ has at least two continuous derivatives everywhere by means of the following constraints.

1. *Interpolation by $S_k$ at both of its endpoints.* Algebraically we require $S_k(t_{k-1}) = y_{k-1}$ and $S_k(t_k) = y_k$ for every $k = 1, \ldots, n$. In terms of (5.3.1), these conditions are

$$a_k = y_{k-1}, \quad (5.3.2)$$

$$a_k + b_k h_k + c_k h_k^2 + d_k h_k^3 = y_k, \quad k = 1, \ldots, n, \quad (5.3.3)$$

where we have used the definition

$$h_k = t_k - t_{k-1}, \quad k = 1, \ldots, n. \quad (5.3.4)$$

The values of $h_k$ are derived from the nodes. Crucially, the unknown coefficients appear only *linearly* in the constraint equations. So we will express the constraints using linear algebra. The left endpoint interpolation constraints (5.3.2) are, in matrix form,

$$\begin{bmatrix} I & 0 & 0 & 0 \end{bmatrix} \begin{bmatrix} a \\ b \\ c \\ d \end{bmatrix} = \begin{bmatrix} y_0 \\ & \ddots \\ & & y_{n-1} \end{bmatrix}, \quad (5.3.5)$$

with $I$ being an $n \times n$ identity. The right endpoint interpolation constraints, given by (5.3.3), become

$$\begin{bmatrix} I & H & H^2 & H^3 \end{bmatrix} \begin{bmatrix} a \\ b \\ c \\ d \end{bmatrix} = \begin{bmatrix} y_1 \\ & \ddots \\ & & y_n \end{bmatrix}, \quad (5.3.6)$$

---

[20] The `cubic` option of `interp1` also produces a piecewise cubic, but one having only one continuous derivative.

## 5.3. Cubic splines

where we have defined the diagonal matrix

$$H = \begin{bmatrix} h_1 & & & \\ & h_2 & & \\ & & \ddots & \\ & & & h_n \end{bmatrix}. \tag{5.3.7}$$

Collectively, (5.3.5) and (5.3.6) express $2n$ scalar constraints on the unknowns.

2. *Continuity of $S'(x)$ at interior nodes.* We do not know what the slope of the interpolant should be at the nodes, but we do want the same slope whether a node is approached from the left or the right. Thus we obtain constraints at the nodes that sit between two neighboring piecewise definitions, so that $S_1'(t_1) = S_2'(t_1)$, and so on. Altogether these are

$$b_k + 2c_k h_k + 3d_k h_k^2 = b_{k+1}, \qquad k = 1,\ldots,n-1. \tag{5.3.8}$$

Moving the unknowns to the left side, as a system these become

$$E\begin{bmatrix} 0 & J & 2H & 3H^2 \end{bmatrix} \begin{bmatrix} a \\ b \\ c \\ d \end{bmatrix} = 0, \tag{5.3.9}$$

where now we have defined

$$J = \begin{bmatrix} 1 & -1 & & & \\ & 1 & -1 & & \\ & & \ddots & \ddots & \\ & & & 1 & -1 \\ & & & & 1 \end{bmatrix}, \tag{5.3.10}$$

and $E$ is the $(n-1)\times n$ matrix resulting from deleting the last row of the identity:

$$E = \begin{bmatrix} 1 & 0 & & & \\ & 1 & 0 & & \\ & & \ddots & \ddots & \\ & & & 1 & 0 \end{bmatrix}. \tag{5.3.11}$$

Left-multiplying by $E$ deletes the last row of any matrix or vector. Hence (5.3.9) represents $n-1$ constraints on the unknowns. (Remember, there are only $n-1$ interior nodes.)

3. *Continuity of $S''(x)$ at interior nodes.* These again apply only at the interior nodes $t_1,\ldots,t_{n-1}$, in the form $S_1''(t_1) = S_2''(t_1)$ and so on. Using (5.3.1) once more, we obtain

$$2c_k + 6d_k h_k = 2c_{k+1}, \qquad k = 1,\ldots,n-1. \tag{5.3.12}$$

In system form (after canceling a factor of 2 from each side) we get

$$E\begin{bmatrix} 0 & 0 & J & 3H \end{bmatrix} \begin{bmatrix} a \\ b \\ c \\ d \end{bmatrix} = 0. \tag{5.3.13}$$

## End constraints

So far the equations (5.3.5), (5.3.6), (5.3.9), and (5.3.13) form $2n + (n-1) + (n-1) = 4n - 2$ linear conditions on the $4n$ unknowns in the piecewise definition (5.3.1). In order to obtain a square system, we must add two more constraints. If the application prescribes values for $S'$ or $S''$ at the endpoints, those may be applied. Otherwise there are two major alternatives:

$$S_1''(t_0) = S_n''(t_n) = 0 \quad \text{(natural spline)}, \tag{5.3.14a}$$

$$S_1'''(t_1) = S_2'''(t_1), \quad S_{n-1}'''(t_{n-1}) = S_n'''(t_{n-1}) \quad \text{(not-a-knot spline)}. \tag{5.3.14b}$$

While natural splines have important theoretical properties, not-a-knot splines give better pointwise approximations, and they are the only type we consider further.

In the not-a-knot spline, the values and first three derivatives of the cubic polynomials $S_1$ and $S_2$ agree at the node $t_1$. Hence they must be the same cubic polynomial! The same is true of $S_{n-1}$ and $S_n$.[21] We could use these facts to eliminate some of the undetermined coefficients from our linear system of constraints. However, rather than rework the algebra we just append two more rows to the system, expressing the conditions

$$d_1 = d_2, \quad d_{n-1} = d_n. \tag{5.3.15}$$

Collectively, (5.3.5), (5.3.6), (5.3.9), (5.3.13), and (5.3.15) comprise a square linear system of size $4n$ which can be solved for the coefficients defining the piecewise cubics in (5.3.1). This is a major difference from the piecewise linear interpolant, for which there is no linear system to solve. Indeed, while it is possible to find a basis for the cubic spline interpolant analogous to the hat functions, it is not possible in closed form to construct a *cardinal* basis, so the solution of a linear system cannot be avoided.

## Implementation

Function 5.3.1 gives an implementation of cubic not-a-knot spline interpolation. For clarity it stays very close to the description given above. There are some possible shortcuts—for example, one could avoid using $E$ and instead directly delete the last row of any matrix it left-multiplies. Observe that the linear system is assembled and solved just once, and the returned evaluation function simply uses the resulting coefficients. This allows us to make multiple calls to evaluate $S$ without unnecessarily repeating the linear algebra.

## Conditioning and convergence

Besides having more smoothness than a piecewise linear interpolant, *the not-a-knot cubic spline improves the order of accuracy to four*.

---

[21] This explains the name of the not-a-knot spline—for splines, "knots" are the points at which different piecewise definitions meet.

## 5.3. Cubic splines

**Function 5.3.1 (spinterp)** Cubic not-a-knot spline interpolation.

```
function S = spinterp(t,y)
% SPINTERP   Cubic not-a-knot spline interpolation.
% Input:
%   t        interpolation nodes (vector, length n+1)
%   y        interpolation values (vector, length n+1)
% Output:
%   S        not-a-knot cubic spline (function)

t = t(:);  y = y(:);  % ensure column vectors
n = length(t)-1;
h = diff(t);           % differences of all adjacent pairs

% Preliminary definitions.
Z = zeros(n);
I = eye(n);   E = I(1:n-1,:);
J = I - diag(ones(n-1,1),1);
H = diag(h);

% Left endpoint interpolation:
AL = [ I, Z, Z, Z ];
vL = y(1:n);

% Right endpoint interpolation:
AR = [ I, H, H^2, H^3 ];
vR = y(2:n+1);

% Continuity of first derivative:
A1 = E*[ Z, J, 2*H, 3*H^2 ];
v1 = zeros(n-1,1);

% Continuity of second derivative:
A2 = E*[ Z, Z, J, 3*H ];
v2 = zeros(n-1,1);

% Not-a-knot conditions:
nakL = [ zeros(1,3*n), [1,-1, zeros(1,n-2)] ];
nakR = [ zeros(1,3*n), [zeros(1,n-2), 1,-1] ];

% Assemble and solve the full system.
A = [ AL; AR; A1; A2; nakL; nakR ];
v = [ vL; vR; v1; v2; 0 ;0 ];
z = A\v;

% Break the coefficients into separate vectors.
rows = 1:n;
a = z(rows);
b = z(n+rows);   c = z(2*n+rows);   d = z(3*n+rows);
S = @evaulate;

    % This function evaluates the spline when called with a value for x.
    function f = evaulate(x)
        f = zeros(size(x));
        for k = 1:n       % iterate over the pieces
            % Evalaute this piece's cubic at the points inside it.
            index = (x>=t(k)) & (x<=t(k+1));
            f(index) = polyval( [d(k),c(k),b(k),a(k)], x(index)-t(k) );
        end
    end

end
```

### Theorem 5.3.1

Suppose that $f(x)$ has four continuous derivatives in $[a,b]$ (i.e., $f \in C^4[a,b]$). Let $S_n(x)$ be the not-a-knot cubic spline interpolant of $(t_i, f(t_i))$ for $i = 0, \ldots, n$, where $t_i = a + ih$ and $h = (b-a)/n$. Then for all sufficiently small $h$ there is a constant $C > 0$ such that

$$\|f - S_n\|_\infty \leq Ch^4. \qquad (5.3.16)$$

### Example 5.3.1

We repeat Example 5.2.3 using cubic splines.

For illustration, here is a spline interpolant using just a few nodes.

```
f = @(x) exp(sin(7*x));
fplot(f,[0,1])
t = [0, 0.075, 0.25, 0.55, 0.7, 1]';   % nodes
y = f(t);                              % values at nodes
hold on, plot(t,y,'.')
S = spinterp(t,y);
fplot(S,[0,1])
```

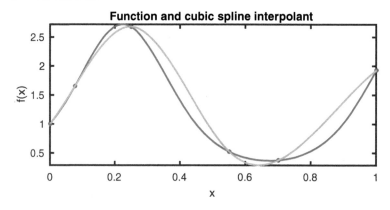

Now we look at the convergence rate as the number of nodes increases.

```
x = linspace(0,1,10001)';   % sample the difference at many
    points
n_ = 2.^(3:0.5:8)';
err_ = 0*n_;
for i = 1:length(n_)
    n = n_(i);
    t = linspace(0,1,n+1)';   % interpolation nodes
    S = spinterp(t,f(t));
    err = norm( f(x) - S(x), inf );
    err_(i) = err;
end
```

Since we expect convergence that is $O(h^4) = O(n^{-4})$, we use a log–log graph of error and expect a straight line of slope $-4$.

```
clf, loglog( n_, err_, '.-' )
hold on, loglog( n_, (n_/n_(1)).^(-4), '--' )
```

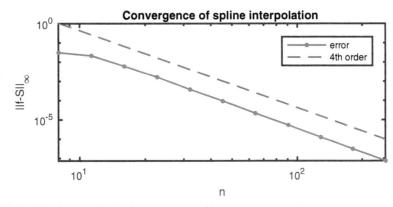

The conditioning of spline interpolation is much more complicated than for the piecewise linear case. First, the fact that the coefficients of all the cubics must be solved for simultaneously implies that each data value in $y$ has an influence on $S$ over the entire interval. Second, $S$ can take on values larger in magnitude than all of the values in $y$ (see Exercise 5.3.5). The details may be found in more advanced texts.

## Exercises

5.3.1. In each case, write out the entries of the matrix and right-hand side of the linear system that determines the coefficients for the cubic not-a-knot spline interpolant of the given function and node vector.

(a) $\cos(\pi^2 x^2)$, $t = [-1, 1, 4]^T$.

(b) $\cos(\pi^2 x^2)$, $t = [0, 1/2, 3/4, 1]^T$.

(c) $\ln(x)$, $t = [1, 2, 3]^T$.

(d) $\sin(x^2)$, $t = [-1, 0, 1]^T$.

5.3.2. (continuation) For each case in Exercise 5.3.1, use MATLAB to set up and solve the linear system you wrote down. Then plot the resulting cubic spline over the interval between the second and third nodes.

5.3.3. For each given function, interval, and value of $n$, define $n+1$ evenly spaced nodes. Then use Function 5.3.1 to plot the cubic spline interpolant at those nodes together with the original function over the given interval.

(a) $\cos(\pi^2 x^2)$, $x \in [0, 1]$, $n = 7$.

(b) $\ln(x)$, $x \in [1, 3]$, $n = 5$.

(c) $\sin(x^2)$, $x \in [0, 2.5]$, $n = 6$.

5.3.4. For each of the functions and intervals in Exercise 5.3.3, use Function 5.3.1 to perform cubic spline interpolation for equispaced nodes with $n = 10, 20, 40$,

80, 160. In each case compute the interpolant at 1600 equally spaced points in the interval and use it to estimate the error

$$E(n) = \|f - S\|_\infty = \max_x |f(x) - S(x)|.$$

Make a log–log plot of $E$ as a function of $n$ and compare it graphically to fourth-order convergence.

5.3.5. ▣ Although the cardinal cubic splines are intractable in closed form, they can be found numerically. Each cardinal spline interpolates the data from one column of an identity matrix. Define the nodes $t = \begin{bmatrix} 0, 0.075, 0.25, 0.55, 1 \end{bmatrix}^T$. Plot over $[0,1]$ the five cardinal functions for this node set over the interval $[0,1]$.

5.3.6. ✎ Suppose you were to define a piecewise quadratic spline that interpolates $n+1$ given values and has a continuous first derivative. Follow the derivation of this section to express all of the interpolation and continuity conditions. How many additional conditions are required to make a square system for the coefficients?

5.3.7. (a) ✎ If $y_0 = y_n$, another possibility for cubic spline end conditions is to make $S(x)$ a periodic function. This implies that $S'$ and $S''$ are also periodic. Write out the two new algebraic equations for these constraints in terms of the piecewise coefficients.

(b) ▣ Modify Function 5.3.1 to compute a periodic spline interpolant. Test by making a plot of the interpolant for $f(x) = \exp(\sin(3x))$ over the interval $[0, 2\pi/3]$ with equally spaced nodes and $n = 8$.

## 5.4 • Finite differences

Much more can be said about interpolation, and we return to it in Chapter 9. But now we turn to one of the most common and important applications of interpolants: finding derivatives. For the moment, we will continue to use $x$ as the independent variable name and $a = t_0, \ldots, t_n = b$ as the interpolation nodes. We also continue to use an equispaced grid, so that $t_i = a + ih$ for $i = 0, \ldots, n$, where $h = (b-a)/n$.

Considering the most common definition of a derivative,

$$f'(x) = \lim_{h \to 0} \frac{f(x+h) - f(x)}{h},$$

it stands to reason that an approximation to $f'$ at a node $t_i$ should depend on the values of $f$ at the nodes closest to $t_i$. Also, because differentiation is a linear operation, we will constrain ourselves to formulas that are linear in these nodal values. Consequently, the sort of formula we seek is the **finite difference formula**

$$f'(t_i) \approx \frac{1}{h} \sum_{k=-p}^{q} a_k f(t_i + kh), \qquad (5.4.1)$$

where $p, q$ are integers, and the $a_k$ are constants known as the **weights** of the formula. Crucially, *the finite difference weights are independent of $f$*, although they do depend

## 5.4. Finite differences

on the nodes. The factor of $h^{-1}$ is present to make the expression more convenient in what follows.

Before deriving some finite difference formulas, we make an important observation about them. Define the new variable $s = x - t_i$ and let $\tilde{f}(s) = f(x - t_i)$. Then it is elementary that

$$\left.\frac{df}{dx}\right|_{x=t_i} = \left.\frac{d\tilde{f}}{ds}\right|_{s=0}.$$

Applying the change of variables to (5.4.1) yields

$$f'(t_i) = \tilde{f}'(0) \approx \frac{1}{h} \sum_{k=i-p}^{i+q} a_k \tilde{f}(kh),$$

with the same constants as before. These manipulations express a property that is simpler than it may appear: we can always derive and write the finite difference formula (5.4.1) with $t_i = 0$ without losing generality. In fact, $t_i$ is just a "dummy" variable that we can replace by $x$, as in

$$f'(x) \approx \frac{1}{h} \sum_{k=-p}^{q} a_k f(x + kh). \tag{5.4.2}$$

This property is *translation invariance*. The formula combines values of the function at points always placed the same way relative to $x$.

An obvious candidate for a finite difference formula is based on the limit definition above:

$$f'(x) \approx \frac{f(x+h) - f(x)}{h}, \tag{5.4.3}$$

which is (5.4.2) with $p = 0$, $q = 1$, $a_0 = -1$, and $a_1 = 1$. This is referred to as a **forward difference formula**, characterized by $p = 0$, because $f$ is evaluated only at points "forward" from $x$. Analogously, we could use the **backward difference formula**

$$f'(x) \approx \frac{f(x) - f(x-h)}{h}, \tag{5.4.4}$$

in which $q = 0$.

Both the forward and backward difference formulas become equalities in the limit $h \to 0$, provided $f$ is differentiable at $x$. However, they are not the only possibilities. One aesthetic objection is the lack of symmetry about the point $x$. In response, we will derive a formula that uses $p = q = 1$, i.e., in which $f(-h)$, $f(0)$, and $f(h)$ are all available.

The formula (5.4.3) (with $x = 0$) is simply the slope of the line through the points $(0, f(0))$ and $(h, f(h))$. A similar observation holds for the backward difference formula. Thus one route to using three function values is to differentiate the quadratic polynomial that interpolates them (see Exercise 5.4.1):

$$Q(x) = \frac{x(x-h)}{2h^2}f(-h) - \frac{x^2-h^2}{h^2}f(0) + \frac{x(x+h)}{2h^2}f(h). \tag{5.4.5}$$

We now use $f'(0) \approx Q'(0)$ to get the **centered finite difference formula**

$$f'(0) \approx \frac{f(h) - f(-h)}{2h}. \tag{5.4.6}$$

This result is equivalent to (5.4.2) with $p = q = 1$ and weights $a_{-1} = -1/2, a_0 = 0$, and $a_1 = 1/2$. Observe that while the value of $f(0)$ was available during the derivation, its weight ends up being zero. We can verify using L'Hôpital's rule that the approximation in (5.4.6) becomes an equality as $h \to 0$. Such an analysis does not, however, reveal a significant accuracy advantage of the centered variant, one that we will take up in Section 5.5.

We can in principle derive any finite difference formula from the same process: *Interpolate the given function values, then differentiate the interpolant exactly.* Some results are given here for two important special cases: $p = q$, or centered differences, in Table 5.1; and $p = 0$, or forward differences, in Table 5.2. For the analogous case of backward differences with $q = 0$, one can use the change of variable $\hat{f}(x) = f(-x)$, which changes the sign and reverses the order of the coefficients in Table 5.2; see Exercise 5.4.2.

Table 5.1. *Weights for centered finite difference formulas ($p = q$ in (5.4.2)). The values given here are for approximating the derivative at zero; the same formulas apply if the independent variable is translated uniformly. The term* order of accuracy *is explained in Section 5.5.*

| Order of accuracy | $-4h$ | $-3h$ | $-2h$ | $-h$ | $0$ | $h$ | $2h$ | $3h$ | $4h$ |
|---|---|---|---|---|---|---|---|---|---|
| 2 | | | | $-\frac{1}{2}$ | $0$ | $\frac{1}{2}$ | | | |
| 4 | | | $\frac{1}{12}$ | $-\frac{2}{3}$ | $0$ | $\frac{2}{3}$ | $-\frac{1}{12}$ | | |
| 6 | | $-\frac{1}{60}$ | $\frac{3}{20}$ | $-\frac{3}{4}$ | $0$ | $\frac{3}{4}$ | $-\frac{3}{20}$ | $\frac{1}{60}$ | |
| 8 | $\frac{1}{280}$ | $-\frac{4}{105}$ | $\frac{1}{5}$ | $-\frac{4}{5}$ | $0$ | $\frac{4}{5}$ | $-\frac{1}{5}$ | $\frac{4}{105}$ | $-\frac{1}{280}$ |

Table 5.2. *Weights for forward finite difference formulas ($p = 0$ in (5.4.2)). The values given here are for approximating the derivative at zero. See the text about the analogous backward differences where $q = 0$. The term* order of accuracy *is explained in Section 5.5.*

| Order of accuracy | $0$ | $h$ | $2h$ | $3h$ | $4h$ |
|---|---|---|---|---|---|
| 1 | $-1$ | $1$ | | | |
| 2 | $-\frac{3}{2}$ | $2$ | $-\frac{1}{2}$ | | |
| 3 | $-\frac{11}{6}$ | $3$ | $-\frac{3}{2}$ | $\frac{1}{3}$ | |
| 4 | $-\frac{25}{12}$ | $4$ | $-3$ | $\frac{4}{3}$ | $-\frac{1}{4}$ |

## 5.4. Finite differences

**Example 5.4.1**

According to the tables, here are two finite difference formulas:

$$f'(0) \approx h^{-1}\left[\tfrac{1}{12}f(-2h) - \tfrac{2}{3}f(-h) + \tfrac{2}{3}f(h) - \tfrac{1}{12}f(2h)\right],$$
$$f'(0) \approx h^{-1}\left[\tfrac{1}{2}f(-2h) - 2f(-h) + \tfrac{3}{2}f(0)\right].$$

### Higher derivatives

Many applications require the second derivative of a function. It's tempting to use the finite difference of a finite difference. For example, applying (5.4.6) twice leads to

$$f''(0) \approx \frac{f(-2h) - 2f(0) + f(2h)}{4h^2}. \tag{5.4.7}$$

This is a valid formula, but it uses values at $\pm 2h$ rather than the closer values at $\pm h$. A better and more generalizable tactic is to return to the quadratic $Q(x)$ in (5.4.5) and use $Q''(0)$ to approximate $f''(0)$. Doing so yields

$$f''(0) \approx \frac{f(-h) - 2f(0) + f(h)}{h^2}, \tag{5.4.8}$$

which is the simplest **centered second difference formula**. As with the first derivative, we can choose larger values of $p$ and $q$ in (5.4.1) to get new formulas, such as

$$f''(0) \approx \frac{f(0) - 2f(h) + f(2h)}{h^2} \tag{5.4.9}$$

and

$$f''(0) \approx \frac{2f(0) - 5f(h) + 4f(2h) - f(3h)}{h^2}. \tag{5.4.10}$$

### Arbitrary nodes

Although function values at equally spaced nodes are a common and convenient situation, the node locations may be arbitrary. The general form of a finite difference formula is

$$f^{(m)}(0) \approx \sum_{k=1}^{r} c_{k,m} f(t_k). \tag{5.4.11}$$

We no longer assume equally spaced nodes, so there is no "$h$" to be used in the formula. As before, the weights may be applied after any translation of the independent variable. The weights again follow from the interpolate/differentiate recipe, but the algebra becomes complicated. Fortunately there is an elegant recursion known as **Fornberg's algorithm** that can calculate these weights for any desired formula. We present it without derivation as Function 5.4.1.

**Function 5.4.1 (fdweights)** Fornberg's algorithm for finite difference weights.

```
function w = fdweights(t,m)
%FDWEIGHTS   Fornberg's algorithm for finite difference weights.
% Input:
%   t    nodes (vector, length r+1)
%   m    order of derivative sought at x=0 (integer scalar)
% Output:
%   w    weights for the approximation to the jth derivative (vector)

% This is a compact implementation, not an efficient one.

r = length(t)-1;
w = zeros(size(t));
for k = 0:r
  w(k+1) = weight(t,m,r,k);
end


function c = weight(t,m,r,k)
% Implement a recursion for the weights.
% Input:
%   t    nodes (vector)
%   m    order of derivative sought
%   r    number of nodes to use from t (<= length(t))
%   k    index of node whose weight is found
% Output:
%   c    finite difference weight

if (m<0) || (m>r)           % undefined coeffs must be zero
  c = 0;
elseif (m==0) && (r==0)     % base case of one-point interpolation
  c = 1;
else                        % generic recursion
  if k<r
    c = (t(r+1)*weight(t,m,r-1,k) - ...
        m*weight(t,m-1,r-1,k))/(t(r+1)-t(k+1));
  else
    beta = prod(t(r)-t(1:r-1)) / prod(t(r+1)-t(1:r));
    c = beta*(m*weight(t,m-1,r-1,r-1) - t(r)*weight(t,m,r-1,r-1));
  end
end
```

**Example 5.4.2**

We try to estimate the derivative of $\cos(x^2)$ at $x = 0.5$ using five nodes.

```
t = [ 0.35 0.5 0.57 0.6 0.75 ]';    % nodes
f = @(x) cos(x.^2);
dfdx = @(x) -2*x.*sin(x.^2);
exact_value = dfdx(0.5)

exact_value =
    -0.2474
```

We have to shift the nodes so that the point of estimation for the derivative is at $x = 0$.

```
w = fdweights(t-0.5,1);
fd_value = w'*f(t)

fd_value =
   -0.2473
```

We can reproduce the weights in the finite difference tables by using equally spaced nodes with $h = 1$. For example, here are two one-sided formulas.

```
format rat
fdweights(0:2,1)
fdweights(-3:0,1)

ans =
   -3/2          2          -1/2
ans =
   -1/3         3/2         -3          11/6
```

## Exercises

5.4.1. ✍ This problem refers to $Q(x)$ defined by (5.4.5).

    (a) Show that $Q(x)$ interpolates the three values of $f$ at $x = -h$, $x = 0$, and $x = h$.

    (b) Show that $Q'(0)$ gives the finite difference formula defined by (5.4.6).

5.4.2. (a) ✍ Table 5.2 lists forward difference formulas in which $p = 0$ in (5.4.2). Show that the change of variable $g(x) = f(-x)$ transforms these formulas into backward difference formulas with $q = 0$, and write out the table analogous to Table 5.2 for backward differences.

    (b) ⌨ Suppose you are given the nodes $t_0 = 0.9$, $t_1 = 1$, and $t_2 = 1.1$, and $f(x) = \sin(2x)$. Using formulas from Table 5.1 and Table 5.2, compute second-order accurate approximations to $f'$ at each of the three nodes.

5.4.3. ⌨ Using Function 5.4.1 to get the necessary weights, find finite difference approximations to the first, second, third, and fourth derivatives of $f(x) = e^{-x}$ at $x = 0.5$. In each case use a centered stencil of minimum possible width. Make a table showing the values and the errors in each case.

5.4.4. ⌨ Use Function 5.4.1 to write out a table analogous to Table 5.1 that lists centered finite difference weights for the second derivative $f''(0)$. (Hint: The `rat` command will let you express the results as exact rational numbers.)

5.4.5. ⌨ For this problem, let $f(x) = \tan(2x)$.

    (a) ⌨ Apply Function 5.4.1 to find a finite difference approximation to $f''(0.3)$ using the five nodes $t_j = 0.03 + jh$ for $j = -2, \ldots, 2$ and $h = 0.05$. Compare to the exact value of $f''(0.3)$.

    (b) ⌨ Repeat part (a) for $f''(0.75)$ and $t_j = 0.75 + jh$. Why is the finite difference result so inaccurate?

5.4.6. (a) ✍ Derive (5.4.7) by applying applying (5.4.6) twice (i.e., apply once to get $f'$ values and then apply again to those values).

(b) ✍ Find the formula for $f''(0)$ that results from applying (5.4.3) and then (5.4.4).

5.4.7. (a) ✍ Show using L'Hôpital's rule that the centered formula approximation (5.4.6) converges to an equality as $h \to 0$.

(b) ✍ Derive two conditions on the finite difference weights in (5.4.2) that arise from requiring convergence as $h \to 0$.

## 5.5 ▪ Convergence of finite differences

All of the finite difference formulas in the previous section based on equally spaced nodes converge as the node spacing $h$ decreases to zero. However, note that to discretize a function over an interval $[a,b]$, we use $h = (b-a)/n$, which implies $n = (b-a)/h = O(h^{-1})$. As $h \to 0$, the total number of nodes needed grows without bound. So we would like to make $h$ as large as possible while still achieving some acceptable accuracy.

To measure this kind of performance, we introduce the **truncation error** of a finite difference formula, defined as the difference between the two sides of (5.4.2):

$$\tau_f(h) = f'(0) - \frac{1}{h} \sum_{k=-p}^{q} a_k f(kh), \tag{5.5.1}$$

where we have used translation invariance to set $x = 0$ for simplicity. In order to make this expression useful, we expand it in a Taylor series about $h = 0$ and ignore all but the leading terms.[22]

---

**Example 5.5.1**

The finite difference formula (5.4.3) implies

$$\begin{aligned}\tau_f(h) &= f'(0) - \frac{f(h)-f(0)}{h} \\ &= f'(0) - h^{-1}\left[\left(f(0) + hf'(0) + \tfrac{1}{2}h^2 f''(0) + \cdots\right) - f(0)\right] \\ &= -\frac{1}{2}h f''(0) + O(h^2).\end{aligned}$$

---

The most important conclusion of Example 5.5.1 is that $\tau_f(h) = O(h)$. The dependence on $h$ raised to the first power leads us to call this a **first-order accurate** formula. In a first-order formula, cutting $h$ in half should reduce the error in the $f'$ estimate by about half as well, when $h$ is small enough.

---

[22] The term "truncation error" is derived from the idea that the finite difference formula, being finite, has to truncate the series representation and thus cannot be exactly correct for all functions.

## 5.5. Convergence of finite differences

**Example 5.5.2**

Let's observe the convergence of the forward difference formula applied to the function $\sin(e^{x+1})$ at $x = 0$.

```
f = @(x) sin( exp(x+1) );
FD1 = [ (f(0.1)-f(0))    /0.1
        (f(0.05)-f(0))   /0.05
        (f(0.025)-f(0))  /0.025 ]

FD1 =
   -2.7379
   -2.6128
   -2.5465
```

It's not clear that the sequence is converging. As predicted, however, the errors are cut approximately by a factor of 2 when $h$ is divided by 2.

```
exact_value = cos(exp(1))*exp(1);
err = exact_value - FD1

err =
    0.2595
    0.1344
    0.0681
```

Asymptotically as $h \to 0$, the error is proportional to $h$.

### Higher-order accuracy

As a rule, including more function values in a finite difference formula (i.e., increasing $p$ and $q$ in (5.4.2)) gives a truncation error that depends on a higher power of $h$ and thus vanishes more quickly as $h \to 0$. The power of $h$ in the leading term of the truncation error is known as the **order of accuracy**.

**Example 5.5.3**

We compute the truncation error of (5.4.6):

$$\begin{aligned}\tau_f(h) &= f'(0) - \frac{f(h) - f(-h)}{2h} \\ &= f'(0) - (2h)^{-1}\Big[\big(f(0) + hf'(0) + \tfrac{1}{2}h^2 f''(0) + \tfrac{1}{6}h^3 f'''(0) + O(h^4)\big) \\ &\quad - \big(f(0) - hf'(0) + \tfrac{1}{2}h^2 f''(0) - \tfrac{1}{6}h^3 f'''(0) + O(h^4)\big)\Big] \\ &= (2h)^{-1}\Big[\tfrac{1}{3}h^3 f'''(0) + O(h^4)\Big] = O(h^2).\end{aligned}$$

Because the lowest-order term is proportional to $h^2$, we say that the method has order of accuracy equal to 2.

Example 5.5.3 implies that (5.4.6) is second-order accurate. Reducing $h$ by a factor of two should cut the error in the $f'$ estimate by a factor of roughly four. This convergence is much more rapid than for the first-order formula (5.4.3).

### Example 5.5.4

We revisit Example 5.5.2 to compare the first-order forward difference formula with the second-order centered formula.

```
f = @(x) sin( exp(x+1) );
exact_value = cos(exp(1))*exp(1);
```

We run both formulas in parallel for a sequence of $h$ values.

```
h = 4.^(-1:-1:-8)';
FD1 = 0*h;   FD2 = 0*h;
for k = 1:length(h)
    FD1(k) = (f(h(k)) - f(0)) / h(k);
    FD2(k) = (f(h(k)) - f(-h(k))) / (2*h(k));
end
```

In each case $h$ is decreased by a factor of 4, so that the error is reduced by a factor of 4 in the first-order method and 16 in the second-order method.

```
error_FD1 = exact_value-FD1;
error_FD2 = exact_value-FD2;
table(h,error_FD1,error_FD2)
```

```
ans =
   8x3 table
         h         error_FD1      error_FD2
     ----------    ----------    -----------
         0.25        0.53167      -0.085897
        0.0625       0.16675      -0.0044438
        0.015625     0.042784     -0.00027403
        0.0039062    0.010751     -1.7112e-05
        0.00097656   0.0026911    -1.0695e-06
        0.00024414   0.00067298   -6.6841e-08
        6.1035e-05   0.00016826   -4.1767e-09
        1.5259e-05   4.2065e-05   -2.7269e-10
```

A graphical comparison can be clearer. On a log–log scale, the error should (roughly) be a straight line whose slope is the order of accuracy. However, it's conventional in convergence plots to show $h$ *decreasing* from left to right, which negates the slopes.

```
loglog(h,[abs(error_FD1),abs(error_FD2)],'.-')
hold on, loglog(h,[h,h.^2],'--')      % perfect 1st and 2nd
    order
set(gca,'xdir','reverse')
```

## 5.5. Convergence of finite differences

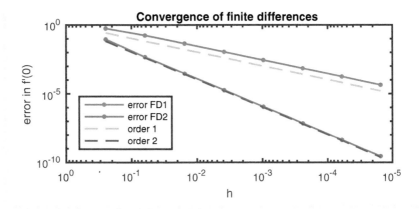

### Conditioning

The truncation error $\tau_f(h)$ of a finite difference formula is dominated by a leading term $O(h^m)$ for an integer $m$. This error decreases as $h \to 0$. However, we have not yet accounted for the effects of roundoff error. To keep matters as simple as possible, let's consider the forward difference

$$\delta(h) = \frac{f(x+h)-f(x)}{h}.$$

As $h \to 0$, the numerator approaches zero even though the values $f(x+h)$ and $f(x)$ are not necessarily near zero. This is the recipe for subtractive cancellation error! In fact, finite difference formulas are inherently ill-conditioned as $h \to 0$ when evaluated in floating point arithmetic.

To be precise, recall that the condition number for the problem of computing $f(x+h) - f(x)$ is

$$\kappa(h) = \frac{\max\{|f(x+h)|, |f(x)|\}}{|f(x+h)-f(x)|},$$

implying a relative error of size $\kappa(h)\varepsilon_{\text{mach}}$ in its computation. Hence the numerical value we actually compute for $\delta$ is

$$\tilde{\delta}(h) = \frac{f(x+h)-f(x)}{h}(1+\kappa(h)\varepsilon_{\text{mach}})$$

$$= \delta(h) + \frac{\max\{|f(x+h)|, |f(x)|\}}{|f(x+h)-f(x)|} \cdot \frac{f(x+h)-f(x)}{h} \cdot \varepsilon_{\text{mach}}$$

$$= \delta(h) \pm \frac{\max\{|f(x+h)|, |f(x)|\}}{h}\varepsilon_{\text{mach}}.$$

As $h \to 0$, $f(x+h) = f(x) + O(h)$, so we can simplify the maximization in the expression to get

$$|\tilde{\delta}(h) - \delta(h)| = |f(x)|\varepsilon_{\text{mach}}h^{-1} + \varepsilon_{\text{mach}} \cdot O(1).$$

Hence as $h \to 0$, the roundoff error is $O(h^{-1})$, which grows without bound.

Combining truncation error with roundoff error leads to

$$|f'(x)-\tilde{\delta}(h)| \leq |\tau_f(h)| + |f(x)|\varepsilon_{\text{mach}} h^{-1}. \tag{5.5.2}$$

Equation (5.5.2) quantifies the contributions of both truncation and roundoff errors. While the truncation error $\tau$ vanishes as $h$ decreases, the roundoff error actually *increases* thanks to the subtractive cancellation. At some value of $h$ the two error contributions will be of roughly equal size. This occurs roughly when

$$|f(x)|\varepsilon_{\text{mach}} h^{-1} \approx Ch, \quad \text{or} \quad h \approx K\sqrt{\varepsilon_{\text{mach}}}$$

for a constant $K$ that depends on $x$ and $f$, but not $h$. For a method of truncation order $m$, the details of the subtractive cancellation are a bit different, but the conclusion generalizes to

$$\varepsilon_{\text{mach}} h^{-1} \approx Ch^m, \quad \text{or} \quad h \approx K\varepsilon_{\text{mach}}^{1/(m+1)}. \tag{5.5.3}$$

Finally, at the optimal $h$ from (5.5.3), both the truncation and roundoff errors are

$$O(h^m) = O\left(\varepsilon_{\text{mach}}^{m/(m+1)}\right). \tag{5.5.4}$$

Hence for a first-order formula ($m = 1$), we can expect only $O\left(\sqrt{\varepsilon_{\text{mach}}}\right)$ error, or about half of the number of accurate machine digits. As $m$ increases, we get ever closer to using the full accuracy available.

**Example 5.5.5**

Let $f(x) = e^{-1.3x}$. We apply finite difference formulas of first, second, and fourth order to estimate $f'(0) = -1.3$.

```
h = 10.^(-1:-1:-12)';
f = @(x) exp(-1.3*x);
for j = 1:length(h)
    nodes = h(j)*(-2:2)';
    vals = f(nodes)/h(j);
    fd1(j) = [   0     0   -1     1      0] * vals;
    fd2(j) = [   0  -1/2    0   1/2      0] * vals;
    fd4(j) = [1/12  -2/3    0   2/3  -1/12] * vals;
end
loglog(h,abs(fd1+1.3),'.-'), hold on
loglog(h,abs(fd2+1.3),'.-')
loglog(h,abs(fd4+1.3),'.-')
loglog(h,0.1*eps./h,'--')
set(gca,'xdir','reverse')
```

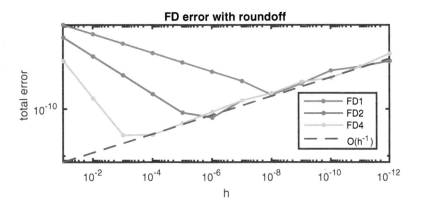

Again the graph is made so that $h$ decreases from left to right. The errors are dominated at first by truncation error, which decreases most rapidly for the fourth-order formula. However, increasing roundoff error eventually equals and then dominates the truncation error as $h$ continues to decrease. As the order of accuracy increases, the crossover point moves to the left (greater efficiency) and down (greater accuracy).

The observations in Example 5.5.5 match the analysis above quite well, with the optimal $h$ becoming larger and the optimal error getting smaller as the order increases. Thus, *higher-order finite difference methods are both more efficient and less vulnerable to roundoff than low-order methods.*

## Exercises

5.5.1. ⌨ Write a code to evaluate the centered second-order finite difference approximation to $f'(\pi/7)$ for $f(x) = \cos(x)$ and $h = 2^{-1}, 2^{-2}, \ldots, 2^{-7}$. On a log–log graph, plot the error as a function of $h$ and compare it graphically to second-order convergence.

5.5.2. ✍ Derive the first two nonzero terms of the Taylor series at $h = 0$ of the truncation error $\tau_f(h)$ for the formula (5.4.4).

5.5.3. ✍ Calculate the first nonzero term in the Taylor series of the truncation error $\tau_f(h)$ for the finite difference formula defined by the second row of Table 5.2.

5.5.4. ✍ Calculate the first nonzero term in the Taylor series of the truncation error $\tau_f(h)$ for the finite difference formula defined by the third row of Table 5.2.

5.5.5. ✍ Using natural extensions of our definitions, show that formula (5.4.8) is second-order accurate.

5.5.6. ✍ A different way to derive finite difference formulas is the *method of undetermined coefficients*. Starting from (5.4.2),

$$f'(x) \approx \frac{1}{h} \sum_{k=-p}^{q} a_k f(x + kh),$$

let each $f(x+kh)$ be expanded in a series around $h = 0$. When the coefficients of powers of $h$ are collected, one obtains

$$\frac{1}{h}\sum_{k=-p}^{q} a_k f(x+kh) = \frac{b_0}{h} + b_1 f'(x) + b_2 f''(x)h + \cdots,$$

where

$$b_i = \sum_{k=-p}^{q} k^i a_k.$$

In order to make the result as close as possible to $f'(x)$, we impose the conditions

$$b_0 = 0, \quad b_1 = 1, \quad b_2 = 0, \quad b_3 = 0, \quad \ldots, \quad b_{p+q} = 0.$$

This provides a system of linear equations for the weights.

(a) For $p = q = 2$, write out the system of equations for $a_{-2}, a_{-1}, a_0, a_1, a_2$.

(b) Verify that the coefficients from the appropriate row of Table 5.1 satisfy the equations you wrote down in part (a).

(c) Derive the finite difference formula for $p = 1, q = 2$ using the method of undetermined coefficients.

## 5.6 • Numerical integration

In calculus you learn that the elegant way to evaluate a definite integral is to apply the Fundamental Theorem of Calculus and find an antiderivative. The connection is so profound and pervasive that it's easy to overlook that a definite integral is a numerical quantity existing independently of antidifferentiation. In fact, most conceivable integrands have no antiderivative in terms of familiar functions.

### Example 5.6.1

The antiderivative of $e^x$ is, of course, itself. That makes evaluation of $\int_0^1 e^x \, dx$ by the Fundamental Theorem trivial.

```
format long, I = exp(1)-1
```

```
I =
    1.718281828459046
```

MATLAB has a built-in `integral` function that estimates the value numerically without finding the antiderivative first. As you can see here, it's often just as accurate.

```
integral(@(x) exp(x),0,1)
```

```
ans =
    1.718281828459045
```

## 5.6. Numerical integration

The numerical approach is far more robust. For example, $e^{\sin x}$ has no useful antiderivative. But numerically it's no more difficult.

```
integral(@(x) exp(sin(x)),0,1)

ans =
    1.631869608418051
```

When you look at the graphs of these functions, what's remarkable is that one of these areas is the most basic calculus, while the other is almost impenetrable analytically. From a numerical standpoint, they are practically the same problem.

```
x = linspace(0,1,201)';
subplot(2,1,1), fill([x;1;0],[exp(x);0;0],[1,0.9,0.9])
subplot(2,1,2), fill([x;1;0],[exp(sin(x));0;0],[1,0.9,0.9])
```

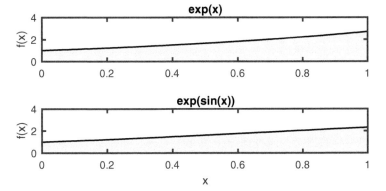

Numerical integration[23] is done by combining values of the integrand sampled at nodes, much like finite differences. In this section we will assume equally spaced nodes using the definitions

$$t_i = a + ih, \quad h = \frac{b-a}{n}, \quad i = 0, \ldots, n. \quad (5.6.1)$$

The integration formulas are expressed as

$$I = \int_a^b f(x)\,dx \approx Q = h \sum_{i=0}^{n} w_i f(t_i)$$
$$= h\big[w_0 f(t_0) + w_1 f(t_1) + \cdots + w_n f(t_n)\big]. \quad (5.6.2)$$

The constants $w_i$ appearing in the formula are called **weights**. As with finite difference formulas, *the weights of numerical integration formulas are chosen independently of the function being integrated, and they determine the formula completely.* We can apply quadrature formulas to sequences of data values even if no function is explicitly known to generate them, but for presentation and implementations we assume that we can evaluate $f(x)$ anywhere.

A straightforward way to derive integration formulas is to mimic the approach taken for finite differences: Find an interpolant and operate exactly on it. If the interpolant is a piecewise polynomial, the result is a **Newton–Cotes formula**.

---
[23] Numerical integration also goes by the older name *quadrature*.

## Trapezoid formula

One of the most important Newton–Cotes formulas results from integration of the piecewise linear interpolant (see Section 5.2). Using the cardinal basis form of the interpolant in (5.2.3), we have

$$I \approx \int_a^b \sum_{i=0}^n f(t_i) H_i(x) \, dx = \sum_{i=0}^n f(t_i) \left[ \int_a^b H_i(x) \right] dx.$$

Thus we can identify the weights as $w_i = h^{-1} \int_a^b H_i(x) \, dx$. Using areas of triangles, it's trivial to derive that

$$w_i = \begin{cases} 1, & i = 1, \ldots, n-1, \\ \frac{1}{2}, & i = 0, n. \end{cases} \quad (5.6.3)$$

Putting everything together, the resulting quadrature formula is

$$I = \int_a^b f(x) \, dx \approx T_f(n) = h \left[ \frac{1}{2} f(t_0) + f(t_1) + f(t_2) + \cdots + f(t_{n-1}) + \frac{1}{2} f(t_n) \right]. \quad (5.6.4)$$

This is called the **trapezoid formula** or trapezoid rule.[24] *The trapezoid formula results from integration of the piecewise linear interpolant*, or equivalently, as illustrated in Figure 5.1, from using the area of approximating trapezoids to estimate the area under a curve. The trapezoid formula is the Swiss Army knife of integration formulas. A short implementation is given as Function 5.6.1.

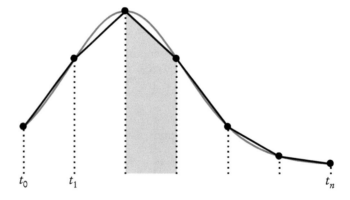

**Figure 5.1.** *Trapezoid formula for integration.*

In Theorem 5.2.2 we stated that the pointwise error in a piecewise linear interpolant with equal node spacing $h$ is bounded by $O(h^2)$ as $h \to 0$. Using $p$ to stand for the piecewise linear interpolant, we obtain

$$I - T_f(n) = I - \int_a^b p(x) \, dx = \int_a^b [f(x) - p(x)] \, dx$$

$$\leq (b-a) \max_{x \in [a,b]} |f(x) - p(x)| = O(h^2).$$

---

[24] Some texts distinguish between a formula for a single subinterval $[t_{k-1}, t_k]$ and a "composite" formula that adds them up over the whole interval to get something like our (5.6.4).

## 5.6. Numerical integration

**Function 5.6.1** (`trapezoid`) Trapezoid formula for numerical integration.

```
function [T,t,y] = trapezoid(f,a,b,n)
%TRAPEZOID   Trapezoid formula for numerical integration.
% Input:
%   f       integrand (function)
%   a,b     interval of integration (scalars)
%   n       number of interval divisions
% Output:
%   T       approximation to the integral of f over (a,b)
%   t       vector of nodes used
%   y       vector of function values at nodes

h = (b-a)/n;
t = a + h*(0:n)';
y = f(t);
T = h * ( sum(y(2:n)) + 0.5*(y(1) + y(n+1)) );
```

Hence the trapezoid formula has second-order error. This fact is embedded rigorously in one of the most remarkable formulas in mathematics, the **Euler–Maclaurin formula**, which may be stated as

$$I = \int_a^b f(x)\,dx = T_f(n) - \frac{h^2}{12}\left[f'(b)-f'(a)\right] + \frac{h^4}{740}\left[f'''(b)-f'''(a)\right] + O(h^6)$$
$$= T_f(n) - \sum_{k=1}^\infty \frac{B_{2k}}{(2k)!}\left[f^{(2k-1)}(b) - f^{(2k-1)}(a)\right],$$

(5.6.5)

where the $B_{2k}$ are constants known as *Bernoulli numbers*. Unless we happen to be fortunate enough to have a function with $f'(b) = f'(a)$, we should expect truncation error at second order and no better.

### Example 5.6.2

We approximate the integral of the function $f(x) = e^{\sin 7x}$ over the interval $[0,2]$.

```
f = @(x) exp(sin(7*x));
a = 0;  b = 2;
```

In lieu of the exact value, we will use the built-in `integral` function to find an accurate result.

```
I = integral(f,a,b,'abstol',1e-14,'reltol',1e-14);
fprintf('Integral = %.15f\n',I)

Integral = 2.663219782761539
```

Here is the error at $n = 40$.

```
T = trapezoid(f,a,b,40);
err = I - T
```

```
err = 
    9.1685e-04
```

In order to check the order of accuracy, we double $n$ a few times and observe how the error decreases.

```
n = 40*2.^(0:5)';
err = zeros(size(n));
for k = 1:length(n)
    T = trapezoid(f,a,b,n(k));
    err(k) = I - T;
end
table(n,err)
```

```
ans =
  6x2 table
     n         err
    ----    ----------
     40     0.00091685
     80     0.00023006
    160     5.7568e-05
    320     1.4395e-05
    640     3.599e-06
   1280     8.9975e-07
```

Each doubling of $n$ cuts the error by a factor of about 4, which is consistent with second-order convergence. Another check: the slope on a log–log graph should be $-2$.

```
loglog(n,abs(err),'.-')
hold on, loglog(n,3e-3*(n/n(1)).^(-2),'--')
```

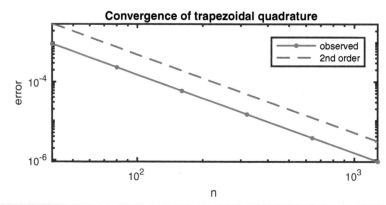

## Extrapolation

If evaluations of $f$ are computationally expensive, we want to get as much accuracy as possible from them by using a higher-order formula. There are many routes for doing so; for example, we could integrate a not-a-knot cubic spline interpolant. However, splines are difficult to compute by hand, and as a result different methods were developed before computers came on the scene.

## 5.6. Numerical integration

*Extrapolation is a general idea that exploits any situation in which we know something about the structure of the error.* Suppose a quantity $A_0$ is approximated by an algorithm $A(h)$ with an error expansion

$$A_0 = A(h) + c_1 h + c_2 h^2 + c_3 h^3 + \cdots. \tag{5.6.6}$$

Crucially, it is not necessary to know the values of the error constants $c_k$, merely that they exist and are independent of $h$. In the case of the trapezoid formula, we have

$$I = T_f(n) + c_2 h^2 + c_4 h^4 + \cdots,$$

as proved by the Euler–Maclaurin formula (5.6.5). The error constants depend on $f$ and can't be evaluated in general, but we know that this expansion holds.

For convenience we recast the error expansion in terms of $n = O(h^{-1})$:

$$I = T_f(n) + c_2 n^{-2} + c_4 n^{-4} + \cdots, \tag{5.6.7}$$

We make the simple observation that

$$I = T_f(2n) + \frac{1}{4} c_2 n^{-2} + \frac{1}{16} c_4 n^{-4} + \cdots. \tag{5.6.8}$$

It follows that if we combine (5.6.7) and (5.6.8) correctly, we can cancel out the second-order term in the error. Specifically, define

$$S_f(2n) = \frac{1}{3}\left[ 4 T_f(2n) - T_f(n) \right]. \tag{5.6.9}$$

(We associate $2n$ rather than $n$ with the extrapolated result because of the total number of nodes needed.) Then

$$I = S_f(2n) + O(n^{-4}) = b_4 n^{-4} + b_6 n^{-6} + \cdots. \tag{5.6.10}$$

The formula (5.6.9) is called **Simpson's rule**. A different presentation and derivation are considered in Exercise 5.6.4.

Equation (5.6.10) is another particular error expansion in the form (5.6.6), so we can extrapolate again! The details change only a little. Considering that

$$I = S_f(4n) = \frac{1}{16} b_4 n^{-4} + \frac{1}{64} b_6 n^{-6} + \cdots,$$

the proper combination this time is

$$R_f(4n) = \frac{1}{15}\left[ 16 S_f(4n) - S_f(2n) \right], \tag{5.6.11}$$

which is sixth-order accurate. Clearly the process can be repeated to get eighth-order accuracy and beyond. Doing so goes by the name of *Romberg integration*, which we will not present in full generality.

## Node doubling

Note in (5.6.11) that $R_f(4n)$ depends on $S_f(2n)$ and $S_f(4n)$, which in turn depend on $T_f(n)$, $T_f(2n)$, and $T_f(4n)$. There is a useful benefit realized by doubling of the nodes in each application of the trapezoid formula. For simplicity, suppose that $[a,b] = [0,1]$ and that $n = 2N$ for some positive integer $N$. The nodes are

$$0, \quad \frac{1}{2N}, \quad \frac{2}{2N}, \quad \frac{3}{2N}, \quad \frac{4}{2N}, \quad \cdots \quad \frac{2N-3}{2N}, \quad \frac{2N-2}{2N}, \quad \frac{2N-3}{2N}, \quad 1.$$

Suppose we delete every other node:

$$0, \quad - \quad \frac{1}{N}, \quad - \quad \frac{2}{N}, \quad \cdots \quad - \quad \frac{N-1}{N}, \quad - \quad 1.$$

What remains are the nodes with $n = N$. That is, if we have computed $T_f(N)$ and want to compute $T_f(2N)$, we begin with half of the evaluations of $f$ already in our pocket. More specifically,

$$\begin{aligned}
T_f(2N) &= \frac{1}{2N}\left[\tfrac{1}{2}f(0) + \tfrac{1}{2}f(1) + \sum_{m=1}^{N-1} f\left(\tfrac{2m-1}{2N}\right) + f\left(\tfrac{2m}{2N}\right)\right] \\
&= \frac{1}{2N}\left[\tfrac{1}{2}f(0) + \tfrac{1}{2}f(1) + \sum_{m=1}^{N-1} f\left(\tfrac{m}{N}\right)\right] + \frac{1}{2N}\sum_{m=1}^{N-1} f\left(\tfrac{2m-1}{2N}\right) \\
&= \frac{1}{2}T_f(N) + \frac{1}{2N}\sum_{m=1}^{N-1} f(t_{2m-1}), \quad (5.6.12)
\end{aligned}$$

where the nodes referenced in the last line are relative to $n = 2N$. To summarize, when $n$ is doubled, new integrand evaluations are needed only at the odd-numbered nodes of the finer grid. Although we derived this result in the particular interval $[0,1]$, it is valid for any interval.

### Example 5.6.3

We estimate $\int_0^2 x^2 e^{-2x}\,dx$ using extrapolation.

```
f = @(x) x.^2.*exp(-2*x);
a = 0;  b = 2;   format short e
I = integral(f,a,b,'abstol',1e-14,'reltol',1e-14);
```

We start with the trapezoid formula on $n = N$ nodes.

```
N = 20;          % the coarsest formula
n = N;   h = (b-a)/n;
t = h*(0:n)';    y = f(t);
```

We can now apply weights to get the estimate $T_f(N)$.

```
T = h*( sum(y(2:N)) + y(1)/2 + y(n+1)/2 );
```

## 5.6. Numerical integration

```
err_2nd = I - T

err_2nd =
   6.2724e-05
```

Now we double to $n = 2N$, but we only need to evaluate $f$ at every other interior node.

```
n = 2*n;  h = h/2;  t = h*(0:n)';
T(2) = T(1)/2 + h*sum( f(t(2:2:n)) );
err_2nd = I - T

err_2nd =
   6.2724e-05   1.5368e-05
```

As expected for a second-order estimate, the error went down by a factor of about 4. We can repeat the same code to double $n$ again.

```
n = 2*n;  h = h/2;  t = h*(0:n)';
T(3) = T(2)/2 + h*sum( f(t(2:2:n)) );
err_2nd = I - T

err_2nd =
   6.2724e-05   1.5368e-05   3.8223e-06
```

Let us now do the first level of extrapolation to get results from Simpson's formula. We combine the elements T(i) and T(i+1) the same way for $i = 1$ and $i = 2$.

```
S = (4*T(2:3) - T(1:2)) / 3;
err_4th = I - S

err_4th =
  -4.1755e-07  -2.6175e-08
```

With the two Simpson values $S_f(N)$ and $S_f(2N)$ in hand, we can do one more level of extrapolation to get a sixth-order accurate result.

```
R = (16*S(2) - S(1)) / 15;
err_6th = I - R

err_6th =
  -8.2748e-11
```

If we consider the computational time to be dominated by evaluations of $f$, then we have obtained a result with twice as many accurate digits as the best trapezoid result, at virtually no extra cost.

## Exercises

5.6.1. ▣ For each integral below, use Function 5.6.1 to estimate the integral for $n = 10 \cdot 2^k$ nodes for $k = 1, 2, \ldots, 10$. Make a log–log plot of the errors and confirm or refute second-order accuracy. (These integrals were taken from [8].)

(a) $\int_0^1 x \log(1+x) \, dx = \dfrac{1}{4}.$

(b) $\int_0^1 x^2 \tan^{-1} x \, dx = \dfrac{\pi - 2 + 2\log 2}{12}.$

(c) $\int_0^{\pi/2} e^x \cos x \, dx = \dfrac{e^{\pi/2} - 1}{2}.$

(d) $\int_0^1 \sqrt{x} \log(x) \, dx = -\dfrac{4}{9}.$ (Note: Although the integrand has the limiting value zero as $x \to 0$, MATLAB cannot compute it there directly. You can start the integral at $x = \varepsilon_{\text{mach}}$ instead.)

(e) $\int_0^1 \sqrt{1 - x^2} \, dx = \dfrac{\pi}{4}.$

5.6.2. ✎ The Euler–Maclaurin error expansion (5.6.5) for the trapezoid formula implies that if we could cancel out the term due to $f'(b) - f'(a)$, we would obtain fourth-order accuracy. We should not assume that $f'$ is available, but approximating it with finite differences can achieve the same goal. Suppose the forward difference formula (5.4.9) is used for $f'(a)$, and its reflected backward difference is used for $f'(b)$. Show that the resulting modified trapezoid formula is

$$G_f(h) = T_f(h) - \dfrac{h}{24}\Big[3\big(f(x_n) + f(x_0)\big) - 4\big(f(x_{n-1}) + f(x_1)\big) + \big(f(x_{n-2}) + f(x_2)\big)\Big], \quad (5.6.13)$$

which is known as a *Gregory quadrature formula*.

5.6.3. ▣ Repeat each integral in Exercise 5.6.1 using Gregory quadrature (5.6.13) instead of the trapezoid formula. Compare the observed errors to fourth-order convergence.

5.6.4. ✎ Simpson's formula can be derived without appealing to extrapolation.

(a) Show that

$$p(x) = \beta + \dfrac{\gamma - \alpha}{2h} x + \dfrac{\alpha - 2\beta + \gamma}{2h^2} x^2$$

interpolates the three points $(-h, \alpha)$, $(0, \beta)$, and $(h, \gamma)$.

(b) Find

$$\int_{-h}^{h} p(s) \, ds,$$

where $p$ is the quadratic polynomial from part (a), in terms of $h$, $\alpha$, $\beta$, and $\gamma$.

(c) Assume equally spaced nodes in the form $t_i = a + ih$ for $h = (b-a)/n$ and $i = 0, \ldots, n$. Suppose $f$ is approximated by $p(x)$ over the subinterval $[t_{i-1}, t_{i+1}]$. Apply the result from (b) to find

$$\int_{t_{i-1}}^{t_{i+1}} f(x)\,dx \approx \frac{h}{3}\bigl[f(t_{i-1}) + 4f(t_i) + f(t_{i+1})\bigr].$$

(Use the change of variable $s = x - t_i$.)

(d) Now also assume that $n = 2m$ for an integer $m$. Derive Simpson's formula,

$$\int_a^b f(x)\,dx \approx \frac{h}{3}\bigl[f(t_0) + 4f(t_1) + 2f(t_2) + 4f(t_3) + 2f(t_4) + \cdots \\ + 2f(t_{n-2}) + 4f(t_{n-1}) + f(t_n)\bigr]. \quad (5.6.14)$$

5.6.5. ✍ Show that the Simpson formula (5.6.14) is equivalent to $S_f(n/2)$, given the definition of $S_f$ in (5.6.9).

5.6.6. ▦ For each integral in Exercise 5.6.1, apply the Simpson formula (5.6.14) and compare the errors to fourth-order convergence.

5.6.7. ▦ For $n = 10, 20, 30, \ldots, 200$, compute the trapezoidal quadrature approximation to

$$\int_0^1 \frac{1}{2.01 + \sin(6\pi x) - \cos(2\pi x)}\,dx \approx 0.9300357672424684.$$

Make two separate plots of the absolute error as a function of $n$, one using log–log scales and the other using log only for the $y$-axis. The graphs suggest that the error asymptotically behaves as $C\alpha^n$ for some $C > 0$ and some $0 < \alpha < 1$. How does this result relate to (5.6.5)?

5.6.8. ▦ For each integral in Exercise 5.6.1, extrapolate the trapezoidal results two levels to get sixth-order accurate results, and compare the expected convergence rate to the observed errors.

5.6.9. ✍ Find a formula like (5.6.11) that extrapolates two values of $R_f$ to obtain an eighth-order accurate one.

## 5.7 • Adaptive integration

To this point, we have used only equally spaced nodes to compute integrals. Yet there are problems in which nonuniformly distributed nodes would clearly be more appropriate.

### Example 5.7.1

This function gets increasingly oscillatory near the right endpoint.

```
f = @(x) (x+1).^2.*cos((2*x+1)./(x-4.3));
fplot(f,[0 4],2000)
```

```
title('A wiggly integrand')
```

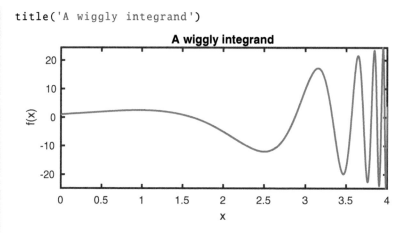

Accordingly, the trapezoid rule is more accurate on the left half of the interval than on the right half.

```
n_ = 50*2.^(0:3)';
T_ = [];
for i = 1:length(n_)
    n = n_(i);
    T_(i,1) = trapezoid(f,0,2,n);
    T_(i,2) = trapezoid(f,2,4,n);
end
left_val = integral(f,0,2,'abstol',1e-14,'reltol',1e-14);
right_val = integral(f,2,4,'abstol',1e-14,'reltol',1e-14);
table(n_,T_(:,1)-left_val,T_(:,2)-right_val,...
    'variablenames',{'n','left_error','right_error'})
```

```
ans =
  4x3 table
     n      left_error    right_error
    ___    _____   _____
    50     -0.0024911      0.50423
    100    -0.00062271     0.096004
    200    -0.00015568     0.022547
    400    -3.8919e-05     0.0055542
```

Both the picture and the numbers suggest that more nodes should be used on the right half of the interval than on the left half.

We would like an algorithm that automatically detects and reacts to a situation like that in Example 5.7.1, a trait known as **adaptivity**.

## Error estimation

Ideally, we would like to make adaptation decisions based on the error of the integration. Knowing the error exactly would be equivalent to knowing the exact answer, but we can estimate it using the extrapolation technique of Section 5.6. Consider the

## 5.7. Adaptive integration

Simpson formula (5.6.10) resulting from one level of extrapolation from trapezoid estimates:

$$S_f(2n) = \frac{1}{3}\Big[4T_f(2n) - T_f(n)\Big]. \tag{5.7.1}$$

We expect this method to be fourth-order accurate, i.e.,

$$\int_a^b f(x)\,dx = S_f(2n) + O(n^{-4}),$$

We can further extrapolate to sixth-order accuracy by (5.6.11),

$$R_f(4n) = \frac{1}{15}\Big[16S_f(4n) - S_f(2n)\Big]. \tag{5.7.2}$$

By virtue of higher order of accuracy, $R_f(4n)$ should be more accurate than $S_f(4n)$. Hence a decent estimate of the error in the better of the two Simpson values is

$$E = R_f(4n) - S_f(4n) = \frac{S_f(4n) - S_f(2n)}{15}. \tag{5.7.3}$$

### Divide and conquer

If $|E|$ is judged to be acceptably small, we are done. This judgment takes some care. For instance, suppose the exact integral is $10^{20}$. Requiring $|E| < \delta \ll 1$ would be fruitless in double precision, since it would require more than 20 accurate digits. Hence checking the absolute size of the error alone is not appropriate. Conversely, consider the integral

$$\int_{10^{-6}}^{2\pi} 2\sin x\,dx \approx -10^{-12}.$$

We are likely to sample values of the integrand that are larger than, say, $1/2$ in absolute value, so obtaining this very small result has to rely on subtractive cancellation. We cannot hope for more than 4–5 accurate digits, so a strict test of the relative error is also not recommended.[25] Typically we use both relative and absolute error, stopping when either one is considered small enough. Algebraically, the test is

$$|E| < \delta_a + \delta_r |S_f(n)|, \tag{5.7.4}$$

where $\delta_a$ and $\delta_r$ are given absolute and relative error tolerances, respectively.

When $|E|$ fails to meet (5.7.4), we bisect the interval $[a,b]$ to exploit the identity

$$\int_a^b f(x)\,dx = \int_a^{(a+b)/2} f(x)\,dx + \int_{(a+b)/2}^b f(x)\,dx,$$

and independently compute estimates to each of the half-length integrals. Each of these half-sized computations recursively applies Simpson's formula and the error estimation criterion, making further bisections as necessary. Such an approach is called **divide and conquer** in computer science: recursively split the problem into easier pieces and glue the results together.

---

[25]In other words, we can have an error that is small relative to the data (the integrand), which is $O(1)$, but not relative to the answer itself.

## Implementation

It is typical to use just the minimal formula $S_f(4)$ and its error estimate $E$ to make decisions about adaptivity. A computation of $S_f(4)$ requires three trapezoid estimates $T_f(1)$, $T_f(2)$, and $T_f(4)$. As observed in (5.6.12) and Example 5.6.3, the five integrand evaluations in $T_f(4)$ are sufficient to compute all of these values. There is one further exploitation of node locations to be found. For simplicity, assume $[a,b] = [0,1]$. The five nodes used in $T_f(4)$ are

$$0 \quad \tfrac{1}{4} \quad \tfrac{1}{2} \quad \tfrac{3}{4} \quad 1.$$

**Function 5.7.1** (intadapt) Adaptive integration with error estimation.

```
function [Q,t] = intadapt(f,a,b,tol)
%INTADAPT   Adaptive integration with error estimation.
% Input:
%   f       integrand (function)
%   a,b     interval of integration (scalars)
%   tol     acceptable error
% Output:
%   Q       approximation to integral(f,a,b)
%   t       vector of nodes used

m = (b+a)/2;
[Q,t] = do_integral(a,f(a),b,f(b),m,f(m),tol);

    % Use error estimation and recursive bisection.
    function [Q,t] = do_integral(a,fa,b,fb,m,fm,tol)

        % These are the two new nodes and their f values.
        xl = (a+m)/2;    fl = f(xl);
        xr = (m+b)/2;    fr = f(xr);
        t = [a;xl;m;xr;b];            % all 5 nodes at this level

        % Compute the trapezoid values iteratively.
        h = (b-a);
        T(1) = h*(fa+fb)/2;
        T(2) = T(1)/2 + (h/2)*fm;
        T(3) = T(2)/2 + (h/4)*(fl+fr);

        S = (4*T(2:3)-T(1:2)) / 3;    % Simpson values
        E = (S(2)-S(1)) / 15;         % error estimate

        if abs(E) < tol*(1+abs(S(2)))  % acceptable error?
            Q = S(2);                  % yes--done
        else
            % Error is too large--bisect and recurse.
            [QL,tL] = do_integral(a,fa,m,fm,xl,fl,tol);
            [QR,tR] = do_integral(m,fm,b,fb,xr,fr,tol);
            Q = QL + QR;
            t = [tL;tR(2:end)];        % merge the nodes w/o duplicate
        end
    end

end
```

## 5.7. Adaptive integration

If we bisect and compute $T_f(4)$ on the subinterval $[0, 1/2]$, we use the nodes

$$0 \quad \tfrac{1}{8} \quad \tfrac{1}{4} \quad \tfrac{3}{8} \quad \tfrac{1}{2}.$$

Only the second and fourth nodes are new. The same is true on the subinterval $[1/2, 1]$, and for every recursive bisection.

Function 5.7.1 shows how to exploit this structure. The subfunction `do_integral` does all of the work. It expects to receive the three nodes and integrand values that it shares with the level above. It adds the two new nodes and uses the set of all five to compute three trapezoid estimates with $n = 1$, $n = 2$, and $n = 4$, using the updating formula (5.6.12) twice. It goes on to find the two Simpson approximations and to estimate the error in the better one by (5.7.3).

If the error estimate passes the test (5.7.4), the better Simpson value is returned as the integral over the given interval. Otherwise, the interval is bisected, the two pieces computed using recursive calls, and those results are added to give the complete integral.

### Example 5.7.2

We revisit the integral from Example 5.7.1.

```
f = @(x) (x+1).^2.*cos((2*x+1)./(x-4.3));
I = integral(f,0,4,'abstol',1e-14,'reltol',1e-14);   % '
   exact' value
```

We perform the integration and show the nodes selected underneath the curve.

```
[Q,t] = intadapt(f,0,4,0.001);
fplot(f,[0 4],2000,'k'), hold on
stem(t,f(t),'.-')
num_nodes = length(t)

num_nodes =
      69
```

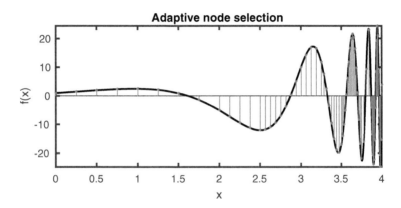

The error turns out to be a bit more than we requested. It's only an estimate, not a guarantee.

```
err = I - Q

err =
   -0.0220
```

Let's see how the number of integrand evaluations and the error vary with the requested tolerance.

```
tol_ = 10.^(-4:-1:-14)';
err_ = 0*tol_;
num_ = 0*tol_;
for i = 1:length(tol_)
    [Q,t] = intadapt(f,0,4,tol_(i));
    err_(i) = I - Q;
    num_(i) = length(t);
end
table(tol_,err_,num_,'variablenames',{'tol','error','
    f_evals'})

ans =
  11x3 table
      tol          error        f_evals
     _____      _____    _____

     0.0001      -0.00041947      113
     1e-05        4.7898e-05      181
     1e-06        6.3144e-06      297
     1e-07       -6.6392e-07      489
     1e-08        7.1808e-08      757
     1e-09        1.2652e-08     1193
     1e-10       -8.4413e-10     2009
     1e-11        2.6126e-11     3157
     1e-12        4.0445e-11     4797
     1e-13       -1.9384e-12     7997
     1e-14        1.6165e-13    12609
```

As you can see, even though the errors are not less than the estimates, the two columns decrease in tandem. If we consider now the convergence not in $h$ (which is poorly defined) but in the number of nodes actually chosen, we come close to the fourth-order accuracy of the underlying Simpson scheme.

```
clf,  loglog(num_,abs(err_),'.-')
hold on
loglog(num_,0.01*(num_/num_(1)).^(-4),'--')
```

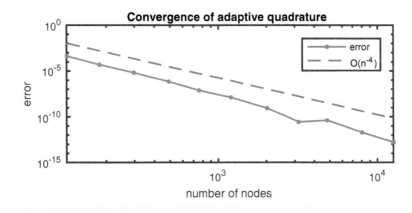

Although adaptivity and the error estimation that goes with it can be very powerful, they come at some cost. The error estimation cannot be universally perfect, so sometimes the answer will not be as accurate as requested (underestimation), and sometimes the function will be evaluated more times than necessary (overestimation). Subtle problems may arise when the integral is a step within a larger computation (see Exercise 5.7.6).

## Exercises

5.7.1. ▦ For each integral below, use Function 5.7.1 with error tolerance $10^{-2}$, $10^{-3}$, ..., $10^{-12}$. Make a table of errors and the number of integrand evaluation nodes used, and use a convergence plot as in Example 5.7.2 to compare to fourth-order accuracy. (These integrals were taken from [8].)

(a) $\int_0^1 x \log(1+x)\,dx = \dfrac{1}{4}$.

(b) $\int_0^1 x^2 \tan^{-1} x\,dx = \dfrac{\pi - 2 + 2\log 2}{12}$.

(c) $\int_0^{\pi/2} e^x \cos x\,dx = \dfrac{e^{\pi/2} - 1}{2}$.

(d) $\int_0^1 \sqrt{x} \log(x)\,dx = -\dfrac{4}{9}$. (Note: Although the integrand has the limiting value zero as $x \to 0$, you have to implement the function carefully to return zero as the value of $f(0)$, or start the integral at $x = \varepsilon_{\text{mach}}$.)

(e) $\int_0^1 \sqrt{1-x^2}\,dx = \dfrac{\pi}{4}$.

5.7.2. ▦ For each integral below, (i) use the built-in **integral** function to find the value to at least 12 digits; (ii) use Function 5.7.1 to evaluate the integral to a tolerance of $10^{-8}$; (iii) compute the absolute error and the number of nodes used; (iv) use the $O(h^2)$ term in the Euler–Maclaurin formula (5.6.5) to estimate how many nodes are required by the fixed-step-size trapezoidal formula to reach an absolute error of $10^{-8}$.

(a) $\int_{0.1}^{3} \operatorname{sech}(\sin(1/x))\,dx.$

(b) $\int_{-0.9}^{9} \ln((x+1)^3))\,dx.$

(c) $\int_{-\pi}^{\pi} \cos(x^3)\,dx.$

5.7.3. ▣ An integral such as $\int_0^1 x^{-\gamma}\,dx$ for $\gamma > 0$, in which the integrand blows up at one or both ends, is known as an *improper* integral. It has a finite value if $\gamma < 1$, despite the singularity. One way to deal with the problem of the infinite value for $f(t_0)$ is to replace the lower limit with a small number $\epsilon$. Using Function 5.7.1 with a small tolerance, make a log–log plot of the error as a function of $\epsilon$ for $\epsilon = 10^{-15}, 10^{-16}, \ldots, 10^{-45}$. (A more robust way to handle improper integrals is discussed in Section 9.7.)

5.7.4. ▣ A curious consequence of our logic in Function 5.7.1 is that the algorithm uses what we believe to be a more accurate, sixth-order answer only for estimating error; the returned value is the supposedly less accurate $S_f(2n)$. The practice of returning the extrapolated $R_f(4n)$ instead is called *local extrapolation*. Modify Function 5.7.1 to use local extrapolation and repeat Exercise 5.7.1. Is the convergence more like fourth order, or sixth order?

5.7.5. ▣ The *sine integral function* is defined by

$$\operatorname{Si}(x) = \int_0^x \frac{\sin z}{z}\,dz.$$

Use Function 5.7.1 to plot Si over the interval $[1, 10]$. Note: You will need to replace the lower bound of integration by $\varepsilon_{\text{mach}}$.

5.7.6. ▣ Adaptive integration can have subtle drawbacks. This exercise is based on the *error function*, a smooth function defined as

$$\operatorname{erf}(x) = \frac{2}{\pi} \int_0^x e^{-s^2}\,ds.$$

(a) Define a function $g$ that approximates erf by applying Function 5.6.1 with $n = 100$. Make a plot of the error $g(x) - \operatorname{erf}(x)$ at 300 points in the interval $[0,3]$.

(b) Define another approximation $h$ that applies Function 5.7.1 with error tolerance $10^{-7}$. Plot the error in $h$ as in part (a). Why does it look so different from the previous case?

(c) Suppose you wished to find $x$ such that $\operatorname{erf}(x) = 0.95$ by using rootfinding on one of your two approximations. Which would be preferable?

## Key ideas in this chapter

1. Polynomial interpolation on equally spaced nodes becomes unstable as the degree increases (page 176).
2. The conditioning of interpolation can be assessed by studying the cardinal basis functions of the interpolant (page 180).

3. Expressing an interpolant using a cardinal basis, such as the hat functions, is trivial (page 184).
4. Piecewise linear interpolation, essentially connecting the data with line segments, is second-order accurate (page 187).
5. A cubic spline is a piecewise cubic function that has two continuous derivatives everywhere (page 190).
6. The not-a-knot cubic spline interpolant is fourth-order accurate (page 192).
7. In finite difference approximations to derivatives, the weights are independent of the function but are dependent on node location (page 196).
8. A finite difference formula can be derived by interpolating the function values and differentiating the interpolant (page 198).
9. Higher-order finite difference methods are both more efficient and less vulnerable to roundoff than low-order methods (page 207).
10. The weights for numerical integration are independent of the function to be integrated but dependent on node location (page 209).
11. The trapezoid formula for integration results from integration of the piecewise linear interpolant (page 210).
12. Extrapolation uses a known series expansion for the truncation error to knock out the leading terms in it (page 212).

## Where to learn more

The algorithmic possibilities for piecewise linear and cubic spline approximation are explored in a different way in Van Loan [75]. In that source, a binary search is used to find the interval for evaluating the piecewise polynomial interpolant.

Further details regarding the derivation of the cubic spline equations, with an emphasis on minimizing memory usage, may be found in a number of sources, e.g., Burden and Faires [14], Cheney and Kincaid [17], and Atkinson and Han [5]. Comprehensive theoretical results can be found in de Boor [24].

On a historical note, Carl De Boor was elected to several National Academies of Science (USA, Poland, and Germany, e.g.) and was awarded the (USA) National Medal of Science in 2003 for his work on splines. Splines continue to be important for computer-aided design and computer graphics, among other applications.

An excellent and pragmatic introduction to finite difference methods can be found in Fornberg [25]. Numerical integration is a large topic unto itself; one longer introduction to it is [23].

# Chapter 6
# Initial-value problems for ODEs

> Without precise calculations we could fly right through a star or bounce too close to a supernova and that'd end your trip real quick, wouldn't it?
> —Han Solo, *Star Wars: A New Hope*

Quantities that change continuously in time or space are often modeled by differential equations. When everything depends on just one independent variable, we call the model an *ordinary differential equation* (ODE). Differential equations need supplemental conditions to define both the modeling situation and the theoretical solutions uniquely. The **initial-value problem**, in which all of the conditions are given at a single value of the independent variable, is the simplest situation. Very often, the independent variable in this case represents time.

Methods for IVPs usually start from the known initial value and iterate or "march" forward from there. There is a large number of them, owing in part to differences in accuracy, stability, and convenience. The most broadly important methods fall into one of two camps: *Runge–Kutta* and *linear multistep* formulas. Each type introduces its own complications, and we will consider them separately.

## 6.1 ▪ Basics of initial-value problems

The general form of a scalar, first-order **initial-value problem** (IVP) is

$$u'(t) = f(t, u(t)), \quad a \le t \le b, \tag{6.1.1a}$$
$$u(a) = u_0. \tag{6.1.1b}$$

We call $t$ the **independent variable** and $u$ the **dependent variable**. When $t$ is in fact meant to be time, sometimes we write $\dot{u}$ (read "u-dot") instead of $u'$. A solution of an initial-value problem is a function $u(t)$ that makes both $u'(t) = f(t, u(t))$ and $u(a) = u_0$ true equations.

> **Example 6.1.1**
>
> Suppose $u(t)$ is the size of a population at time $t$. We idealize by allowing $u$ to take any real (not just integer) value. If we assume a constant per capita birth rate (births per unit population per unit time), then
>
> $$\frac{du}{dt} = ku, \qquad u(0) = u_0$$
>
> for some $k > 0$. The solution is $u(t) = e^{kt} u_0$, which is exponential growth.
>
> A more realistic model would cap the growth due to finite resources. Suppose the death rate is proportional to the size of the population, indicating competition. Then
>
> $$\frac{du}{dt} = ku - ru^2, \qquad u(0) = u_0. \tag{6.1.2}$$
>
> This is the **logistic equation**. Although crude, it is still useful in population models. The solution relevant for population models has the form
>
> $$u(t) = \frac{k/r}{1 + \left(\frac{k}{ru_0} - 1\right)e^{-kt}}.$$
>
> For $k, r, u_0 > 0$, the solution smoothly varies from the initial population $u_0$ to a finite population, equal to $k/r$, that has been limited by competition.

If $u' = f(t, u) = g(t) + u h(t)$, the differential equation is **linear**. Linear problems can be solved in terms of standard integrals. Defining the *integrating factor* $\rho(t) = \exp\left[\int -h(t)\, dt\right]$, the solution is derived from

$$\rho(t) u(t) = u_0 + \int_a^t \rho(s) g(s)\, ds.$$

In many cases, however, the integrals cannot be done in closed form. When the differential equation is nonlinear, there is often no analytic formula available for its solution.

An ODE may have higher derivatives of the unknown solution present. For example, a **second-order ordinary differential equation** is often given in the form $u''(t) = f(t, u, u')$. A second-order IVP requires two conditions at the initial time in order to specify a solution completely. As we make clear in Section 6.3, we are always able to reformulate higher-order IVPs in a first-order form, so we will deal with first-order problems exclusively.[26]

---

[26]There are, however, some numerical methods for second-order problems that are important to certain application areas.

## 6.1. Basics of initial-value problems

### Numerical solutions

MATLAB has many built-in solvers for IVPs, all expecting the same input and returning similar types of output. One of the best-known all-purpose solvers is called **ode45**.

#### Example 6.1.2

The equation $u' = \sin[(u + t)^2]$ also has a solution that can be found numerically with ease, even though no formula exists for its solution.

```
f = @(t,u) sin( (t+u).^2 );
[t,u] = ode45(f,[0,4],-1);
plot(t,u)
```

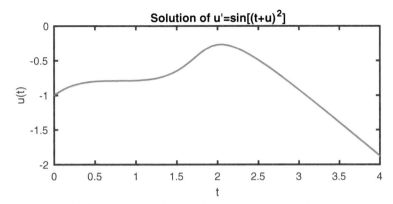

In some cases you may want solution values at times other than those that were automatically selected. While we could construct an interpolant using the methods of Chapter 5, there are two alternative ways to use the MATLAB solvers that give you this control and save you the extra work.

#### Example 6.1.3

We return to the equation $u' = \sin[(u + t)^2]$ to see two variations on how to obtain numerical solutions for it. In the first one, we can supply a vector of time nodes and receive the solution at exactly those times.

```
f = @(t,u) sin( (t+u).^2 );
t = linspace(0,4,60)';
[t,u] = ode45(f,t,-1);
plot(t,u,'.')
```

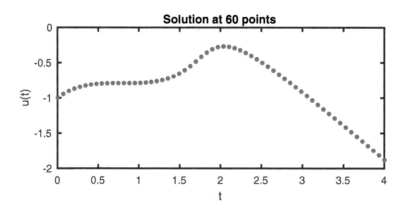

In the second variation, we see that it's also possible to create a callable function for the solution. Essentially this is a high-quality interpolant of the computed particular values.

```
u = @(t) deval( ode45(f,[0,4],-1), t );
fplot(u,[0,4])
```

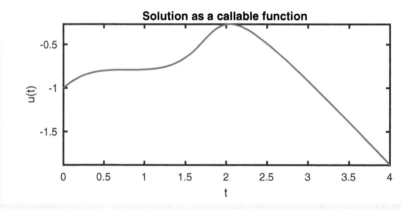

## Existence and uniqueness

As with linear systems of equations, there are cases where an IVP has no solution or has more than one solution, though the details are much different.

> **Example 6.1.4**
>
> The functions $u(t) = u^2$ and $u(t) \equiv 0$ both satisfy the differential equation $u' = 2\sqrt{u}$ and the initial condition $u(0) = 0$. Thus the corresponding IVP has more than one solution.

## 6.1. Basics of initial-value problems

### Example 6.1.5

The equation $u' = (u+t)^2$ gives us some trouble.

```
f = @(t,u) (t+u).^2;
[t,u] = ode45(f,[0,1],1);
semilogy(t,u)
xlabel('t'), ylabel('u(t)'), title('Finite time blowup')
```

```
Warning: Failure at t=7.853789e-01.  Unable to meet
    integration tolerances
without reducing the step size below the smallest value
    allowed (1.776357e-15)
at time t.
```

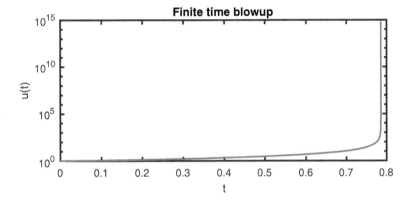

The warning message we received can mean that there is a bug in the formulation of the problem. But if everything has been done correctly, it suggests that the solution does not exist past the indicated time.

The following standard theorem gives us a condition that is easy to check and guarantees that a unique solution exists. But it is not the most general possible such condition, so there are problems with a unique solution that it cannot detect. We state the theorem without proof.

### Theorem 6.1.1

If the derivative $\frac{\partial f}{\partial u}$ exists and $\left|\frac{\partial f}{\partial u}\right|$ is bounded by a constant $L$ for all $a \le t \le b$ and all $u$, then the IVP (6.1.1) has a unique solution for $t \in [a,b]$.

## Conditioning of first-order IVPs

In a numerical context we have to be concerned about the conditioning of the IVP. There are two key items in (6.1.1) that we might consider to be the data of the initial-value ODE problem: the function $f(t,u)$, and the initial value $u_0$. It's easier to discuss perturbations to numbers than to functions, so we will focus on the effect of $u_0$ on the solution, using the following theorem that we give without proof. Happily, its conditions are identical to those in Theorem 6.1.1.

> **Theorem 6.1.2**
>
> If the derivative $\frac{\partial f}{\partial u}$ exists and $\left|\frac{\partial f}{\partial u}\right|$ is bounded by a constant $L$ for all $a \leq t \leq b$ and all $u$, then the solution $u(t; u_0 + \delta)$ of $u' = f(t,u)$ with initial condition $u(0) = u_0 + \delta$ satisfies
>
> $$\|u(t; u_0 + \delta) - u(t; u_0)\|_\infty \leq |\delta| e^{L(b-a)} \qquad (6.1.3)$$
>
> for all sufficiently small $|\delta|$.

Numerical solutions of IVPs have errors, and those errors can be seen as perturbations to the solution. Theorem 6.1.2 gives an upper bound of $e^{L(b-a)}$ on the infinity norm (i.e., pointwise) absolute condition number of the solution with respect to perturbations at an initial time. However, the upper bound may be a terrible overestimate of the actual sensitivity for a particular problem.

> **Example 6.1.6**
>
> Consider the ODEs $u' = u$ and $u' = -u$. In each case we compute $\partial f / \partial u = \pm 1$, so the condition number bound is $e^{(b-a)}$ in both problems. However, they behave quite differently. In the case of exponential growth, $u' = u$, the bound is the actual condition number.
>
> ```
> for u0 = [0.7 1 1.3]
>     fplot(@(t) exp(t)*u0,[0 3]), hold on
> end
> ```
>
>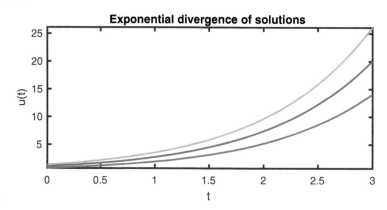

But with $u' = -u$, solutions actually get closer together with time.

```
clf
for u0 = [0.7 1 1.3]
    fplot(@(t) exp(-t)*u0,[0 3]), hold on
end
```

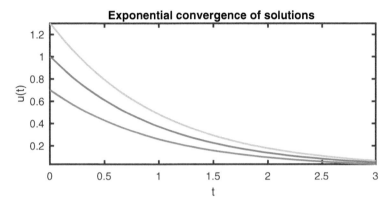

In this case the actual condition number is one, due to the difference of solutions at the initial time.

In general, solutions can diverge from, converge to, or oscillate around the original trajectory in response to perturbations. We won't fully consider these behaviors and their implications for numerical methods again until Section 11.4.

## Exercises

6.1.1. ✍ For each IVP, determine whether the problem satisfies the conditions of Theorems 6.1.1 and 6.1.2. If so, determine the smallest possible value for $L$.

(a) $f(t,u) = 3u$, $0 \le t \le 1$.

(b) $f(t,u) = -t \sin(u)$, $0 \le t \le 5$.

(c) $f(t,u) = -(1+t^2)u^2$, $1 \le t \le 3$.

(d) $f(t,u) = \sqrt{u}$, $0 \le t \le 1$.

6.1.2. ⌨ Solve each IVP in the preceding problem with MATLAB's built-in function ode45 using default settings, and make a plot of the solution.

6.1.3. ✍ Use an integrating factor to find the solution of each problem in analytic form.

(a) $u' = -tu$, $0 \le t \le 5$, $u(0) = 2$.

(b) $u' - 3u = e^{-2t}$, $0 \le t \le 1$, $u(0) = 5$.

6.1.4. ✍ Consider the IVP $u' = u^2$, $u(0) = \alpha$.

(a) Does Theorem 6.1.1 apply to this problem?

(b) Show that $u(t) = \alpha/(1-\alpha t)$ is a solution of the IVP.

(c) Does this solution necessarily exist for all $t \in [0,1]$?

6.1.5. Using **ode45** in MATLAB, compute solutions $x(t)$ to the logistic equation with harvesting,

$$x' = k(S-x)(x-M), \qquad 0 \le t \le 10,$$

using $k = S = 1$ and $M = 0.25$ and for the initial conditions $x(0) = 0.9M, 1.1M, 1.5M, 0.9S, 1.1S, 3S$. Show all the solutions together on one plot. (Note: One of the solutions will throw a warning and fail to reach $t = 10$. You can plot it anyway, and make it look better with **ylim**; use ylim([0,3.5]) at the end to create a reasonable scale for the vertical axis.)

6.1.6. Using **ode45** in MATLAB, solve the IVP $u' = u\cos(u) + \cos(4t)$, $0 \le t \le 10$, $u(0) = u_0$ for $u_0 = -2, -1.5, -1, \ldots, 1.5, 2$. Plot all the solutions on a single graph. You should find that they all settle into one of two periodic oscillations. To two digits of accuracy, find the value of $u_0$ in $(-2, 2)$ at which the attracting solution changes.

6.1.7. Experimental evidence (see [49]) shows that a 300 mg oral dose of caffeine, such as might be found in a large mug of drip-brewed coffee, creates a concentration of about 8 $\mu$g/mL in blood plasma. This boost is followed by first-order kinetics with a half-life of about 6 hours (although this rate can vary a great deal from person to person). We can model the caffeine concentration due to one drink taken over half an hour via

$$x'(t) = -kx + C(t), \quad x(0) = 0,$$

where $k = \log(2)/6$ and

$$C(t) = \begin{cases} 16, & 0 \le t \le 0.5, \\ 0, & t > 0.5. \end{cases}$$

Use **ode45** to make a plot of the caffeine concentration for 12 hours. Then change $k = \log(2)/8$ (half-life of 8 hours) and plot the solution again.

6.1.8. A reasonable model of the velocity $v(t)$ of a skydiver is

$$\frac{dv}{dt} = -g + \frac{k}{m}v^2, \qquad v(0) = 0,$$

where $g = 9.8$ m/sec$^2$ is gravitational acceleration, $m$ is the mass of the skydiver with parachute, and $k$ quantifies the effect of air resistance. At the U.S. Air Force Academy, a training jump starts at about 1200 m and has $k = 0.4875$ for $t < 13$ and $k = 29.16$ for $t \ge 13$. (This is an oversimplification; see [45].) Find the time at which the skydiver reaches the ground. Keep in mind that the distance fallen up to time $t$ is $\int_0^t v(s)\,ds$, and use the output form of **ode45** that enables you to use **deval** as shown in Example 6.1.3.

## 6.2 • Euler's method

Let a first-order IVP be given in the form

$$u'(t) = f(t, u(t)), \quad a \leq t \leq b,$$
$$u(a) = u_0.$$

*We represent a numerical solution of an IVP by its values at a finite collection of nodes,* which for now we require to be equally spaced:

$$t_i = a + ih, \quad h = \frac{b-a}{n}, \quad i = 0, \ldots, n.$$

The number $h$ is called the **step size**.

Because we don't get exactly correct values of the solution at the nodes, we need to take some care with the notation. From now on we let $\hat{u}(t)$ denote the exact solution of the IVP. The approximate value at $t_i$ computed at the nodes by our numerical methods will be denoted by $u_i \approx \hat{u}(t_i)$. Because we are given the initial value $u(a) = u_0$ exactly, there is no need to distinguish whether we mean $u_0$ as the exact or the numerical solution.

Consider a piecewise linear interpolant to the (as yet unknown) values $u_0, u_1, \ldots, u_n$. Its derivative is piecewise constant with values

$$\frac{u_{i+1} - u_i}{t_{i+1} - t_i} = \frac{u_{i+1} - u_i}{h}, \quad t_i < t < t_{i+1},$$

where $i = 0, \ldots, n-1$. We can connect this derivative to the differential equation by following the model of $u' = f(t, u)$:

$$\frac{u_{i+1} - u_i}{h} = f(t_i, u_i), \quad i = 0, \ldots, n-1.$$

We could also view the left-hand side as a forward difference approximation to $u'(t)$ at $t = t_i$. Either way, we can rearrange to get

$$u_{i+1} = u_i + h f(t_i, u_i), \quad i = 0, \ldots, n-1. \tag{6.2.1}$$

Together with the starting value $u_0$ from the initial condition, this formula defines an iteration known as **Euler's method**. It is an explicit method because the answer at the new time level is explicitly given in terms of the older time level(s).

A basic implementation of Euler's method is shown in Function 6.2.1. Like the built-in integrator `ode45` that we encountered in Example 6.1.2, it expects the problem to be specified in the form of a function $f$ of two arguments, an interval defining the time domain, and an initial condition. It also requires the number of intervals $n$ defined by the nodes (or, equivalently, the number of steps in the iteration). The output of Function 6.2.1 is a vector of the nodes and a vector of approximate solution values at those nodes, again like `ode45`.

**Function 6.2.1 (eulerivp)** Euler's method for a scalar initial-value problem.

```
function [t,u] = eulerivp(dudt,tspan,u0,n)
% EULERIVP   Euler's method for a scalar initial-value problem.
% Input:
%    dudt     defines f in u'(t)=f(t,u).  (function)
%    tspan    endpoints of time interval (2-vector)
%    u0       initial value (scalar)
%    n        number of time steps (integer)
% Output:
%    t        selected nodes   (vector, length n+1)
%    u        solution values  (vector, length n+1)

a = tspan(1);  b = tspan(2);
h = (b-a)/n;
t = a + (0:n)'*h;
u = zeros(n+1,1);
u(1) = u0;
for i = 1:n
  u(i+1) = u(i) + h*dudt(t(i),u(i));
end
```

### Example 6.2.1

We consider the IVP $u' = \sin[(u+t)^2]$ over $0 \le t \le 4$, with $u(0) = -1$.

```
f = @(t,u) sin( (t+u).^2 );
a = 0;  b = 4;
u0 = -1;
[t,u] = eulerivp(f,[a,b],u0,20);
plot(t,u,'.-')
```

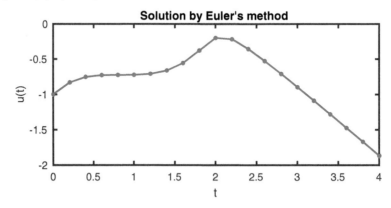

We could define a different interpolant to get a smoother picture above, but the derivation assumed the piecewise linear interpolant, so it is the most meaningful one. We can instead request more steps to make the interpolant look smoother.

```
[t,u] = eulerivp(f,[a,b],u0,200);
hold on, plot(t,u,'-')
```

## 6.2. Euler's method

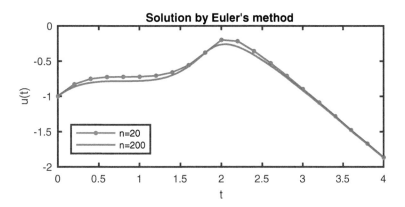

Increasing $n$ changed the solution noticeably. Since we know that interpolants and finite differences become more accurate as $h \to 0$, we should expect that from Euler's method too.

We don't have an exact solution to compare to, so we will use the built-in solver ode113 with settings chosen to construct an accurate solution.

```
opt = odeset('abstol',5e-14,'reltol',5e-14);
uhat = ode113(f,[a,b],u0,opt);
u_exact = @(t) deval(uhat,t)';
fplot(u_exact,[a,b],'k-')
```

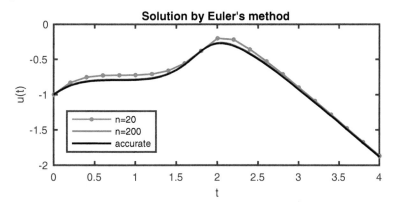

Now we can perform a convergence study.

```
n = 50*2.^(0:5)';
err = 0*n;
for j = 1:length(n)
    [t,u] = eulerivp(f,[a,b],u0,n(j));
    err(j) = max(abs(u_exact(t)-u));
end
table(n,err)

ans =
  6x2 table
     n         err
    ____    _____
     50      0.029996
```

```
  100    0.014229
  200    0.0069443
  400    0.0034295
  800    0.0017041
 1600    0.00084942
```

The error is almost perfectly halved at each step, so we expect that a log–log plot will reveal first-order convergence.

```
clf
loglog(n,err,'.-'), hold on
loglog(n,0.05*(n/n(1)).^(-1),'--')
```

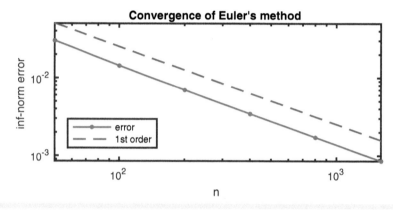

## Local truncation error

It should be clear that Euler's method cannot get the exact solution unless it happens to be piecewise linear. More generally, suppose that an oracle has granted us the exact solution at $t = t_i$, so that $u_i = \hat{u}(t_i)$. How would we then fare in obtaining $u_{i+1}$ as an approximation to $\hat{u}(t_{i+1})$? The answer is revealed through a Taylor series:

$$\begin{aligned}
\hat{u}(t_{i+1}) - \left[u_i + hf(t_i, u_i)\right] &= \hat{u}(t_{i+1}) - \left[\hat{u}(t_i) + hf(t_i, \hat{u}(t_i))\right] \\
&= \left[\hat{u}(t_i) + h\hat{u}'(t_i) + \tfrac{1}{2}h^2\hat{u}''(t_i) + \cdots\right] \\
&\quad - \left[\hat{u}(t_i) + h\hat{u}'(t_i)\right] \\
&= \tfrac{1}{2}h^2\hat{u}''(t_i) + O(h^3),
\end{aligned} \qquad (6.2.2)$$

where we used the fact that $\hat{u}$ satisfies the differential equation.

We formalize this calculation as follows. Euler's method may be written in the abstract form

$$u_{i+1} = u_i + h\phi(t_i, u_i, h), \qquad i = 0, \ldots, n-1, \qquad (6.2.3)$$

which we call a general **one-step method**. Euler's method is the particular case $\phi(t, u, h) = f(t, u)$, but we will see other one-step methods in future sections. *When we substitute the exact solution at $t = t_i$, calculate the resulting error at $t_{i+1}$, and divide by $h$, we get a quantity called the **local truncation error** (LTE) of the one-step formula.* In

## 6.2. Euler's method

the general one-step formula this is

$$\tau_{i+1}(h) := \frac{\hat{u}(t_{i+1}) - \hat{u}(t_i)}{h} - \phi(t_i, \hat{u}(t_i), h). \qquad (6.2.4)$$

Compared to (6.2.2) there is an extra division by $h$, which we explain below. First, though, note that in the limit $h \to 0$ in (6.2.4), we obtain

$$\hat{u}'(t_i) - \phi(t_i, \hat{u}(t_i), 0),$$

which through the ODE is the same as

$$f(t_i, \hat{u}(t_i)) - \phi(t_i, \hat{u}(t_i), 0).$$

It seems very reasonable to expect the LTE to vanish as the step size goes to zero for any ODE, which implies that $\phi(t, u, 0) = f(t, u)$ for any function $u$ whatsoever. This condition on the one-step formula is called **consistency**. It is trivially true for Euler's method.

### Convergence

The LTE measures the effect of a single step of the numerical method. It's straightforward to calculate from the formula, but the practical quantity of interest is the **global error**, $\hat{u}(t_i) - u_i$, over the entire time interval. By (6.2.4), $h\tau_{i+1}(h)$ describes how much error is made by taking a single step, starting from the exact value. If there were no other sources of or effects on the error, we would add up all of those local errors to get the global error.

To reach the time $t = b$ from $t = a$ with step size $h$, we need to take $n = (b-a)/h$ steps. If we want to reach, say, $t = (a+b)/2$, then we would have to take $n/2$ steps, and so on. The point is that to reach any fixed time in the interval, we need to take $O(n) = O(h^{-1})$ steps. That is why we express the error made in one step as $h\tau_{i+1}(h)$, with that extra factor of $h$ taken out. By this reasoning, for instance, the LTE of Euler computed in (6.2.2) implies a global error that is $O(h)$.

However, global error is not just a simple sum of local errors. As each step causes a perturbation of the solution, we jump from one solution curve to a new one. The new curve will have its own trajectory, i.e., the error will propagate through the ODE (see Example 6.1.6). This phenomenon is precisely the subject of Theorem 6.1.2: jumping to a different solution curve incurs a condition number at time $t > t_i$ of $e^{L(t-t_i)}$, which is constant at fixed time as $h \to 0$.

The following theorem puts our above observations on a rigorous footing.

> **Theorem 6.2.1**
>
> Suppose that the unit LTE of the one-step method (6.2.3) satisfies
>
> $$|\tau_{i+1}(h)| \leq Ch^p \qquad (6.2.5)$$

and that
$$\left|\frac{\partial \phi}{\partial u}\right| \leq L \tag{6.2.6}$$
for all $t \in [a,b]$, all $u$, and all $h > 0$. Then the global error satisfies
$$|\hat{u}(t_i) - u_i| \leq \frac{Ch^p}{L}\left[e^{L(t_i-a)} - 1\right] = O(h^p), \tag{6.2.7}$$
as $h \to 0$.

*Proof.* Define the global error sequence $E_i = \hat{u}(t_i) - u_i$. Using (6.2.3), we obtain
$$E_{i+1} - E_i = \hat{u}(t_{i+1}) - \hat{u}(t_i) - (u_{i+1} - u_i) = \hat{u}(t_{i+1}) - \hat{u}(t_i) - h\phi(t_i, u_i, h)$$
or
$$E_{i+1} = E_i + \hat{u}(t_{i+1}) - \hat{u}(t_i) - h\phi(t_i, \hat{u}(t_i), h) + h[\phi(t_i, \hat{u}(t_i), h) - \phi(t_i, u_i, h)].$$
We apply the triangle inequality, (6.2.4), and (6.2.5) to find
$$|E_{i+1}| \leq |E_i| + Ch^{p+1} + h|\phi(t_i, \hat{u}(t_i), h) - \phi(t_i, u_i, h)|.$$
The Fundamental Theorem of Calculus implies that
$$|\phi(t_i, \hat{u}(t_i), h) - \phi(t_i, u_i, h)| = \left|\int_{u_i}^{\hat{u}(t_i)} \frac{\partial \phi}{\partial u} du\right|$$
$$\leq \int_{u_i}^{\hat{u}(t_i)} \left|\frac{\partial \phi}{\partial u}\right| du$$
$$= \leq L|\hat{u}(t_i) - u_i| = L|E_i|.$$
Thus
$$|E_{i+1}| \leq Ch^{p+1} + (1+hL)|E_i|$$
$$\leq Ch^{p+1} + (1+hL)\left[Ch^{p+1} + (1+hL)|E_{i-1}|\right]$$
$$\vdots$$
$$\leq Ch^{p+1}\left[1 + (1+hL) + (1+hL)^2 + \cdots + (1+hL)^i\right].$$
To get the last line we applied the inequality recursively until reaching $E_0$, which is zero. Replacing $i+1$ by $i$ and simplifying the geometric sum, we get
$$|E_i| \leq Ch^{p+1}\frac{(1+hL)^i - 1}{(1+hL) - 1} = \frac{Ch^p}{L}\left[(1+hL)^i - 1\right].$$
We observe that $1 + x \leq e^x$ for $x \geq 0$ (see Exercise 6.2.6). Hence $(1+hL)^i \leq e^{ihL}$, which completes the proof. □

The theorem justifies a general definition of **order of accuracy** as the leading exponent of $h$ in $\tau_{i+i}(h)$: *the local truncation error of a one-step method has the same order of* *accuracy as the global error.* This agrees with the first-order convergence we observed experimentally for Euler in Example 6.2.1. Note, however, that the $O(h^p)$ convergence hides a leading constant that grows exponentially in time. When the time interval is bounded as $h \to 0$, this does not interfere with the conclusion, but the behavior as $t \to \infty$ contains no such guarantee.

## Exercises

6.2.1. ✍ Do two steps of Euler's method for the following problems using the given step size $h$. Then compute the error using the given exact solution.

(a) $u' = -2tu$, $u(0) = 2$; $h = 0.1$; $\hat{u}(t) = 2e^{-t^2}$.

(b) $u' = u + t$, $u(0) = 2$; $h = 0.2$; $\hat{u}(t) = -1 - t + 3e^t$.

(c) $tu' + u = 1$, $u(1) = 6$, $h = 0.25$; $\hat{u}(t) = 1 + 5/t$.

(d) $u' - 2u(1-u) = 0$, $u(0) = 1/2$, $h = 0.25$; $\hat{u}(t) = 1/(1 + e^{-2t})$.

6.2.2. 💻 For each IVP, solve the problem using Function 6.2.1. (i) Plot the solution for $n = 320$. (ii) For $n = 10 \cdot 2^k$, $k = 2, 3, \ldots, 10$, compute the error at the final time and make a log–log convergence plot, including a reference line for first-order convergence.

(a) $u' = -2tu$, $0 \le t \le 2$, $u(0) = 2$; $\hat{u}(t) = 2e^{-t^2}$.

(b) $u' = u + t$, $0 \le t \le 1$, $u(0) = 2$; $\hat{u}(t) = -1 - t + 3e^t$.

(c) $(1 + t^3)uu' = t^2$, $0 \le xt \le 3$, $u(0) = 1$; $\hat{u}(t) = [1 + (2/3)\ln(1 + xt^3)]^{1/2}$.

(d) $u' - 2u(1-u) = 0$, $0 \le t \le 2$, $u(0) = 1/2$; $\hat{u}(t) = 1/(1 + e^{-2t})$.

(e) $v' - (1 + x^2)v = 0$, $1 \le x \le 3$, $v(1) = 1$, $\hat{v}(x) = e^{-(\pi/4) + \arctan(x)}$.

(f) $v' + (1 + x^2)v^2 = 0$, $0 \le x \le 2$, $v(0) = 2$, $\hat{v}(x) = 1/(0.5 + \arctan(x))$.

(g) $u' = 2(1 + t)(1 + u^2)$, $0 \le t \le 0.5$, $u(0) = 0$, $\hat{u}(t) = \tan(2t + t^2)$.

6.2.3. 💻 For each IVP, compute the error at the final time in the numerical solution obtained by Euler's method with $n = 32$. Compare this error to the bound (6.2.7) with the smallest allowable value of $L$.

(a) $y' = -2ty$, $0 \le t \le 2$, $y(0) = 2$; $\hat{y}(t) = 2e^{-t^2}$.

(b) $y' = y + t$, $0 \le t \le 1$, $y(0) = 2$; $\hat{y}(t) = -1 - t + 3e^t$.

6.2.4. ✍ Here is an alternative to Euler's method:

$$v_{i+1} = u_i + hf(t_i, u_i),$$
$$u_{i+1} = u_i + hf(t_i + h, v_{i+1}).$$

(a) Write out the method explicitly in the general one-step form (6.2.3) (i.e., clarify what $\phi$ is for this method).

(b) Show that the method is consistent.

6.2.5. ✍ Consider the problem $u' = ku$, $u(0) = 1$ for constant $k$ and $t > 0$.

(a) Find an explicit formula in terms of $h$, $k$, and $i$ for the Euler solution $u_i$ at $t = ih$.

(b) Find values of $k$ and $h$ such that $|u_i| \to \infty$ as $i \to \infty$ while the exact solution $\hat{u}(t)$ is bounded as $t \to \infty$.

6.2.6. ⚠ Prove the fact, used in the proof of Theorem 6.2.1, that $1 + x \le e^x$ for all $x \ge 0$.

6.2.7. ⚠ Suppose that the error in making a step is also subject to roundoff error $\epsilon_{i+1}$, so that $\tau_{i+1}(h) = Ch^p + \epsilon_{i+1}h^{-1}$; assume that $|\epsilon_{i+1}| \le \epsilon$ is the largest roundoff error in the computation and that the initial condition is known exactly. Generalize Theorem 6.2.1 for this case.

## 6.3 • Systems of differential equations

Before we improve upon Euler's method, we will address a straightforward but critically important generalization of the problem. Very few applications involve an IVP with just a single dependent variable. Most of the time there are multiple unknowns and a system of equations to define them.

---

**Example 6.3.1**

Variations of the following model are commonly seen in ecology and epidemiology:

$$\begin{aligned} \frac{dy}{dt} &= y(1-\alpha y) - \frac{yz}{1+\beta y}, \\ \frac{dz}{dt} &= -z + \frac{yz}{1+\beta y}, \end{aligned} \quad (6.3.1)$$

where $\alpha$ and $\beta$ are positive constants. This model is a system of two differential equations for the unknown functions $y(t)$, which could represent a prey species or susceptible host, and $z(t)$, which could represent a predator species or infected population. We refer to this as a **predator–prey model**. Both of the equations involve both of the unknowns, with no clear way to separate them.

We can pack the two dependent variables $y$ and $z$ into a vector-valued function of time, $u(t)$, writing

$$\begin{aligned} u_1'(t) &= f_1(t, u) = u_1(1 - a u_1) - \frac{u_1 u_2}{1 + b u_1}, \\ u_2'(t) &= f_2(t, u) = -u_2 + \frac{u_1 u_2}{1 + b u_1}, \end{aligned}$$

and identifying $u_1 = y$, $u_2 = z$.

---

The generic form of a first-order system IVP is

$$u'(t) = f(t, u(t)), \quad a \le t \le b, \quad u(a) = u_0, \quad (6.3.2)$$

where all of the boldface quantities are vectors in $\mathbb{R}^m$ or $\mathbb{C}^m$. This is an IVP for a **first-order system of ODEs**, and the number $m$ of dependent variables (and equations) is called the **dimension** of the system. In particular, note that $f$ represents $m$ scalar functions in $m + 1$ scalar variables, including time.

## 6.3. Systems of differential equations

**Example 6.3.2**

Let $A(t)$ be an $m \times m$ matrix whose entries depend on $t$. Then

$$\frac{du}{dt} = A(t)u$$

is a **linear system** of differential equations. If the matrix $A$ is independent of time, it is a **linear, constant-coefficient system**.

The solution to a linear, constant-coefficient IVP, where also $u(0) = u_0$, is formally

$$u(t) = e^{tA} u_0, \qquad (6.3.3)$$

where $e^{tA}$ is a **matrix exponential**, which can be defined using Taylor series or by other means. This result is a seamless generalization of the scalar case, $m = 1$. MATLAB has a built-in function called **expm** for the matrix exponential. (*Be careful!* While exp(A) is a valid command, it's very different from expm(A).)

```
A = [ -2  5; -1  0 ]

A =
    -2     5
    -1     0


u0 = [1;0];
t = linspace(0,6,600);    % times for plotting
u = zeros(length(t),length(u0));
for j=1:length(t)
    ut = expm(t(j)*A)*u0;
    u(j,:) = ut';
end

plot(t,u)
```

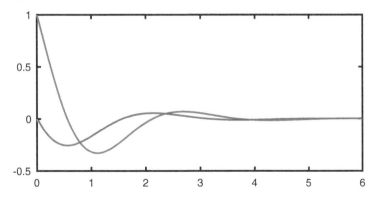

While the matrix exponential is a vital theoretical tool, computing it is too slow to be a practical numerical method.

The built-in IVP solvers all handle first-order systems. You must provide them with a function (either named or anonymous) that accepts the two arguments $t$ and $u$ and returns $f(t, u)$.

## Example 6.3.3

We encode the predator-prey equations via an included function. (This entire example was written in a function, not a script, in order to make this definition possible.)

```
alpha = 0.1;   beta = 0.25;
function dudt = predprey(t,u)
    y = u(1);   z = u(2);
    s = (y*z) / (1+beta*y);    % appears in both equations
    dudt = [ y*(1-alpha*y) - s;   -z + s ];
end
```

Note that the function must accept both t and u inputs, even though there is no explicit dependence on t. To solve the IVP we must also provide the initial condition, which is a 2-vector here, and the interval for the independent variable.

```
u0 = [1; 0.01];
t = linspace(0,80,2001)';        % specify where the time
    outputs will be
[t,u] = ode45(@predprey,t,u0);
```

Each row of the output u is the value of the solution vector at a single node in time. Each column of u is the entire history of one component of the solution.

```
size_u = size(u)
y = u(:,1);   z = u(:,2);
plot(t,y,t,z)
```

```
size_u =
        2001         2
```

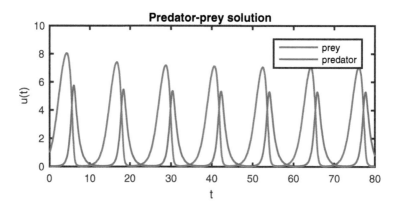

When there are just two components, it's common to plot the solution in the *phase plane*, i.e., with $u_1$ and $u_2$ along the axes and time as a parameterization of the curve.

```
plot(y,z)
```

## 6.3. Systems of differential equations

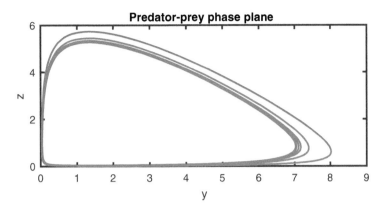

From this plot we can deduce that the solution approaches a periodic one, which in the phase plane is represented by a closed loop.

## Transformation of high-order systems

Fortunately the ability to solve first-order ODE systems implies the ability to solve (most practical) systems of higher differential order, too. The reason is that *there is a systematic way to turn a higher-order problem into a first-order one of higher dimension.*

### Example 6.3.4

Consider the nonlinear IVP

$$y'' + (1+y')^3 y = 0, \qquad y(0) = y_0, \quad y'(0) = 0.$$

In order to write this problem as a first-order system we define two scalar unknown functions, $u_1 = y$ and $u_2 = y'$. With these definitions, we have the two differential equations

$$u_1' = u_2,$$
$$u_2' = -(1+u_2)^3 u_1,$$

which is a first-order system in two dimensions. The initial condition of the system is

$$u_1(0) = y_0, \quad u_2(0) = 0.$$

### Example 6.3.5

Two identical pendula suspended from the same rod and swinging in parallel planes can be modeled as the second-order system

$$\theta_1''(t) + \gamma \theta_1' + \frac{g}{L} \sin\theta_1 + k(\theta_1 - \theta_2) = 0,$$
$$\theta_2''(t) + \gamma \theta_2' + \frac{g}{L} \sin\theta_2 + k(\theta_2 - \theta_1) = 0,$$

where $\theta_1$ and $\theta_2$ are angles made by the two pendula, $L$ is the length of each pendulum, $\gamma$ is a frictional parameter, and $k$ is a parameter describing a torque produced by the rod when it is twisted. We can convert this problem into a first-order system using the substitutions

$$u_1 = \theta_1, \quad u_2 = \theta_2, \quad u_3 = \theta_1', \quad u_4 = \theta_2'.$$

With these definitions the system becomes

$$u_1' = u_3,$$
$$u_2' = u_4,$$
$$u_3' = -\gamma u_3 - \frac{g}{L} \sin u_1 + k(u_2 - u_1),$$
$$u_4' = -\gamma u_4 - \frac{g}{L} \sin u_2 + k(u_1 - u_2),$$

which is a first-order system in four dimensions. To complete the description of the problem, you would need to specify values for $\theta_1(0)$, $\theta_1'(0)$, $\theta_2(0)$, and $\theta_2'(0)$.

The trick illustrated in the preceding examples is always available. Specifically, one introduces a new variable (that is, a component of $u$) for all but the highest derivative appearing for every variable of the original formulation. The surest way to get the transformation correct is to define one component of the new vector variable for each scalar initial condition given. Many equations for the first-order system then come from the trivial relationships among all the lower derivatives. The remaining equations for the system come from the original, high-order equations. In the end, there must be as many scalar component equations as unknown first-order variables.

## Methods for IVP systems

The generalization of a scalar IVP solver to handle systems is straightforward. Consider Euler's method, which in system form becomes

$$\boldsymbol{u}_{i+1} = \boldsymbol{u}_i + h\boldsymbol{f}(t_i, \boldsymbol{u}_i), \qquad i = 0, \ldots, n-1. \tag{6.3.4}$$

The vector difference equation (6.3.4) is just Euler's formula applied simultaneously to each component of the ODE system. The method is still explicit for the solution at the new time level.

Note here that as always in this book, an indexed boldface quantity such as $\boldsymbol{u}_i$ is a vector. If we want to refer to component $j$ of that vector, we would write $u_{i,j}$. We could also follow our convention and define an $m \times (n+1)$ matrix $\boldsymbol{U}$ whose columns are the $\boldsymbol{u}_i$. This matrix is the transpose of the one that is returned by the built-in IVP solvers, and our solvers follow the same convention. Function 6.3.1 shows our implementation of Euler's method for systems—it barely differs from Function 6.2.1 and completely replaces it.

In the rest of this chapter we present methods as though they are for scalar equations, but their application to systems is straightforward.[27] Our MATLAB functions, on the other hand, are written to accept systems.

---

[27] The generalization of error analysis can be more complicated, but our statements about order of accuracy and other properties are true for systems as well as scalars.

**Function 6.3.1 (eulersys)** Euler's method for a first-order IVP system.

```
function [t,u] = eulersys(dudt,tspan,u0,n)
% EULERSYS   Euler's method for a first-order IVP system.
% Input:
%   dudt     defines f in u'(t)=f(t,u) (function)
%   tspan    endpoints of time interval (2-vector)
%   u0       initial value (vector, length m)
%   n        number of time steps (integer)
% Output:
%   t        selected nodes   (vector, length n+1)
%   u        solution values  (array, (n+1)-by-m)

% Time discretization.
a = tspan(1);  b = tspan(2);
h = (b-a)/n;
t = a + (0:n)'*h;

% Initial condition and output setup.
m = length(u0);
u = zeros(m,n+1);
u(:,1) = u0(:);

% The time stepping iteration.
for i = 1:n
    u(:,i+1) = u(:,i) + h*dudt(t(i),u(:,i));
end

% This line makes the output conform to MATLAB conventions.
u = u.';
```

## Exercises

6.3.1. ⚠ Rewrite the given higher-order problems as first-order systems.

(a) $y''' - 3y'' + 3y' - y = t$, $y(0) = 1$, $y'(0) = 2$, $y''(0) = 3$.

(b) $y'' + 4(x^2 - 1)y' + y = 0$, $y(0) = 2$, $y'(0) = -1$.

(c) For a given constant $a$,

$$x'' + \frac{ax}{(x^2 + y^2)^{3/2}} = 0,$$

$$y'' + \frac{ay}{(x^2 + y^2)^{3/2}} = 0,$$

with initial values $x(0) = 1$, $x'(0) = y(0) = 0$, $y'(0) = 3$.

(d) $y^{(4)} - y = e^{-t}$, $y(0) = 0$, $y'(0) = 0$, $y''(0) = 1$, $y'''(0) = 0$.

(e) $y''' - y'' + y' - y = t$, $y(0) = 1$, $y'(0) = 2$, $y''(0) = 3$.

6.3.2. ⚠ Write the given IVP as a system. Then do two steps of Euler's method by hand (perhaps with a calculator) with the indicated step size $h$. Using the given exact solution, compute the error after the second step.

(a) $y'' + 4y = 4t$, $y(0) = 1$, $y'(0) = 1$; $\hat{y}(t) = t + \cos(2t)$, $h = 0.1$.

(b) $y'' - 4y = 4t$, $y(0) = 2$, $y'(0) = -1$; $\hat{y}(t) = e^{2t} + e^{-2t} - t$, $h = 0.1$.

(c) $2x^2y'' + 3xy' - y = 0$, $y(2) = 1$, $y'(2) = -1/2$, $\hat{y}(x) = 2/x$, $h = 1/8$.

(d) $2x^2y'' + 3xy' - y = 0$, $y(1) = 4$, $y'(1) = -1$, $\hat{y}(x) = 2(x^{1/2} + x^{-1})$, $h = 1/4$.

6.3.3. ▣ Solve the following IVPs using Function 6.3.1 using $n = 100$ steps. Plot $y(t)$ and $y'(t)$ as functions of time together on one plot, and plot the error in each component as functions of time on another.

(a) $y'' + 4y = 4t$, $0 < t < 2\pi$, $y(0) = 1$, $y'(0) = 1$; $y(t) = t + \cos(2t)$.

(b) $y'' + 9y = \sin(2t)$, $0 < t < 2\pi$, $y(0) = 2$, $y'(0) = 1$; $\hat{y}(t) = (1/5)\sin(3t) + 2\cos(3t) + (1/5)\sin(2t)$.

(c) $y'' - 4y = 4t$ $0 < t < \pi$, $y(0) = 2$, $y'(0) = -1$; $\hat{y}(t) = e^{2t} + e^{-2t} - t$.

(d) $y'' + 4y' + 4y = t$, $0 < t < 4$, $y(0) = 1$, $y'(0) = 3/4$; $\hat{y}(t) = (3t + 5/4)e^{-2t} + (t-1)/4$.

(e) $x^2y'' + 5xy' + 4y = 0$, $1 < x < e^2$, $y(1) = 0$, $y'(1) = 2$, $\hat{y}(x) = (2/x^2)\ln x$.

(f) $x^2y'' + 5xy' + 4y = 0$, $1 < x < e^2$, $y(1) = 1$, $y'(1) = -1$, $\hat{y}(x) = x^{-2}(1 + \ln x)$.

(g) $2x^2y'' + 3xy' - y = 0$, $2 < x < 20$, $y(2) = 1$, $y'(2) = -1/2$, $\hat{y}(x) = 2/x$.

(h) $2x^2y'' + 3xy' - y = 0$, $1 < x < 16$, $y(1) = 4$, $y'(1) = -1$, $\hat{y}(x) = 2(x^{1/2} + x^{-1})$.

(i) $x^2y'' - xy' + 2y = 0$, $1 < x < e^\pi$, $y(1) = 3$, $y'(1) = 4$, $\hat{y}(x) = x[3\cos(\ln x) + \sin(\ln x)]$.

(j) $x^2y'' + 3xy' + 4y = 0$, $e^{\pi/12} < x < e^\pi$, $y(e^{\pi/12}) = 0$, $y'(e^{\pi/12}) = -6$, $\hat{y}(x) = x^{-1}[3\cos(3\ln x) + \sin(3\ln x)]$.

6.3.4. ▣ A disease that is endemic to a population can be modeled by tracking the fraction of the population that is susceptible to infection, $v(t)$, and the fraction that is infectious, $w(t)$. (The rest of the population is considered to be recovered and immune.) A simple example model is (see [12])

$$\frac{dv}{dt} = 0.2(1-v) - 3vw, \qquad \frac{dw}{dt} = (3v-1)w.$$

Starting with $v(0) = 0.95$ and $w(0) = 0.05$, use `ode45` to find the long-term steady values of $v(t)$ and $w(t)$. Plot both components of the solution as functions of time.

6.3.5. ▣ Compute the solution to the systems for the given initial conditions using `ode45`. Plot your results as a curve in the phase plane (that is, with $x$ and $y$ as the axes of the plot).

(a) Using initial conditions with $x(0)^2 + y(0)^2$ both smaller and larger than 1 (inside and outside the unit circle), solve

$$x'(t) = -4y + x(1 - x^2 - y^2),$$
$$y'(t) = 4x + y(1 - x^2 - y^2),$$

starting at $0 < t < 10$. What is the final state of the system?

(b) Using initial conditions with $x(0)^2 + y(0)^2$ both inside and outside circles of radii 1 and 2, solve

$$x'(t) = -4y + x(1 - x^2 - y^2)(4 - x^2 - y^2),$$
$$y'(t) = 4x + y(1 - x^2 - y^2)(4 - x^2 - y^2),$$

starting at $0 < t < 10$. What is the final state of the system? Justify your answer with a plot showing the trajectories in the $(x,y)$ plane for a few different initial conditions.

6.3.6. The *Fitzhugh–Nagumo equations* are a simple model of the repeated firing of a neuron. They are given by

$$\frac{dv_1}{dt} = -v_1(v_1-1)(v_1-a) - v_2 + I,$$
$$\frac{dv_2}{dt} = \epsilon(v_1 - \gamma v_2).$$

Assume $v_1(0) = 0.5$, $v_2(0) = 0.1$, $a = 0.1$, $\epsilon = 0.008$, $\gamma = 1$. For each value of $I$ below, find and plot the solution using `ode45` for $0 \le t \le 600$. The solutions are highly sensitive to $I$, and you need to read the documentation on `odeset` to change the requested absolute and relative error tolerances to $10^{-9}$. Each time, the solution quickly approaches a periodic oscillation.

(a) $I = 0.05527$, (b) $I = 0.05683$, (c) $I = 0.0568385$, (d) $I = 0.05740$.

This exploration was carried out by Baer and Erneux [7].

## 6.4 • Runge–Kutta methods

We come now to one of the major and most-used types of methods for IVPs: **Runge–Kutta** (RK) methods.[28] They are one-step methods in the sense of (6.2.3), though they are not often written in that form. *RK methods boost the accuracy past first order by evaluating the ODE function $f(t,u)$ more than once per time step.*

### A second-order method

Consider a series expansion of the exact solution to $u' = f(t,u)$,

$$\hat{u}(t_{i+1}) = \hat{u}(t_i) + h\hat{u}'(t_i) + \frac{1}{2}h^2 \hat{u}''(t_i) + O(h^3). \tag{6.4.1}$$

If we replace $\hat{u}'$ by $f$ and keep only the first two terms on the right-hand side, we would obtain the Euler method. To get more accuracy we will need to compute or estimate the third term too. Note that

$$\hat{u}'' = f' = \frac{df}{dt} = \frac{\partial f}{\partial t} + \frac{\partial f}{\partial u}\frac{du}{dt} = f_t + f_u f,$$

where we have applied the multidimensional chain rule to the derivative, because both of the arguments to $f$ depend on $t$. Using this expression in (6.4.1), we obtain

$$\hat{u}(t_{i+1}) = \hat{u}(t_i) + h\left[ f(t_i, \hat{u}(t_i)) + \frac{h}{2} f_t(t_i, \hat{u}(t_i)) + \frac{h}{2} f(t_i, \hat{u}(t_i)) f_u(t_i, \hat{u}(t_i)) \right]$$
$$+ O(h^3). \tag{6.4.2}$$

---

[28] Americans tend to pronounce these German names as "run-ghuh kut-tah."

We have no desire to calculate and then code those partial derivatives of $f$ directly; an approximate approximation is called for. Observe that

$$f(t_i+\alpha,\hat{u}(t_i)+\beta) = f(t_i,\hat{u}(t_i))+\alpha f_t(t_i,\hat{u}(t_i))+\beta f_u(t_i,\hat{u}(t_i))+O(\alpha^2+|\alpha\beta|+\beta^2). \tag{6.4.3}$$

Matching this expression to the term in brackets in (6.4.2), it seems natural to select $\alpha = h/2$ and $\beta = \tfrac{1}{2}hf(t_i,\hat{u}(t_i))$. Doing so, we find

$$\hat{u}(t_{i+1}) = \hat{u}(t_i)+h\left[f(t_i+\alpha,\hat{u}(t_i)+\beta)\right]+O(h\alpha^2+h|\alpha\beta|+h\beta^2+h^3).$$

Truncating results in the one-step formula

$$u_{i+1} = u_i + hf\left(t_i+\tfrac{1}{2}h, u_i+\tfrac{1}{2}hf(t_i,u_i)\right), \qquad i=0,\ldots,n-1. \tag{6.4.4}$$

We will call this the **improved Euler** method. Thanks to the definitions above of $\alpha$ and $\beta$, the omitted terms are of size

$$O(h\alpha^2+h|\alpha\beta|+h\beta^2+h^3) = O(h^3).$$

Therefore $h\tau_{i+1} = O(h^3)$, and the order of accuracy of improved Euler is two. We will refer to it as **IE2**.

## Implementation

RK methods are called **multistage** methods. We can see why if we interpret (6.4.4) from the inside out. In the first stage, the method takes an Euler half-step to time $t_i + h/2$:

$$k_1 = hf(t_i, u_i),$$
$$v = u_i + \tfrac{1}{2}k_1.$$

The second stage employs an Euler-style strategy over the whole time step, but using the value from the first stage to get the slope, rather than using $f(t_i, w_i)$:

$$k_2 = hf\left(t_i+\tfrac{1}{2}h, v\right),$$
$$u_{i+1} = u_i + k_2.$$

A MATLAB implementation of IE2 is shown in Function 6.4.1.

## More Runge–Kutta methods

The idea of matching Taylor expansions can be generalized to higher orders of accuracy. To do so, however, we must introduce additional stages, each having free parameters so that more terms in the series may be matched. The amount of algebra grows rapidly in size and complexity, though there is a sophisticated theory for keeping track of it. We do not give the derivation details.

## 6.4. Runge–Kutta methods

**Function 6.4.1** (ie2) Improved Euler method for an IVP.

```
function [t,u] = ie2(dudt,tspan,u0,n)
% IE2     Improved Euler method for an IVP.
% Input:
%   dudt     defines f in u'(t)=f(t,u) (function)
%   tspan    endpoints of time interval (2-vector)
%   u0       initial value (vector, length m)
%   n        number of time steps (integer)
% Output:
%   t        selected nodes (vector, length N+1)
%   u        solution values (array, (n+1)-by-m)

% Time discretization.
a = tspan(1);  b = tspan(2);
h = (b-a)/n;
t = a + h*(0:n)';

% Initialize solution array.
u = zeros(length(u0),n+1);
u(:,1) = u0;

% Time stepping.
for i = 1:n
    uhalf = u(:,i) + h/2 * dudt(t(i),u(:,i));
    u(:,i+1) = u(:,i) + h * dudt(t(i)+h/2,uhalf);
end

u = u.';     % conform with MATLAB output convention
```

There are many known RK methods. We present a generic $s$-stage method in the form

$$
\begin{aligned}
k_1 &= hf(t_i, u_i), \\
k_2 &= hf(t_i + c_1 h, u_i + a_{11} k_1), \\
k_3 &= hf(t_i + c_2 h, u_i + a_{21} k_1 + a_{22} k_2), \\
&\vdots \\
k_s &= hf(t_i + c_{s-1} h, u_i + a_{s-1,1} k_1 + \cdots + a_{s-1,s-1} k_{s-1}), \\
u_{i+1} &= u_i + b_1 k_1 + \cdots + b_s k_s.
\end{aligned}
\tag{6.4.5}
$$

This recipe is completely determined by the number of stages $s$ and the constants $a_{ij}$, $b_j$, and $c_i$. Often an RK method is presented as just a table of these numbers, as in

| | | | | | |
|---:|---|---|---|---|---|
| 0 | | | | | |
| $c_1$ | $a_{11}$ | | | | |
| $c_2$ | $a_{21}$ | $a_{22}$ | | | |
| $\vdots$ | $\vdots$ | | $\ddots$ | | |
| $c_{s-1}$ | $a_{s-1,1}$ | $\cdots$ | | $a_{s-1,s-1}$ | |
| | $b_1$ | $b_2$ | $\cdots$ | $b_{s-1}$ | $b_s$ |

For example, IE2 is given by

$$\begin{array}{c|cc} 0 & & \\ \frac{1}{2} & \frac{1}{2} & \\ \hline & 0 & 1 \end{array}$$

Here are two more 2-stage, second-order methods, **modified Euler** and **Heun's**, respectively:

$$\begin{array}{c|cc} 0 & & \\ 1 & 1 & \\ \hline & \frac{1}{2} & \frac{1}{2} \end{array} \qquad \begin{array}{c|cc} 0 & & \\ \frac{2}{3} & \frac{2}{3} & \\ \hline & \frac{1}{4} & \frac{3}{4} \end{array}$$

The most commonly used RK method, and perhaps the most popular IVP method of

---

**Function 6.4.2 (rk4)** Fourth-order Runge–Kutta for an IVP.

```
function [t,u] = rk4(dudt,tspan,u0,n)
% RK4     Fourth-order Runge-Kutta for an IVP.
% Input:
%   dudt     defines f in u'(t)=f(t,u)   (function)
%   tspan    endpoints of time interval (2-vector)
%   u0       initial value (vector, length m)
%   n        number of time steps (integer)
% Output:
%   t        selected nodes (vector, length n+1)
%   u        solution values (array, (n+1) by m)

% Define time discretization.
a = tspan(1);  b = tspan(2);
h = (b-a)/n;
t = a + (0:n)'*h;

% Initialize solution array.
u = zeros(length(u0),n+1);
u(:,1) = u0(:);

% Time stepping.
for i = 1:n
  k1 = h*dudt( t(i),     u(:,i) );
  k2 = h*dudt( t(i)+h/2, u(:,i)+k1/2 );
  k3 = h*dudt( t(i)+h/2, u(:,i)+k2/2 );
  k4 = h*dudt( t(i)+h,   u(:,i)+k3 );
  u(:,i+1) = u(:,i) + (k1 + 2*(k2 + k3) + k4)/6;
end

u = u.';    % conform to MATLAB output convention
```

## 6.4. Runge–Kutta methods

all, is the fourth-order one given by

$$
\begin{array}{c|cccc}
0 & & & & \\
\frac{1}{2} & \frac{1}{2} & & & \\
\frac{1}{2} & 0 & \frac{1}{2} & & \\
1 & 0 & 0 & 1 & \\
\hline
& \frac{1}{6} & \frac{1}{3} & \frac{1}{3} & \frac{1}{6}
\end{array}
\tag{6.4.6}
$$

This formula is often called "the" fourth-order RK method—even though there are others—and we shall refer to it as RK4. Written out, the recipe is

$$
\begin{aligned}
k_1 &= hf(t_i, u_i), \\
k_2 &= hf(t_i + h/2, u_i + k_1/2), \\
k_3 &= hf(t_i + h/2, u_i + k_2/2), \\
k_4 &= hf(t_i + h, u_i + k_3), \\
u_{i+1} &= u_i + \frac{1}{6}k_1 + \frac{1}{3}k_2 + \frac{1}{3}k_3 + \frac{1}{6}k_4.
\end{aligned}
\tag{6.4.7}
$$

An implementation is given in Function 6.4.2.

### Example 6.4.1

We consider the IVP $u' = \sin[(u+t)^2]$ over $0 \le t \le 4$, with $u(0) = -1$.

```
f = @(t,u) sin( (t+u).^2 );
a = 0;   b = 4;
u0 = -1;
```

We use the built-in solver ode113 to construct an accurate approximation to the exact solution.

```
opt = odeset('abstol',5e-14,'reltol',5e-14);
uhat = ode113(f,[a,b],u0,opt);
u_exact = @(t) deval(uhat,t)';
```

Now we perform a convergence study of our two Runge–Kutta implementations.

```
n = 50*2.^(0:5)';
err_IE2 = 0*n;
err_RK4 = 0*n;
for j = 1:length(n)
    [t,u] = ie2(f,[a,b],u0,n(j));
    err_IE2(j) = max(abs(u_exact(t)-u));
    [t,u] = rk4(f,[a,b],u0,n(j));
    err_RK4(j) = max(abs(u_exact(t)-u));
end
```

The amount of computational work at each time step is assumed to be proportional to the number of stages. Let's compare on an apples-to-apples basis by using the number of $f$-evaluations on the horizontal axis.

```
loglog(2*n,err_IE2,'.-')
hold on, loglog(4*n,err_RK4,'.-')
loglog(2*n,0.01*(n/n(1)).^(-2),'--')
loglog(4*n,1e-6*(n/n(1)).^(-4),'--')
```

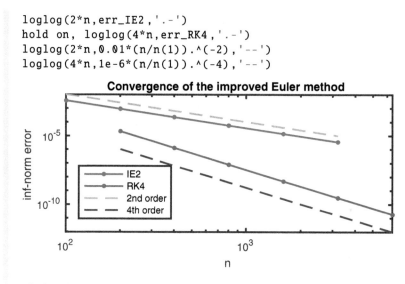

The fourth-order variant is more efficient in this problem over a wide range of accuracy.

## Efficiency

As with rootfinding and integration, the usual point of view is that evaluations of $f$ are the only significant computations and are therefore to be minimized in number. One of the most important characteristics of a multistage method is that each stage requires an evaluation of $f$; that is, a single time step of an $s$-stage method requires $s$ evaluations of $f$.

The error decreases *geometrically* as the number of stages grows algebraically, so trading a stage for an increase in order is a good deal. But $s = 5, 6,$ or $7$ gives a maximal order of accuracy of $s - 1$; this decreases to $s - 2$ for $s = 8$ and $s = 9$, etc. Fourth order is considered adequate and the "sweet spot" for many applications.

## Exercises

6.4.1. ⚠ For each IVP, write out (possibly using a calculator) the first time step of the improved Euler method with $h = 0.2$.

    (a) $u' = -2tu$, $0 \le t \le 2$, $u(0) = 2$; $\hat{u}(t) = 2e^{-t^2}$.

    (b) $u' = u + t$, $0 \le t \le 1$, $u(0) = 2$; $\hat{u}(t) = -1 - t + 3e^t$.

    (c) $(1 + x^3)uu' = x^2$, $0 \le x \le 3$, $u(0) = 1$; $\hat{u}(x) = [1 + (2/3)\ln(1 + x^3)]^{1/2}$.

6.4.2. ⚠ Use the modified Euler method to solve Exercise 6.4.1.

6.4.3. ⚠ Use Heun's method to solve Exercise 6.4.1.

6.4.4. ⚠ Use RK4 to solve Exercise 6.4.1.

6.4.5. ⚠ Using (6.4.2) and (6.4.3), show that Heun's method has order of accuracy at least two.

6.4.6. ✍ Using (6.4.2) and (6.4.3), show that the modified Euler method has order of accuracy at least two.

6.4.7. 💻 For each IVP, compute the solution using Function 6.4.2. (i) Plot the solution for $n = 300$. (ii) For $n = 100, 200, 300, \ldots, 1000$, compute the error at the final time and make a log–log convergence plot, including a reference line for fourth-order convergence.

(a) $u'' + 9u = 9t$, $0 < t < 2\pi$, $u(0) = 1$, $u'(0) = 1$; $\hat{u}(t) = t + \cos(3t)$.

(b) $u'' + 9u = \sin(2t)$, $0 < t < 2\pi$, $u(0) = 2$, $u'(0) = 1$; $\hat{u}(t) = (1/5)\sin(3t) + 2\cos(3t) + (1/5)\sin(2t)$.

(c) $u'' - 9u = 9t$, $0 < t < 1$, $u(0) = 2$, $u'(0) = -1$; $\hat{u}(t) = e^{3t} + e^{-3t} - t$.

(d) $u'' + 4u' + 4u = t$, $0 < t < 4$, $u(0) = 1$, $u'(0) = 3/4$; $\hat{u}(t) = (3t + 5/4)e^{-2t} + (t - 1)/4$.

(e) $x^2 y'' + 5xy' + 4y = 0$, $1 < x < e^2$, $y(1) = 1$, $y'(1) = -1$, $\hat{y}(x) = x^{-2}(1 + \ln x)$.

(f) $2x^2 y'' + 3xy' - y = 0$, $1 < x < 16$, $y(1) = 4$, $y'(1) = -1$, $\hat{y}(x) = 2(x^{1/2} + x^{-1})$.

(g) $x^2 y'' - xy' + 2y = 0$, $1 < x < e^\pi$, $y(1) = 3$, $y'(1) = 4$, $\hat{y}(x) = x[3\cos(\ln x) + \sin(\ln x)]$.

(h) $x^2 y'' + 3xy' + 4y = 0$, $e^{\pi/12} < x < e^\pi$, $y(e^{\pi/12}) = 0$, $y'(e^{\pi/12}) = -6$, $\hat{y}(x) = x^{-1}[3\cos(3\ln x) + \sin(3\ln x)]$.

6.4.8. 💻 Do Exercise 6.3.4, but using Function 6.4.2 instead of ode45.

6.4.9. 💻 Do Exercise 6.3.6, but using Function 6.4.2 instead of ode45.

6.4.10. ✍ Consider the problem $u' = ku$, $u(0) = 1$ for constant $k$ and $t > 0$.

(a) Find an explicit formula in terms of $h$ and $k$ for $u_{i+1}/u_i$ in the modified Euler solution.

(b) Find values of $k$ and $h$ such that $|u_i| \to \infty$ as $i \to \infty$ while the exact solution $\hat{u}(t)$ is bounded as $t \to \infty$.

6.4.11. 💻 Modify Function 6.4.2 to implement Heun's method. Test your function on the problem in Exercise 6.4.7(f), showing that the error at $x = 16$ converges at second order.

## 6.5 • Adaptive Runge–Kutta

The derivation and analysis of methods for IVPs usually assumes a fixed step size $h$. While the error behavior $O(h^p)$ is guaranteed by Theorem 6.2.1 as $h \to 0$, this bound comes with an unknowable constant, and it is not very useful as a guide to the numerical value of the error at any particular value of $h$. Furthermore, as we saw in Section 5.7 with numerical integration, in many problems a fixed value of $h$ throughout $a \le t \le b$ is far from the most efficient strategy.

In response we will employ the basic strategy of Section 5.7: *adapt the time step size in order to reach an accuracy goal, as measured by an error estimate formed from computing multiple approximations.* The details are quite different, however.

## Error estimation

Suppose that, starting from a given value $u_i$ and using a step size $h$, we run one step of *two* RK methods simultaneously: one method with order $p$, producing $u_{i+1}$, and the other method with order $p+1$, producing $\tilde{u}_{i+1}$. In most circumstances, we can expect that $\tilde{u}_{i+1}$ is a much better approximation to the solution than $u_{i+1}$ is. So it seems reasonable to use $E_i(h) = |\tilde{u}_{i+1} - u_{i+1}|$ (in the vector case, a norm) as an estimate of the actual local error made by the $p$th-order method. If our goal is to keep error less than some predetermined value $\epsilon$, we could decide to accept the new solution value if $E_i < \epsilon$ and otherwise reject it. (Even though the estimate $E_i$ is meant to go with the *less* accurate proposed value $u_{i+1}$, it's hard to resist the temptation to keep the more accurate value $\tilde{u}_{i+1}$ instead, and this is common in practice.)

Regardless of whether $E_i < \epsilon$ and we accept the step, we now ask a question: looking back, what step size *should* we have taken to just meet our error target $\epsilon$? Let's speculate that $E_i(h) \approx Ch^{p+1}$ for an unknown constant $C$, given the behavior of local truncation error as $h \to 0$. If we had used a step size $qh$ for some $q > 0$, then, trivially, $E_i(qh) \approx Cq^{p+1}h^{p+1}$ is what we would expect to get. Our best guess for $q$ would be to set $E_i(qh) \approx \epsilon$, or

$$q \approx \left(\frac{\epsilon}{E_i}\right)^{1/(p+1)}. \tag{6.5.1}$$

Whether or not we accepted the value proposed for $t = t_{i+1}$, we will adjust the step size to $qh$ for the next attempted step.

Given what we know about the connection between local and global errors, we might instead decide that controlling the normalized contribution to *global* error, which is closer to $E_i(qh)/(qh)$, is more reasonable. Then we end up with

$$q \leq \left(\frac{\epsilon}{E_i}\right)^{1/p}. \tag{6.5.2}$$

Expert authors have different recommendations about whether to use (6.5.1) or (6.5.2). Even though (6.5.2) appears to be more in keeping with our assumptions about global errors, modern practice seems to favor (6.5.1).

## Embedded formulas

We have derived two useful pieces of information: a reasonable estimate of the actual value of the local (or global) error, and a prediction how the step size will affect that error. Together they can be used to adapt step size and keep errors near some target level. But there remains one more important twist to the story.

At first glance, it would seem that to use (for example) any pair of second- and third-order RK methods to get the $u_{i+1}$ and $\tilde{u}_{i+1}$ needed for adaptive error control, we need at least $2 + 3 = 5$ evaluations of $f(t, y)$ for each attempted time step. This is more than double the computational work needed by the second-order method without

## 6.5. Adaptive Runge–Kutta

adaptivity. Fortunately, the marginal cost of adaptation can be substantially reduced by using **embedded Runge–Kutta** formulas. *Embedded RK formulas are a pair of RK methods whose stages share the same internal f evaluations, combining them differently in order to get estimates of two different orders of accuracy.*

A good example of an embedded method is the **Bogacki–Shampine** (BS23) formula, given by the table

$$
\begin{array}{c|cccc}
0 & & & & \\
\frac{1}{2} & \frac{1}{2} & & & \\
\frac{3}{4} & 0 & \frac{3}{4} & & \\
1 & \frac{2}{9} & \frac{1}{3} & \frac{4}{9} & \\
\hline
 & \frac{2}{9} & \frac{1}{3} & \frac{4}{9} & 0 \\
\hline
 & \frac{7}{24} & \frac{1}{4} & \frac{1}{3} & \frac{1}{8}
\end{array}
\tag{6.5.3}
$$

The top part of the table describes four stages in the usual RK fashion. The last two rows describe how to construct a second-order estimate $u_{i+1}$ and a third-order estimate $\tilde{u}_{i+1}$ by taking different combinations of those stages.

Both the `ode23` and the `ode45` solvers built into MATLAB are based on embedded RK formulas.

## Implementation

Our implementation of an embedded second-/third-order (RK23) code is given in Function 6.5.1. It has a few details that are worth explaining.

First, as in (5.7.4), we use a combination of absolute and relative tolerances to judge the acceptability of a solution value. Second, we have a check whether $t_i + h$ equals $t_i$, which looks odd. This check is purely about roundoff error, because $h$ can become so small that it no longer changes the floating point value of $t_i$. When this happens, it's often a sign that the underlying exact solution has a singularity near $t = t_i$. Third, some adjustments are made to the step size prediction factor $q$. We use a smaller value than (6.5.1), to be conservative about the many assumptions that were made to derive it. We also prevent a huge jump in the step size for the same reason. And we make sure that our final step doesn't take us past the requested end of the domain.

Finally, there is some careful programming done to avoid redundant evaluations of $f$. As written in (6.5.3), there seem to be four stages needed to find the paired second- and third-order estimates. This is unfortunate, since there are three-stage formulas of order three. But BS23 has a special property called "first same as last" (FSAL). If the proposed step is accepted, the final stage computed in stepping from $t_i$ to $t_{i+1}$ is identical to the *first* stage needed to step from $t_{i+1}$ to $t_{i+2}$, so in that sense one of the stage evaluations comes at no cost. This detail is addressed in our code.

**Function 6.5.1 (rk23)** Adaptive IVP solver based on embedded RK formulas.

```
1   function [t,u] = rk23(dudt,tspan,u0,tol)
2   % RK23    Adaptive IVP solver based on embedded RK formulas.
3   % Input:
4   %    dudt     defines f in u'(t)=f(t,u)  (function)
5   %    tspan    endpoints of time interval (2-vector)
6   %    u0       initial value (vector, length m)
7   %    tol      global error target (positive scalar)
8   % Output:
9   %    t        selected nodes (vector, length n+1)
10  %    u        solution values (array, size m by n+1)
11
12  % Initialize for the first time step.
13  t = tspan(1);
14  u(:,1) = u0(:);    i = 1;
15  h = 0.5*tol^(1/3);
16  s1 = dudt(t(1), u(:,1));
17
18  % Time stepping.
19  while t(i) < tspan(2)
20
21      % Detect underflow of the step size.
22      if t(i)+h == t(i)
23          warning('Stepsize too small near t=%.6g.',t(i))
24          break  % quit time stepping loop
25      end
26
27      % New RK stages.
28      s2 = dudt( t(i)+h/2,   u(:,i)+(h/2)*s1   );
29      s3 = dudt( t(i)+3*h/4, u(:,i)+(3*h/4)*s2 );
30      unew2 = u(:,i) + h*(2*s1 + 3*s2 + 4*s3)/9;   % 2rd order solution
31      s4 = dudt( t(i)+h,     unew2 );
32      err = h*(-5*s1/72 + s2/12 + s3/9 - s4/8);    % 2nd/3rd order
                difference
33      E = norm(err,inf);                           % error estimate
34      maxerr = tol*(1 + norm(u(:,i),inf));  % relative/absolute blend
35
36      % Accept the proposed step?
37      if E < maxerr      % yes
38          t(i+1) = t(i) + h;
39          u(:,i+1) = unew2;
40          i = i+1;
41          s1 = s4;        % use FSAL property
42      end
43
44      % Adjust step size.
45      q = 0.8*(maxerr/E)^(1/3);      % conservative optimal step factor
46      q = min(q,4);                  % limit step size growth
47      h = min(q*h,tspan(2)-t(i));    % don't step past the end
48  end
49
50  t = t';  u = u.';   % conform to MATLAB output convention
```

## 6.5. Adaptive Runge–Kutta

**Example 6.5.1**

Let's run adaptive RK on $u' = e^{t-u\sin u}$.

```
f = @(t,u) exp(t-u*sin(u));
[t,u] = rk23(f,[0,5],0,1e-5);
plot(t,u,'.')
```

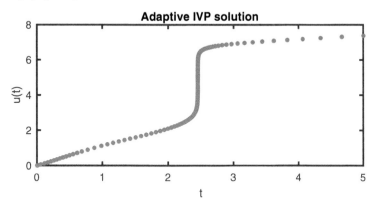

The solution makes a very abrupt change near $t = 2.4$. The resulting time steps vary over three orders of magnitude.

```
semilogy(t(1:end-1),diff(t))
```

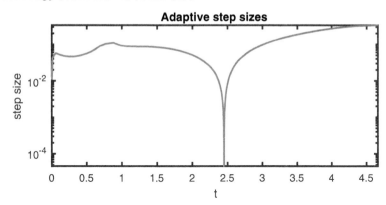

If we had to run with a uniform step size to get this accuracy, it would be

```
h_min = min(diff(t))
```

```
h_min =
    4.6097e-05
```

On the other hand, the average step size that was actually taken was

```
h_avg = mean(diff(t))
```

```
h_avg =
    0.0321
```

We took fewer steps by a factor of almost 1000! Even accounting for the extra stage per step and the occasional rejected step, the savings are clear.

### Example 6.5.2

In Example 6.1.5 we found an IVP that appears to blow up in a finite amount of time. Because the solution increases so rapidly as it approaches the blowup, adaptive stepping is required to even get close. In fact it's the failure of adaptivity that is used to get an idea of when the singularity occurs.

```
f = @(t,u) (t+u).^2;
warning on
[t,u] = rk23(f,[0,1],1,1e-5);
semilogy(t,u)
```

```
Warning: Stepsize too small near t=0.785409.
```

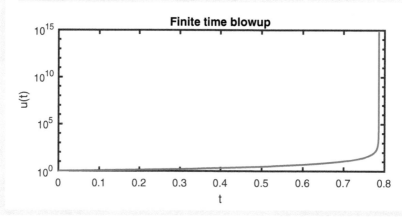

Often, the steps chosen adaptively clearly correspond to identifiable features of the solution. However, there are so-called *stiff problems* in which the time steps seem unreasonably small in relation to the observable behavior of the solution. These problems benefit from a particular type of solver and will be taken up in Section 6.7 and, in more mathematical detail, in Section 11.4.

## Exercises

6.5.1. ▣ Using Function 6.5.1, solve $y'' + (1+y')^3 y = 0$ over $0 \le t \le 4\pi$ with the indicated initial conditions. Plot $y(t)$ and $y'(t)$ as a function of $t$ and separately plot the solution curve parametrically in the phase plane—that is, the $(y(t), y'(t))$ plane.

(a) $y(0) = 0.1$, $y'(0) = 0$.
(b) $y(0) = 0.5$, $y'(0) = 0$.
(c) $y(0) = 0.75$, $y'(0) = 0$.
(d) $y(0) = 0.95$, $y'(0) = 0$.

6.5.2. Solve Exercise 6.1.7 using Function 6.5.1 with an error tolerance of $10^{-5}$. Plot the solution so that you can see the individual points. What is the smallest time step taken, and at what time does it occur?

6.5.3. Solve Exercise 6.3.6 using Function 6.5.1. Let the error tolerance be $10^{-k}$, increasing the integer $k$ until the graph of the solutions no longer changes. (This illustrates that the error tolerance is a request, not a guarantee!)

6.5.4. Derive equation (6.5.2) using the stated assumption about controlling global rather than local error.

6.5.5. Solve the problem $u' = u^2 - u^3$, $u(0) = 0.001$, $0 \le t \le 2000$ and make plots as in Example 6.5.1 that show both the solution and the time steps taken. Does the step size selection seem to be entirely explained by the local variability of the solution?

## 6.6 • Multistep methods

In RK methods we start at $u_i$ to find $u_{i+1}$, taking multiple $f$-evaluations (stages) to achieve high accuracy. In contrast, *multistep methods boost accuracy by employing more of the history of the solution*, taking information from time steps $i-1$, $i-2$, etc. For the discussion in this and following sections, we introduce the shorthand notation

$$f_i = f(t_i, u_i).$$

A $k$-step **multistep** (or linear multistep) method is given by the difference equation

$$u_{i+1} = a_{k-1}u_i + \cdots + a_0 u_{i-k+1} + h(b_k f_{i+1} + \cdots + b_0 f_{i-k+1}), \quad i = k-1, \ldots, n-1, \quad (6.6.1)$$

where the $a_j$ and the $b_j$ are constants. If $b_k = 0$, the method is **explicit**; otherwise, it is **implicit**. In order to use (6.6.1), we also need some way of generating the so-called **starting values**

$$u_1 = \alpha_1, \quad \ldots, \quad u_{k-1} = \alpha_{k-1},$$

which are otherwise undefined. In practice the starting values are often found using an RK formula.[29]

The difference formula (6.6.1) defines $u_{i+1}$ in terms of known values of the solution and its derivative from the past. In the explicit case with $b_k = 0$, equation (6.6.1) immediately gives a formula for the unknown quantity $u_{i+1}$ in terms of values at time level $t_i$ and earlier. Thus only one new evaluation of $f$ is needed to make a time step, provided that we store the recent history. For an implicit method, however, $b_k \ne 0$ and (6.6.1) has the form

$$u_{i+1} - h b_k f(t_{i+1}, u_{i+1}) = F(u_i, u_{i-1}, \ldots, u_{i-k+1}).$$

---
[29]This begs a question: if we must use an RK method anyway, why bother with multistep formulas at all? The answer is that multistep methods can be more efficient, even at the same order of accuracy.

**Table 6.1.** *Coefficients of Adams multistep formulas. All have $a_{k-1} = 1$ and $a_{k-2} = \cdots = a_0 = 0$. Alternate names are given in parentheses.*

| name/order | steps $k$ | $b_k$ | $b_{k-1}$ | $b_{k-2}$ | $b_{k-3}$ | $b_{k-4}$ |
|---|---|---|---|---|---|---|
| AB1 | 1 | 0 | 1 | (Euler) | | |
| AB2 | 2 | 0 | $\frac{3}{2}$ | $-\frac{1}{2}$ | | |
| AB3 | 3 | 0 | $\frac{23}{12}$ | $-\frac{16}{12}$ | $\frac{5}{12}$ | |
| AB4 | 4 | 0 | $\frac{55}{24}$ | $-\frac{59}{24}$ | $\frac{37}{24}$ | $-\frac{9}{24}$ |
| AM1 | 1 | 1 | (Backward Euler) | | | |
| AM2 | 1 | $\frac{1}{2}$ | $\frac{1}{2}$ | (Trapezoid) | | |
| AM3 | 2 | $\frac{5}{12}$ | $\frac{8}{12}$ | $-\frac{1}{12}$ | | |
| AM4 | 3 | $\frac{9}{24}$ | $\frac{19}{24}$ | $-\frac{5}{24}$ | $\frac{1}{24}$ | |
| AM5 | 4 | $\frac{251}{720}$ | $\frac{646}{720}$ | $-\frac{264}{720}$ | $\frac{106}{720}$ | $-\frac{19}{720}$ |

**Table 6.2.** *Coefficients of backward differentiation formulas. All have $b_k \neq 0$ and $b_{k-1} = \cdots = b_0 = 0$.*

| name/order | steps $k$ | $a_{k-1}$ | $a_{k-2}$ | $a_{k-3}$ | $a_{k-4}$ | $b_k$ |
|---|---|---|---|---|---|---|
| BD1 | 1 | 1 | (Backward Euler) | | | 1 |
| BD2 | 2 | $\frac{4}{3}$ | $-\frac{1}{3}$ | | | $\frac{2}{3}$ |
| BD3 | 3 | $\frac{18}{11}$ | $-\frac{9}{11}$ | $\frac{2}{11}$ | | $\frac{6}{11}$ |
| BD4 | 4 | $\frac{48}{25}$ | $-\frac{36}{25}$ | $\frac{16}{25}$ | $-\frac{3}{25}$ | $\frac{12}{25}$ |

Now the unknown $u_{i+1}$ that we seek appears inside the function $f$. In general this equation is a nonlinear rootfinding problem for $u_{i+1}$ and is not solvable in a finite number of steps by a formula. The implementation of both explicit and implicit multistep formulas is discussed in detail in Section 6.7.

As with RK formulas, a multistep method is entirely specified by the values of a few constants. Tables 6.1 and 6.2 present some of the most well-known and important formulas. The **Adams–Bashforth** (AB) methods are explicit, while the **Adams–Moulton** (AM) and **backward differentiation** (BD) formulas are implicit. The tables also list the methods' order of accuracy, to be defined shortly. We adopt the convention of referring to a multistep method by appending its order of accuracy to a two-letter name abbreviation, e.g., the "AB3 method."

There is a simple shorthand notation for a multistep method, the **generating polynomials**

$$\rho(z) = z^k - a_{k-1} z^{k-1} - \cdots - a_0, \tag{6.6.2}$$

$$\sigma(z) = b_k z^k + b_{k-1} z^{k-1} + \cdots + b_0. \tag{6.6.3}$$

For example, the AB3 method is completely specified by

$$\rho(z) = z^3 - z^2, \qquad \sigma(z) = \tfrac{1}{12}(23z^2 - 16z + 5).$$

In general, the polynomial $\rho(z)$ is monic (i.e., its leading term has a unit coefficient), and the degree of $\rho$ is the number of steps $k$. Furthermore, $\deg \sigma(z) = k$ for an implicit

## 6.6. Multistep methods

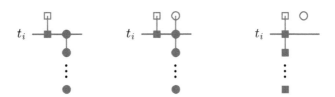

**Figure 6.1.** *Stencils of Adams and backward differentiation methods. Open shapes are unknowns for a single step, and the filled shapes are known quantities with nonzero coefficients. Boxes are values of the solution, while circles are values of its derivative.*

method and $\deg \sigma(z) < k$ for an explicit method. The connection with polynomials is straightforward, if a bit abstract. Let $\mathscr{Z}$ be a **forward-shift operator**, so that, for example, $\mathscr{Z} t_i = t_{i+1}$, $\mathscr{Z}^3 u_{i-1} = u_{i+2}$, etc. With this, the difference formula (6.6.1) can be written concisely as

$$\rho(\mathscr{Z})u_{i-k+1} = h\sigma(\mathscr{Z})f_{i-k+1}. \tag{6.6.4}$$

We can also draw a **stencil** to clarify what information is being used in each formula, as shown in Figure 6.1.

### Truncation and global error

The definition of local truncation error (LTE) is easily extended to multistep methods. As with RK, we plug the exact solution $\hat{u}$ into the difference formula and see what is left over, dividing by $h$ to account for the order difference between local and global errors. Thus the **local truncation error of a $k$-step multistep formula** is defined as

$$\tau_{i+1}(h) = \frac{\hat{u}(t_{i+1}) - a_{k-1}\hat{u}(t_i) - \cdots - a_0 \hat{u}(t_{i-k+1})}{h} \\ - \left[ b_k f(t_{i+1}, \hat{u}(t_{i+1})) + \cdots + b_0 f(t_{i-k+1}, \hat{u}(t_{i-k+1})) \right]. \tag{6.6.5}$$

The **order of accuracy** of the method is the leading (lowest) exponent of $h$ in the series expansion of $\tau_{i+1}(h)$ around $h = 0$. Although we shall not present the analysis, the conclusion for the multistep methods in this section is the same as for one-step methods: *the order of accuracy in the global error is the same as for the local truncation error.*

#### Example 6.6.1

The first-order Adams–Moulton method is also known as **backward Euler**, because its difference equation is

$$u_{i+1} = u_i + h f_{i+1},$$

which is equivalent to a backward difference approximation to $u'(t_{i+1})$. AM1 is characterized by $\rho(z) = z - 1$ and $\sigma(z) = z$.

To derive the LTE, we use the definition

$$h\tau_{i+1}(h) = \hat{u}(t_{i+1}) - \hat{u}(t_i) - hf\bigl(t_{i+1}, \hat{u}(t_{i+1})\bigr)$$

$$= \hat{u}(t_i) + h\hat{u}'(t_i) + \frac{h^2}{2}\hat{u}''(t_i) + O(h^3) - \hat{u}(t_i) - h\hat{u}'(t_{i+1})$$

$$= h\hat{u}'(t_i) + \frac{h^2}{2}\hat{u}''(t_i) + O(h^3) - h[\hat{u}'(t_i) + h\hat{u}''(t_i) + O(h^2)]$$

$$= -\frac{h^2}{2}\hat{u}''(t_i) + O(h^3).$$

Thus $\tau_{i+1}(h) = O(h)$ and AM1 (backward Euler) is a first-order method.

**Example 6.6.2**

The AB2 method has the formula

$$u_{i+1} = u_i + h\left(\frac{3}{2}f_i - \frac{1}{2}f_{i-1}\right).$$

The generating polynomials are $\rho(z) = z^2 - z$ and $\sigma(z) = (3z-1)/2$. We find that the method is second order from the LTE:

$$h\tau_{i+1}(h) = \hat{u}(t_{i+1}) - \hat{u}(t_i) - h\left[\frac{3}{2}f(t_i, \hat{u}(t_i)) - \frac{1}{2}f(t_{i-1}, \hat{u}(t_{i-1}))\right]$$

$$= \hat{u}(t_i) + h\hat{u}'(t_i) + \frac{h^2}{2}\hat{u}''(t_i) + \frac{h^3}{6}\hat{u}'''(t_i) + O(h^4)$$

$$\quad - \hat{u}(t_i) - \frac{3h}{2}\hat{u}'(t_i)$$

$$\quad + \frac{h}{2}[\hat{u}'(t_i) - h\hat{u}''(t_i) + \frac{h^2}{2}\hat{u}'''(t_i) + O(h^3)]$$

$$= \frac{5h^3}{12}\hat{u}'''(t_i) + O(h^4),$$

so that $\tau_{i+1}(h) = O(h^2)$.

## Derivation of the formulas

Where do coefficients like those in Table 6.1 come from? There are different ways to answer that question, but Adams and BD methods have distinctive stories to tell. The derivation of Adams methods begins with the observation that

$$\hat{u}(t_{i+1}) = \hat{u}(t_i) + \int_{t_i}^{t_{i+1}} \hat{u}'(t)\,dt = \hat{u}(t_i) + \int_{t_i}^{t_{i+1}} f(t, \hat{u}(t))\,dt. \qquad (6.6.6)$$

The integrand is unknown over the interval of integration. But we can approximate it by a polynomial interpolant by using the solution history. The polynomial can be integrated analytically, leading to a derivation of the coefficients $b_0, \ldots, b_k$.

### Example 6.6.3

Let's derive a one-step AM method using the two values $(t_i, f_i)$ and $(t_{i+1}, f_{i+1})$. The interpolating polynomial is the linear function

$$p(t) = f_i \frac{t_{i+1} - t}{t_{i+1} - t_i} + f_{i+1} \frac{t - t_i}{t_{i+1} - t_i}.$$

Things become a little clearer with the change of variable $s = t - t_i$, and using $h = t_{i+1} - t_i$:

$$\int_{t_i}^{t_{i+1}} p(t)\,dt = \int_0^h p(t_i + s)\,ds = h^{-1} \int_0^h [(h-s)f_i + s f_{i+1}]\,ds = \frac{h}{2}(f_i + f_{i+1}),$$

which explains the entries for AM2 in Table 6.1. The derivation also points out why this method is commonly called "trapezoid," because like the trapezoid formula for a definite integral, we compute the exact integral of a piecewise linear interpolant.

In AB methods, the interpolating polynomial has degree $k-1$, which means that its interpolation error is $O(h^k)$. Upon integrating we get a local error of $O(h^{k+1})$, which reduces to a global error of $O(h^k)$. The AM interpolating polynomial is one degree larger, so its order of accuracy is one higher for the same number of steps.

The idea behind backward difference formulas is complementary to that for Adams: Interpolate solution values $u_{i+1}, \ldots, u_{i-k+1}$ by a polynomial $q$, and then, motivated by $f(t, \hat{u}) = \hat{u}'(t)$, set

$$f_{i+1} = q'(t_{i+1}). \tag{6.6.7}$$

With some algebra, the quantity $q'(t_{i+1})$ can be expressed as a linear combination of the past solution values to get the coefficients in Table 6.2. In summary, Adams methods are based on local integration, and BD methods are based on local differentiation (i.e., finite differences).

## Exercises

6.6.1. ✍ For each method, write out the generating polynomials $\rho(z)$ and $\sigma(z)$, and draw the stencil of the method.

(a) AM2.  (b) AB2.  (c) BD2.  (d) AM3.  (e) AB3.

6.6.2. ✍ Write out by hand an equation that defines the first solution value $u_1$ produced by AM1 (backward Euler) for each IVP. (Reminder: This is an implicit formula.)

(a) $u' = -2tu$, $\quad 0 \le t \le 2$, $\quad u_0 = 2$, $\quad h = 0.2$.

(b) $u' = u + t$, $\quad 0 \le t \le 1$, $\quad u_0 = 2$, $\quad h = 0.1$.

(c) $(1 + x^3)uu' = x^2$, $\quad 0 \le x \le 3$, $\quad u_0 = 1,$, $\quad h = 0.5$.

6.6.3. ✍ Do the preceding problem for AM2 (trapezoid) instead of backward Euler.

6.6.4. ✍ For each method, find the leading term in the local truncation error using (6.6.5).

(a) AM2.  (b) AB2.  (c) BD2.

6.6.5. ✍ / 🖥 For each method, find the leading term in the local truncation error using (6.6.5). (Computer algebra is recommended.)

(a) AM3.  (b) AB3,.  (c) BD4.

6.6.6. ✍ A formula for the quadratic polynomial interpolant through the points $(s_1, y_1)$, $(s_2, y_2)$, and $(s_3, y_3)$ is

$$p(x) = \frac{(x-s_2)(x-s_3)}{(s_1-s_2)(s_1-s_3)} y_1 + \frac{(x-s_1)(x-s_3)}{(s_2-s_1)(s_2-s_3)} y_2 + \frac{(x-s_1)(x-s_2)}{(s_3-s_1)(s_3-s_2)} y_3.$$

(a) Use (6.6.6) and a polynomial interpolant through three points to derive the coefficients of the AM3 method.

(b) Use (6.6.7) and a polynomial interpolant through three points to derive the coefficients of the BD2 method.

6.6.7. ✍ By doing series expansion about the point $z = 1$, show for BD2 that

$$\frac{\rho(z)}{\sigma(z)} = \log(z-1) + O((z-1)^3).$$

6.6.8. ✍ / 🖥 By doing series expansion about the point $z = 1$, show for AB3 and AM3 that

$$\frac{\rho(z)}{\sigma(z)} = \log(z-1) + O((z-1)^4).$$

(Computer algebra is recommended.)

## 6.7 ▪ Implementation of multistep methods

We now consider some of the practical issues that arise when multistep formulas are used to solve IVPs. Implementation of the explicit case is relatively straightforward. In what follows we use boldface for the vector form of the problem, $\mathbf{u}' = \mathbf{f}(t, \mathbf{u})$. For instance, the explicit AB4 method is defined by the formula

$$\mathbf{u}_{i+1} = \mathbf{u}_i + \frac{h}{24}(55\mathbf{f}_i - 59\mathbf{f}_{i-1} + 37\mathbf{f}_{i-2} - 9\mathbf{f}_{i-3}), \quad i = 3, \ldots, n-1. \quad (6.7.1)$$

Function 6.7.1 shows a basic implementation of this formula. Observe that Function 6.4.2 is used to find the starting values $\mathbf{u}_1, \mathbf{u}_2, \mathbf{u}_3$ that are needed before the iteration formula takes over. As far as RK4 is concerned, it needs to solve the IVP over the time interval $a \leq t \leq a + 3h$, using a step size $h$ (the same step size as in the AB4 iteration). These values are then used by Function 6.7.1 to find $\mathbf{f}_0, \ldots, \mathbf{f}_3$ and get the main iteration started.

For each value of $i$ the formula uses the four most recently known values of the solution's derivative in order to advance by one step. In Function 6.7.1 only these values

## 6.7. Implementation of multistep methods

**Function 6.7.1 (ab4)** Fourth-order Adams–Bashforth formula for an IVP.

```
function [t,u] = ab4(dudt,tspan,u0,n)
%AB4    4th-order Adams-Bashforth formula for an IVP.
% Input:
%   dudt    defines f in u'(t)=f(t,u) (function)
%   tspan   endpoints of time interval (2-vector)
%   u0      initial value (m-vector)
%   n       number of time steps (integer)
% Output:
%   t       selected nodes (vector, length n+1)
%   u       solution values (array, size n+1 by m)

% Discretize time.
a = tspan(1);   b = tspan(2);
h = (b-a)/n;
t = tspan(1) + (0:n)'*h;

% Constants in the AB4 method.
k = 4;
sigma = [55; -59; 37; -9]/24;

% Find starting values by RK4.
u = zeros(length(u0),n+1);
[ts,us] = rk4(dudt,[a, a+(k-1)*h],u0,k-1);
u(:,1:k) = us(1:k,:).';

% Compute history of u' values, from oldest to newest.
f = zeros(length(u0),k);
for i = 1:k-1
  f(:,k-i) = dudt(t(i),u(:,i));
end

% Time stepping.
for i = k:n
  f = [dudt(t(i),u(:,i)), f(:,1:k-1)];    % new value of du/dt
  u(:,i+1) = u(:,i) + h*(f*sigma);        % advance one step
end

u = u.';    % conform to MATLAB output convention
```

of $f$ are stored, and a matrix-vector product is used for the linear combination implied in (6.7.1):

$$u_{i+1} = u_i + h \begin{bmatrix} f_i & f_{i-1} & f_{i-2} & f_{i-3} \end{bmatrix} \begin{bmatrix} 55/24 \\ -59/24 \\ 37/24 \\ -9/24 \end{bmatrix}. \qquad (6.7.2)$$

We have distributed the factor of $1/24$ in order to point out that the $4 \times 1$ constant vector is just the vector of coefficients of the generating polynomial $\sigma(z)$ from (6.6.3). At the start of an iteration, the value of $f$ at the most recent solution step is unknown, so a call is made to evaluate it, and the other columns are shifted to the right (i.e., into the past).

### Example 6.7.1

We consider the IVP $u' = \sin[(u+t)^2]$ over $0 \le t \le 4$, with $u(0) = -1$.

```
f = @(t,u) sin( (t+u).^2 );
a = 0;   b = 4;
u0 = -1;
```

We use the built-in solver ode113 to construct an accurate approximation to the exact solution.

```
opt = odeset('abstol',5e-14,'reltol',5e-14);
uhat = ode113(f,[a,b],u0,opt);
u_exact = @(t) deval(uhat,t)';   % callable function
```

Now we perform a convergence study of the AB4 code.

```
n = 10*2.^(0:5)';
err = 0*n;
for j = 1:length(n)
    [t,u] = ab4(f,[a,b],u0,n(j));
    err(j) = max(abs(u_exact(t)-u));
end
```

The method should converge as $O(h^4)$, so a log–log scale is appropriate for the errors.

```
loglog(n,err,'.-')
hold on, loglog(n,0.1*(n/n(1)).^(-4),'--')
```

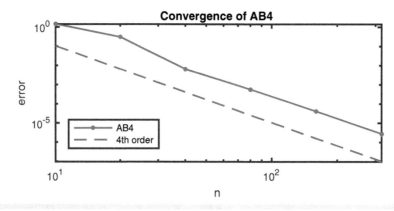

## Implicit methods

The implementation of an implicit multistep method is a bit more involved. Consider the second-order implicit formula AM2, also known as the trapezoid method. To advance from step $i$ to $i+1$, we need to solve

$$z - u_i - \tfrac{1}{2}h\big[f(t_i,u_i) + f(t_{i+1},z)\big] = 0 \qquad (6.7.3)$$

for $z$, and then set $u_{i+1} = z$. This equation takes the form $g(z) = 0$, so we have a rootfinding problem as in Chapter 4. An implementation of AM2 using Function 4.6.2

## 6.7. Implementation of multistep methods

**Function 6.7.2** (am2) Second-order Adams–Moulton (trapezoid) formula for an IVP.

```
function [t,u] = am2(dudt,tspan,u0,n)
% AM2     2nd-order Adams-Moulton (trapezoid) formula for an IVP.
% Input:
%   dudt     f(t,y) for the ODE (function)
%   tspan    endpoints of time interval (2-vector)
%   u0       initial value (m-vector)
%   n        number of time steps (integer)
% Output:
%   t        vector of times (vector, length n+1)
%   u        solution (array, size n+1 by m)

% Discretize time.
a = tspan(1);  b = tspan(2);
h = (b-a)/n;
t = tspan(1) + (0:n)'*h;

m = numel(u0);
u = zeros(m,n+1);
u(:,1) = u0(:);

% Time stepping.
for i = 1:n
  % Data that does not depend on the new value.
  known = u(:,i) + h/2*dudt(t(i),u(:,i));
  % Find a root for the new value.
  unew = levenberg(@trapzero,known);
  u(:,i+1) = unew(:,end);
end

u = u.';   % conform to MATLAB output convention

    % This function defines the rootfinding problem at each step.
    function F = trapzero(z)
        F = z - h/2*dudt(t(i+1),z) - known;
    end

end  % main function
```

from Section 4.6 is shown in Function 6.7.2. It defines a nested function called trapzero that evaluates the left-hand side of (6.7.3), given any value of $z$. The time stepping iteration calls levenberg at each step, starting from the value $u_i + \frac{1}{2}h f_i$ that is halfway between $u_i$ and the Euler step $u_i + h f_i$. A robust code would have to intercept the case where Function 4.6.2 fails to converge, but we have ignored this issue for the sake of simplicity.

## Stiff problems

At each time step in Function 6.7.2 (or any implicit IVP solver), a rootfinding iteration of unknown length is needed. This fact makes the cost of an implicit method much greater on a per-step basis than for an explicit one. Given this drawback, you are justified to wonder whether implicit methods are ever competitive! The answer is emphatically yes, as the following simple example shows.

## Example 6.7.2

The following simple ODE uncovers a surprise.

```
f = @(t,u) u.^2 - u.^3;
u0 = 0.005;
```

We will solve the problem first with the implicit AM2 method using $n = 200$ steps.

```
[tI,uI] = am2(f,[0 400],u0,200);
plot(tI,uI)
```

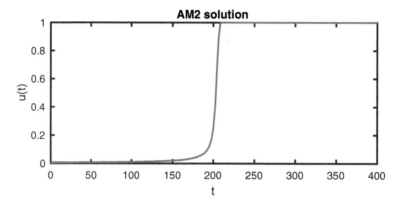

Now we repeat the process using the explicit AB4 method.

```
[tE,uE] = ab4(f,[0 400],u0,200);
hold on, plot(tE,uE), ylim([-1 2])
```

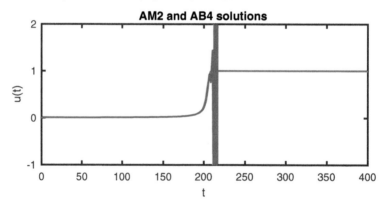

Once the solution starts to take off, the AB4 result goes catastrophically wrong.

```
format short e, uE(105:111)
```

```
ans =
   7.5539e-01
   1.4373e+00
  -3.2890e+00
   2.1418e+02
  -4.4821e+07
```

## 6.7. Implementation of multistep methods

```
    4.1269e+23
    -3.2214e+71
```

We hope that AB4 will converge in the limit $h \to 0$, so let's try using more steps.

```
for n = [1000 1600]
    [tE,uE] = ab4(f,[0 400],u0,n);
    plot(tE,uE)
end
```

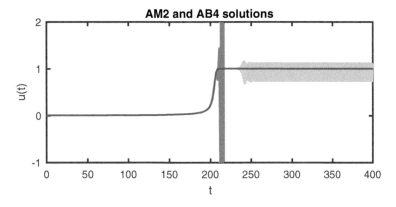

So AB4, which is supposed to be *more* accurate than AM2, actually needs something like 8 times as many steps to get a reasonable-looking answer!

Although the result of Example 6.7.2 may seem strange, there is no contradiction: a fourth-order explicit formula is indeed more accurate than a second-order implicit one, in the limit $h \to 0$. But there is another limit to consider, $t \to \infty$ with $h$ fixed, and in this one the implicit method wins. Such problems are called **stiff**. A complete mathematical description lies in Section 11.4, but a sure sign of stiffness is the presence of phenomena on widely different time scales. In the example, there is "slow time," where the solution changes very little, and "fast time," when it suddenly jumps from zero to one. For stiff problems, implicit methods are usually preferred, because they can take far fewer steps than an explicit method, more than offsetting the extra work required per step.

## Adaptivity

As with RK methods, we can run two time stepping methods simultaneously in order to estimate the error and adjust the step size accordingly. For example, we could pair AB3 with AB4 at practically no cost, because the methods differ only in how they include known information from the recent past. The more accurate AB4 value should allow an accurate estimate of the local error in the AB3 value, and so on.

Because multistep methods rely on the solution history, though, changing the step size is more complicated than for RK methods. If $h$ is changed, then the historical values $u_{i-1}, u_{i-2} \ldots$ and $f_{i-1}, f_{i-2} \ldots$ are no longer given at the right moments in time to apply the iteration formula. A typical remedy is to use interpolation to re-evaluate the historical values at the appropriate times. The details are important but not especially illuminating, and we do not give them here.

## Exercises

6.7.1. For each IVP, use Function 6.7.1 to find the solution over the indicated time interval for $n = 250$. Plot the computed solution $(t_i, u_i)$ for $i = 0, \ldots, n$, and separately plot the error $(t_i, u_i - \hat{u}(t_i))$.

(a) $u' = -2tu$, $0 \le t \le 2$, $u(0) = 2$; $\hat{u}(t) = 2e^{-t^2}$.

(b) $u' = u + t$, $0 \le t \le 1$, $u(0) = 2$; $\hat{u}(t) = 1 - t + e^t$.

(c) $u' = x^2/[u(1+x^3)]$, $0 \le x \le 3$, $u(0) = 1$; $\hat{u}(x) = [1 + (2/3)\ln(1+x^3)]^{1/2}$.

(d) $u'' + 9u = 9t$, $0 < t < 2\pi$, $u(0) = 1$, $u'(0) = 1$; $\hat{u}(t) = t + \cos(3t)$.

(e) $u'' + 9u = \sin(2t)$, $0 < t < 2\pi$, $u(0) = 2$, $u'(0) = 1$; $\hat{u}(t) = (1/5)\sin(3t) + 2\cos(3t) + (1/5)\sin(2t)$.

(f) $u'' - 9u = 9t$ $0 < t < 1$, $u(0) = 2$, $u'(0) = -1$; $\hat{u}(t) = e^{3t} + e^{-3t} - t$.

(g) $u'' + 4u' + 4u = t$, $0 < t < 4$, $u(0) = 1$, $u'(0) = 3/4$; $\hat{u}(t) = (3t + 5/4)e^{-2t} + (t-1)/4$.

(h) $x^2 u'' + 5xu' + 4u = 0$, $1 < x < e^2$, $u(1) = 1$, $u'(1) = -1$, $\hat{u}(x) = x^{-2}(1 + \ln x)$.

(i) $2x^2 u'' + 3xu' - u = 0$, $1 < x < 16$, $u(1) = 4$, $u'(1) = -1$, $\hat{u}(x) = 2(x^{1/2} + x^{-1})$.

(j) $x^2 u'' - xu' + 2u = 0$, $1 < x < e^\pi$, $u(1) = 3$, $u'(1) = 4$, $\hat{u}(x) = x[3\cos(\ln x) + \sin(\ln x)]$.

6.7.2. For each IVP in Exercise 6.7.1, use Function 6.7.1 for $n = 10 \cdot 2^d$ and $d = 1, \ldots, 10$. Make a log–log convergence plot for the final time error $|u_n - \hat{u}(t_n)|$ versus $n$, and add a straight line indicating fourth-order convergence.

6.7.3. Line 35 of Function 6.7.1 reads

```
u(:,i+1) = u(:,i) + h*(f*sigma);
```

Explain carefully why this is preferable to

```
u(:,i+1) = u(:,i) + h*f*sigma;
```

6.7.4. Repeat Exercise 6.7.1 using Function 6.7.2.

6.7.5. Repeat Exercise 6.7.2 using Function 6.7.2 and comparing to second-order rather than fourth-order convergence.

6.7.6. Using Function 6.7.2 as a model, write a function **bd2** that applies the BD2 method to solve an IVP. Test the convergence of your function on one of the IVPs in Exercise 6.7.1.

6.7.7. 📷 For numerical purposes, the exact solution of the IVP in Example 6.7.2 satisfies $\hat{u}(400) = 1$.
  (a) Use Function 6.7.1 with $n = 200, 400, 600, \ldots, 2000$ and make a log–log convergence plot of the error $|u_n - 1|$ as a function of $n$.
  (b) Repeat part (a) using Function 6.7.2.

6.7.8. Consider the IVP
$$u'(t) = Au(t), \quad A = \begin{bmatrix} 0 & -4 \\ 4 & 0 \end{bmatrix}, \quad u(0) = \begin{bmatrix} 1 \\ 0 \end{bmatrix}.$$
  (a) ✏️ Define $E(t) = \|u(t)\|_2^2$. Show that $E(t)$ is constant. (Differentiate $u^T u$ with respect to time and show that it simplifies to zero.)
  (b) 📷 Use Function 6.7.1 to solve the IVP for $t \in [0, 20]$ with $n = 100$ and $n = 150$. On a single graph using a log scale on the $y$-axis, plot $|E(t) - E(0)|$ versus time for both solutions. You should see exponential growth in time.
  (c) 📷 Repeat part (b) with $n = 400$ and $n = 600$, but use a linear scale on the $y$-axis. Now you should see only linear growth of $|E(t) - E(0)|$.

6.7.9. 📷
  (a) Modify Function 6.7.1 to implement the AB2 method.
  (b) Repeat part (b) of Exercise 6.7.8, using AB2 in place of AB4.
  (c) Repeat part (c) of Exercise 6.7.8, using AB2 in place of AB4.

## 6.8 • Zero-stability of multistep methods

For one-step methods such as RK, Theorem 6.2.1 guarantees that the method converges and that the global error is of the same order as the LTE. For multistep methods, however, a new wrinkle is introduced. As an example, it can be checked that the 2-step method "LIAF," defined by

$$u_{i+1} = -4u_i + 5u_{i-1} + h(4f_i + 2f_{i-1}), \tag{6.8.1}$$

is third-order accurate. Yet it is not very useful.

### Example 6.8.1

Consider the ridiculously simple IVP $u' = u$, $u(0) = 1$, whose solution is $e^t$.

```
dudt = @(t,u) u;
u_exact = @exp;
a = 0;   b = 1;
```

Let's apply the LIAF method to this problem for varying fixed step sizes. We'll measure the error at the time $t = 1$.

```
n = [5;10;20;40;60];
err = 0*n;
```

```
for j = 1:length(n)
    h = (b-a)/n(j);
    t = a + h*(0:n(j))';
    u = [1; u_exact(h); zeros(n(j)-1,1)];
    f = [dudt(t(1),u(1)); zeros(n(j)-2,1)];
    for i = 2:n(j)
        f(i) = dudt(t(i),u(i));
        u(i+1) = -4*u(i) + 5*u(i-1) + h*(4*f(i)+2*f(i-1));
    end
    err(j) = abs(u_exact(b) - u(end));
end
h = (b-a)./n;
table(n,h,err)
```

```
ans =
  5x3 table
    n         h          err
    __    _____    _____

     5         0.2       0.016045
    10         0.1         2.8455
    20        0.05      1.6225e+06
    40       0.025      9.3442e+18
    60    0.016667      1.7401e+32
```

The error starts out promisingly, but things explode from there. A graph of the last numerical attempt yields a clue.

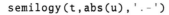

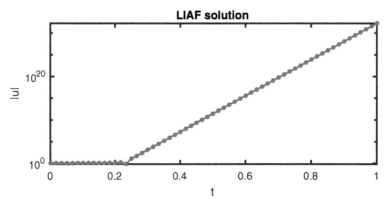

It's clear that the solution is growing exponentially in time.

The source of the instability in Example 6.8.1 is not hard to identify. First, though, we're going to encounter the possibility of complex numbers shortly. Because it's easy to confuse the node index $i$ with the imaginary unit,[30] we'll switch to using $m$ as the step index in this section.

---

[30] We refuse to use $j$ to mean $\sqrt{-1}$, because the name for it is not "jmaginary."

## 6.8. Zero-stability of multistep methods

Let's recall that we can rewrite (6.8.1) as $\rho(\mathscr{Z})u_{m-1} = h\sigma(\mathscr{Z})u_{m-1}$ using the forward shift operator $\mathscr{Z}$:

$$(\mathscr{Z}^2 + 4\mathscr{Z} - 5)u_{m-1} = h(4\mathscr{Z} + 2)f_{m-1}. \tag{6.8.2}$$

(See (6.6.4).) Next, suppose that $h$ is negligible in (6.8.2). Then the numerical solution of LIAF is roughly defined by

$$(\mathscr{Z}^2 + 4\mathscr{Z} - 5)u_{m-1} = 0. \tag{6.8.3}$$

The graph in Example 6.8.1 strongly suggests that for small $h$, $|u_m| \approx c\alpha^m$ for some $\alpha > 1$ as $m$ gets large. So we are motivated to posit $u_m = cz^m$ for all $m$ and see if we can prove that it is an exact solution. The beauty of this choice is that $\mathscr{Z}u_m = zu_m$; that is, the "shift ahead" operator on the sequence $u_0, u_1, \ldots$ is identical to "multiply by $z$." This observation implies that we can solve (6.8.3) with $u_m = cz^m$ for all $m$ simply by finding a numerical value of $z$ such that

$$z^2 + 4z - 5 = 0. \tag{6.8.4}$$

We've arrived at a polynomial rootfinding problem! The roots for this method are $z = 1$ and $z = -5$.

Now we see why exponentially large numbers were observed in Example 6.8.1: the sequence $u_m = c(-5)^m$ is approximately a solution of the multistep formula for small values of $h$. This causes exponential growth that drowns out the solution we were trying to find. In the general case, whenever there is a root $r$ of $\rho(z) = 0$ such that $|r| > 1$, we should expect exponentially growing solutions as $h \to 0$ and $m \to \infty$.

### The root condition

The property we lacked in Example 6.8.1 is called **zero-stability**. To state it precisely, *zero-stability requires that as $h \to 0$, every numerical solution produced by the multistep formula remains bounded throughout $a \le t_m \le b$.* Without this property, any kind of error, whether from truncation or roundoff, will get exponentially amplified and overwhelm convergence to the exact solution. The following theorem concisely summarizes when we can expect zero-stability.

> **Theorem 6.8.1: Root condition**
>
> A linear multistep method is zero-stable (i.e., has only bounded solutions as $h \to 0$) if and only if every root $r$ of the generating polynomial $\rho(z)$ satisfies $|r| \le 1$, and any root $r$ with $|r| = 1$ is simple (that is, $\rho'(r) \ne 0$).

*Partial proof.* As described above, the values produced by the numerical method approach solutions of the difference equation $\rho(\mathscr{Z})u_m = 0$. We consider only the case where the roots $r_1, \ldots, r_k$ of $\rho(z)$ are all distinct. Then $u_m = (r_j)^m$ is a solution of

$\rho(\mathcal{Z})u_m = 0$ for each $j = 1, \ldots, k$. By linearity,

$$u_m = c_1(r_1)^m + c_2(r_2)^m + \cdots + c_k(r_k)^m$$

is a solution for any values of $c_1, \ldots, c_k$. These constants are determined uniquely by the starting values $u_0, \ldots, u_{k-1}$ (we omit the proof). Now, if all the roots satisfy $|r_j| \leq 1$, then

$$|u_m| \leq \sum_{j=1}^{k} |c_j||r_j|^m \leq \sum_{j=1}^{k} |c_j|,$$

independently of $h$ and $m$. This proves zero-stability. Conversely, if some $|r_j| > 1$, then $|u_m|$ cannot be bounded above by a constant independent of $m$. Since $b = t_m$, $m \to \infty$ at $t = b$ as $h \to 0$, and so zero-stability cannot hold. □

### Example 6.8.2

A $k$-step Adams method has $\rho(z) = z^k - z^{k-1} = z^{k-1}(z-1)$. Hence 1 is a simple root and 0 is a root of multiplicity $k-1$. So the Adams methods are all stable.

## Dahlquist theorems

It turns out that lacking zero-stability is the only thing that can go wrong for a consistent multistep method. (Recall that a method is consistent if its LTE is $O(h)$ as $h \to 0$.)

### Theorem 6.8.2: Dahlquist equivalence

A linear multistep method is convergent if and only if it is consistent and zero-stable.

The Dahlquist Equivalence Theorem is one of the most important and celebrated in the history of numerical analysis. It can be proved more precisely that a zero-stable, consistent method is convergent in the same sense as Theorem 6.2.1, with the error between numerical and exact solutions being of the same order as the local truncation error, for a wide class of problems.

Now have a look back at the stencils of the Adams and BD formulas in Figure 6.1. In each case only about half of the available data from the past $k$ steps is used. For instance, a $k$-step AB method uses only the $f_j$ values and has order $k$. The order could be made higher by also using $u_j$ values, like the LIAF method does for $k = 2$. Also like the LIAF method, however, such attempts are doomed by instability.

> **Theorem 6.8.3: First Dahlquist stability barrier**
>
> The order of accuracy $p$ of a stable $k$-step linear multistep method satisfies
> $$p \le \begin{cases} k+2 & \text{if } k \text{ is even,} \\ k+1 & \text{if } k \text{ is odd,} \\ k & \text{if the method is explicit.} \end{cases}$$

## Exercises

6.8.1. ⚞ Show that the LIAF method (6.8.1) has order of accuracy equal to three.

6.8.2. ⚞ / 🖥 Verify that the order of accuracy of the given multistep method is at least one. Then apply Theorem 6.8.1 to determine whether it is zero-stable.

   (a) BD2.

   (b) BD3.

   (c) $u_{i+1} = u_{i-1} + 2hf_i$.

   (d) $u_{i+1} = -u_i + u_{i-1} + u_{i-2} + \frac{2h}{3}(4f_i + f_{i-1} + f_{i-2})$.

   (e) $u_{i+1} = u_{i-3} + \frac{4h}{3}(2f_i - f_{i-1} + 2f_{i-2})$.

   (f) $u_{i+1} = -2u_i + 3u_{i-1} + h(f_{i+1} + 2f_i + f_{i-1})$.

6.8.3. ⚞ A Fibonacci sequence is defined by $u_{i+1} = u_i + u_{i-1}$, where $u_0$ and $u_1$ are seed values. Using the proof of Theorem 6.8.1, find $r_1$ and $r_2$ such that $u_i = c_1(r_1)^i + c_2(r_2)^i$ for all $i$.

6.8.4. ⚞ Suppose that $r$ is a root of multiplicity at least two for $\rho(z)$. Show that $u_m = cmr^m$ is a solution of the difference equation $\rho(\mathscr{Z})u_m = 0$.

## Key ideas in this chapter

1. A numerical solution of an IVP is represented by approximate values of the solution at selected nodes (page 235).

2. The error introduced by a method in one step, divided by $h$, is called the local truncation error (page 238).

3. The leading power of $h$ in the local truncation error is the same as the leading power of the global error (page 240).

4. Systems can be solved using the same methods as scalar problems, because high-order equations can always be reduced to first-order systems (page 245).

5. Runge–Kutta (RK) methods achieve high order of accuracy by multiple evaluations of $f$ at each time step (page 249).

6. Practical methods compute two or more approximations per time step to estimate the error in the resulting solution and take the most appropriate step size (page 255). There are embedded RK formulas to do this efficiently (page 257).

7. Multistep methods use the history of the solution to achieve high order (page 261).
8. Zero-stability ensures that numerical solutions do not grow unboundedly over the time interval in the limit $h \to 0$ and is necessary for convergence (page 275).

## Where to learn more

We use RK methods as a class to represent what are also called single-step or one-step methods. A gentle introduction to these and other kinds of IVP methods can be found in [4]. More advanced introductions are given in [21] and [34]. The most definitive reference is [30].

An interesting article about the built-in functions for solving IVPs is [58]. The capabilities of these functions go well beyond the ODE IVPs described here. For example, ode15s can solve implicit differential equations and differential-algebraic systems; a relevant text on this subject is [11].

Interesting history of IVP methods can be found at history.siam.org, where C. W. Gear gives both an oral history (history.siam.org/oralhistories/gear.htm) and an article (history.siam.org/pdf/cwgear.pdf, reprinted from [48]).

# Part II

*The techniques of the first six chapters are excellent for a variety of problems. However, they all face major limitations when pushed into problems that are too large, high-dimensional, demanding in accuracy, or otherwise difficult in some way.*

*In the next seven chapters we introduce some of the techniques developed in response to some of these challenges. These techniques and the concepts they stem from have proved useful in many problems of serious interest. Moreover, they also connect in deep ways to topics in complex variables, statistics, probability, function theory, graph theory, the analysis of differential equations, and more. Since we don't assume that you have much prior experience in any of those areas, we can only occasionally offer glimpses into these connections. But numerical computation plays a uniquely central role in combining fascinating theoretical mathematics with applications of interest within math and well beyond.*

# Chapter 7
# Matrix analysis

> Judge me by my size, do you?
> — Yoda, *The Empire Strikes Back*

In previous chapters we have seen how matrices that represent square or overdetermined linear systems of equations can be manipulated into LU and QR factorizations. But matrices have other factorizations that are more intrinsic to their nature as mathematical linear transformations. The most fundamental of these are the eigenvalue and singular value decompositions.

These decompositions can be used to solve linear and least squares systems, but they have greater value in how they represent the matrix itself. They lead to critical and quantitative insights about the structure of the underlying transformation and suggest ways to approximate it efficiently. In this chapter we will look at both of these fundamental decompositions and hint at just a few of their computational applications.

## 7.1 ▪ From matrix to insight

*Any two-dimensional array of numbers may be interpreted as a matrix.* Whether or not this is the only point of view that matters to a particular application, it does lead to particular types of analysis. The related mathematical and computational tools are universally applicable and find diverse uses.

### Tables as matrices

Tables are used to represent variation of a quantity with respect to two variables. These variables may be encoded as the rows and columns of a matrix.

### Example 7.1.1

Suppose we have a *corpus*, or collection of text documents. A *term-document matrix* has one column for each document and one row for each unique term appearing in the corpus. The $(i, j)$ entry of the matrix is the number of times term $i$ appears in document $j$. That is, column $j$ of the matrix is a term-frequency vector quantifying all occurrences of the indexed terms. A new document could be represented by its term-frequency vector, which is then comparable to the columns of the matrix. Or a new term could be represented by counting its appearances in all of the documents and be compared to the rows of the matrix.

It turns out that by finding the singular value decomposition (Section 7.3) of the term-document matrix, the strongest patterns within the corpus can be isolated, frequently corresponding to what we interpret as textual meaning. This is known as *latent semantic analysis*.

### Example 7.1.2

The website www.congress.gov/roll-call-votes offers data on all the votes cast in each session of the U.S. Congress. We can put members of Congress along the columns of a matrix and bills along the rows, recording a number that codes for "yea," "nay," "none," etc. The singular value decomposition (Section 7.3) can reveal an objective, reproducible analysis of the partisanship and cooperation of individual members.

### Example 7.1.3

In 2006 the online video service Netflix started an open competition for a $1 million prize. They provided a data set of 100,480,507 ratings (one to five stars) made by 480,189 users for 17,770 movies. Each rating is implicitly an entry in a 17,770×480,189 matrix. The object of the prize was to predict a user's ratings for movies they had not rated. This is known as a *matrix completion problem*. (It took 6 days for a contestant to improve on Netflix's private algorithm, and in 2009 the $1 million prize was awarded to a team that had improved the performance by over 10%.)

## Graphs as matrices

An important concept in modern mathematics is that of a **graph**. A graph consists of a set $V$ of **nodes** and a set $E$ of **edges**, each of which is an ordered pair of nodes. The natural interpretation is that the edge $(v_i, v_j)$ denotes a link from node $i$ to node $j$, in which case we say that node $i$ is adjacent to node $j$. One usually visualizes small graphs by drawing points for nodes and arrows or lines for the edges.

Graphs are useful because they are the simplest way to represent link structure—of social networks, airline routes, power grids, sports teams, and web pages, to name a few examples. They also have close ties to linear algebra. Prominent among these is

## 7.1. From matrix to insight

the **adjacency matrix** of the graph. If the graph has $n$ nodes, then its $n \times n$ adjacency matrix $A$ has elements

$$A_{ij} = \begin{cases} 1 & \text{if } (v_i, v_j) \in E \text{ (i.e., there is an edge from node } i \text{ to node } j), \\ 0 & \text{otherwise.} \end{cases} \quad (7.1.1)$$

**Example 7.1.4**

Here is the adjacency matrix of a graph with 60 nodes from a built-in MATLAB function.

```
[A,v] = bucky;
size(A)

ans =
    60    60
```

The extra vector $v$ gives particular coordinates for each node on the unit sphere. Plotting the nodes and edges with these coordinates reveals beautiful structure.

```
gplot(A,v), axis equal
```

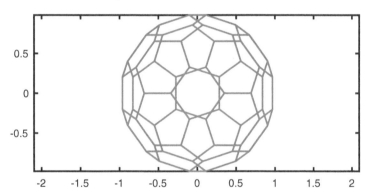

As you can see, the edges resemble those on a soccer ball. The same structure, when made of carbon atoms at the nodes, is called a *buckyball* in materials science.

The representation of a graph by its adjacency matrix opens up the possibility for many kinds of analysis of the graph. One might ask whether the nodes admit a natural partition into clusters, for example. Or one might ask to rank the nodes in order of importance to the network as determined by some objective criteria—an application made famous by Google's PageRank algorithm, and one which is mathematically stated as an eigenvalue problem (Section 7.2).

## Images as matrices

Computers most often represent images as rectangular arrays of pixels, each of which is colored according to numerical values for red (R), green (G), and blue (B) components of white light. Typically these are given as integers in the range from zero (no color) to 255 (full color). Thus, an image that is $m \times n$ pixels can be stored as an $m \times n \times 3$ array of integer values.

We will simplify the representation by considering images represented using pixels that can take only shades of gray. We will also use floating point numbers rather than integers, so that we can operate on them using real arithmetic, though we will stay with the convention that values should be in the range $[0, 255]$. (Pixels below zero or above 255 will be colored pure black or pure white, respectively.)

### Example 7.1.5

MATLAB reads an image using the `imread` command. We'll use an image that ships with MATLAB for demonstrations.

```
A = imread('peppers.png');
size(A)

ans =
    384   512     3
```

As stated above, the image has three "layers" for red, green, and blue. We can deal with each layer as a matrix, or (as below) convert it to a single matrix indicating shades of gray from black (0) to white (255). Then we have to tell MATLAB to represent the entries as floating point values rather than integers.

```
A = rgb2gray(A);   % collapse from 3 dimensions to 2
A = double(A);     % convert to floating point
[m,n] = size(A)

m =
    384
n =
    512
```

We now have a matrix. To show it as an image, we use the `imshow` command. We can tell it explicitly which values correspond to black and white.

```
imshow(A,[0 255])
```

Representation of an image as a matrix allows us to describe some common image operations in terms of linear algebra. Furthermore, the singular value decomposition (Section 7.3) can be used to compress the information.

## Exercises

7.1.1. ✎ Consider the terms *numerical*, *analysis*, and *fun*. Write out the term-document matrix for the following statements:

- *Numerical analysis is the most fun type of analysis.*
- *It's fun to produce numerical values for the digits of pi.*
- *Complex analysis is a beautiful branch of mathematics.*

7.1.2. ✎ Write out the adjacency matrix for the following graph on six nodes:

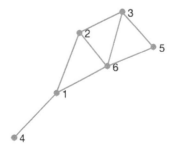

7.1.3. ✍ Here is a graph adjacency matrix:

$$\begin{bmatrix} 0 & 1 & 0 & 1 & 0 & 1 & 0 \\ 1 & 0 & 1 & 0 & 0 & 1 & 0 \\ 0 & 1 & 0 & 0 & 1 & 0 & 1 \\ 1 & 0 & 0 & 0 & 0 & 0 & 0 \\ 0 & 0 & 1 & 0 & 0 & 1 & 0 \\ 1 & 1 & 0 & 0 & 1 & 0 & 0 \\ 0 & 0 & 1 & 0 & 0 & 0 & 0 \end{bmatrix}.$$

(a) How many vertices are adjacent to vertex 5?

(b) How many edges are in the graph?

(c) Draw the graph.

7.1.4. ⌨ Refer to Example 7.1.5.

(a) Display the peppers.png image upside-down.

(b) Display it mirror-reversed from left to right.

(c) Display the image so that it is cropped to include only the garlic bulb in the lower right of the original picture.

## 7.2 • Eigenvalue decomposition

To this point we have dealt frequently with the solution of the linear system $Ax = b$. Alongside this problem in its importance to linear algebra is the eigenvalue problem,

$$Ax = \lambda x, \qquad (7.2.1)$$

for a scalar **eigenvalue** $\lambda$ and an associated nonzero **eigenvector** $x$.

### Complex matrices

A matrix with real entries can have complex eigenvalues. Therefore we assume all matrices, vectors, and scalars may be complex in what follows. Recall that a complex number can be represented as $a + ib$ for real $a$ and $b$ and where $i^2 = -1$. The **complex conjugate** of $x = a + ib$ is denoted $\bar{x}$ and is given by $\bar{x} = a - ib$. The **hermitian** or conjugate transpose of a matrix $A$ is denoted $A^*$ and is given by $A^* = (\overline{A})^T = \overline{A^T}$.

For the most part, "hermitian" replaces "transpose" when dealing with complex matrices. For instance, the inner product of complex vectors $u$ and $v$ is

$$u^* v = \sum_{k=1}^{n} \overline{u}_k v_k, \qquad (7.2.2)$$

which in turn defines the 2-norm for complex vectors, and thereby matrices as well. The definitions of orthogonal and orthonormal sets of complex-valued vectors use $*$ instead of $^T$. The analog of an orthogonal matrix in the complex case—that is, a square matrix whose columns are orthonormal in the complex sense—is said to be **unitary**. A unitary matrix $U$ satisfies $U^{-1} = U^*$ and $\|Ux\|_2 = \|x\|_2$ for any complex vector $x \in \mathbb{C}^n$.

## Eigenvalue decomposition

The eigenvalue equation $Ax = \lambda x$ is equivalent to $(\lambda I - A)x = 0$, which, since $x$ is nonzero in order to be an eigenvector, implies that $\lambda I - A$ is a singular matrix. This observation leads to the familiar property of an eigenvalue being a root of the **characteristic polynomial** $\det(\lambda I - A)$. From here one concludes that an $n \times n$ matrix has $n$ eigenvalues, counting multiplicity.

Hence suppose that $Av_k = \lambda_k v_k$ for $k = 1, \ldots, n$. We can summarize these as

$$\begin{bmatrix} Av_1 & Av_2 & \cdots & Av_n \end{bmatrix} = \begin{bmatrix} \lambda_1 v_1 & \lambda_2 v_2 & \cdots & \lambda_n v_n \end{bmatrix},$$

$$A \begin{bmatrix} v_1 & v_2 & \cdots & v_n \end{bmatrix} = \begin{bmatrix} v_1 & v_2 & \cdots & v_n \end{bmatrix} \begin{bmatrix} \lambda_1 & & & \\ & \lambda_2 & & \\ & & \ddots & \\ & & & \lambda_n \end{bmatrix},$$

$$AV = VD. \tag{7.2.3}$$

So far $A$ could be any square matrix. But if we also assume that $V$ is a nonsingular matrix, we can rewrite (7.2.3) as

$$A = VDV^{-1}. \tag{7.2.4}$$

*Equation (7.2.4) is called an* **eigenvalue decomposition (EVD)** *of $A$.* If $A$ has an EVD, we say that $A$ is **diagonalizable**; otherwise $A$ is **nondiagonalizable** (or *defective*). One simple example of a nondiagonalizable matrix is

$$B = \begin{bmatrix} 1 & 1 \\ 0 & 1 \end{bmatrix}. \tag{7.2.5}$$

There is a common circumstance in which we can guarantee an EVD exists; the proof of the following theorem can be found in many elementary texts on linear algebra.

### Theorem 7.2.1

If the $n \times n$ matrix $A$ has $n$ distinct eigenvalues, then $A$ is diagonalizable.

The `eig` command can be used to compute the eigenvalue decomposition of a given matrix.

### Example 7.2.1

The `eig` command will give just the eigenvalues if one output is given, or both $V$ and $D$ if two are given.

```
A = pi*ones(2,2);
lambda = eig(A)
```

```
lambda =
         0
    6.2832

[V,D] = eig(A)

V =
   -0.7071    0.7071
    0.7071    0.7071
D =
         0         0
         0    6.2832
```

We can check the fact that this is an EVD.

```
norm( A - V*D/V )    % /V is like *inv(V)

ans =
   8.8818e-16
```

Even if the matrix is not diagonalizable, `eig` will run successfully, but the matrix $V$ will not be invertible.

```
[V,D] = eig([1 1;0 1])
rankV = rank(V)

V =
    1.0000   -1.0000
         0    0.0000
D =
    1    0
    0    1
rankV =
    1
```

Observe that if $Av = \lambda v$ for nonzero $v$, then the equation remains true for any nonzero multiple of $v$, so eigenvectors are not unique, and neither is an EVD.

## Similarity and change of basis

The particular relationship between matrices $A$ and $D$ in (7.2.4) is important. If $S$ is any nonsingular matrix, we say that $B = SAS^{-1}$ is **similar** to $A$. A similarity transformation does not change eigenvalues, a fact that is typically proved in elementary linear algebra texts.

### Theorem 7.2.2

If $X$ is an nonsingular matrix, then $XAX^{-1}$ has the same eigenvalues as $A$.

## 7.2. Eigenvalue decomposition

Similarity transformation has a relatively simple interpretation. First, consider the product of a nonsingular $X$ with any vector:

$$y = Xz = z_1 x_1 + \cdots + z_n x_n.$$

We call $z_1, \ldots, z_n$ the *coordinates* of the vector $y$ with respect to the columns of $X$. That is, $z$ is a representation of $y$ relative to the basis implied by the columns of $X$. But then $z = X^{-1} y$, so left-multiplication by $X^{-1}$ converts the vector $y$ into that representation. In other words, *multiplication by the inverse of a matrix performs a change of basis into the coordinates associated with the matrix.*

So now consider the EVD (7.2.4) and the product $u = Ax$, or $(V^{-1}u) = D(V^{-1}x)$. This equation says that if you convert the input $x$ and the output $u$ into the coordinates of the $V$-basis, then the relationship between them is diagonal. That is, the EVD is about finding a basis for $\mathbb{C}^n$ in which the map $x \mapsto Ax$ is a diagonal one in which the coordinates are independently rescaled.

The fact that the EVD represents a change of basis in both the domain and range spaces makes it useful for matrix powers:

$$A^2 = (VDV^{-1})(VDV^{-1}) = VD(V^{-1}V)DV^{-1} = VD^2V^{-1},$$

and so on. Because $D$ is diagonal, its power $D^k$ is just the diagonal matrix of the powers of the eigenvalues.

### Conditioning of eigenvalues

Just as linear systems have condition numbers that quantify the effect of fixed precision, eigenvalue problems may be poorly conditioned too. While many possible results can be derived, we will use just one, the **Bauer–Fike theorem**.[31]

> **Theorem 7.2.3**
>
> Let $A \in \mathbb{C}^{n \times n}$ be diagonalizable, $A = VDV^{-1}$, with eigenvalues $\lambda_1, \ldots, \lambda_n$. If $\mu$ is an eigenvalue of $A + E$ for a complex matrix $E$, then
>
> $$\min_{j=1,\ldots,n} |\mu - \lambda_j| \leq \kappa(V) \|E\|, \qquad (7.2.6)$$
>
> where $\|\cdot\|$ and $\kappa$ are in the 2-norm.

The Bauer–Fike theorem tells us that eigenvalues can be perturbed by an amount that is $\kappa(V)$ times larger than perturbations to the matrix. This result is a bit less straightforward than it might seem—eigenvectors are not unique, so there are multiple possible values for $\kappa(V)$. Still, the theorem indicates caution when a matrix has eigenvectors that form an ill-conditioned matrix. The limiting case of $\kappa(V) = \infty$ might be interpreted as indicating a nondiagonalizable matrix $A$.

---

[31] We will apply it only in the 2-norm, though it is more generally true.

At the other extreme, if a unitary eigenvector matrix $V$ can be found, then $\kappa(V) = 1$ and (7.2.6) guarantees that eigenvalues are robust under perturbations to the original matrix $A$. Such matrices are called **normal**, and they include the hermitian (or real symmetric) matrices. We consider them again in Section 7.4.

> **Example 7.2.2**
>
> We will confirm the Bauer–Fike theorem on a triangular matrix. These tend to be far from normal.
>
> ```
> n = 15;
> lambda = (1:n)';
> A = triu( ones(n,1)*lambda' );
> ```
>
> The Bauer–Fike theorem provides an upper bound on the condition number of these eigenvalues.
>
> ```
> [V,D] = eig(A);
> kappa = cond(V)
> ```
>
> ```
> kappa =
>    7.1978e+07
> ```
>
> The theorem suggests that eigenvalue changes may be up to 7 orders of magnitude larger than a perturbation to the matrix. A few random experiments show that effects of nearly that size are not hard to observe.
>
> ```
> for k = 1:3
>     E = randn(n);  E = 1e-7*E/norm(E);
>     mu = eig(A+E);
>     max_change = norm( sort(mu)-lambda, inf )
> end
> ```
>
> ```
> max_change =
>    0.2407
> max_change =
>    0.4492
> max_change =
>    0.2737
> ```

## Computing the EVD

In elementary linear algebra you use the characteristic polynomial to compute the eigenvalues of small matrices. However, computing polynomial roots in finite time is impossible for degree 5 and over. In principle one could use Newton-like methods to find all of the roots, but doing so is relatively slow and difficult. Furthermore we know that polynomial roots tend to become poorly conditioned when roots get close to one another (see Example 1.2.4).

## 7.2. Eigenvalue decomposition

Practical algorithms for computing the EVD go beyond the scope of this book. The essence of the matter is the connection to matrix powers, that is, $A^k = VD^kV^{-1}$. (We will see much more about the importance of matrix powers in Chapter 8.) If the eigenvalues have different complex magnitudes, then as $k \to \infty$ the entries on the diagonal of $D^k$ become increasingly well separated and easy to pick out. It turns out that there is an astonishingly easy and elegant way to accomplish this separation without explicitly computing the matrix powers.

### Example 7.2.3

Let's start with a known set of eigenvalues and an orthogonal eigenvector basis.

```
D = diag([-6 -1 2 4 5]);
[V,R] = qr(randn(5));
A = V*D*V';    % note that V' = inv(V)
```

Now we will take the QR factorization and just reverse the factors.

```
[Q,R] = qr(A);
A = R*Q;
```

It turns out that this is a similarity transformation, so the eigenvalues are unchanged.

```
eig(A)
```

```
ans =
   -6.0000
   -1.0000
    5.0000
    4.0000
    2.0000
```

What's remarkable is that if we repeat the transformation many times, the process converges to $D$.

```
for k = 1:15
    [Q,R] = qr(A);
    A = R*Q;
end
A
```

```
A =
   -5.9984   -0.1336    0.0100   -0.0000    0.0000
   -0.1336    4.9960   -0.0491    0.0000   -0.0000
    0.0100   -0.0491    4.0024   -0.0001   -0.0000
   -0.0000    0.0000   -0.0001    2.0000   -0.0001
    0.0000   -0.0000   -0.0000   -0.0001   -1.0000
```

The process demonstrated in Example 7.2.3 is known as the *Francis QR iteration*, and it can be formulated as an $O(n^3)$ algorithm for finding the EVD. Such an algorithm is what the `eig` command uses.

## Exercises

**7.2.1.** (a) 🖉 Suppose that matrix $A$ has an eigenvalue $\lambda$. Show that for any induced matrix norm, $\|A\| \geq |\lambda|$.

(b) 🖉 Find a matrix $A$ such that $\|A\|_2$ is strictly larger than $|\lambda|$ for all eigenvalues $\lambda$. (Proof-by-computer isn't allowed here. You don't need to compute $\|A\|_2$, just a lower bound for it.)

**7.2.2.** 🖉 Prove that the matrix $B$ in (7.2.5) does not have two independent eigenvectors.

**7.2.3.** 💻 Use `eig` to find the EVD of each matrix. Then for each eigenvalue $\lambda$, use the `rank` command to verify that $\lambda I - A$ is singular.

(a) $A = \begin{bmatrix} 2 & -1 & 0 \\ -1 & 2 & -1 \\ 0 & -1 & 2 \end{bmatrix}$, (b) $B = \begin{bmatrix} 2 & -1 & -1 \\ -2 & 2 & -1 \\ -1 & -2 & 2 \end{bmatrix}$, (c) $C = \begin{bmatrix} 2 & -1 & -1 \\ -1 & 2 & -1 \\ -1 & -1 & 2 \end{bmatrix}$,

(d) $D = \begin{bmatrix} 3 & 1 & 0 & 0 \\ 1 & 3 & 1 & 0 \\ 0 & 1 & 3 & 1 \\ 0 & 0 & 1 & 3 \end{bmatrix}$, (e) $E = \begin{bmatrix} 4 & -3 & -2 & -1 \\ -2 & 4 & -2 & -1 \\ -1 & -2 & 4 & -1 \\ -1 & -2 & -1 & 4 \end{bmatrix}$.

**7.2.4.** (a) 🖉 Show that the eigenvalues of a diagonal $n \times n$ matrix $D$ are the diagonal entries of $D$. (That is, produce the associated eigenvectors.)

(b) 🖉 The eigenvalues of a triangular matrix are its diagonal entries. Prove this in the $3 \times 3$ case

$$T = \begin{bmatrix} t_{11} & t_{12} & t_{13} \\ 0 & t_{22} & t_{23} \\ 0 & 0 & t_{33} \end{bmatrix}$$

by finding the eigenvectors. (Start by showing that $[1,0,0]^T$ is an eigenvector. Then show how to make $[a,1,0]^T$ an eigenvector, except for one case that does not change the outcome. Continue the same logic for $[a,b,1]^T$.)

**7.2.5.** 🖉 If $p(z) = c_0 + c_1 z + \cdots + c_k z^k$ is a polynomial, then its value for a matrix argument is defined as

$$p(A) = c_0 I + c_1 A + \cdots + c_k A^k.$$

Show that if an EVD of $A$ is available, then $p(A)$ can be found using only evaluations of $p$ at the eigenvalues, plus two matrix multiplications.

**7.2.6.** 💻 In Exercise 2.6.3, you showed that the displacements of point masses placed along a string satisfy a linear system $Aq = f$ for an $(n-1) \times (n-1)$ matrix $A$. The eigenvalues and eigenvectors of $A$ correspond to resonant frequencies and modes of vibration of the string. For $n = 40$ and the physical parameters given in part (b) of that exercise, find the eigenvalue decomposition of $A$. Report the four eigenvalues with smallest absolute value, and plot all four associated eigenvectors on a single graph (as functions of the vector index number).

**7.2.7.** 💻 The eigenvalues of *Toeplitz* matrices, which have a constant value on each diagonal, have beautiful connections to complex analysis. (This example is adapted from Chapter 7 of [73].)

(a) Define a $200 \times 200$ matrix using

## 7.3. Singular value decomposition

```
z = zeros(1,160);
A = toeplitz( [0,0,-4,-2i,z], [0,2i,-1,2,z] );
```

Display the upper left $6 \times 6$ block of $A$ so that you get the idea of how it's defined.

(b) Plot the eigenvalues of $A$ as black dots in the complex plane.

(c) Let $E$ and $F$ be $164 \times 164$ random matrices generated by `randn`. On top of the plot from (b), plot the eigenvalues of the matrix $A + 10^{-3}E + 10^{-3}iF$ as blue dots.

(d) Repeat part (c) 100 more times (generating a single plot).

(e) Compute $\kappa(V)$ for an eigenvector matrix $V$ and relate your picture to the conclusion of Theorem 7.2.3.

7.2.8. ▭ Eigenvalues of random matrices and their perturbations can be very interesting.

(a) Let `A=randn(36,36)`. Plot its eigenvalues as black dots in the complex plane.

(b) Let $E$ be another random $36 \times 36$ matrix, and on top of the previous graph, plot the eigenvalues of $A + 0.01E$ as blue dots. Repeat this for 100 different values of $E$.

(c) Let `T=triu(A)`. On a new graph, plot the eigenvalues of $T$ as black dots in the complex plane (they are all real numbers).

(d) Repeat part (b) with $T$ in place of $A$.

(e) Compute two condition numbers and use Theorem 7.2.3 to explain your graphs.

## 7.3 · Singular value decomposition

We now introduce another factorization that is as fundamental as the EVD.

---

**Theorem 7.3.1**

Let $A \in \mathbb{C}^{m \times n}$. Then $A$ can be written as

$$A = USV^*, \qquad (7.3.1)$$

where $U \in \mathbb{C}^{m \times m}$ and $V \in \mathbb{C}^{n \times n}$ are unitary and $S \in \mathbb{R}^{m \times n}$ is real and diagonal with nonnegative entries. If $A$ is real, then so are $U$ and $V$ (which are then orthogonal matrices).

---

*Equation (7.3.1) is called a* **singular value decomposition**, *or SVD, of $A$.* The columns of $U$ and $V$ are called **left** and **right singular vectors**, respectively. The diagonal entries of $S$, written $\sigma_1, \ldots, \sigma_r$, for $r = \min\{m, n\}$, are called the **singular values** of $A$. By convention the singular values are ordered so that

$$\sigma_1 \geq \sigma_2 \geq \cdots \geq \sigma_r \geq 0, \qquad r = \min\{m, n\}. \qquad (7.3.2)$$

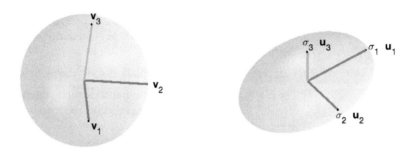

**Figure 7.1.** *Visual interpretation of the SVD for $3 \times 3$ real matrices. Under multiplication by the matrix, a unit sphere, expressed in the right singular vector coordinates (shown at left), maps to an ellipsoid, given in the left singular vector coordinates (shown at right).*

We call $\sigma_1$ the **principal singular value** and $u_1$ and $v_1$ the **principal singular vectors**. The matrix $S$ in the SVD is uniquely defined when the ordering is imposed, but the singular vectors are not—one could replace both $U$ and $V$ by their negatives, for example.

---

**Example 7.3.1**

Suppose $A$ is a real matrix and that $A = USV^T$ is an SVD. Then $A^T = VS^TU^T$ meets all the requirements of an SVD for $A^T$: the first and last matrices are orthogonal, and the middle matrix is diagonal with nonnegative entries. Hence $A$ and $A^T$ have the same singular values.

---

## Interpreting the SVD

Another way to write $A = USV^*$ is as $AV = US$. Taken columnwise, this means

$$Av_k = \sigma_k u_k, \qquad k = 1, \ldots, r = \min\{m, n\}. \tag{7.3.3}$$

In words, each right singular vector is mapped by $A$ to a scaled version of its corresponding left singular vector; the magnitude of scaling is its singular value. Together with the orthonormality of the columns of $U$ and $V$, these equations provide a simple graphical interpretation of the SVD. A representative $3 \times 3$ case is shown in Figure 7.1.

Both the SVD and the EVD describe a matrix in terms of some special vectors and a small number of scalars. Table 7.1 summarizes the key differences. The SVD sacrifices having the same basis in both source and image spaces—after all, they may not even have the same dimension—but gains orthogonality in both spaces.

## 7.3. Singular value decomposition

**Table 7.1.** *Differences between the EVD and the SVD.*

| EVD | SVD |
|---|---|
| most square matrices | all rectangular and square matrices |
| $Ax_k = \lambda_k x_k$ | $Av_k = \sigma_k u_k$ |
| same basis for domain and range of $A$ | two orthogonal bases |
| may have poor conditioning | perfectly conditioned |

### SVD and the 2-norm

*The SVD is intimately connected to the 2-norm,* as the following theorem describes.

---

**Theorem 7.3.2**

Let $A \in \mathbb{C}^{m \times n}$ have an SVD $A = USV^*$ in which (7.3.2) holds. Then the following hold:

1. The 2-norm satisfies
$$\|A\|_2 = \sigma_1. \tag{7.3.4}$$
2. The rank of $A$ is the number of nonzero singular values.
3. Let $r = \min\{m, n\}$. Then
$$\kappa_2(A) = \|A\|_2 \|A^+\|_2 = \frac{\sigma_1}{\sigma_r}, \tag{7.3.5}$$
where a division by zero implies that $A$ does not have full rank.

---

The conclusion (7.3.4) is intuitively clear from Figure 7.1, and some straightforward vector calculus provides a proof (see Exercise 7.3.10). In the square case $m = n$, $A$ having full rank is identical to being nonsingular. The SVD is the usual means for computing the 2-norm and condition number of a matrix. We demonstrate this using MATLAB's **svd** command.

---

**Example 7.3.2**

We verify some of the fundamental SVD properties using the built-in MATLAB command svd.

```
A = vander(1:5);
A = A(:,1:4)
```

```
A =
         1       1       1       1
        16       8       4       2
        81      27       9       3
       256      64      16       4
       625     125      25       5
```

```
[U,S,V] = svd(A);
norm(U'*U - eye(5))

ans =
    1.4850e-15

norm(V'*V - eye(4))

ans =
    1.2551e-15

sigma = diag(S)

sigma =
  695.8403
   18.2045
    1.6989
    0.2120

[ norm(A) sigma(1) ]

ans =
  695.8403  695.8403

[ cond(A) sigma(1)/sigma(4) ]

ans =
   1.0e+03 *
    3.2820    3.2820
```

## Connections to the EVD

Let $A = USV^*$ be $m \times n$, and consider the square hermitian matrix $B = A^*A$:

$$B = (VS^*U^*)(USV^*) = VS^*SV^* = V(S^TS)V^{-1}.$$

Note that $S^TS$ is a diagonal $n \times n$ matrix. There are two cases to consider:

$m \geq n$:

$$S^TS = \begin{bmatrix} \sigma_1^2 & & \\ & \ddots & \\ & & \sigma_n^2 \end{bmatrix},$$

$m < n$:

$$S^TS = \begin{bmatrix} \sigma_1^2 & & & \\ & \ddots & & \\ & & \sigma_m^2 & \\ & & & 0 \end{bmatrix},$$

where the lower-right zero in the last matrix is $n - m$ square. In both cases we may conclude that the squares of the singular values of $A$ are all eigenvalues of $B$. Conversely, an EVD of $B$ reveals the singular values and a set of right singular vectors of $A$. The left singular vectors could then be deduced from the identity $AV = US$.

Another close connection between EVD and SVD comes via the $(m+n) \times (m+n)$ matrix

$$C = \begin{bmatrix} 0 & A^* \\ A & 0 \end{bmatrix}. \tag{7.3.6}$$

If $\sigma$ is a singular value of $B$, then $\sigma$ and $-\sigma$ are eigenvalues of $C$, and the associated eigenvector immediately reveals a left and a right singular vector (see Exercise 7.3.11). This connection is implicitly exploited by software to compute the SVD.

### Thin form

In Section 3.3 we saw that a matrix has both a "full" and a "thin," or reduced, form of the QR factorization. A similar situation holds with the SVD. Suppose $A$ is $m \times n$ with $m > n$, and let $A = USV^*$ be an SVD. The last $m - n$ rows of $S$ are all zero due to the fact that $S$ is diagonal. Hence

$$US = \begin{bmatrix} u_1 & \cdots & u_n & u_{n+1} & \cdots & u_m \end{bmatrix} \begin{bmatrix} \sigma_1 & & & \\ & \ddots & & \\ & & \sigma_n & \\ & & 0 & \end{bmatrix}$$

$$= \begin{bmatrix} u_1 & \cdots & u_n \end{bmatrix} \begin{bmatrix} \sigma_1 & & \\ & \ddots & \\ & & \sigma_n \end{bmatrix} = \widehat{U}\widehat{S}, \tag{7.3.7}$$

in which $\widehat{U}$ is $m \times n$ and $\widehat{S}$ is $n \times n$. This allows us to define the **thin SVD** $A = \widehat{U}\widehat{S}V^*$, in which $\widehat{S}$ is square and diagonal and $\widehat{U}$ is ONC but not unitary. This form is computationally preferable when $m \gg n$, since it requires far less storage and contains the same information about $A$. In MATLAB one uses svd(A,0) to get the thin form. In the dual case where $m < n$, you can use the reduced SVD of $A^*$ and then take the hermitian to get a reduced SVD for $A$.

## Exercises

7.3.1. ✍ Each factorization below is algebraically correct. In each case determine whether it is an SVD. If it is, write down $\sigma_1$, $u_1$, and $v_1$. The notation $I_n$ means an $n \times n$ identity.

(a) $\begin{bmatrix} 0 & 0 \\ 0 & -1 \end{bmatrix} = \begin{bmatrix} 0 & 1 \\ 1 & 0 \end{bmatrix} \begin{bmatrix} 1 & 0 \\ 0 & 0 \end{bmatrix} \begin{bmatrix} 0 & 1 \\ -1 & 0 \end{bmatrix}$.

(b) $\begin{bmatrix} 0 & 0 \\ 0 & -1 \end{bmatrix} = I_2 \begin{bmatrix} 0 & 0 \\ 0 & -1 \end{bmatrix} I_2$.

(c) $\begin{bmatrix} 1 & 0 \\ 0 & \sqrt{2} \\ 1 & 0 \end{bmatrix} = \begin{bmatrix} \alpha & 0 & -\alpha \\ 0 & 1 & 0 \\ \alpha & 0 & -\alpha \end{bmatrix} \begin{bmatrix} \sqrt{2} & 0 \\ 0 & \sqrt{2} \\ 0 & 0 \end{bmatrix} \begin{bmatrix} 0 & 1 \\ 1 & 0 \end{bmatrix}$ ($\alpha = 1/\sqrt{2}$).

(d) $\begin{bmatrix} \sqrt{2} & \sqrt{2} \\ -1 & 1 \\ 0 & 0 \end{bmatrix} = I_3 \begin{bmatrix} 2 & 0 \\ 0 & \sqrt{2} \\ 0 & 0 \end{bmatrix} \begin{bmatrix} \alpha & \alpha \\ -\alpha & \alpha \end{bmatrix}$ ($\alpha = 1/\sqrt{2}$).

7.3.2. ✍ Solve a 2 × 2 eigenvalue problem to find the singular values of

$$A = \begin{bmatrix} 1 & 0 \\ 0 & 0 \\ 0 & 1 \\ -1 & -1 \end{bmatrix}.$$

7.3.3. 🖥 Let $x$ be a vector of 1000 equally spaced points between 0 and 1, and let $A_n$ be the 1000 × $n$ Vandermonde matrix whose $(i, j)$ entry is $x_i^{j-1}$ for $j = 1, \ldots, n$.

   (a) Print out the singular values of $A_1$, $A_2$, and $A_3$.

   (b) Make a semi-log plot of the singular values of $A_{25}$.

   (c) Use rank to find the rank of $A_{25}$. How does this relate to the graph from part (b)? You may want to use the online help for the rank command to understand what it does.

7.3.4. 🖥 MATLAB ships with some sample images for trying out ideas. You can get one of these by using

```
load mandrill
imshow(X,map)
```

Make a semi-log plot of the singular values of $X$. (The shape of this graph is surprisingly similar across a wide range of images.)

7.3.5. ✍ Prove that for a square real matrix $A$, $\|A\|_2 = \|A^T\|_2$.

7.3.6. ✍ Prove (7.3.5) of Theorem 7.3.2, given that (7.3.4) is true. (Hint: If the SVD of $A$ is known, what is the SVD of $A^+$?)

7.3.7. ✍ Suppose $A \in \mathbb{R}^{m \times n}$, for $m > n$, has the reduced SVD $A = \widehat{U}\widehat{S}V^T$. Show that the orthogonal projector $AA^+$ is equal to $\widehat{U}\widehat{U}^T$. (You must be careful with matrix sizes in this derivation.)

7.3.8. ✍ In (3.2.3) we defined the 2-norm condition number of a rectangular matrix as $\kappa(A) = \|A\| \cdot \|A^+\|$, and then claimed (in the real case) that $\kappa(A^*A) = \kappa(A)^2$. Prove this assertion using the SVD.

7.3.9. ✍ Show that the square of each singular value of $A$ is an eigenvalue of the matrix $AA^*$ for any $m \times n$ matrix $A$. (You should consider the cases $m > n$ and $m \leq n$ separately.)

7.3.10. ✍ In this problem you will see how (7.3.4) is proved.

   (a) Use the technique of Lagrange multipliers to show that among vectors that satisfy $\|x\|_2^2 = 1$, any vector that maximizes $\|Ax\|_2^2$ must be an eigenvector of $A^*A$. It will help to know that if $B$ is any hermitian matrix, the gradient of the scalar function $x^*Bx$ with respect to $x$ is $2Bx$.

   (b) Use the result of part (a) to prove (7.3.4).

7.3.11. ✍ Suppose $A \in \mathbb{R}^{n \times n}$, and define $C$ as in (7.3.6).

(a) Suppose that $v = \begin{bmatrix} x \\ y \end{bmatrix}$, and write the block equation $Cv = \lambda v$ as two individual equations involving both $x$ and $y$.

(b) By applying some substitutions, rewrite the equations from part (a) as one in which $x$ was eliminated and another in which $y$ was eliminated.

(c) Substitute the SVD $A = USV^T$ and explain why $\lambda^2 = \sigma_k^2$ for some singular value $\sigma_k$.

(d) As a more advanced variation, modify the argument to show that $\lambda = 0$ is another possibility if $A$ is not square.

## 7.4 • Symmetry and definiteness

As we saw in Section 2.9, symmetry can simplify the LU factorization into a symmetric form, $A = LDL^T$. Certain specializations occur too for the eigenvalue and singular value factorizations. In this section we stay with complex-valued matrices, so we are interested in the case when $A^* = A$, or $A$ is hermitian. However, we often loosely speak of "symmetry" to mean this property in the complex case. All of the statements in this section easily specialize to the real case.

### Unitary diagonalization

Suppose now that $A^* = A$ and that $A = USV^*$ is an SVD. Since $S$ is real and square, we have
$$A^* = VS^*U^* = VSU^*,$$
and it's tempting to conclude that $U = V$. Happily, this is nearly true. The following theorem is typically proved in an advanced linear algebra course.

> **Theorem 7.4.1: Spectral decomposition**
>
> If $A = A^*$, then $A$ has a diagonalization $A = VDV^{-1}$ in which $V$ is unitary and $D$ is diagonal and real.

Another way to state the result of Theorem 7.4.1 is that *a hermitian matrix has a complete set of orthonormal eigenvectors—that is, a* unitary diagonalization—*and real eigenvalues.* In this case, the EVD $A = VDV^{-1} = VDV^*$ is almost an SVD.

> **Theorem 7.4.2**
>
> If $A^* = A$ and $A = VDV^{-1}$ is a unitary diagonalization, then
> $$A = (VT) \cdot |D| \cdot V^* \qquad (7.4.1)$$
> is an SVD, where $|D|$ is the elementwise absolute value and $T$ is diagonal with $|T_{ii}| = 1$ for all $i$.

*Proof.* Let $T_{ii} = \text{sign}(D_{ii})$. Then $T^2 = I$ and $|D| = TD$. The result follows. □

The converse of Theorem 7.4.1 is also true: every matrix with a unitary diagonalization and real eigenvalues is hermitian. Moreover, there are nonhermitian matrices that meet just the requirement of a unitary EVD; any such matrix is called **normal**.

### Example 7.4.1

The following matrix is not hermitian.

```
A = [0 2; -2 0]

A =
      0     2
     -2     0
```

It has an EVD with a unitary matrix of eigenvectors, though, so it is normal.

```
[V,D] = eig(A);
norm( V'*V - eye(2) )

ans =
   2.2204e-16
```

The eigenvalues are pure imaginary.

```
lambda = diag(D)

lambda =
   0.0000 + 2.0000i
   0.0000 - 2.0000i
```

The singular values are the complex magnitudes of the eigenvalues.

```
svd(A)

ans =
     2
     2
```

Now consider again Theorem 7.2.3, which says that the condition number of the eigenvalues is bounded above by $\kappa(V)$, where $V$ is an eigenvector matrix. Because $\kappa = 1$ for any unitary or orthogonal matrix, Theorem 7.4.1 then implies that the condition number of the eigenvalues of a hermitian or any normal matrix is one. That is, *eigenvalues of a normal matrix can be changed by no more than the norm of the perturbation to the matrix.*

## 7.4. Symmetry and definiteness

**Example 7.4.2**

We construct a real symmetric matrix with known eigenvalues by using the QR factorization to produce a random orthogonal set of eigenvectors.

```
n = 30;
lambda = (1:n)';
D = diag(lambda);
[V,R] = qr(randn(n));    % get a random orthogonal V
A = V*D*V';
```

The condition number of these eigenvalues is one. Thus the effect on them is bounded by the norm of the perturbation to $A$.

```
for k = 1:3
    E = randn(n); E = 1e-4*E/norm(E);
    mu = sort(eig(A+E));
    max_change = norm(mu-lambda,inf)
end
```

```
max_change =
   2.5564e-05
max_change =
   2.0501e-05
max_change =
   2.3712e-05
```

## Rayleigh quotient

Recall that for a matrix $A$ and compatible vector $x$, the quadratic form $x^*Ax$ is a scalar. With a suitable normalization, it becomes the **Rayleigh quotient**

$$R_A(x) = \frac{x^*Ax}{x^*x}. \tag{7.4.2}$$

If $v$ is an eigenvector such that $Av = \lambda v$, then one easily calculates that $R_A(v) = \lambda$. That is, *the Rayleigh quotient maps an eigenvector into its associated eigenvalue.*

If $A^* = A$, then the Rayleigh quotient has another interesting property: $\nabla R_A(v) = 0$ if $v$ is an eigenvector. By a multidimensional Taylor series, then,

$$R_A(v + \epsilon z) = R_A(v) + 0 + O(\epsilon^2) = \lambda + O(\epsilon^2), \tag{7.4.3}$$

as $\epsilon \to 0$. The conclusion is that a good estimate of an eigenvector becomes an even better estimate of an eigenvalue.

## Example 7.4.3

We construct a symmetric matrix with a known EVD.

```
n = 20;
lambda = (1:n)';   D = diag(lambda);
[V,~] = qr(randn(n));    % get a random orthogonal V
A = V*D*V';
```

The Rayleigh quotient of an eigenvector is its eigenvalue.

```
R = @(x) (x'*A*x)/(x'*x);
format long, R(V(:,7))
```

```
ans =
    7.000000000000001
```

The Rayleigh quotient's value is much closer to an eigenvalue than its input is to an eigenvector. In this experiment, each additional digit of accuracy in the eigenvector estimate gives two more digits to the eigenvalue estimate.

```
delta = 1./10.^(1:4)';
quotient = 0*delta;
for k = 1:4
    e = randn(n,1);  e = delta(k)*e/norm(e);
    x = V(:,7)+e;
    quotient(k) = R(x);
end
table(delta,quotient)
```

```
ans =
  4x2 table
    delta            quotient
    _____        _____

     0.1          7.05738940427937
     0.01         7.00066684894918
     0.001        7.00000278235035
     0.0001       7.00000005557751
```

## Definite and indefinite matrices

In the real case, we called a symmetric matrix $A$ *symmetric positive definite* (SPD) if $x^T A x > 0$ for all nonzero vectors $x$. In the complex case the relevant property is **hermitian positive definite** (HPD), meaning that $A^* = A$ and $x^* A x > 0$ for all nonzero complex vectors $x$. Putting this property together with the Rayleigh quotient leads to the following theorem.

## Theorem 7.4.3

If $A^* = A$, then the following statements are equivalent:

1. $A$ is HPD.
2. The eigenvalues of $A$ are positive numbers.
3. Any unitary EVD of $A$ is also an SVD of $A$.

Naturally, a hermitian matrix with all negative eigenvalues is called *negative definite*, and one with eigenvalues of different signs is *indefinite*. Finally, if one or more eigenvalues is zero and the rest have one sign, it is *semidefinite*.

## Exercises

7.4.1. Each line below is an EVD for a hermitian matrix. State whether the matrix is definite, indefinite, or semidefinite. Then state whether the given factorization is also an SVD, and if it is not, modify it to find an SVD.

(a) $\begin{bmatrix} 0 & 0 \\ 0 & -1 \end{bmatrix} = \begin{bmatrix} 0 & 1 \\ 1 & 0 \end{bmatrix} \begin{bmatrix} -1 & 0 \\ 0 & 0 \end{bmatrix} \begin{bmatrix} 0 & 1 \\ 1 & 0 \end{bmatrix}$.

(b) $\begin{bmatrix} 4 & -2 \\ -2 & 1 \end{bmatrix} = \begin{bmatrix} 1 & -0.5 \\ -0.5 & -1 \end{bmatrix} \begin{bmatrix} 5 & 0 \\ 0 & 0 \end{bmatrix} \begin{bmatrix} 0.8 & -0.4 \\ -0.4 & -0.8 \end{bmatrix}$.

(c) $\begin{bmatrix} -5 & 3 \\ 3 & -5 \end{bmatrix} = \begin{bmatrix} \alpha & \alpha \\ \alpha & -\alpha \end{bmatrix} \begin{bmatrix} -2 & 0 \\ 0 & -8 \end{bmatrix} \begin{bmatrix} \alpha & \alpha \\ \alpha & -\alpha \end{bmatrix}$ $(\alpha = 1/\sqrt{2})$.

7.4.2. Prove true, or give a counterexample: If $A$ and $B$ are hermitian matrices of the same size, then

$$R_{A+B}(x) = R_A(x) + R_B(x).$$

7.4.3. The range of the function $R_A(x)$ is a subset of the complex plane known as the *field of values* of the matrix $A$. Use 500 random vectors to plot points in the field of values of

$$A = \begin{bmatrix} 1 & 0 & -2 \\ 0 & 2 & 0 \\ -2 & 0 & 1 \end{bmatrix}.$$

Then compute its eigenvalues and guess what the exact field of values is.

7.4.4. Let $A = \begin{bmatrix} 3 & -2 \\ -2 & 0 \end{bmatrix}$.

(a) Write out $R_A(x)$ explicitly as a function of $x_1$ and $x_2$.
(b) Find $R_A(x)$ for $x_1 = 1$, $x_2 = 2$.
(c) Find the gradient vector $\nabla R_A(x)$.
(d) Show that the gradient vector is zero when $x_1 = 1$, $x_2 = 2$.

7.4.5. A *skew-Hermitian* matrix is one that satisfies $A^* = -A$. Show that if $A$ is skew-Hermitian, then $R_A$ is imaginary-valued.

## 7.5 • Dimension reduction

The SVD has another important property that proves very useful in a variety of applications. Let $A$ be a real $m \times n$ matrix with SVD $A = USV^T$ and (momentarily) $m \geq n$. Another way of writing the thin form of the SVD is

$$A = \widehat{U}\widehat{S}V^T = \begin{bmatrix} u_1 & u_2 & \cdots & u_n \end{bmatrix} \begin{bmatrix} \sigma_1 & & \\ & \ddots & \\ & & \sigma_n \end{bmatrix} \begin{bmatrix} v_1^T \\ \vdots \\ v_n^T \end{bmatrix}$$

$$= \begin{bmatrix} \sigma_1 u_1 & \cdots & \sigma_n u_n \end{bmatrix} \begin{bmatrix} v_1^T \\ \vdots \\ v_n^T \end{bmatrix}$$

$$= \sigma_1 u_1 v_1^T + \cdots + \sigma_r u_r v_r^T = \sum_{i=1}^{r} \sigma_i u_i v_i^T, \qquad (7.5.1)$$

where $r$ is the rank of $A$. The final formula also holds for the case $m < n$.

Each outer product $u_i v_i^T$ is a rank-1 matrix of unit norm. Thanks to the ordering of singular values, then, equation (7.5.1) expresses $A$ as a sum of decreasingly important contributions. This motivates the definition, for $1 \leq k \leq r$,

$$A_k = \sum_{i=1}^{k} \sigma_i u_i v_i^T = U_k S_k V_k^T. \qquad (7.5.2)$$

where $U_k$ and $V_k$ are the first $k$ columns of $U$ and $V$, respectively, and $S_k$ is the upper-left $k \times k$ submatrix of $S$.

The rank of a sum of matrices is always less than or equal to the sum of the ranks, so $A_k$ is a rank-$k$ approximation to $A$. It turns out that $A_k$ *is the* best *rank-k approximation of $A$,* as measured in the matrix 2-norm.

> **Theorem 7.5.1**
>
> Suppose $A$ has rank $r$ and let $A = USV^T$ be an SVD. Let $A_k$ be as in (7.5.2) for $1 \leq k < r$. Then
>
> 1. $\|A - A_k\|_2 = \sigma_{k+1}$, $k = 1, \ldots, r-1$.
> 2. If the rank of $B$ is $k$ or less, then $\|A - B\|_2 \geq \sigma_{k+1}$.

*Proof of part* 1. Note that (7.5.2) is identical to (7.5.1) with $\sigma_{k+1}, \ldots, \sigma_r$ all set to zero. This implies that

$$A - A_k = U(S - \widehat{S})V^T,$$

where $\widehat{S}$ has those same values of $\sigma_i$ replaced by zero. But that makes the above an SVD of $A - A_k$, with singular values $0, \ldots, 0, \sigma_{k+1}, \ldots, \sigma_r$, the largest of which is $\sigma_{k+1}$. That proves the first claim. □

## 7.5. Dimension reduction

If the singular values of $A$ decrease sufficiently rapidly, then $A_k$ may capture the "most significant" behavior of the matrix for a reasonably small value of $k$.

### Example 7.5.1

We make an image from some text, then reload it as a matrix.

```
tobj = text(0,0,'Hello world','fontsize',44);
saveas(gcf,'hello.png')
A = imread('hello.png');
A = double(rgb2gray(A));
imagesc(A), colormap gray
[m,n] = size(A)

m =
    300
n =
    675
```

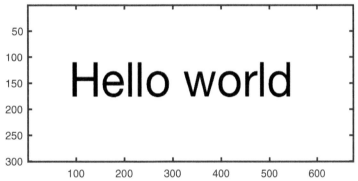

Next we show that the singular values decrease exponentially until they reach zero (more precisely, are about $\sigma_1 \varepsilon_{\text{mach}}$). For all numerical purposes, this determines the rank of the matrix.

```
[U,S,V] = svd(A);
sigma = diag(S);
semilogy(sigma,'.')
r = find(sigma/sigma(1) > 10*eps,1,'last')

r =
    55
```

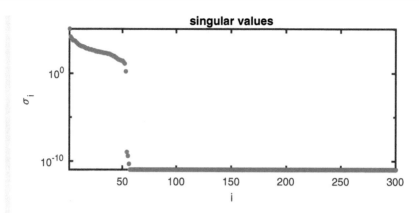

The rapid decrease suggests that we can get fairly good low-rank approximations.

```
for i = 1:4
    subplot(2,2,i)
    k = 2*i;
    Ak = U(:,1:k)*S(1:k,1:k)*V(:,1:k)';
    imshow(Ak,[0 255])
    title(sprintf('rank = %d',k))
end
```

**rank = 2**

**rank = 4**

Hollo world

**rank = 6**

Hello world

**rank = 8**

Hello world

Consider how little data is needed to reconstruct these images. For rank 8, for instance, we have 8 left and right singular vectors plus 8 singular values, for a compression ratio of better than 25:1.

```
compression = 8*(m+n+1) / (m*n)

compression =
    0.0386
```

## Capturing major trends

The use of dimension reduction offered by low-rank SVD approximation goes well beyond simply reducing computation time. By isolating the most important contributions to the matrix, *dimension reduction can uncover deep connections and trends that are otherwise obscured by lower-order effects and noise.*

## 7.5. Dimension reduction

One useful way to quantify the decay in the singular values is to compute

$$\tau_k = \frac{\sum_{i=1}^k \sigma_i^2}{\sum_{i=1}^r \sigma_i^2}, \quad k=1,\ldots,r. \tag{7.5.3}$$

Clearly $0 \le \tau_k \le 1$ and $\tau_k$ is nondecreasing as a function of $k$. We can think of $\tau_k$ as the fraction of "energy" contained in the singular values up to and including the $k$th.[32]

### Example 7.5.2

This matrix describes the votes on bills in the 111th session of the United States Senate. (The data set was obtained from voteview.com.) Each row is one senator and each column is a vote item.

```
load voting      % get from the book's website
[m,n] = size(A);
```

If we visualize the votes (white is "yea," black is "nay," and gray is anything else), we can see great similarity between many rows, reflecting party unity.

```
imagesc(A)
colormap gray
```

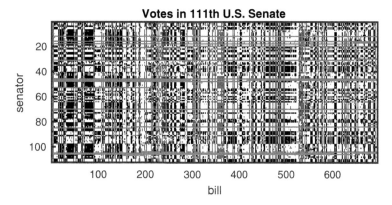

We use singular value "energy" to quantify the decay rate of the values.

```
[U,S,V] = svd(A);
sigma = diag(S);
tau = cumsum(sigma.^2) / sum(sigma.^2);
plot(tau(1:16),'.')
```

---

[32] In statistics this quantity may be interpreted as the fraction of explained variance.

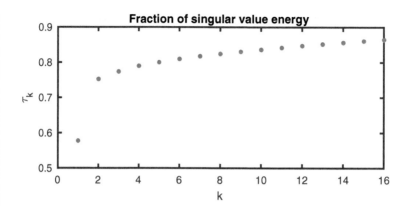

The first and second singular triples contain about 58% and 17%, respectively, of the energy of the matrix. All others have far less effect, suggesting that the information is primarily two-dimensional. The first left and right singular vectors also contain interesting structure.

```
subplot(211), plot(U(:,1),'.')
subplot(212), plot(V(:,1),'.')
```

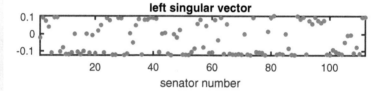

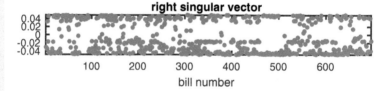

Both vectors have values greatly clustered near $\pm C$ for a constant $C$. These can be roughly interpreted as how partisan a particular senator or bill was, and for which political party.

Projecting the senators' vectors into the first two $V$-coordinates gives a particularly nice way to reduce them to two dimensions. Political scientists label these dimensions "partisanship" and "bipartisanship." Here we color them by actual party affiliation (also given in the data file): red for Republican, blue for Democrat, and black for independent.

```
clf
x1 = V(:,1)'*A';    x2 = V(:,2)'*A';
scatter(x1(Dem),x2(Dem),20,'b'),   hold on
scatter(x1(Rep),x2(Rep),20,'r')
scatter(x1(Ind),x2(Ind),20,'k')
title('111th US Senate in 2D')
```

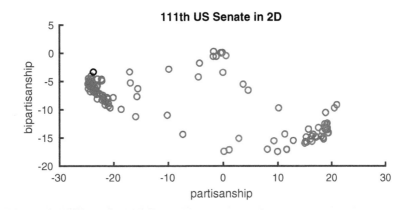

Not all data sets can be reduced effectively to a small number of dimensions, but as Example 7.5.2 shows, in some cases reduction reveals information that may correspond to real-world understanding.

## Exercises

7.5.1. ✍ Suppose that $A$ is an $n \times n$ matrix. Explain why $\sigma_n$ is the distance (in the 2-norm) from $A$ to the set of all singular matrices.

7.5.2. ✍ Suppose $A$ is a $7 \times 4$ matrix and the eigenvalues of $A^*A$ are 3, 4, 7, and 10. How close is $A$ (in the 2-norm) to (a) a rank-3 matrix? (b) a rank-2 matrix?

7.5.3. (a) ⌨ Find the rank-1 matrix closest to $A = \begin{bmatrix} 1 & 5 \\ 5 & 1 \end{bmatrix}$, as measured in the 2-norm.

(b) ⌨ Repeat part (a) for $A = \begin{bmatrix} 1 & 5 \\ 0 & 1 \end{bmatrix}$.

7.5.4. ✍ Find the rank-1 matrix closest to $A = \begin{bmatrix} 1 & b \\ b & 1 \end{bmatrix}$, as measured in the 2-norm, where $b > 0$.

7.5.5. ⌨ Following Example 7.5.1 as a guide, load peppers.png and convert it to a matrix of floating point pixel intensities. Using the SVD, display as images the best approximations of rank 5, 10, 15, and 20.

## Key ideas in this chapter

1. Tables, graphs, and images are all examples of two-dimensional data that can be usefully interpreted as matrices (page 281).
2. A square matrix may have an eigenvalue decomposition $A = VDV^{-1}$, where $D$ is diagonal (page 287).
3. Multiplication by the inverse of a matrix performs a change of basis into the coordinates associated with the matrix. An EVD changes basis to one in which the matrix acts diagonally (page 289).

4. Any matrix has a singular value decomposition (SVD) $A = USV^*$, where $S$ is diagonal and nonnegative, and $U$ and $V$ are unitary (page 293).
5. The SVD is intimately connected to the 2-norm (page 295).
6. A hermitian matrix has a complete set of orthonormal eigenvectors and real eigenvalues (page 299).
7. The eigenvalues of a normal matrix are perturbed by no more than the size of the perturbation to the matrix (page 300).
8. The Rayleigh quotient is a function mapping an eigenvector into its associated eigenvalue, and an eigenvector estimate into an eigenvalue estimate (page 301).
9. Truncating the SVD provides the best low-rank approximation to a matrix (page 304).
10. Dimension reduction via the SVD can uncover connections that are otherwise obscured by lower-order effects and noise (page 306).

## Where to learn more

Details on the computation of the EVD and SVD are presented at length in [60] and more briefly in Chapters 7 and 8 of [27]. A classic reference on the particulars of the symmetric case is [53], while [73] focuses on the nonnormal case. Dimension reduction via the SVD often goes by the name *principal component analysis*, which is the subject of [35].

# Chapter 8
# Krylov methods in linear algebra

> I warn you not to underestimate my powers.
> —Luke Skywalker, *Return of the Jedi*

What are the implications of the $O(n^3)$ work requirements for solving linear systems? Suppose tomorrow your computer became a thousand times faster. (Historically this has taken about 15 years in the real world.) Assuming you are willing to wait just as long then as you are today, the size of the linear system you can solve will go up only by a factor of 10. Nice, but not nearly the jump that you get in hardware power. In fact, there is an odd paradox: faster computers make faster algorithms *more* important, not less—because they demand that you work at larger values of $n$, where asymptotic differences are large.

In practice the only reasonable way to deal with large matrices (at this writing, $n > 10^4$ or so) is if they are sparse, or can be approximated sparsely. But LU factorization of a sparse matrix does not necessarily lead to sparse factors, particularly when row pivoting is required. The algorithm can be improved to be more sparse-aware, but we will not go into the details.

Instead, we will replace LU factorization with an iterative algorithm. Unlike the LU factorization, iteration gives useful intermediate and continually improving results before the exact solution is found, allowing us to stop well before the nominal exact termination. More importantly, though, these iterations, based on an idea called *Krylov subspaces*, allow us to fully exploit sparsity.

Krylov subspace methods have two other advantages that are subtle but critically relevant to applications. One is that they allow us to do linear algebra *even without having the relevant matrix*. This may sound undesirable or even impossible, but it exploits the connection between matrix-vector multiplication and a linear transformation, as discussed in Section 8.7. The other major unique advantage of Krylov subspace iterations is that they can exploit "approximate inverses" when they are available. These two features are among the most powerful ideas behind scientific computation today.

## 8.1 · Sparsity and structure

Very large matrices cannot be stored all within the primary memory of a computer unless they are sparse. *A sparse matrix has structural zeros, meaning entries that are known to be exactly zero.* For instance, the adjacency matrix of a graph has zeros where there are no links in the graph. To store and operate with a sparse matrix efficiently, it is not represented as an array of all of its values. There is a variety of sparse formats available; MATLAB chooses *compressed sparse column* format. For the most part, you can imagine that the matrix is stored as triples $(i, j, A_{ij})$ for all the nonzero $(i, j)$ locations.

### Computing with sparse matrices

Most graphs with real applications have many fewer edges than the maximum possible $n^2$ for $n$ nodes. Accordingly, their adjacency matrices have mostly zero elements and should be represented sparsely.

**Example 8.1.1**

Here we load the adjacency matrix of a graph with 2790 nodes. Each node is a web page referring to Roswell, NM, and the edges represent links between web pages.

```
load roswelladj    % get from the book's website
a = whos('A')

a =
  struct with fields:
         name: 'A'
         size: [2790 2790]
        bytes: 158120
        class: 'double'
       global: 0
       sparse: 1
      complex: 0
      nesting: [1x1 struct]
   persistent: 0
```

We may define the density of $A$ as the number of nonzeros divided by the total number of entries.

```
sz = size(A);  n = sz(1);
density = nnz(A) / prod(sz)

density =
    0.0011
```

We can compare the storage space needed for the sparse $A$ with the space needed for its dense or full counterpart. This ratio can never be as small as the density of nonzeros,

## 8.1. Sparsity and structure

because of the need to store locations as well as data. However, it's still quite small here, even though the matrix is not really large.

```
F = full(A);
f = whos('F');
a.bytes/f.bytes
```

```
ans =
    0.0025
```

Matrix-vector products are also much faster using the sparse form, because operations with structural zeros are skipped.

```
x = randn(n,1);
tic, for i = 1:200, A*x; end
sparse_time = toc
```

```
sparse_time =
    0.0038
```

```
tic, for i = 1:200, F*x; end
dense_time = toc
```

```
dense_time =
    0.6262
```

However, the sparse storage format in MATLAB is column-oriented. Operations on rows may take a lot longer than similar ones on columns.

```
v = A(:,1000);
tic, for i = 1:n, A(:,i)=v; end
column_time = toc
r = v';
tic, for i = 1:n, A(i,:)=r; end
row_time = toc
```

```
column_time =
    0.0066
row_time =
    0.0666
```

Arithmetic operations such as +, -, *, and ^ respect and exploit sparsity, if the matrix operands are sparse. However, *matrix operations may substantially decrease the amount of sparsity, a phenomenon known as fill-in.*

In the case of an adjacency matrix $A$, for example, the $(i,j)$ entry of matrix $A^k$ for positive integer $k$ is the number of paths of length $k$ from node $i$ to node $j$.

### Example 8.1.2

Here is the buckyball adjacency matrix again.

```
[A,v] = bucky;
```

The number of vertex pairs on a soccer ball connected by a path of length $k > 1$ grows with $k$, as can be seen here for $k = 3$.

```
subplot(1,2,1), spy(A)
subplot(1,2,2), spy(A^3)
```

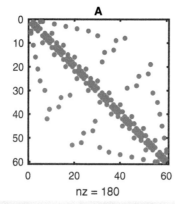
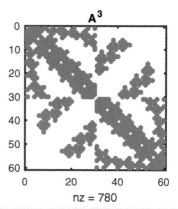

## Banded matrices

A particularly important type of sparse matrix is a banded matrix. Recall from Section 2.9 that $A$ has **upper bandwidth** $p$ if $j-i > p$ implies $A_{ij} = 0$, and **lower bandwidth** $q$ if $i-j > q$ implies $A_{ij} = 0$; we say the total **bandwidth** is $p+q+1$. Banded matrices appear naturally in many applications where each node interacts directly with only its closest neighbors. In MATLAB, they are often straightforward to construct using the command **spdiags**. *Without pivoting, an LU factorization preserves bandwidth, but pivoting can change or destroy bandedness.*

### Example 8.1.3

Here is a matrix with both lower and upper bandwidth equal to one. Such a matrix is called **tridiagonal**. The `spdiags` command creates a sparse matrix given its diagonal elements. The main or central diagonal is numbered zero, above and to the right of that is positive, and below and to the left is negative.

```
n = 50;
d = [ n*ones(n,1), ones(n,1), -(1:n)'];  % diagonal entries
pos = [-3 0 1];                           % which diagonals
A = spdiags(d,pos,n,n);
full( A(1:7,1:7) )
```

## 8.1. Sparsity and structure

```
ans =
     1    -2     0     0     0     0     0
     0     1    -3     0     0     0     0
     0     0     1    -4     0     0     0
    50     0     0     1    -5     0     0
     0    50     0     0     1    -6     0
     0     0    50     0     0     1    -7
     0     0     0    50     0     0     1
```

Without pivoting, the LU factors have the same lower and upper bandwidth as the original matrix.

```
[L,U] = lufact(A);
subplot(1,2,1), spy(L), title('L')
subplot(1,2,2), spy(U), title('U')
```

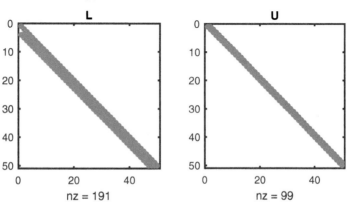

However, if we introduce row pivoting, bandedness may be expanded or destroyed.

```
[L,U,P] = lu(A);
subplot(1,2,1), spy(L), title('L')
subplot(1,2,2), spy(U), title('U')
```

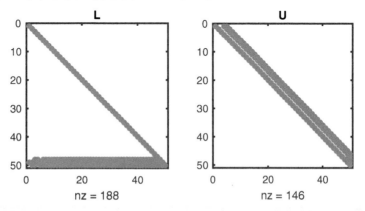

## Linear systems and eigenvalues

If given a sparse matrix, the backslash operator will automatically try a form of sparse-aware Cholesky or pivoted LU factorization. Depending on the sparsity pattern of the matrix, the time taken to solve the linear system may be well below the $O(n^3)$ needed in the general case.

For very large matrices, it's unlikely that you will want to find all of its eigenvalues and eigenvectors. The **eigs** command will find a selected number of eigenvalues of largest magnitude, eigenvalues lying to the extreme left or right, or eigenvalues nearest a given complex number. The algorithm that powers **eigs** will be described in Section 8.4. A similar command **svds** exists for singular values.

### Example 8.1.4

The sprandsym command generates a random sparse matrix with prescribed eigenvalues.

```
n = 3000;
density = 1.23e-3;
lambda = 1./(1:n);
A = sprandsym(n,density,lambda);
spy(A)
```

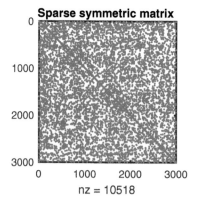

```
eigs(A,5)      % largest magnitude

ans =
    1.0000
    0.5000
    0.3333
    0.2500
    0.2000

eigs(A,5,0)    % closest to zero

ans =
   1.0e-03 *
```

```
   0.3338
   0.3337
   0.3336
   0.3334
   0.3333
```

The scaling of time to solve a sparse linear system is not easy to predict unless you have some more information about the matrix (such as bandedness). But it will typically be a great deal faster than the dense or full matrix case.

```
x = 1./(1:n)';  b = A*x;
tic, sparse_err = norm(x - A\b), sparse_time = toc

sparse_err =
   2.1074e-14
sparse_time =
   0.0016

A = full(A);
tic, dense_err = norm(x - A\b), dense_time = toc

dense_err =
   7.2871e-14
dense_time =
   0.1735
```

## Exercises

8.1.1. Use **spdiags** to build the 200 × 200 matrices

$$A = \begin{bmatrix} -2 & 1 & & & \\ 1 & -2 & 1 & & \\ & \ddots & \ddots & \ddots & \\ & & 1 & -2 & 1 \\ & & & 1 & -2 \end{bmatrix}, \quad B = \begin{bmatrix} -2 & 1 & & & 1 \\ 1 & -2 & 1 & & \\ & \ddots & \ddots & \ddots & \\ & & 1 & -2 & 1 \\ 1 & & & 1 & -2 \end{bmatrix}.$$

Use **spy** and an inspection of the 5 × 5 submatrices in the corners of $A$ to verify your matrices.

8.1.2. This problem uses **bucky** as introduced in Example 7.1.4 to generate an adjacency matrix $A$.

(a) Find the density of $A$ (number of nonzeros divided by total number of elements), $A^2$, $A^4$, and $A^8$. You should find that it increases with the power of $A$.

(b) The LU factors tend to at least partially retain sparsity. Find the density of the $L$ and $U$ factors of $A$ using the built-in **lu** (with two outputs).

(c) The QR factors have no particular relationship to sparsity in general. Repeat part (b) for the QR factorization using **qr**.

8.1.3. ▣ The matrix A = gallery('poisson',n); arises in the numerical solution of elliptic partial differential equations using finite difference methods. The resulting $n^2 \times n^2$ matrix $A$ is sparse, banded, and SPD.

  (a) For $n = 10, 20, \ldots, 80$, compute [L,U] = lu(A);. Make a table with $n$, the number of elements in $A$, the number of nonzeros in $A$ (using nnz), and the number of nonzeros in $L$. In the case $n = 10$, use subplot to make side-by-side spy plots of $A$ and $L$.

  (b) Repeat part (a) using [L,U,P,Q] = lu(A);, which uses a better sparse algorithm, and add the number of nonzeros in this version of $L$ to your table. Also make a spy plot for $n = 10$.

8.1.4. ✍ Prove the statement illustrated by Example 8.1.2: if $A$ is an adjacency matrix, the $(i, j)$ entry of matrix $A^k$ for positive integer $k$ is the number of paths of length $k$ from node $i$ to node $j$.

8.1.5. ▣ A common practical use of adjacency matrices is to analyze the links between members of a collection. Obtain the adjacency matrix $A$ by loading the file roswelladj.mat, available from the book website. (Credit goes to Panayiotis Tsaparas and the University of Toronto for making this data public.) The matrix catalogs the links between websites related to the town of Roswell, NM, with $A_{ij} = 1$ if and only if site $i$ points to site $j$.

  (a) Verify numerically that the matrix does not include self-links (links from a site to itself).

  (b) Note that $A$ does not need to be symmetric, because web links point in one direction. Verify numerically that $A$ is not symmetric.

  (c) How many sites in the group are not pointed to by any other sites in the group?

  (d) Which site points to the most other sites?

  (e) Which site is pointed to the most by the other sites? This is a crude way to establish the "most important" site.

  (f) There are $2790^2$ possible ways to connect ordered pairs of sites. What fraction of these pairs is connected by a path of links that is no greater than three in length?

8.1.6. ▣ The *graph Laplacian matrix* is $L = D - A$, where $A$ is the adjacency matrix and $D$ is the *degree matrix*, a diagonal matrix with diagonal entries $d_{jj} = \sum_{i=1}^{n} a_{ij}$. Load the file roswelladj.mat, available from the book website, to get an adjacency matrix $A$. Then find the five eigenvalues of $L$ having largest magnitude.

8.1.7. ▣ The file actors.mat (available at the book site and based on data provided by the Self-Organized Networks Database at the University of Notre Dame) contains information about the appearances of 392,400 actors in 127,823 movies, as given by the Internet Movie Database. The matrix $A$ has $A_{ij} = 1$ if actor $j$ appeared in movie $i$ and zero elsewhere.

  (a) What is the maximum number of actors appearing in any one movie?

  (b) How many actors appeared in exactly three movies?

  (c) Define $C = A^T A$. How many nonzero entries does $C$ have? What is the simple meaning of $C_{ij}$?

## 8.2 • Power iteration

Given that matrix-vector multiplication is so fast for sparse matrices, let's see what we might accomplish with just that at our disposal.

### Example 8.2.1

Here we let $A$ be a $5 \times 5$ matrix. We also choose a random 5-vector.

```
A = magic(5)/65;
x = randn(5,1)

x =
    1.7491
    0.1326
    0.3252
   -0.7938
    0.3149
```

Applying matrix-vector multiplication once doesn't do anything recognizable.

```
y = A*x

y =
    0.4864
    0.5707
    0.0473
    0.1467
    0.4770
```

Repeating the multiplication still doesn't do anything obvious.

```
z = A*y

z =
    0.4668
    0.3701
    0.2987
    0.2634
    0.3291
```

But if we keep repeating the matrix-vector multiplication, something remarkable happens: $Ax \approx x$.

```
for j = 1:8,  x = A*x;  end
[x,A*x]

ans =
    0.3457    0.3457
    0.3457    0.3456
```

```
        0.3455       0.3456
        0.3455       0.3456
        0.3456       0.3456
```

This seems to occur regardless of the starting value of $x$.

```
x = randn(5,1)
for j = 1:8,   x = A*x;   end
[x,A*x]
```

```
x =
    -0.5273
     0.9323
     1.1647
    -2.0457
    -0.6444
ans =
    -0.2240    -0.2241
    -0.2239    -0.2241
    -0.2239    -0.2241
    -0.2242    -0.2241
    -0.2243    -0.2240
```

Example 8.2.1 leads to some speculation. If it holds exactly, the equation $Ax = x$ is a special case of the eigenvalue equation $Ax = \lambda x$. We can use `eig` to determine that the eigenvalues of $A$ from the example are about 1, $\pm 0.327$, and $\pm 0.202$. So it appears that starting from an arbitrary $x$, repeated multiplication by $A$ leads to an eigenvector associated with the eigenvalue $\lambda = 1$.

## Dominant eigenvalue

Analysis of Example 8.2.1 is fairly straightforward. Let $A$ be any diagonalizable $n \times n$ matrix having eigenvalues $\lambda_1, \ldots, \lambda_n$ and corresponding linearly independent eigenvectors $v_1, \ldots, v_n$. Furthermore, suppose the eigenvalues (possibly after renumbering) are such that

$$|\lambda_1| > |\lambda_2| \geq |\lambda_3| \geq \cdots \geq |\lambda_n|. \tag{8.2.1}$$

Given (8.2.1) we say that $\lambda_1$ is the **dominant eigenvalue**. This was the case with $\lambda_1 = 1$ for $A$ in Example 8.2.1.

Now let the initially chosen $x$ be expressed as a linear combination in the eigenvector basis:

$$x = c_1 v_1 + c_2 v_2 + \cdots + c_n v_n.$$

Observe that

$$Ax = c_1 A v_1 + c_2 A v_2 + \cdots + c_n A v_n$$
$$= c_1 \lambda_1 v_1 + c_2 \lambda_2 v_2 + \cdots + c_n \lambda_n v_n.$$

## 8.2. Power iteration

If we apply $A$ repeatedly, then

$$A^k x = \lambda_1^k c_1 v_1 + \lambda_2^k c_2 v_2 + \cdots + \lambda_n^k c_n v_n$$
$$= \lambda_1^k \left[ c_1 v_1 + \left(\frac{\lambda_2}{\lambda_1}\right)^k c_2 v_2 + \cdots + \left(\frac{\lambda_n}{\lambda_1}\right)^k c_n v_n \right]. \quad (8.2.2)$$

Since $\lambda_1$ is dominant, we see that in any norm,

$$\left\| \frac{A^k x}{\lambda_1^k} - c_1 v_1 \right\| \leq |c_2| \cdot \left|\frac{\lambda_2}{\lambda_1}\right|^k \|v_2\| + \cdots + |c_n| \cdot \left|\frac{\lambda_n}{\lambda_1}\right|^k \|v_n\| \to 0 \quad \text{as } k \to \infty. \quad (8.2.3)$$

As long as $c_1 \neq 0$, $A^k x$ eventually is almost parallel to the dominant eigenvector.[33]

For algorithmic purposes, it is important to interpret $A^k x$ as $A(\cdots(A(Ax))\cdots)$, i.e., as repeated applications of $A$ to a vector, as implemented in Example 8.2.1. This interpretation allows us to fully exploit any sparsity of $A$, something which is not preserved by taking a matrix power $A^k$ explicitly.

### Power iteration

An important technicality separates us from an algorithm: unless $|\lambda_1| = 1$, the factor $\lambda_1^k$ tends to make $\|A^k x\|$ either very large or very small. *To make a practical algorithm, we alternate matrix-vector multiplication with a renormalization of the vector.* This algorithm is known as the **power iteration**.

1. Choose $x_1$.
2. For $k = 1, 2, \ldots$

$$y_k = A x_k, \quad (8.2.4a)$$

$$\alpha_k = \frac{1}{y_{k,m}}, \quad \text{where } |y_{k,m}| = \|y_k\|_\infty, \quad (8.2.4b)$$

$$x_{k+1} = \alpha_k y_k. \quad (8.2.4c)$$

Our notation here uses $y_{k,m}$ to mean the $m$th component of $y_k$, or $e_m^T y_k$. Note that now $\|x_{k+1}\|_\infty = 1$. We can write

$$x_k = (\alpha_1 \alpha_2 \cdots \alpha_k) A^k x_1. \quad (8.2.5)$$

Thus the renormalization step modifies (8.2.2) and (8.2.3) only slightly.

So far we have discussed only eigenvector estimation. However, if $x_k$ is nearly a dominant eigenvector of $A$, then $A x_k$ is nearly $\lambda_1 x_k$, and we can take the ratio $\gamma_k = y_{k,m}/x_{k,m}$ as an eigenvalue estimate. In fact, revisiting (8.2.2), the extra $\alpha_j$ normalization factors cancel in the ratio, and, after some simplification, we get

$$\gamma_k = \frac{y_{k,m}}{x_{k,m}} = \lambda_1 \frac{1 + r_2^{k+1} b_2 + \cdots + r_n^{k+1} b_n}{1 + r_2^k b_2 + \cdots + r_n^k b_n}, \quad (8.2.6)$$

---

[33] If $x$ is chosen randomly, the odds that $c_1 = 0$ are mathematically zero.

**Function 8.2.1 (poweriter)** Power iteration for the dominant eigenvalue.

```
1  function [gamma,x] = poweriter(A,numiter)
2  % POWERITER   Power iteration for the dominant eigenvalue.
3  % Input:
4  %    A          square matrix
5  %    numiter    number of iterations
6  % Output:
7  %    gamma      sequence of eigenvalue approximations (vector)
8  %    x          final eigenvector approximation
9
10 n = length(A);
11 x = randn(n,1);
12 x = x/norm(x,inf);
13 for k = 1:numiter
14   y = A*x;
15   [normy,m] = max(abs(y));
16   gamma(k) = y(m)/x(m);
17   x = y/y(m);
18 end
```

where $r_j = \lambda_j/\lambda_1$ and the $b_j$ are constants. By assumption (8.2.1), each $r_j$ satisfies $|r_j| < 1$, so we see that $\gamma_k \to \lambda_1$ as $k \to \infty$.

An implementation of power iteration is shown in Function 8.2.1. Line 15 of the code uses the syntax of **max** to find both the largest element of a vector and its location in the vector. Observe that the only use of $A$ is to find the matrix-vector product $Ax$, which makes exploitation of the sparsity of $A$ automatic.

## Convergence

Let's examine the terms in the numerator and denominator of (8.2.6) more carefully:

$$r_2^k b_2 + \cdots + r_n^k b_n = r_2^k \left[ b_2 + \left(\frac{r_3}{r_2}\right)^k b_3 + \cdots + \left(\frac{r_n}{r_2}\right)^k b_n \right]$$
$$= r_2^k \left[ b_2 + \left(\frac{\lambda_3}{\lambda_2}\right)^k b_3 + \cdots + \left(\frac{\lambda_n}{\lambda_2}\right)^k b_n \right]. \tag{8.2.7}$$

At this point we'll introduce an extra assumption,

$$|\lambda_2| > |\lambda_3| \geq \cdots \geq |\lambda_n|. \tag{8.2.8}$$

This condition isn't strictly necessary, but it simplifies the following statements considerably, because now it's clear that the quantity in (8.2.7) approaches $b_2 r_2^k$ as $k \to \infty$.

Next we estimate (8.2.6) for large $k$, using a geometric series expansion for the denominator to get

$$\gamma_k \to \lambda_1 \left(1 + b_2 r_2^{k+1}\right)\left(1 - b_2 r_2^k + O(r_2^{2k})\right),$$
$$\gamma_k - \lambda_1 \to \lambda_1 b_2 (r_2 - 1) r_2^k. \tag{8.2.9}$$

## 8.2. Power iteration

This is linear convergence with factor $r_2$. That is,

$$\frac{\gamma_{k+1} - \lambda_1}{\gamma_k - \lambda_1} \to r_2 = \frac{\lambda_2}{\lambda_1} \quad \text{as } k \to \infty. \tag{8.2.10}$$

*The error in the eigenvalue estimates $\gamma_k$ of power iteration is reduced asymptotically by a constant factor $\lambda_2/\lambda_1$ on each iteration.*

### Example 8.2.2

We set up a 5 × 5 matrix with prescribed eigenvalues, then apply the power iteration.

```
lambda = [1 -0.75 0.6 -0.4 0];
A = triu(ones(5),1) + diag(lambda)    % triangular matrix

A =
    1.0000    1.0000    1.0000    1.0000    1.0000
         0   -0.7500    1.0000    1.0000    1.0000
         0         0    0.6000    1.0000    1.0000
         0         0         0   -0.4000    1.0000
         0         0         0         0         0
```

We run the power iteration 60 times. The best estimate of the dominant eigenvalue is the last entry of gamma.

```
[gamma,x] = poweriter(A,60);
eigval = gamma(end)

eigval =
    1.0000
```

We check linear convergence using a log-linear plot of the error. We use our best estimate in order to compute the error at each step.

```
err = eigval - gamma;
semilogy(abs(err),'.-')
```

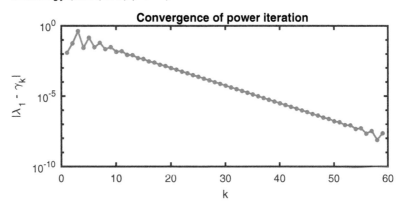

The trend is clearly a straight line asymptotically. We can get a refined estimate of the error reduction in each step by using the exact eigenvalues.

```
theory = lambda(2)/lambda(1)
observed = err(40)/err(39)

theory =
   -0.7500
observed =
   -0.7454
```

Note that the error is supposed to change sign on each iteration. An effect of these alternating signs is that estimates oscillate around the exact value.

```
format long
gamma(36:40)'

ans =
   0.999990221596137
   1.000007341824289
   0.999994498441981
   1.000004129053093
   0.999996904940087
```

The practical utility of (8.2.10) is limited, because if we knew $\lambda_1$ and $\lambda_2$, we wouldn't be running the power iteration in the first place! Sometimes it's possible to find estimates or bounds of the ratio. But for the most part we just find it a valuable theoretical statement of how power iteration should converge.

## Exercises

8.2.1. Use Function 8.2.1 to perform 20 iterations of the power method for the following matrices. Quantitatively compare the observed convergence to (8.2.10).

(a)
$$A = \begin{bmatrix} 1.1 & 1 \\ 0 & 2.1 \end{bmatrix}.$$

(b)
$$A = \begin{bmatrix} 2 & 1 \\ 1 & 0 \end{bmatrix}.$$

(c)
$$A = \begin{bmatrix} 6 & 5 & 4 \\ 5 & 4 & 3 \\ 4 & 3 & 2 \end{bmatrix}.$$

(d)
$$A = \begin{bmatrix} 2 & -1 & 0 \\ -1 & 2 & -1 \\ 0 & -1 & 2 \end{bmatrix}.$$

8.2.2. 🖳 Let
$$A = \begin{bmatrix} 6 & 3 & 3 \\ 1 & 10 & 1 \\ 2 & 5 & 5 \end{bmatrix}.$$

Use `eig` to find an eigenvalue decomposition of $A$.

(a) Modify Function 8.2.1 so that instead of choosing $x$ as a random vector initially, choose x=[3;-1;1]. Run the new `poweriter` for 100 iterations on $A$. Explain why power iteration does *not* find the dominant eigenvalue in this case.

(b) Now make the starting vector be x=[2;-1;2] and again run for 100 iterations. Plot $|\gamma_k - 6|$ and $|\gamma_k - 12|$ versus $k$ on a single semi-log graph.

(c) To the starting vector in part (b) add $10^{-8}$ elementwise and run again, adding the results to the graph.

(d) Explain what you observed in parts (b) and (c).

8.2.3. ✍ Describe what happens during power iteration using the matrix $A = \begin{bmatrix} 0 & 1 \\ 1 & 0 \end{bmatrix}$ and initial vector $x = \begin{bmatrix} 0.4 \\ 0.7 \end{bmatrix}$. Does the algorithm converge to an eigenvector? How does this relate to (8.2.2)?

8.2.4. 🖳 In Exercise 2.6.3 on page 73 we considered a mass-lumped model of a hanging string that led to a tridiagonal system of linear equations. The same process can be applied to an idealized membrane hanging from a square frame. Now we use a cartesian grid of masses, each mass directly interacting with the four neighbors immediately to the north, south, east, and west. If $n$ masses are used in each coordinate direction, we get a sparse matrix $A$ of size $n^2 \times n^2$, which can be found in MATLAB simply by using

```
A = n^2*gallery('poisson',n);
```

(a) Let $n = 10$ and make a `spy` plot of $A$. What is the density of $A$? Most rows all have the same number of nonzeros; find this number and explain it.

(b) The eigenvalues of $A$ are approximately squares of the frequencies of vibration for the membrane. Find the dominant $\lambda_1$ using `eig` for $n = 10, 15, 20, 25$.

(c) For each case of $n$ in part (b), apply 100 steps of Function 8.2.1. On one graph plot the four convergence curves $|\gamma_k - \lambda_1|$ using a semi-log scale. (They will not be smooth curves, because the matrix has many repeated eigenvalues that complicate our convergence analysis.)

8.2.5. 🖳 The matrix $A$ in the file `actors.mat` available on the book site is described in Exercise 8.1.7. Use Function 8.2.1 to find the leading eigenvalue of $A^T A$ to at least 6 significant digits.

8.2.6. 📖 For symmetric matrices, the Rayleigh quotient (7.4.2) converts an $O(\epsilon)$ eigenvector estimate into an $O(\epsilon^2)$ eigenvalue estimate. Duplicate Function 8.2.1 and rename it `powitersym`. Modify the new function to use the Rayleigh quotient to produce the entries of `gamma`. Test the original Function 8.2.1 and the new `powitersym` on a 50 × 50 symmetric matrix, and observe that the convergence rate for the new function is the square of that for the original one.

## 8.3 • Inverse iteration

Power iteration finds only the dominant eigenvalue. We next show that it can be adapted to find any eigenvalue, provided you start with a reasonably good estimate of it. Some simple linear algebra is all that is needed.

> **Theorem 8.3.1**
>
> Let $A$ be an $n \times n$ matrix with eigenvalues $\lambda_1, \ldots, \lambda_n$ (possibly with repeats), and let $s$ be a complex scalar. Then the following hold:
>
> 1. The eigenvalues of the matrix $A - sI$ are $\lambda_1 - s, \ldots, \lambda_n - s$.
> 2. If $s$ is not an eigenvalue of $A$, the eigenvalues of the matrix $(A - sI)^{-1}$ are $(\lambda_1 - s)^{-1}, \ldots, (\lambda_n - s)^{-1}$.
> 3. The eigenvectors associated with the eigenvalues in the first two parts are the same as those of $A$.

*Proof.* The equation $Av = \lambda v$ implies that $(A - sI)v = Av - sIv = \lambda v - sv = (\lambda - s)v$. That proves the first part of the theorem. For the second part, we note that, by assumption, $(A - sI)$ is nonsingular, so $(A - sI)v = (\lambda - s)v$ implies that $v = (\lambda - s)(A - sI)^{-1}v$, or $(\lambda - s)^{-1}v = (A - sI)^{-1}v$. The discussion above also proves the third part of the theorem. □

Consider first part 2 of the theorem with $s = 0$, and suppose that $A$ has a *smallest* eigenvalue,
$$|\lambda_n| \geq |\lambda_{n-1}| \geq \cdots > |\lambda_1|.$$
Then clearly
$$|\lambda_1^{-1}| > |\lambda_2^{-1}| \geq \cdots \geq |\lambda_n^{-1}|,$$
and $A^{-1}$ has a dominant eigenvalue. Hence power iteration on $A^{-1}$ can be used to find the eigenvalue of $A$ closest to zero. This is called **inverse iteration**. Comparing to (8.2.1) and (8.2.10), it is clear that the linear convergence rate of inverse iteration is the ratio
$$\frac{\lambda_2^{-1}}{\lambda_1^{-1}} = \frac{\lambda_1}{\lambda_2}.$$

These observations generalize easily to nonzero values of $s$. Specifically, if we suppose that the eigenvalues are ordered by their distance to $s$, i.e.,
$$|\lambda_n - s| \geq \cdots \geq |\lambda_2 - s| > |\lambda_1 - s|, \tag{8.3.1}$$

## 8.3. Inverse iteration

**Function 8.3.1 (inviter)** Shifted inverse iteration for the closest eigenvalue.

```
function [gamma,x] = inviter(A,s,numiter)
% INVITER    Shifted inverse iteration for the closest eigenvalue.
% Input:
%    A           square matrix
%    s           value close to targeted eigenvalue (complex scalar)
%    numiter     number of iterations
% Output:
%    gamma       sequence of eigenvalue approximations (vector)
%    x           final eigenvector approximation

n = length(A);
x = randn(n,1);
x = x/norm(x,inf);
B = A - s*eye(n);
[L,U] = lu(B);
for k = 1:numiter
    y = U \ (L\x);
    [normy,m] = max(abs(y));
    gamma(k) = x(m)/y(m) + s;
    x = y/y(m);
end
```

then it follows that

$$|\lambda_1 - s|^{-1} > |\lambda_2 - s|^{-1} \geq \cdots \geq |\lambda_n - s|^{-1}.$$

Hence *power iteration on the matrix* $(A - sI)^{-1}$ *converges to* $(\lambda_1 - s)^{-1}$. We call this **shifted inverse iteration** with shift $s$, though the term "inverse iteration" is often used to refer to the shifted variety too.

## Algorithm

Shifted inverse iteration introduces a major new algorithmic wrinkle. The key step to consider is the counterpart of (8.2.4a),

$$y_k = (A - sI)^{-1} x_k. \tag{8.3.2}$$

We do not want to explicitly find the inverse of a matrix. Instead we should write this step as

$$\text{Solve } (A - sI)y_k = x_k \text{ for } y_k. \tag{8.3.3}$$

Each step of inverse iteration therefore requires the solution of a linear system of equations with the matrix $B = A - sI$. The discussion becomes awkward for us at this moment, because the solution of the large, sparse linear system $By = x$ is something we consider later on in this chapter. In order to get things working, here we use (sparse) PLU factorization for this system and hope for the best. Note that the matrix of the linear system is constant, so the factorization needs to be done only once for all iterations, with only triangular solves being done repeatedly. The details are in Function 8.3.1. Note the use of **speye**, the sparse identity matrix, to ensure that $B$ is sparse when $A$ is.

One additional detail is worth mentioning. Suppose the power iteration is applied to the matrix $(A - sI)^{-1}$, producing estimates $\beta_k$. These are actually converging to $(\lambda_1 - s)^{-1}$, so to recover $\lambda_1$ we compute $\gamma_k = s + \beta_k^{-1}$.

## Convergence

The convergence rate is found by interpreting (8.2.10) from the power iteration in the new context:

$$\frac{\gamma_{k+1} - \lambda_1}{\gamma_k - \lambda_1} \to \frac{\lambda_1 - s}{\lambda_2 - s} \quad \text{as } k \to \infty. \tag{8.3.4}$$

We observe that the convergence is best when the shift $s$ is close to the target eigenvalue $\lambda_1$, specifically when it is much closer to that eigenvalue than to any other.

### Example 8.3.1

We set up a $5 \times 5$ triangular matrix with prescribed eigenvalues on its diagonal.

```
lambda = [1 -0.75 0.6 -0.4 0];
A = triu(ones(5),1) + diag(lambda);
format long
```

We run inverse iteration with the shift $s = 0.7$ and take the final estimate as our "exact" answer to observe the convergence.

```
[gamma,x] = inviter(A,0.7,30);
eigval = gamma(end)
```

```
eigval =
   0.599999999999998
```

As expected, the eigenvalue that was found is the one closest to 0.7. The convergence is again linear.

```
err = eigval - gamma;
semilogy(abs(err),'.-')
```

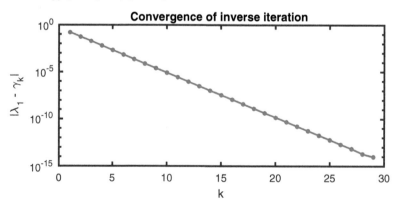

## 8.3. Inverse iteration

The observed linear convergence rate is found from the data.

```
observed_rate = err(26)/err(25)

observed_rate =
  -0.327983951855567
```

In the numbering of this example, the eigenvalue closest to $s = 0.7$ is $\lambda_3$ and the next closest is $\lambda_1$.

```
theoretical_rate = (lambda(3)-0.7) / (lambda(1)-0.7)

theoretical_rate =
  -0.333333333333333
```

### Dynamic shifting

There is an obvious opportunity for positive feedback in Function 8.3.1. The convergence rate of inverse iteration improves as the shift gets closer to the true eigenvalue—and the output of the algorithm is a sequence of improving eigenvalue estimates! If we update the shift to $s = \gamma_k$ after each iteration, the convergence accelerates. You are asked to implement this algorithm in Exercise 8.3.6.

Let's estimate the resulting convergence. If the eigenvalues are ordered by distance to $s$, then the convergence is linear with rate $|\lambda_1 - s|/|\lambda_2 - s|$. As $s \to \lambda_1$, the change in the denominator is negligible. So if the error $(\lambda_1 - s)$ is $\epsilon$, then the error in the next estimate is reduced by a factor $O(\epsilon)$. That is, $\epsilon$ becomes $O(\epsilon^2)$, which is *quadratic convergence*.

#### Example 8.3.2

We return to the matrix from Example 8.3.1.

```
lambda = [1 -0.75 0.6 -0.4 0];
A = triu(ones(5),1) + diag(lambda);
format long
```

We begin with a shift $s = 0.7$, which is closest to the eigenvalue 0.6.

```
I = eye(5);   s = 0.7;
x = ones(5,1);
y = (A-s*I)\x;   gamma = x(1)/y(1) + s

gamma =
  0.703481392557023
```

Note that the result is not yet any closer to the targeted 0.6. But we proceed (without being too picky about normalization here).

```
s = gamma;
x = y/y(1);
y = (A-s*I)\x;   gamma = x(1)/y(1) + s

gamma =
   0.561276140617300
```

Still not much apparent progress. However, in just a few more iterations the results are dramatically better.

```
for k = 1:4
    s = gamma;   x = y/y(1);
    y = (A-s*I)\x;   gamma = x(1)/y(1) + s
end

gamma =
   0.596431288475387
gamma =
   0.599971709182010
gamma =
   0.599999997855635
gamma =
   0.600000000000000
```

There is a price to pay for this improvement. The matrix of the linear system to be solved, $(A - sI)y = x$, now changes with each iteration. That means that we can no longer do just one LU factorization to do the entire iteration. The speedup in convergence usually makes this trade-off worthwhile, however.

In practice power and inverse iteration are not as effective as the algorithms used by `eigs` and based on the mathematics described in the rest of this chapter. However, inverse iteration can be useful for turning an eigenvalue estimate into an eigenvector estimate.

## Exercises

8.3.1. Use Function 8.3.1 to perform 20 iterations for the given matrix and shift. Compare the results quantitatively to the convergence given by (8.3.4).

(a)
$$A = \begin{bmatrix} 1.1 & 1 \\ 0 & 2.1 \end{bmatrix}, \quad s = 1.$$

(b)
$$A = \begin{bmatrix} 1.1 & 1 \\ 0 & 2.1 \end{bmatrix}, \quad s = 2.$$

(c)
$$A = \begin{bmatrix} 1.1 & 1 \\ 0 & 2.1 \end{bmatrix}, \quad s = 1.6.$$

(d)
$$A = \begin{bmatrix} 2 & 1 \\ 1 & 0 \end{bmatrix}, \quad s = -0.4.$$

(e)
$$A = \begin{bmatrix} 6 & 5 & 4 \\ 5 & 4 & 3 \\ 4 & 3 & 2 \end{bmatrix}, \quad s = 0.1.$$

8.3.2. ✍ Given the starting vector $x_1 = [1,1]^T$, find the vector $x_2$ for the matrix and each of the shifts in Exercise 8.3.1(a)–(c).

8.3.3. ✍ Why is it a bad idea to use unshifted inverse iteration with the matrix $\begin{bmatrix} 0 & 1 \\ -1 & 0 \end{bmatrix}$? Does the shift $s = -1$ improve matters?

8.3.4. ✍ When the shift $s$ is very close to an eigenvalue of $A$, the matrix $A - sI$ is close to a singular matrix. But then (8.3.3) is a linear system with a badly conditioned matrix, which should create a lot of error in the numerical solution for $y_k$. However, it happens that the error is mostly in the direction of the eigenvector we are looking for, as the following toy example demonstrates.

Prove that $\begin{bmatrix} 1 & 1 \\ 0 & 0 \end{bmatrix}$ has an eigenvalue at zero with associated eigenvector $v = [-1, 1]^T$. Suppose this matrix is perturbed slightly to $A = \begin{bmatrix} 1 & 1 \\ 0 & \epsilon \end{bmatrix}$, and that $x_k = [1, 1]^T$ in (8.3.3). Show that once $y_k$ is normalized by its infinity norm, the result is within $\epsilon$ of a multiple of $v$.

8.3.5. ⌨ (Continuation of Exercise 8.2.4.) This problem concerns the $n^2 \times n^2$ sparse matrix defined in MATLAB by

```
A = n^2*gallery('poisson',n);
```

(a) The eigenvalues of $A$ are approximately squares of the frequencies of vibration for the membrane. Using eig, find the eigenvalue $\lambda_m$ closest to zero for $n = 10, 15, 20, 25$.

(b) For each case of $n$ in part (a), apply 50 steps of Function 8.3.1. On one graph plot the four convergence curves $|\gamma_k - \lambda_m|$ using a semi-log scale.

(c) Let v be the eigenvector (second output) found by Function 8.3.1; one can visualize using the **mesh** command. Use mesh(reshape(v,n,n)) with $n = 25$ to see the physical vibration mode associated with this lowest frequency of the membrane.

8.3.6. ⌨

(a) Modify Function 8.3.1 to use a dynamic shift. That is, it should change the value of the shift $s$ to be the most recent value in gamma. Note that the matrix **B** must also change with each iteration, and the LU factorization cannot be done just once.

(b) Let A=magic(99)/99 and compute its eigenvalues using eig. Using an initial shift of $s = 100$, apply the dynamic inverse iteration. Determine which eigenvalue was found using min, and make a table of the log10 of the errors in the iteration as a function of iteration number. (These should approximately double, until machine precision is reached, due to quadratic convergence.)

(c) Repeat part (b) using an initial shift of your choice.

## 8.4 ▪ Krylov subspaces

The power and inverse iterations have a flaw that seems obvious once it is pointed out. Given a seed vector $u$, they produce a sequence of vectors $u_1, u_2, \ldots$ that are scalar multiples of $u, Au, A^2 u, \ldots$, but only the most recent vector is used to produce an eigenvector estimate. It stands to reason that we could do no worse, and perhaps much better, if we searched among all linear combinations of the vectors seen in the past. In other words, we seek a solution in the range (column space) of the $m \times n$ **Krylov matrix**

$$K_m = \begin{bmatrix} u & Au & A^2 u & \cdots & A^{m-1} u \end{bmatrix}. \qquad (8.4.1)$$

Such a space is called the $m$th **Krylov subspace** $\mathcal{K}_m$ of $\mathbb{C}^n$.[34] Implicitly we understand that $K_m$ and $\mathcal{K}_m$ depend on both $A$ and the initial vector $u$, but we rarely express the dependence notationally. In general, we expect that the dimension of $\mathcal{K}_m$, which is the rank of $K_m$, equals $m$, though it may be smaller.

### Properties

As we have seen with the power iteration, part of the appeal of the Krylov matrix is that it can be generated in a way that fully exploits the sparsity of $A$, simply through repeated matrix-vector multiplication. Furthermore, we have some important mathematical properties.

> **Lemma 8.4.1**
>
> Suppose $A$ is $n \times n$, $0 < m < n$, and a vector $u$ is used to generate Krylov subspaces. If $x \in \mathcal{K}_m$, then the following hold:
>
> 1. $x = K_m z$ for some $z \in \mathbb{C}^m$.
> 2. $x \in \mathcal{K}_{m+1}$.
> 3. $Ax \in \mathcal{K}_{m+1}$.

*Proof.* If $x \in \mathcal{K}_m$, then for some coefficients $c_1, \ldots, c_m$,

$$x = c_1 u + c_2 A u + \cdots + c_m A^{m-1} u.$$

---

[34] A proper pronunciation of "Krylov" is something like "kree-luv," but American English speakers often say "kreye-lahv."

## 8.4. Krylov subspaces

Thus let $z = \begin{bmatrix} c_1 & \cdots & c_m \end{bmatrix}^T$. Also $x \in \mathcal{K}_{m+1}$, as we can add zero times $A^m u$ to the sum. Finally,

$$Ax = c_1 Au + c_2 A^2 u + \cdots + c_m A^m u \in \mathcal{K}_{m+1}. \qquad \square$$

**Reducing dimension**

The problems $Ax = b$ and $Ax = \lambda x$ are statements about a very high-dimensional space $\mathbb{C}^n$. One way to approximate them is to *replace the full n-dimensional space with a much lower-dimensional $\mathcal{K}_m$ for $m \ll n$.* This is the essence of the Krylov subspace approach.

For instance, we can interpret $Ax_m \approx b$ in the sense of linear least squares—that is, using Lemma 8.4.1 to let $x = K_m z$,

$$\min_{x \in \mathcal{K}_m} \|Ax - b\| = \min_{z \in \mathbb{C}^m} \|A(K_m z) - b\| = \min_{z \in \mathbb{C}^m} \|(AK_m)z - b\|. \qquad (8.4.2)$$

The natural seed vector for $\mathcal{K}_m$ in this case is the vector $b$. In the next example we try to implement (8.4.2). We do take one precaution: because the vectors $A^k b$ may become very large or small in norm, we normalize after each multiplication by $A$, just as we did in the power iteration.

**Example 8.4.1**

First we define a triangular matrix with known eigenvalues and a random vector $b$.

```
lambda = 10 + (1:100);
A = diag(lambda) + triu(rand(100),1);
b = rand(100,1);
```

Next we build up the first ten Krylov matrices iteratively, using renormalization after each matrix-vector multiplication.

```
Km = b;
for m = 1:29
    v = A*Km(:,m);
    Km(:,m+1) = v/norm(v);
end
```

Now we solve a least squares problem for Krylov matrices of increasing dimension.

```
resid = zeros(30,1);
for m = 1:30
    z = (A*Km(:,1:m))\b;
    x = Km(:,1:m)*z;
    resid(m) = norm(b-A*x);
end
```

The linear system approximations show smooth linear convergence at first, but the convergence stagnates after only a few digits have been found.

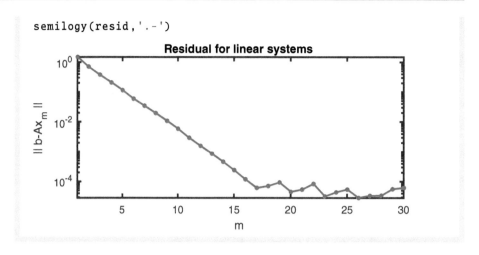

## The Arnoldi iteration

The breakdown of convergence in Example 8.4.1 is due to a critical numerical defect in our approach: the columns of the Krylov matrix (8.4.1) increasingly become parallel to one another (and the dominant eigenvector), as (8.2.3) predicts. As we saw in Section 3.3, near-parallel vectors create the potential for numerical cancellation. This manifests as a large condition number for $\boldsymbol{K}_m$ as $m$ grows, eventually creating excessive error when solving the least squares system.

The polar opposite of an ill-conditioned basis for $\mathcal{K}_m$ is an orthonormal one. Suppose we had a thin QR factorization of $\boldsymbol{K}_m$:

$$\boldsymbol{K}_m = \boldsymbol{Q}_m \boldsymbol{R}_m = \begin{bmatrix} \boldsymbol{q}_1 & \boldsymbol{q}_2 & \cdots & \boldsymbol{q}_m \end{bmatrix} \begin{bmatrix} R_{11} & R_{12} & \cdots & R_{1m} \\ 0 & R_{22} & \cdots & R_{2m} \\ \vdots & & \ddots & \\ 0 & 0 & \cdots & R_{mm} \end{bmatrix}.$$

Then the vectors $\boldsymbol{q}_1, \ldots, \boldsymbol{q}_m$ are the orthonormal basis we seek for $\mathcal{K}_m$. By Lemma 8.4.1, we know that $\boldsymbol{A}\boldsymbol{q}_m \in \mathcal{K}_{m+1}$, and therefore

$$\boldsymbol{A}\boldsymbol{q}_m = H_{1m}\boldsymbol{q}_1 + H_{2m}\boldsymbol{q}_2 + \cdots + H_{m+1,m}\boldsymbol{q}_{m+1} \tag{8.4.3}$$

for some choice of the $H_{ij}$. Note that by using orthonormality, we have

$$\boldsymbol{q}_i^*(\boldsymbol{A}\boldsymbol{q}_m) = H_{im}, \qquad i = 1, \ldots, m. \tag{8.4.4}$$

Since we started by assuming that we know $\boldsymbol{q}_1, \ldots, \boldsymbol{q}_m$, the only unknowns in (8.4.3) are $H_{m+1,m}$ and $\boldsymbol{q}_{m+1}$. But they appear only as a product, and we know that $\boldsymbol{q}_{m+1}$ is a *unit* vector, so they are uniquely defined (up to sign) by the other terms in the equation.

We can now proceed iteratively. If $\boldsymbol{u}$ is the Krylov seed vector, then do the following:

1. Let $\boldsymbol{q}_1 = \boldsymbol{u}/\|\boldsymbol{u}\|$.

## 8.4. Krylov subspaces

2. For $m = 1, 2, \ldots$
   (i) use (8.4.4) to find $H_{im}$ for $i = 1, \ldots, m$;
   (ii) let
   $$v = (Aq_m) - H_{1m}q_1 - H_{2m}q_2 - \cdots - H_{mm}q_m; \qquad (8.4.5)$$
   (iii) let $H_{m+1,m} = \|v\|$;
   (iv) let $q_{m+1} = v/H_{m+1,m}$.

We have just described the **Arnoldi iteration**. *The Arnoldi iteration finds an orthonormal basis for a Krylov subspace.*

### Example 8.4.2

We illustrate a few steps of the Arnoldi iteration for a small matrix.

```
A = magic(6);
```

The seed vector determines the first member of the orthonormal basis.

```
u = randn(6,1);
Q = u/norm(u);
```

Multiplication by $A$ gives us a new vector in $\mathcal{K}_2$.

```
Aq = A*Q(:,1);
```

We subtract off its projection in the previous direction. The remainder is rescaled to give us the next orthonormal column.

```
v = Aq - (Q(:,1)'*Aq)*Q(:,1);
Q(:,2) = v/norm(v);
```

On the next pass, we have to subtract off the projections in two previous directions.

```
Aq = A*Q(:,2);
v = Aq - (Q(:,1)'*Aq)*Q(:,1) - (Q(:,2)'*Aq)*Q(:,2);
Q(:,3) = v/norm(v);
```

At every step, $Q_m$ is an ONC matrix.

```
norm( Q'*Q - eye(3) )

ans =
   3.5209e-16
```

And $Q_m$ spans the same space as the three-dimensional Krylov matrix.

```
K = [ u A*u A*A*u ];
rank( [Q,K] )

ans =
   3
```

## Key identity

Up to now we have focused only on finding the orthonormal basis that lies in the columns of $Q_m$. But the $H_{ij}$ values found during the iteration are also very important. Taking $j = 1, 2, \ldots, m$ in (8.4.3) leads to

$$AQ_m = \begin{bmatrix} Aq_1 & \cdots & Aq_m \end{bmatrix}$$

$$= \begin{bmatrix} q_1 & q_2 & \cdots & q_{m+1} \end{bmatrix} \begin{bmatrix} H_{11} & H_{12} & \cdots & H_{1m} \\ H_{21} & H_{22} & \cdots & H_{2m} \\ & H_{32} & \ddots & \vdots \\ & & \ddots & H_{mm} \\ & & & H_{m+1,m} \end{bmatrix} = Q_{m+1} H_m, \quad (8.4.6)$$

where the matrix $H_m$ has an *upper Hessenberg* structure. Equation (8.4.6) is a fundamental identity of Krylov subspace methods.

## Implementation

An implementation of the Arnoldi iteration is given in Function 8.4.1. A careful inspection shows that the loop at line 18 does not exactly implement (8.4.4) and (8.4.5). The reason is numerical stability. Though the described and implemented versions

---

**Function 8.4.1 (arnoldi)** Arnoldi iteration for Krylov subspaces.

```
function [Q,H] = arnoldi(A,u,m)
% ARNOLDI   Arnoldi iteration for Krylov subspaces.
% Input:
%   A    square matrix (n by n)
%   u    initial vector
%   m    number of iterations
% Output:
%   Q    orthonormal basis of Krylov space (n by m+1)
%   H    upper Hessenberg matrix, A*Q(:,1:m)=Q*H (m+1 by m)

n = length(A);
Q = zeros(n,m+1);
H = zeros(m+1,m);
Q(:,1) = u/norm(u);
for j = 1:m
   % Find the new direction that extends the Krylov subspace.
   v = A*Q(:,j);
   % Remove the projections onto the previous vectors.
   for i = 1:j
      H(i,j) = Q(:,i)'*v;
      v = v - H(i,j)*Q(:,i);
   end
   % Normalize and store the new basis vector.
   H(j+1,j) = norm(v);
   Q(:,j+1) = v/H(j+1,j);
end
```

are mathematically equivalent in exact arithmetic (see Exercise 8.4.6), the approach in Function 8.4.1 is much more stable to roundoff.

In the next section we revisit the idea of approximately solving $Ax = b$ over a Krylov subspace $\mathcal{K}_m$, using the ONC matrix $Q_m$ in place of $K_m$. A related idea is used to approximate the eigenvalue problem for $A$ (see Exercise 8.4.7); this is the approach that underlies `eigs` in MATLAB.

## Exercises

8.4.1. ✍ Consider the matrix

$$A = \begin{bmatrix} 0 & 1 & 0 & 0 \\ 0 & 0 & 1 & 0 \\ 0 & 0 & 0 & 1 \\ 1 & 0 & 0 & 0 \end{bmatrix}.$$

(a) Find the Krylov matrix $K_3$ for the seed vector $u = e_1$.

(b) Find $K_3$ for the seed vector $u = \begin{bmatrix} 1 & 1 & 1 & 1 \end{bmatrix}^T$.

8.4.2. 🖳 For each matrix, make a table of the 2-norm condition numbers $\kappa(K_m)$ for $m = 1, \ldots, 10$. Use a vector of all ones as the Krylov seed.

(a) Matrix from Example 8.4.1.

(b) $\begin{bmatrix} -2 & 1 & & & \\ 1 & -2 & 1 & & \\ & \ddots & \ddots & \ddots & \\ & & 1 & -2 & 1 \\ & & & 1 & -2 \end{bmatrix}$ $(100 \times 100)$.

(c) $\begin{bmatrix} -2 & 1 & & & 1 \\ 1 & -2 & 1 & & \\ & \ddots & \ddots & \ddots & \\ & & 1 & -2 & 1 \\ 1 & & & 1 & -2 \end{bmatrix}$ $(200 \times 200)$.

8.4.3. ✍ (See also Exercise 7.2.5.) If $p(z) = c_0 + c_1 z + \cdots + c_k z^k$ is a polynomial, then its value for a matrix argument is defined as

$$p(A) = c_0 I + c_1 A + \cdots + c_k A^k.$$

Show that if $x \in \mathcal{K}_m$, then $x = p(A)u$ for a polynomial $p$ of degree at most $m - 1$.

8.4.4. ✍ Compute the asymptotic flop requirements for Function 8.4.1. Assume that, due to sparsity, a matrix-vector multiplication $Au$ requires only $cn$ flops for a constant $c$, rather than the usual $O(n^2)$.

8.4.5. ■ When Arnoldi iteration is performed on the Krylov subspace generated using the matrix

$$A = \begin{bmatrix} 2 & 1 & 1 & 0 \\ 1 & 3 & 1 & 0 \\ 0 & 1 & 3 & 1 \\ 0 & 1 & 1 & 2 \end{bmatrix},$$

the results can depend strongly on the initial vector $u$.

(a) Compare the results for Q and H using the following initial guesses: (i) u=[1; 0; 0; 0], (ii) u=[1; 1; 1; 1], (iii) u=rand(4,1).

(b) Can you explain what is different about case (ii)? You may have to show the output of Function 8.4.1 to see what is happening.

8.4.6. ✍ As mentioned in the text, Function 8.4.1 does not compute $H_{ij}$ as defined by (8.4.4), but rather

$$S_{ij} = q_i^*(Aq_j - S_{1j}q_1 - \cdots - S_{i-1,j}q_{i-1})$$

for $i = 1, \ldots, j$. Show, however, that $S_{ij} = H_{ij}$ (hence Function 8.4.1 is mathematically equivalent to our Arnoldi formulas).

8.4.7. How should we approximate the eigenvalue problem $Ax = \lambda x$ over $\mathcal{K}_m$? One simple idea is to restrict $x$ to $\mathcal{K}_m$, so that $x = Q_m z$. Then multiply the approximate equality $AQ_m z \approx \lambda Q_m z$ on the left by $Q_m^*$.

(a) ✍ Starting from (8.4.6), show that

$$Q_m^* A Q_m = \widetilde{H}_m,$$

where $\widetilde{H}_m$ is the upper Hessenberg matrix resulting from deleting the last row of $H_m$. What is the size of this matrix?

(b) ✍ Show that the reasoning above leads to the approximate eigenvalue problem $\widetilde{H}_m z \approx \lambda z$. (That is, the exact eigenvalues of $\widetilde{H}_m$ ought to approximate some of the eigenvalues of $A$.)

(c) ■ Using the matrix of Example 8.4.1 and Function 8.4.1, compute eigenvalues of $\widetilde{H}_m$ for $m = 1, \ldots, 40$, keeping track of the error between the largest of those values and the largest eigenvalue of $A$. Make a semi-log graph of the error as a function of $m$.

## 8.5 ▪ GMRES

The most important use of the Arnoldi iteration is to solve the square linear system $Ax = b$. The iteration's generation of an orthogonal basis is the key to fixing the trouble we encountered in Example 8.4.1. Recall that we replaced the linear system $Ax = b$ by

$$\min_{x \in \mathcal{K}_m} \|Ax - b\| = \min_{z \in \mathbb{C}^m} \|AK_m z - b\|,$$

where $K_m$ is the Krylov matrix generated using $A$ and the seed vector $b$. This method was unstable due to the poor conditioning of $K_m$, which is a numerically poor basis of

## 8.5. GMRES

$\mathcal{K}_m$. If we use the columns of $Q_m$ instead as a basis, then we set $x = Q_m z$ and obtain

$$\min_{z \in \mathbb{C}^m} \|AQ_m z - b\|. \tag{8.5.1}$$

From the fundamental Arnoldi identity (8.4.6), this is equivalent to

$$\min_{z \in \mathbb{C}^m} \|Q_{m+1} H_m z - b\|. \tag{8.5.2}$$

Note that $q_1$ is a unit multiple of $b$, so $b = \|b\| Q_{m+1} e_1$. Thus (8.5.2) becomes

$$\min_{z \in \mathbb{C}^m} \|Q_{m+1}(H_m z - \|b\| e_1)\|. \tag{8.5.3}$$

The least squares problems (8.5.1), (8.5.2), and (8.5.3) are all $n \times m$. But observe that for any $w \in \mathbb{C}^{m+1}$,

$$\|Q_{m+1} w\|^2 = w^* Q_{m+1}^* Q_{m+1} w = w^* w = \|w\|^2.$$

The first norm in that equation is on $\mathbb{C}^n$, while the last is on the much smaller space $\mathbb{C}^{m+1}$. Hence the least squares problem (8.5.3) is equivalent to

$$\min_{z \in \mathbb{C}^m} \|H_m z - \|b\| e_1\|, \tag{8.5.4}$$

which is of size $(m+1) \times m$. We call the solution of this minimization $z_m$, and then $x_m = Q_m z_m$ is the $m$th approximation to the solution of $Ax = b$.

The algorithm resulting from this discussion is known as **GMRES** (for Generalized Minimum RESidual). *GMRES uses the output of the Arnoldi iteration to minimize the residual of $Ax = b$ over successive Krylov subspaces.*

### Example 8.5.1

We define a triangular matrix with known eigenvalues and a random vector $b$.

```
lambda = 10 + (1:100);
A = diag(lambda) + triu(rand(100),1);
b = rand(100,1);
```

Instead of building up the Krylov matrices, we use the Arnoldi iteration to generate equivalent orthonormal vectors.

```
[Q,H] = arnoldi(A,b,60);
```

The Arnoldi bases are used to solve the least squares problems defining the GMRES iterates.

```
for m = 1:60
    s = [norm(b); zeros(m,1)];
    z = H(1:m+1,1:m)\s;
    x = Q(:,1:m)*z;
    resid(m) = norm(b-A*x);
end
```

The approximations converge smoothly, practically all the way to machine epsilon.

```
semilogy(resid,'.-')
```

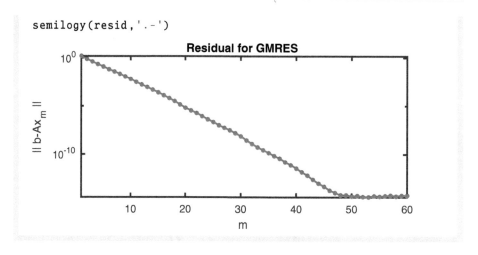

Compare the graph in Example 8.5.1 to that in Example 8.4.1. Both start with the same linear convergence, but only the version using Arnoldi avoids the instability created by the poor Krylov basis.

A basic implementation of GMRES is given in Function 8.5.1. It performs the same computations as were done in Example 8.5.1. For practical use, MATLAB has a built-in **gmres** that is more robust and full-featured than our simple version.

## Convergence and restarting

Thanks to Lemma 8.4.1, minimization of $\|b - Ax\|$ over $\mathcal{K}_{m+1}$ includes minimization over $\mathcal{K}_m$. Hence the norm of the residual $r_m = b - Ax_m$ (being the minimized quantity) cannot increase as the iteration unfolds.

Unfortunately, making other conclusive statements about the convergence of GMRES is neither easy nor simple. Example 8.5.1 shows the cleanest behavior: nearly linear convergence down to the range of machine epsilon. But it is possible for the convergence to go through phases of sublinear and superlinear convergence as well. There is a strong dependence on the spectrum of the matrix, a fact we state with more precision and detail in Section 8.6.

One of the practical challenges in GMRES is that as the dimension of the Krylov subspace grows, the number of new entries to be found in $H_m$, and the total number of columns in $Q$, also grow. Thus both the work and the storage requirements are quadratic in $m$, which can become intolerable in some applications. For this reason, GMRES is often used with **restarting**.

Suppose $\hat{x}$ is an approximate solution of $Ax = b$. Then if we set $x = u + \hat{x}$, we have $A(u + \hat{x}) = b$, or $Au = b - A\hat{x}$. The conclusion is that if we get an approximate solution and compute its residual $r = b - A\hat{x}$, then we need only to solve $Au = r$ in order to get a "correction" to $\hat{x}$.[35]

Restarting guarantees a fixed upper bound on the per-iteration cost of GMRES.

---

[35] The new problem needs to be solved for accuracy relative to $\|b\|$, *not* relative to $\|r\|$.

## 8.5. GMRES

**Function 8.5.1 (arngmres)** GMRES for a linear system.

```
function [x,residual] = arngmres(A,b,M)
% ARNGMRES   GMRES for a linear system (demo only).
% Input:
%   A         square matrix (n by n)
%   b         right-hand side (n by 1)
%   M         number of iterations
% Output:
%   x         approximate solution (n by 1)
%   r         history of norms of the residuals

n = length(A);
Q = zeros(n,M);
Q(:,1) = b/norm(b);
H = zeros(M,M-1);

% Initial "solution" is zero.
residual(1) = norm(b);

for m = 1:M

  % Next step of Arnoldi iteration.
  v = A*Q(:,m);
  for i = 1:m
      H(i,m) = Q(:,i)'*v;
      v = v - H(i,m)*Q(:,i);
  end
  H(m+1,m) = norm(v);
  Q(:,m+1) = v/H(m+1,m);

  % Solve the minimum residual problem.
  r = norm(b)*eye(m+1,1);
  z = H(1:m+1,1:m) \ r;
  x = Q(:,1:m)*z;
  residual(m+1) = norm( A*x - b );

end
```

However, this bound comes at a price. Even though restarting preserves progress made in previous iterations, the Krylov space information is discarded and the residual minimization process starts again over low-dimensional choices. That can significantly retard or even stagnate the convergence. The implementation of `gmres` in MATLAB has an option for using restarts.

### Example 8.5.2

The following experiments are based on a matrix resulting from discretization of a partial differential equation.

```
maxit = 120;   rtol = 1e-8;
d = 50;
A = d^2*gallery('poisson',d);
n = size(A,1)
b = ones(n,1);
```

```
n =
        2500
```

We compare unrestarted GMRES with three different thresholds for restarting.

```
rest = [maxit 20 40 60];
for j = 1:4
    [~,~,~,~,rv] = gmres(A,b,rest(j),rtol,maxit/rest(j));
    semilogy(0:length(rv)-1,rv,'-'), hold on
end
```

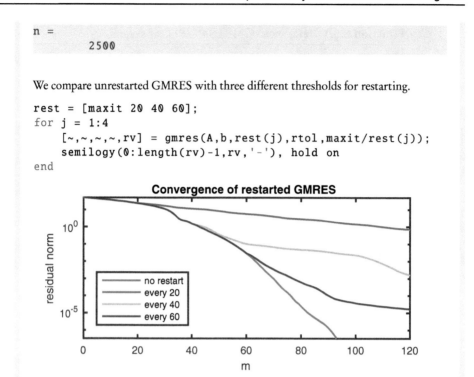

The "pure" curve is the lowest one. All of the other curves agree with it until they encounter their first restart.

As you can see from Example 8.5.2, if the restart takes place before GMRES has entered a rapidly converging phase, the restarted GMRES can converge a great deal more slowly. However, the later iterations for the restarted versions should be much faster than those of pure GMRES, making an apples-to-apples comparison difficult.

There are other ways to avoid the growth in computational effort as the GMRES/Arnoldi iteration proceeds. Three of the more popular variations are abbreviated CGS, BiCGSTAB, and QMR, and these are also implemented in MATLAB. We do not describe them in this book.

## Exercises

8.5.1. ⚐ (See also Exercise 8.4.1.) Consider the linear system with

$$A = \begin{bmatrix} 0 & 1 & 0 & 0 \\ 0 & 0 & 1 & 0 \\ 0 & 0 & 0 & 1 \\ 1 & 0 & 0 & 0 \end{bmatrix}, \quad b = e_1.$$

(a) Find the exact solution by inspection.

(b) Find the GMRES approximate solutions $x_m$ for $m = 1, 2, 3, 4$.

8.5.2. ✍ (Continuation of Exercise 8.4.3.) Show that if $x_m \in \mathcal{K}_m$, then the residual $b - Ax_m$ is equal to $q(A)b$, where $q$ is a polynomial of degree at most $m$ and $q(0) = 1$. (This is a key fact for many convergence results.)

8.5.3. ✍ Explain why GMRES, in exact arithmetic, converges to the true solution in $n$ iterations for an $n \times n$ matrix if rank$(K_n) = n$. (Hint: Consider how the algorithm is defined from first principles.)

8.5.4. 💻 Let $A$ be the $n \times n$ tridiagonal matrix

$$\begin{bmatrix} -4 & 1 & & & \\ 1 & -4 & 1 & & \\ & \ddots & \ddots & \ddots & \\ & & 1 & -4 & 1 \\ & & & 1 & -4 \end{bmatrix},$$

and let the $n$-vector $b$ have entries $b_i = i/n$. For $n = 8, 16, 32, 64$, run Function 8.5.1 for $m = n/2$ iterations. On one graph plot $\|r_k\|/\|b\|$ for all the cases. How does the convergence rate of GMRES seem to depend on $n$?

8.5.5. 💻 In this problem you will see the strong effect the eigenvalues of the matrix may have on GMRES convergence. Let

$$B = \begin{bmatrix} 1 & & & \\ & 2 & & \\ & & \ddots & \\ & & & 100 \end{bmatrix},$$

let $I$ be a $100 \times 100$ identity, and let $Z$ be a $100 \times 100$ matrix of zeros. Also let $b$ be a $200 \times 1$ vector of ones.

(a) Let $A = \begin{bmatrix} B & I \\ Z & B \end{bmatrix}$. What are its eigenvalues (no computer required here)?

Apply gmres with tolerance $10^{-10}$ for 100 iterations without restarts, and plot the residual convergence.

(b) Repeat part (a) with restarts every 20 iterations.

(c) Now let $A = \begin{bmatrix} B & I \\ Z & -B \end{bmatrix}$. What are its eigenvalues? Repeat part (a). Which matrix is more difficult for GMRES?

8.5.6. 💻 (Continuation of Exercise 8.2.5.) We again consider the $n^2 \times n^2$ sparse matrix defined in MATLAB by

```
A = n^2*gallery('poisson',n);
```

The solution of $Ax = b$ may be interpreted as the deflection of a lumped membrane in response to a constant load represented by $b$.

(a) For $n = 10, 15, 20, 25$, let $b$ be the vector of $n^2$ ones and apply Function 8.5.1 for 50 iterations. On one semi-log graph, plot the four convergence curves $\|r_m\|/\|b\|$.

(b) For the case $n = 25$ use mesh(reshape(x,25,25)) to plot the solution, which should look physically plausible.

## 8.6 • MINRES and conjugate gradients

We have seen before that certain matrix properties enhance solutions to linear algebra problems. One of the most important of these is when $A^* = A$; i.e., $A$ is hermitian. The Arnoldi iteration has a particularly useful specialization to this case. Starting from (8.4.6), we left-multiply by $Q_m^*$ to get

$$Q_m^* A Q_m = Q_m^* Q_{m+1} H_m = \widetilde{H}_m,$$

where $\widetilde{H}_m$ is rows 1 through $m$ of $H_m$. If $A$ is hermitian, then so is the left side of this equation, hence $\widetilde{H}_m$ is hermitian too. But it is also upper Hessenberg, so it must be lower Hessenberg as well. The conclusion is that $\widetilde{H}_m$ is tridiagonal.

Equation (8.4.3) of the Arnoldi iteration now simplifies to a much shorter expression:

$$A q_m = H_{m-1,m} q_{m-1} + H_{mm} q_m + H_{m+1,m} q_{m+1}. \tag{8.6.1}$$

As before in deriving the Arnoldi iteration, when given the first $m$ vectors we can solve for the entries in column $m$ of $H$ and then for $q_{m+1}$. The resulting process is known as the **Lanczos iteration**. In a simple implementation, it needs just a single minor change from Arnoldi, but numerical stability requires some extra effort. We do not present the details. The most important practical difference is that *while Arnoldi needs $O(m)$ steps to get $q_{m+1}$ from the previous vectors, Lanczos needs only $O(1)$ steps, so restarting isn't required for symmetric matrices.*

### MINRES

When $A$ is hermitian and the Arnoldi iteration is reduced to Lanczos, the analog of GMRES is known as MINRES. MATLAB has a native implementation as the function `minres`. Like GMRES, MINRES minimizes the residual $\|b - Ax\|$ over increasingly large Krylov spaces.

MINRES is also more theoretically tractable than GMRES. Recall that the eigenvalues of a hermitian matrix are real. Of the eigenvalues that are positive, let $\kappa_+$ be the ratio of the one farthest from the origin (largest) to the one closest to the origin (smallest). Similarly, let $\kappa_-$ be the ratio of the negative eigenvalue farthest from the origin to the negative eigenvalue closest to the origin. Then there is a rigorous upper bound on the residual:

$$\frac{\|r_m\|_2}{\|b\|_2} \leq \left( \frac{\sqrt{\kappa_+ \kappa_-} - 1}{\sqrt{\kappa_+ \kappa_-} + 1} \right)^{\lfloor m/2 \rfloor}, \tag{8.6.2}$$

where $\lfloor m/2 \rfloor$ means to round $m/2$ down to the nearest integer. This bound (though not necessarily MINRES itself) obeys a linear convergence rate. As the product $\kappa_+ \kappa_-$ grows, the rate of this convergence approaches one, i.e., is slower. Hence the convergence of MINRES may depend strongly on the eigenvalues of the matrix, with eigenvalues close to the origin (relative to the max eigenvalues) predicted to force a slower convergence.

## Conjugate gradients

With another property in addition to symmetry, we arrive at perhaps the most famous Krylov subspace method for $Ax = b$, called **conjugate gradients**. Suppose now that $A$ is hermitian and positive definite (HPD). We know that $A$ has a Cholesky factorization, which in the complex case is $A = R^*R$, and therefore for any vector $u$,

$$u^*Au = (Ru)^*(Ru) = \|Ru\|^2,$$

which is nonnegative and zero only when $u = 0$, provided $A$ (and therefore $R$) is nonsingular. Hence we can define a special vector norm relative to $A$:

$$\|u\|_A = (u^*Au)^{1/2}. \tag{8.6.3}$$

*The conjugate gradients algorithm minimizes the error, as measured in the A-norm, over the sequence of Krylov subspaces.* That is, $x_m$ makes $\|x_m - x\|_A$ as small as possible over $\mathcal{K}_m$. We do not show any details or code for the resulting algorithm, instead relying on the built-in command `pcg` to perform conjugate gradients for us.

## Convergence

The convergence of CG and MINRES is dependent on the eigenvalues of $A$. In the HPD case the eigenvalues are real and positive, and they equal the singular values. Hence the condition number $\kappa$ is equal to the ratio of the largest eigenvalue to the smallest one. The following theorem suggests that MINRES and CG are not so different in convergence.

> **Theorem 8.6.1**
> 
> Let $A$ be real and SPD with 2-norm condition number $\kappa$. For MINRES define $R(m) = \|r_m\|_2/\|b\|_2$, and for CG define $R(m) = \|x_m - x\|_A/\|x\|_A$, where $r_m$ and $x_m$ are the residual and solution approximation associated with the space $\mathcal{K}_m$. Then
> 
> $$R(m) \leq 2\left(\frac{\sqrt{\kappa}-1}{\sqrt{\kappa}+1}\right)^m. \tag{8.6.4}$$

As in the indefinite case with MINRES, *larger eigenvalue ratio (condition number) is associated with slower convergence in the positive definite case for both MINRES and CG.* That is, a large condition number is a double penalty, increasing both the time it takes to obtain a solution and the effects of numerical errors. Specifically, to make the bound of (8.6.4) less than a number $\epsilon$ requires

$$2\left(\frac{\sqrt{\kappa}-1}{\sqrt{\kappa}+1}\right)^m \approx \epsilon,$$

$$m \log\left(\frac{\sqrt{\kappa}-1}{\sqrt{\kappa}+1}\right) \approx \log\left(\frac{\epsilon}{2}\right).$$

We estimate

$$\frac{\sqrt{\kappa}-1}{\sqrt{\kappa}+1} = (1-\kappa^{-1/2})(1+\kappa^{-1/2})^{-1}$$
$$= (1-\kappa^{-1/2})(1-\kappa^{-1/2}+\kappa^{-1}+\cdots)$$
$$= 1-2\kappa^{-1/2}+O(\kappa^{-1}), \quad \text{as } \kappa \to \infty.$$

With the Taylor expansion $\log(1+x) = x - (x^2/2) + \cdots$, we finally conclude

$$2m\kappa^{-1/2} \approx \log\left(\frac{\epsilon}{2}\right), \quad \text{or} \quad m = O(\sqrt{\kappa}),$$

as an estimate of the number of iterations needed to achieve a fixed accuracy. (This estimate fails for very large $\kappa$, however.)

### Example 8.6.1

In this example we compare MINRES and CG on some pseudorandom SPD problems. The first matrix has a condition number of 100.

```
n = 1000;
density = 0.008;
A = sprandsym(n,density,1e-2,2);
```

We cook up a linear system whose solution we happen to know exactly.

```
x = (1:n)'/n;
b = A*x;
```

Now we apply both methods and compare the convergence of the system residuals, using the built-in function pcg in the latter case.

```
[xMR,~,~,~,residMR] = minres(A,b,1e-12,100);
[xCG,~,~,~,residCG] = pcg(A,b,1e-12,100);
semilogy(0:100,residMR/norm(b),'.-')
hold on, semilogy(0:100,residCG/norm(b),'.-')
```

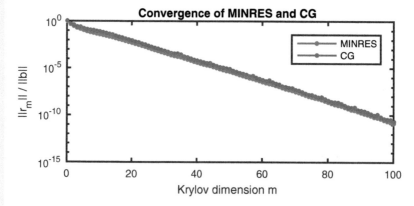

There is virtually no difference between the two methods here when measuring the residual. We see little difference in the errors as well.

## 8.6. MINRES and conjugate gradients

```
errorMR = norm( xMR - x ) / norm(x)
errorCG = norm( xCG - x) / norm(x)

errorMR =
   1.1820e-10
errorCG =
   8.0067e-11
```

Next we use a system matrix whose condition number is $10^4$.

```
A = sprandsym(n,density,1e-4,2);
```

Now we find that the CG residual jumps up initially, and then both methods converge at about the same linear rate. Note that both methods have actually made very little progress after 100 iterations, though.

```
[xMR,~,~,~,residMR] = minres(A,b,1e-12,100);
[xCG,~,~,~,residCG] = pcg(A,b,1e-12,100);
clf
semilogy(0:100,residMR/norm(b),'.-')
hold on, semilogy(0:100,residCG/norm(b),'.-')
```

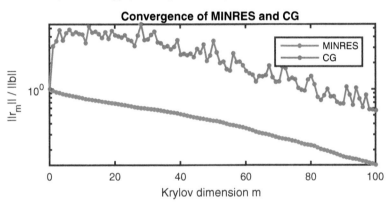

The errors confirm that we are nowhere near the correct solution in either case.

```
errorMR = norm( xMR - x ) / norm(x)
errorCG = norm( xCG - x) / norm(x)

errorMR =
  922.4364
errorCG =
   1.0080e+03
```

We can explain some of the behavior in Example 8.6.1. The first matrix has a condition number of $10^2$, whereas the second has $\kappa = 10^4$. The linear convergence bounds of the two cases after 100 iterations have values $(9/11)^{100} \approx 2 \times 10^{-9}$ and $(99/101)^{100} \approx 0.14$, respectively, which agrees fairly well with the observed reductions in the residual norms. The major practical difference between MINRES and CG lies in the interpretation of minimization of the residual versus minimization of the error in the $\boldsymbol{A}$-norm.

## Exercises

8.6.1. ✍ For each part, the eigenvalues of $A$ are as given. Suppose MINRES is applied to solve $Ax = b$. Find a numerical value for the upper bound in (8.6.2) or (8.6.4), whichever is most appropriate. Then determine which of the cases gives the slowest and which gives the fastest convergence.

(a) $-100, -99, \ldots, -1, 1, 2, \ldots, 100$.

(b) $-100, 1, 2, \ldots, 100$.

(c) $1, 2, \ldots, 100$.

8.6.2. 💻 Define the matrix

```
u = linspace(-200,-1); v = linspace(10,100);
A = diag([u v]);
```

Let $b$ be a random unit vector of length 200.

(a) Apply 120 iterations of `minres` to solve $Ax = b$. Compute the relative norm of the answer. Plot the norm of the residual as a function of $m$.

(b) Add to your graph the line representing the upper bound (8.6.2). (Ignore the rounding in the exponent.) This line should stay strictly on or above the convergence curve.

8.6.3. 💻 Define the matrix

```
A = diag(linspace(4,10000,500));
```

Let $b$ be a random unit vector of length 500.

(a) Apply 80 iterations of `minres` to solve $Ax = b$. Compute the relative norm of the answer. Plot the norm of the residual as a function of $m$.

(b) Add to your graph the line representing the upper bound (8.6.4). This line should stay strictly on or above the convergence curve.

(c) Add a convergence curve for 80 iterations of `pcg`.

8.6.4. ✍ Given real $n \times n$ symmetric $A$ and vector $b = Ax$, we can define the scalar-valued function
$$\varphi(u) = u^T A u - 2 u^T b, \qquad u \in \mathbb{R}^n.$$

(a) Expand and simplify the expression $\varphi(x+v) - \varphi(x)$, keeping in mind that $Ax = b$.

(b) Using the result of (a), prove that if $A$ is an SPD matrix, $\varphi$ has a global minimum at $x$.

(c) Show that for any vector $u$, $\|u - x\|_A^2 - \varphi(u)$ is constant.

(d) Using the result of (c), prove that CG minimizes $\varphi(u)$ over Krylov subspaces.

8.6.5. 💻 The following linear system arises from the Helmholtz equation for wave propagation:

```
A = n^2*gallery('poisson',n) - k^2*speye(n^2);
b = -ones(n^2,1);
```

(a) Repeat Example 8.6.1 using this linear system with $n = 50$ and $k = 1.3\pi$.

(b) Repeat Example 8.6.1 using this linear system with $n = 50$ and $k = 1.5\pi$. Use `eig` to explain why CG fails in this case.

## 8.7 • Matrix-free iterations

A primary reason for our interest in matrices is their relationship to linear transformations. If we define $f(x) = Ax$, then for all $x$, $y$, and $\alpha$,

$$f(\mathbf{x}+\mathbf{y}) = f(\mathbf{x}) + f(\mathbf{y}), \qquad (8.7.1\text{a})$$
$$f(\alpha \mathbf{x}) = \alpha f(\mathbf{x}). \qquad (8.7.1\text{b})$$

This is what defines a linear transformation. Moreover, *every* linear transformation between finite-dimensional vector spaces can be represented the same way, as a matrix-vector multiplication.

In scientific computing one often has a procedure for computing a transformation at any given input. That is, for some multidimensional function $f$ you have the ability to compute $f(x)$ at any given value of $x$. In Chapter 4 we solved the rootfinding problem $f(x) = 0$ with secant and quasi-Newton methods that needed exactly this capability. By repeatedly evaluating $f$ at cleverly chosen points, these algorithms were able to return a value for $f^{-1}(0)$.

A close examination reveals that Krylov subspace methods have a similar feel when the transformation is a linear map $f(x) = Ax$. While we presumed that we had access to $A$ directly, in fact the only use of it was to compute $Ax$ for a sequence of iteratively chosen $x$ values. This detail goes back to where we started with the power iteration: what can we compute about $A$ by using only matrix-vector multiplications? In the cases of GMRES, MINRES, and CG, the answer is that we can find $A^{-1}b$.

Bringing all of these observations together leads us to a cornerstone of modern scientific computation: *matrix-free iterations*. *Krylov subspace methods can be used to invert a linear transformation if one provides code for the transformation, even if its associated matrix is not known explicitly.* That is, it's possible to solve $Ax = b$ without even knowing $A$! We look at an extended example of this capability in the rest of this section.

### Blurring images

In Section 7.1 we saw that a grayscale image can be stored as an $m \times n$ matrix $X$ of pixel intensity values. Now consider a simple model for blurring the image. Define $B$ as the $m \times m$ tridiagonal matrix

$$B_{ij} = \begin{cases} \frac{1}{2} & \text{if } i = j, \\ \frac{1}{4} & \text{if } |i-j| = 1, \\ 0 & \text{otherwise.} \end{cases} \qquad (8.7.2)$$

Now $BX$ applies $B$ to each column of $X$. Within that column it does a weighted average of the values of each pixel and its two neighbors. That has the effect of blurring the image vertically. We can increase the amount of blur by applying $B$ repeatedly.

In order to blur horizontally, we can transpose the image and apply blurring in the same way. We need a blurring matrix defined as in (8.7.2) but with size $n \times n$. We call this matrix $C$. Altogether the horizontal blurring is done by transposing, applying $C$, and transposing back to the original orientation. That is,

$$\left(CX^T\right)^T = XC^T = XC,$$

using the symmetry of $C$. So we can describe blur in both directions as the function

$$\text{blur}(X) = B^k X C^k \qquad (8.7.3)$$

for a positive integer $k$.

### Example 8.7.1

We use an image that is built into MATLAB.

```
load mandrill
[m,n] = size(X)
image(X), colormap(gray(256))
```

```
m =
    480
n =
    500
```

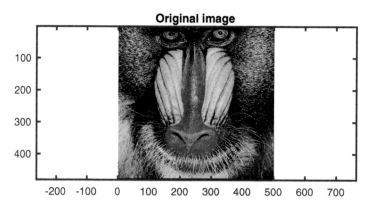

We define the one-dimensional tridiagonal blurring matrices.

```
v = [1/4 1/2 1/4];
B = spdiags( repmat(v,m,1), -1:1, m,m);
C = spdiags( repmat(v,n,1), -1:1, n,n);
```

Finally, we show the results of using $k = 12$ repetitions of the blur in each direction.

```
blur = @(X) B^12 * X * C^12;
image(blur(X))
```

## 8.7. Matrix-free iterations

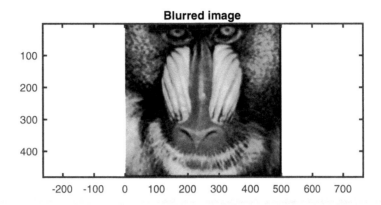
**Blurred image**

## Deblurring

A more interesting operation is *deblurring*: given an image blurred by poor focus, can we reconstruct the true image? Conceptually, we want to invert the function blur($X$).

It's easy to see from (8.7.3) that the blur operation is a linear transformation on image matrices. But an $m \times n$ image matrix is equivalent to a length-$mn$ vector—it's just a matter of interpreting the shape of the same data. Let vec($X$) = $x$ and unvec($x$) = $X$ be the mathematical statements of such reshaping operations. Now say $X$ is the original image and $Z$ = blur($X$) is the blurred one. Then by linearity there is some matrix $A$ such that

$$A \operatorname{vec}(X) = \operatorname{vec}(Z),$$

or $Ax = z$.

The matrix $A$ is $mn \times mn$; for a 12-megapixel image, it would have $1.4 \times 10^{14}$ entries! It is true that the matrix is extremely sparse. But more to our point, it's unnecessary. Instead, given any vector $u$ we can compute $v = Au$ through the steps

$$U = \operatorname{unvec}(u),$$
$$V = \operatorname{blur}(U),$$
$$v = \operatorname{vec}(V).$$

The following example shows how to put these ideas into practice with GMRES.

**Example 8.7.2**

We repeat the earlier process to blur the original image $X$ to get $Z$.

```
load mandrill
[m,n] = size(X)
v = [1/4 1/2 1/4];
B = spdiags( repmat(v,m,1), -1:1, m,m);
C = spdiags( repmat(v,n,1), -1:1, n,n);
blur = @(X) B^12 * X * C^12;
```

```
Z = blur(X);

m =
    480
n =
    500
```

Now we imagine that $X$ is unknown and that the blurred $Z$ is given. We want to invert the blur transformation using the transformation itself. But we have to translate between vectors and images each time.

```
vec = @(X) reshape(X,m*n,1);
unvec = @(x) reshape(x,m,n);
T = @(x) vec( blur(unvec(x)) );
```

Now we apply gmres to the composite blurring transformation T.

```
y = gmres(T,vec(Z),50,1e-5);
Y = unvec(y);
subplot(121)
image(X), colormap(gray(256))
subplot(122)
image(Y), colormap(gray(256))
```

```
gmres(50) converged at outer iteration 2 (inner iteration
    45) to a solution with relative residual 1e-05.
```

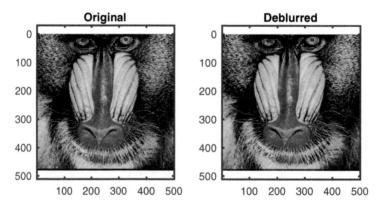

The reconstruction isn't perfect because the condition number of repeated blurring happens to be very large.

## Exercises

8.7.1. ✍ Show using (8.7.1) and (8.7.3) that the blur operation is a linear transformation.

8.7.2. ✍ In each case, state with reasons whether the given transformation on $n$-vectors is linear.

$$\text{(a) } f(x) = \begin{bmatrix} x_2 \\ x_3 \\ \vdots \\ x_n \\ x_1 \end{bmatrix}. \quad \text{(b) } f(x) = \begin{bmatrix} x_1 \\ x_1 + x_2 \\ x_1 + x_2 + x_3 \\ \vdots \\ x_1 + \cdots + x_n \end{bmatrix}. \quad \text{(c) } f(x) = \begin{bmatrix} x_1 + 1 \\ x_2 + 2 \\ x_3 + 3 \\ \vdots \\ x_n + n \end{bmatrix}.$$

8.7.3. ◢ Suppose that code for the linear transformation $f(x) = Ax$ is given for an unknown matrix $A$. Explain carefully how one could construct $A$.

8.7.4. ⌨ The matrix of the blur operation happens to be symmetric and positive definite. Repeat Example 8.7.2, adding lines to do MINRES and CG (setting the maximum number of iterations to get convergence to tolerance $10^{-5}$). Report the running time required by each of the three Krylov methods.

8.7.5. The condition number of the unknown matrix associated with blurring is plausibly related to the condition numbers of the single-dimension matrices $B^k$ and $C^k$ in (8.7.3).

 (a) ⌨ Let $m = 50$. Show that $B$ has a Cholesky factorization and thus is SPD. Find $\kappa(B)$.

 (b) ◢ Explain why part (a) implies $\kappa(B^k) = \kappa(B)^k$.

 (c) ◢ Explain two important effects of the severity of blur (that is, as $k \to \infty$) on deblurring by Krylov methods.

## 8.8 ▪ Preconditioning

As we saw in Section 8.6, the convergence of a Krylov method can be expected to deteriorate as the condition number of the matrix increases. Even moderately large condition numbers can make the convergence impractically slow. Therefore it's common for these methods to be used with a technique known as **preconditioning** to reduce the relevant condition number.

A problem $Ax = b$ with a difficult $A$ can be made more tractable in the mathematically equivalent form

$$(M^{-1}A)x = M^{-1}b \qquad (8.8.1)$$

for a matrix $M$ of our choosing. One goal in this choice is to make $M^{-1}A \approx I$, which makes (8.8.1) easy to solve by Krylov iteration. In a loose sense, this means $M \approx A$. On the other hand, there is an important constraint on $M$. As usual, we do not wish to actually compute $M^{-1}$. Instead, we have a linear system with the matrix $M^{-1}A$, and we take a two-step process to compute any $y = M^{-1}Av$ within the Krylov iteration:

1. Set $u = Av$.

2. Solve $My = u$ for $y$.

Hence we desire that solving the system $My = u$ be relatively fast. In short, *preconditioning is a matter of looking for an inexpensive—that is, easily inverted—approximation of the original matrix.*

Methods for deriving a good preconditioner are numerous and often problem dependent. Certain generic algebraic tricks are available. One of these is an **incomplete LU factorization**. Since true factorization of a sparse matrix usually leads to an undesirable amount of fill-in, incomplete LU prohibits or limits the fill-in in exchange for not getting an exact factorization.

### Example 8.8.1

Here is a 1000 × 1000 matrix of density around 0.5%.

```
A = 0.6*speye(1000) + sprand(1000,1000,0.005,1/10000);
```

Without a preconditioner, GMRES takes a large number of iterations.

```
b = rand(1000,1);
[x,~,~,~,resid_plain] = gmres(A,b,50,1e-10,6);   % restart
    at 50
clf, semilogy(resid_plain,'-')
```

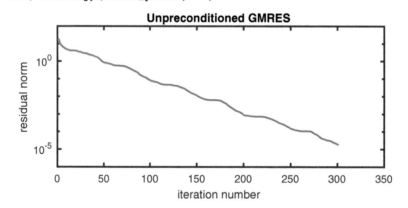

This version of incomplete LU factorization simply prohibits fill-in for the factors, freezing the sparsity pattern of the approximate factors.

```
[L,U] = ilu(A);
subplot(121), spy(A)
subplot(122), spy(L)
```

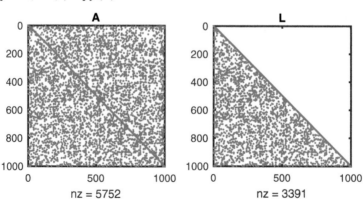

## 8.8. Preconditioning

It does *not* produce a true factorization of A.

```
norm( full(A - L*U) )

ans =
   1.5199
```

The actual preconditioning matrix is $M = LU$. However, the gmres function allows setting the preconditioner by giving the factors independently.

```
[x,~,~,~,resid_prec] = gmres(A,b,[],1e-10,300,L,U);
```

The preconditioning is fairly successful in this case.

```
clf, semilogy(resid_prec,'-')
```

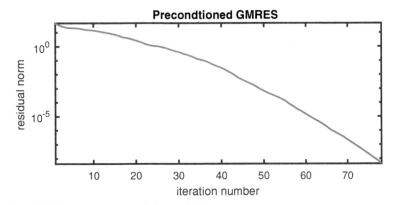

### Symmetric preconditioning

A major drawback to replacing $Ax = b$ by the preconditioned $M^{-1}Ax = M^{-1}b$ is that the new system matrix $M^{-1}A$ need not be symmetric, even if $M$ and $A$ are. In the SPD case where we have a Cholesky factorization $M = R^T R$, a symmetric form of preconditioning is

$$(R^{-T}AR^{-1})(Rx) = R^{-T}b, \qquad (8.8.2)$$

in which the system matrix is SPD if $A$ is (see Exercise 8.8.1). When using minres or pcg, one gets symmetric preconditioning automatically. Corresponding to the incomplete LU factorization for any square matrix, there is an **incomplete Cholesky factorization** for an SPD matrix, leading to a natural preconditioner.

#### Example 8.8.2

First we create a 1000 × 1000 SPD matrix with density 0.5%.

```
A = sprandsym(1000,0.005,1/1000,1);
```

Now we solve a linear system with a random right-hand side, without preconditioner.

```
b = rand(1000,1);
[x,~,~,~,resid_plain] = minres(A,b,1e-10,400);
semilogy(resid_plain,'-')
```

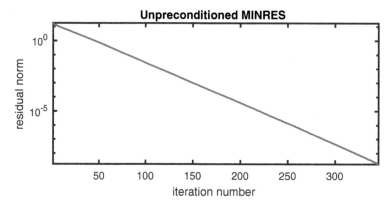

For an SPD matrix we can use an incomplete Cholesky factorization. (It returns a lower triangular $L = R^T$ rather than an upper triangular $R$.) However, it can fail, so here we add a shift to the eigenvalues of $A$ to make it "more positive."

```
L = ichol(A+0.05*speye(1000));
[x,~,~,~,residPrec] = minres(A,b,1e-10,400,L,L');
hold on, semilogy(residPrec,'-')
```

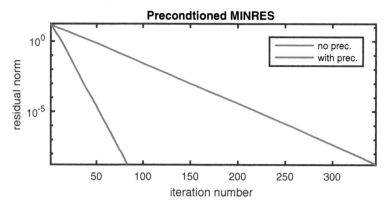

The preconditioning was moderately successful.

In many applications the problem $Ax = b$ has a known structure that may be exploited. It may be some approximation of a continuous mathematical model, and then $M$ can be derived by using a cruder form of the approximation. Another important idea is to distinguish near-field and far-field influences in a physically motivated problem and make a fast approximation of the far field. These are advanced topics and require appreciation of the underlying problem that produces the linear system to be solved.

## Exercises

8.8.1. Show that the matrix $R^{-T}AR^{-1}$ in (8.8.2) is SPD, given that $A$ is SPD and $R$ is nonsingular.

8.8.2. Suppose $M = R^T R$. Show that the eigenvalues of $R^{-T}AR^{-1}$ are the same as the eigenvalues of $M^{-1}A$.

8.8.3. Using the definitions in Example 8.8.1, with $M = LU$, plot the eigenvalues of $A$ and of $M^{-1}A$ in the complex plane. Do they support the notion that $M^{-1}A$ is "more like" an identity matrix than $A$ is? (You have to convert a sparse matrix to full form in order to apply `eig` to it, as in `eig(full(A))`.)

8.8.4. (Continuation of Exercise 8.5.5.) Let

$$B = \begin{bmatrix} 1 & & & \\ & 2 & & \\ & & \ddots & \\ & & & 100 \end{bmatrix},$$

let $I$ be a 100 × 100 identity, and let $Z$ be a 100 × 100 matrix of zeros. Define $A = \begin{bmatrix} B & I \\ Z & -B \end{bmatrix}$ and let $b$ be a 200 × 1 vector of ones. The matrix $A$ is difficult for GMRES.

(a) Design a diagonal preconditioner $M$ such that $M^{-1}A$ has all positive eigenvalues. Apply `gmres` without restarts using this preconditioner and a tolerance of $10^{-10}$ for 100 iterations. Plot the convergence curve.

(b) Now design another diagonal preconditioner such that all the eigenvalues of $M^{-1}A$ are 1 and apply preconditioned `gmres` again. How many iterations are apparently needed for full convergence?

## Key ideas in this chapter

1. A sparse matrix has many structural zeros, meaning entries that are known to be exactly zero (page 312).
2. Matrix operations may substantially decrease the amount of sparsity, a phenomenon known as fill-in (page 313).
3. Without pivoting, an LU factorization preserves bandwidth, but pivoting can change or destroy bandedness (page 314).
4. The power iteration alternates matrix-vector multiplication with renormalization to converge to the dominant eigenvalue (page 321).
5. The power iteration converges linearly with rate $\lambda_2/\lambda_1$, where the eigenvalues are ordered by decreasing magnitude (page 323).
6. Power iteration on a shifted and inverted matrix converges to the eigenvalue closest to a given complex scalar (page 327).
7. Krylov subspace methods replace a high-dimensional problem with one posed on lower-dimensional Krylov subspaces (page 333).
8. The Arnoldi iteration finds orthonormal bases for Krylov subspaces of increasing dimension (page 335).

9. GMRES uses the output of the Arnoldi iteration to minimize the residual of $Ax = b$ over successive Krylov subspaces (page 339).
10. While Arnoldi needs $O(m)$ steps to get the $m$th new basis vector, Lanczos needs only $O(1)$ steps. Consequently MINRES, which is GMRES for symmetric matrices, does not need restarts (page 344).
11. The conjugate gradients algorithm for positive definite matrices minimizes the error, as measured in the $A$-norm, over the Krylov subspaces (page 345).
12. Larger eigenvalue ratio (condition number) is associated with slower convergence for MINRES and CG (page 345).
13. Krylov subspace methods can be used to invert a linear transformation if one provides code for the transformation, even if its associated matrix is not known explicitly (page 349).
14. Preconditioning is the use of a cheap—that is, easily inverted—approximation of the original matrix to speed Krylov iterations (page 353).

## Where to learn more

The iterative solution of large linear systems is a vast and difficult subject. A broad yet detailed introduction to the subject, including classical topics such as Jacobi and Gauss–Seidel methods not mentioned in this chapter, is [57]. A more focused introduction to Krylov methods is given in [74].

The conjugate gradient method was originally intended to be a direct method. Theoretically, the answer is found in $n$ steps if there are $n$ unknowns if the arithmetic is perfect. However, for floating point arithmetic this result no longer holds. The trouble is that as the method progresses, the succeeding search directions become closer to being dependent, and this causes problems for conditioning and floating point computation. The method was not successful until it came to be viewed as an iterative method that could be stopped once a reasonable approximation was reached. The method was discovered by M. Hestenes and E. Stiefel independently, but they joined forces to publish a widely cited paper [32] as part of an early research program in computing run by what was then called the (U.S.) National Bureau of Standards (now called the National Institute of Standards and Technology). It took until the 1970s for the method to catch on as a computational method [26]. The interested reader can visit the SIAM History Project's articles at history.siam.org/pdf/mhestenes.pdf to find an article by Hestenes that recounts the discovery (reprinted from Nash [48]).

For those not experienced with preconditioning, it can seem like something of an art. The approach that works best very often depends on the application. Summaries of some approaches can be found in Quarteroni, Sacco, and Saleri [55] and Trefethen and Bau [72].

# Chapter 9
# Global function approximation

> Not entirely stable? I'm glad you're here to tell us these things.
> — Han Solo, *The Empire Strikes Back*

In Chapter 5 we considered a few ways to map data values to functions via interpolation. The methods we deemed successful were piecewise low-degree polynomials. In this chapter we deal with approximations that are globally defined over the entire interval, not piecewise. As shown in Section 5.1, the conditioning of a global polynomial is unacceptable for high-degree interpolants of equally spaced data. We'll remedy that issue by changing how the interpolation nodes are distributed. With that change, polynomial interpolation becomes extremely accurate and fast. Then we will look beyond interpolation and beyond polynomials a bit and consider the application of these global methods to numerical integration.

## 9.1 • Polynomial interpolation

In Sections 2.1 and 5.1 we encountered polynomial interpolation for the $n+1$ data points $(t_0, y_0), \ldots, (t_n, y_n)$.[36] Theoretically at least, *we can always construct an interpolating polynomial, and the result is unique among polynomials whose degree is less than $n+1$.*

> **Theorem 9.1.1**
>
> If the nodes $t_0, \ldots, t_n$ are all distinct, there exists a unique polynomial $p$ of degree at most $n$ that satisfies $p(t_k) = y_k$ for all $k = 0, \ldots, n$.

*Proof.* We defer the existence part to equation (9.1.3). As for uniqueness, if $p$ and $q$ are two interpolating polynomials, then $p - q$ is a polynomial of degree at most $n$

---
[36] As in Chapter 5, we use $t_i$ to denote interpolation or data nodes, and $x$ to denote the independent variable.

that is zero at the $n+1$ points $t_0, \ldots, t_n$. By the Fundamental Theorem of Algebra, which states that a $k$th-degree polynomial has no more than $k$ roots, we conclude that $p - q \equiv 0$, so $p = q$. $\square$

## Lagrange formula

In our earlier encounters with polynomial interpolation, we found the interpolant by solving a linear system of equations with a Vandermonde matrix. The first step was to express the polynomial in the natural monomial basis $1, x, x^2, \ldots$. However, as we saw in Section 5.2, no basis is more convenient than a cardinal basis, in which each member is one at a single node and zero at all of the other nodes. It is surprisingly easy to construct a cardinal basis for global polynomial interpolation. By definition, each member $\ell_k$ of the basis, for $k = 0, \ldots, n$, is an $n$th-degree polynomial satisfying the cardinality conditions

$$\ell_k(t_j) = \begin{cases} 1 & \text{if } j = k, \\ 0 & \text{otherwise.} \end{cases} \tag{9.1.1}$$

Recall that any polynomial of degree $n$ can be expressed as

$$c(x - r_1)(x - r_2) \cdots (x - r_n) = c \prod_{k=1}^{n} (x - r_k),$$

where $r_1, \ldots, r_n$ are the roots of the polynomial and $c$ is a constant. The conditions (9.1.1) give all $n$ roots of $\ell_k$, and the normalization $\ell_k(t_k) = 1$ tells us how to find $c$. The result is

$$\ell_k(x) = \frac{(x - t_0) \cdots (x - t_{k-1})(x - t_{k+1}) \cdots (x - t_n)}{(t_k - t_0) \cdots (t_k - t_{k-1})(t_k - t_{k+1}) \cdots (t_k - t_n)} = \prod_{\substack{i=0 \\ i \neq k}}^{n} \frac{(x - t_i)}{(t_k - t_i)}, \tag{9.1.2}$$

which is called a **Lagrange polynomial**.

---

**Example 9.1.1**

We plot a cardinal Lagrange polynomial for $n = 5$ and $k = 2$.

```
t = [ 1, 1.5, 2, 2.25, 2.75, 3 ];
n = 5;  k = 2;
not_k = [0:k-1 k+1:n];    % all except the kth node
```

Whenever we index into the vector t, we have to add 1 since our mathematical index starts at zero.

```
phi = @(x) prod(x-t(not_k+1));
ell_k = @(x) phi(x) ./ phi(t(k+1));
fplot(ell_k,[1 3])
hold on, grid on
plot(t(not_k+1),0*t(not_k+1),'.')
plot(t(k+1),1,'.')
```

## 9.1. Polynomial interpolation

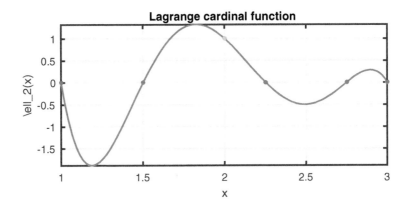
**Lagrange cardinal function**

Observe that $\ell_k$ is *not* between zero and one everywhere between the nodes.

Because they are a cardinal basis, *the Lagrange polynomials lead to a simple expression for the polynomial interpolating the $(t_k, y_k)$ points*:

$$p(x) = \sum_{k=0}^{n} y_k \ell_k(x). \tag{9.1.3}$$

This is called the **Lagrange formula** for the interpolating polynomial. At this point we can say that we have completed the proof of Theorem 9.1.1.

### Example 9.1.2

We construct the Lagrange interpolating polynomials of degrees $n = 1$ and 2 to interpolate samples of $f(x) = \tan(x)$. For $n = 1$, we use $t_0 = 0$ and $t_1 = \pi/3$; the Lagrange formula then gives

$$\begin{aligned} P_1(x) &= y_0 \ell_0(x) + y_1 \ell_1(x) \\ &= y_0 \frac{x - t_1}{t_0 - t_1} + y_1 \frac{x - t_0}{t_1 - t_0} \\ &= 0 \cdot \frac{x - \frac{\pi}{3}}{0 - \frac{\pi}{3}} + \sqrt{3} \cdot \frac{x - 0}{\frac{\pi}{3} - 0} \\ &= \frac{3\sqrt{3}}{\pi} x. \end{aligned}$$

This is the unique linear function passing through $(0,0)$ and $(\pi/3, \sqrt{3})$.

For $n = 2$, we use $t_0 = 0$, $t_1 = \pi/6$ and $t_2 = \pi/3$. We now have

$$\begin{aligned} P_2(x) &= y_0 \ell_0(x) + y_1 \ell_1(x) + y_2 \ell_2(x) \\ &= y_0 \frac{(x-t_1)(x-t_2)}{(t_0-t_1)(t_0-t_2)} + y_1 \frac{(x-t_0)(x-t_2)}{(t_1-t_0)(t_1-t_2)} + y_2 \frac{(x-t_0)(x-t_1)}{(t_2-t_0)(t_2-t_1)} \\ &= 0 + \frac{1}{\sqrt{3}} \frac{(x-0)(x-\frac{\pi}{3})}{(\frac{\pi}{6}-0)(\frac{\pi}{6}-\frac{\pi}{3})} + \sqrt{3} \frac{(x-0)(x-\frac{\pi}{6})}{(\frac{\pi}{3}-0)(\frac{\pi}{3}-\frac{\pi}{6})} = \frac{6\sqrt{3}}{\pi^2} x^2 + \frac{\sqrt{3}}{\pi} x. \end{aligned}$$

## Error formula

In addition to existence, uniqueness, and the constructive Lagrange formula, *we have a useful formula for the error in a polynomial interpolant when the data are samples of a smooth function.*

> **Theorem 9.1.2: Polynomial interpolation error**
>
> Let $t_0,\ldots,t_n$ be distinct points in $[a,b]$, and suppose $f$ has at least $n+1$ continuous derivatives in that interval. Let $p(x)$ be the unique polynomial of degree at most $n$ interpolating $f$ at $t_0,\ldots,t_n$. Then for each $x \in [a,b]$ there exists a number $\xi(x) \in (a,b)$ such that
>
> $$f(x) - p(x) = \frac{f^{(n+1)}(\xi)}{(n+1)!} \prod_{i=0}^{n}(x - t_i). \qquad (9.1.4)$$

*Proof.* Define

$$\Phi(x) = \prod_{i=0}^{n}(x - t_i). \qquad (9.1.5)$$

If $x = t_i$ for some $i$, the statement of the theorem is trivially true. Otherwise, we define a new function $g(s)$ by

$$g_x(s) = \Phi(s)[f(x) - p(x)] - \Phi(x)[f(s) - p(s)].$$

(Note that $x$ is now arbitrary but fixed.) Clearly $g_x(t_i) = 0$ for each $i = 0,\ldots,n$, because both $\Phi$ and the error $f - p$ have that property. Also, $g_x(x) = 0$. So $g_x$ has at least $n+2$ zeros in $[a,b]$. This is possible only if $g_x$ has at least $n+1$ local minima in $(a,b)$; i.e., $g_x'$ has at least $n+1$ zeros. But that implies that $g_x''$ must have at least $n$ zeros, etc. Eventually we conclude that $g_x^{(n+1)}$ has at least one zero in $(a,b)$.[37] Let $\xi(x)$ be such a zero.

Observe that $\Phi$ is a monic polynomial (i.e., its leading coefficient is 1) of degree $n+1$. Hence $\Phi^{(n+1)}(t) = (n+1)!$. Since $p$ has degree at most $n$, $p^{(n+1)} = 0$. Finally, we write

$$0 = g_x^{(n+1)}(\xi) = \Phi^{(n+1)}(\xi)[f(x) - p(x)] - \Phi(x)[f^{(n+1)}(\xi) - p^{(n+1)}(\xi)]$$
$$= (n+1)![f(x) - p(x)] - \Phi(x)f^{(n+1)}(\xi),$$

which is a restatement of (9.1.4). $\square$

Usually $f^{(n+1)}$ and the function $\xi(x)$ are unknown. The importance of formula (9.1.4) is how it helps to express the error as a function of $x$, and its dependence on the nodes $t_0,\ldots,t_n$. We will exploit this knowledge in Section 9.3.

---

[37]This deduction on $g_x^{(n+1)}$ is known as Rolle's Theorem in calculus.

## 9.1. Polynomial interpolation

**Example 9.1.3**

Consider the problem of interpolating $\log(x)$ at these nodes:

```
t = [ 1, 1.6, 1.9, 2.7, 3 ];
```

Here $n = 4$ and $f^{(5)}(\xi) = 4!/\xi^5$. For $\xi \in [1,3]$ we can say that $|f^{(5)}(\xi)| \leq 4!$. Hence

$$|f(x) - p(x)| \leq \frac{1}{5}\Phi(x).$$

```
Phi = @(x) prod(x-t);
fplot(@(x) Phi(x)/5,[1 3])
hold on, plot(t,0*t,'.')
```

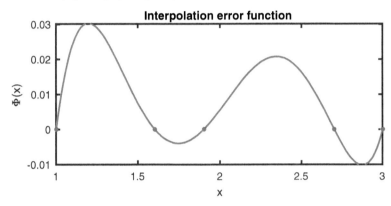

The error bound has one local extreme point between each consecutive pair of nodes.

For equispaced nodes, Theorem 9.1.1 has an immediate consequence that we already referred to in Chapter 5.

---

**Theorem 9.1.3**

Suppose $t_i = ih$ for constant step size $h$ and all $i = 0, 1, \ldots, n$, and that $f$ has $n+1$ continuous derivatives in $(t_0, t_n)$. If $x \in [t_0, t_n]$, then there exist $\xi(x) \in (t_0, t_n)$ and $C$ independent of $x$ such that

$$|f(x) - p(x)| \leq C f^{(n+1)}(\xi) h^{n+1}. \tag{9.1.6}$$

In particular, $|f(x) - p(x)| = O(h^{n+1})$ as $h \to 0$.

---

*Proof.* If $x \in [t_0, t_n]$, then $|x - t_i| < nh$ for all $i$, and (9.1.4) implies (9.1.6). As $h \to 0$, $\xi \to x$, and the continuity of $f^{(n+1)}$ allows us to make the asymptotic conclusion. □

Thus, linear (and piecewise linear) interpolation on an interval of width $O(h)$ is $O(h^2)$, and a finite difference method based on $n+1$ nodes is $O(h^n)$ (because of division by $h$ in the finite difference formula).

## Instability

As presented in (9.1.3), the Lagrange formula is not a good choice for numerical computation, because it is unstable (see Exercise 9.1.7). In the next section we derive an algebraically equivalent formula that is numerically stable and faster to apply as well.

---

## Exercises

9.1.1. ✎ Write out the Lagrange form of the interpolating polynomial of degree $n$ for the given functions and nodes. Using a calculator, evaluate the polynomial at $x = \pi/4$ and compute the error there.

   (a) $f(x) = \sin(x)$, $n = 1$, $t_0 = 0, t_1 = \pi/2$.
   (b) $f(x) = \sin(x)$, $n = 2$, $t_0 = 0, t_1 = \pi/6, t_2 = \pi/2$.
   (c) $f(x) = \cos(x)$, $n = 2$, $t_0 = 0, t_1 = \pi/3, t_2 = \pi/2$.
   (d) $f(x) = \tanh(x)$, $n = 2$, $t_0 = 0, t_1 = \pi/3, t_2 = \pi/2$.

9.1.2. 🖥 For each case, plot the requested Lagrange cardinal polynomial for the given set of nodes over the interval $[t_0, t_n]$. Superimpose dots or circles for the points represented by the cardinal conditions (9.1.1). (If you use `fplot`, force it to use at least 500 points in the plot, as described in the help text.)

   (a) $n = 2$, $t_0 = -1, t_1 = -0.2, t_2 = 0$, $\ell_2(x)$.
   (b) $n = 4$, $t_0 = 0, t_1 = 1, t_2 = 1.5, t_3 = 2.5, t_4 = 3$, $\ell_3(x)$.
   (c) $n = 20$, $t_i = i/n$ for $i = 0, \ldots, n$, $\ell_0(x)$.
   (d) $n = 20$, $t_i = i/n$ for $i = 0, \ldots, n$, $\ell_{10}(x)$.
   (e) $n = 40$, $t_i = i/n$ for $i = 0, \ldots, n$, $\ell_{20}(x)$.

9.1.3. ✎ Suppose $p$ is the quadratic polynomial interpolating the points $(-2, 12)$, $(1, 3a)$, and $(2, 0)$. Use (9.1.3) to compute $p'(0)$.

9.1.4. ✎ Explain carefully why using (9.1.3) to compute $p(x)$ at a single value of $x$ takes $O(n^2)$ floating point operations.

9.1.5. ✎ Explain why, for any distribution of nodes and all $x$,

$$1 = \sum_{k=0}^{n} \ell_k(x).$$

(Hint: This problem does not require any computation or formula manipulation.)

9.1.6. ✎ Show that

$$\ell_k(x) = \frac{\Phi(x)}{(x - t_k)\Phi'(t_k)},$$

where $\Phi$ is the function defined in (9.1.5).

9.1.7. ✎ Consider the nodes $t_0 = 0, t_1 = 1, t_2 = \beta > 1$.

   (a) Write out the Lagrange cardinal polynomials $\ell_0, \ell_1,$ and $\ell_2$.

(b) Suppose the data are $y_0 = y_1 = y_2 = 1$. What is the unique interpolating polynomial of degree no greater than 2?

(c) By letting $x = 1/2$ and $\beta \to 1$, explain how (a) and (b) demonstrate a numerical instability in the Lagrange formula.

## 9.2 • The barycentric formula

The Lagrange formula (9.1.3) is useful theoretically but not ideal for computation. For each new value of $x$, all of the cardinal functions $\ell_k$ must be evaluated at $x$, which requires a product of $n$ terms. Thus the total work is $O(n^2)$ for every value of $x$. Moreover, the formula is numerically unstable (see Exercise 9.1.7). An alternative version of the formula improves both issues.

### Derivation

Define, as in the proof of Theorem 9.1.2,

$$\Phi(x) = \prod_{j=0}^{n}(x-t_j). \tag{9.2.1}$$

Also define the **barycentric weights**

$$w_k = \frac{1}{\prod_{\substack{j=0 \\ j \neq k}}^{n}(t_k - t_j)}, \quad k = 0, \ldots, n. \tag{9.2.2}$$

Then (9.1.2) can be written as

$$\ell_k(x) = \Phi(x) \frac{w_k}{x - t_k}, \tag{9.2.3}$$

and thus the interpolating polynomial for $f(x)$ is

$$p(x) = \Phi(x) \sum_{k=0}^{n} \frac{w_k}{x - t_k} y_k. \tag{9.2.4}$$

We are still one step away from the most useful formula. Obviously, the constant function $p(x) \equiv 1$ is its own polynomial interpolant on any set of nodes. The uniqueness of the interpolating polynomial, as proved in Theorem 9.1.1, allows us to plug $y_k = 1$ for all $k$ into (9.2.4) to obtain

$$1 = \Phi(x) \sum_{k=0}^{n} \frac{w_k}{x - t_k}.$$

This is solved for $\Phi(x)$ and put back into (9.2.4) to get

$$p(x) = \frac{\displaystyle\sum_{k=0}^{n} y_k \frac{w_k}{x - t_k}}{\displaystyle\sum_{k=0}^{n} \frac{w_k}{x - t_k}}, \tag{9.2.5}$$

which is the **barycentric formula** for the interpolating polynomial.[38] Equation (9.2.5) is certainly an odd-looking way to write a polynomial! But *the barycentric formula is the key to efficient and stable evaluation of a polynomial interpolant.*

---

**Example 9.2.1**

We now write the barycentric formula for the interpolating polynomial for the quadratic case ($n = 2$) for Example 9.1.2. The weights are computed from (9.2.2),

$$w_0 = \frac{1}{(t_0 - t_1)(t_0 - t_2)} = \frac{1}{\left(0 - \frac{\pi}{6}\right)\left(0 - \frac{\pi}{3}\right)} = \frac{18}{\pi^2}, \qquad (9.2.6)$$

and similarly, $w_1 = -36/\pi^2$ and $w_2 = 18/\pi^2$. Note that in (9.2.5) any common factor in the weights cancels out without affecting the results. Hence it's a lot easier to use $w_0 = w_2 = 1$ and $w_1 = -2$. Then

$$p(x) = \frac{\dfrac{w_0}{x - t_0} y_0 + \dfrac{w_1}{x - t_1} y_1 + \dfrac{w_2}{x - t_2} y_2}{\dfrac{w_0}{x - t_0} + \dfrac{w_1}{x - t_1} + \dfrac{w_2}{x - t_2}}$$

$$= \frac{\left(\dfrac{1}{x}\right) 0 - \left(\dfrac{2}{x - \dfrac{\pi}{6}}\right) \dfrac{1}{\sqrt{3}} + \left(\dfrac{1}{x - \dfrac{\pi}{3}}\right) \sqrt{3}}{\dfrac{1}{x} - \dfrac{2}{x - \dfrac{\pi}{6}} + \dfrac{1}{x - \dfrac{\pi}{3}}}.$$

Further algebraic manipulation could return this expression to the classical Lagrange form derived in Example 9.1.2.

---

For certain canonical node distributions, simple formulas for the weights $w_k$ are known. Otherwise, computing all $n + 1$ weights from (9.2.2) takes $O(n^2)$ operations. However, the weights depend only on the nodes, not the data—and once they are known, computing $p(x)$ from (9.2.4) for any set of data at a particular value of $x$ takes just $O(n)$ operations.

## Implementation

In Function 9.2.1 we show an implementation of the barycentric formula for polynomial interpolation. The first phase is to compute the weights $w_k$, or, more conveniently, $w_k^{-1}$. As noted in Example 9.2.1, a common scaling factor in the weights does not affect the barycentric formula (9.2.5). In our code this fact is used to rescale the nodes so as to avoid arriving at very small or very large numbers.

The weight computation begins with the singleton node set $\{t_0\}$, for which one gets the single weight $w_0 = 1$. The idea is to grow this single node into the set of all the

---

[38] More precisely, (9.2.4) and (9.2.5) are the "first" and "second" barycentric formulas, respectively.

## 9.2. The barycentric formula

**Function 9.2.1** (`polyinterp`) Polynomial interpolation by the barycentric formula.

```
function p = polyinterp(t,y)
% POLYINTERP Polynomial interpolation by the barycentric formula.
% Input:
%   t     interpolation nodes (vector, length n+1)
%   y     interpolation values (vector, length n+1)
% Output:
%   p     polynomial interpolant (function)

t = t(:);                    % column vector
n = length(t)-1;
C = (t(end)-t(1)) / 4;       % scaling factor to ensure stability
tc = t/C;

% Adding one node at a time, compute inverses of the weights.
omega = ones(n+1,1);
for m = 1:n
    d = (tc(1:m) - tc(m+1));      % vector of node differences
    omega(1:m) = omega(1:m).*d;   % update previous
    omega(m+1) = prod( -d );      % compute the new one
end
w = 1./omega;                     % go from inverses to weights

p = @evaluate;

    function f = evaluate(x)
        % % Compute interpolant, one value of x at a time.
        f = zeros(size(x));
        for j = 1:numel(x)
            terms = w ./ (x(j) - t );
            f(j) = sum(y.*terms) / sum(terms);
        end

        % Apply L'Hopital's rule exactly.
        for j = find( isnan(f(:)) )'    % divided by zero here
            [~,idx] = min( abs(x(j)-t) ); % node closest to x(j)
            f(j) = y(idx);                % value at node
        end
    end

end
```

nodes through a recursive formula. Define $\omega_{k,m-1}$ (for $k < m$) as the inverse of the weight for node $k$ using the set $\{t_0, \ldots, t_{m-1}\}$. Then

$$\omega_{k,m} = \prod_{\substack{j=0 \\ j \neq k}}^{m}(t_k - t_j) = \omega_{k,m-1} \cdot (t_k - t_m), \qquad k = 0, 1, \ldots, m-1.$$

A direct application of (9.2.2) can be used to find $\omega_{m,m}$. This process is iterated over $m = 1, \ldots, n$ to find $w_k = \omega_{k,n}^{-1}$.

Once the weights are computed, the function loops over the interpolation nodes to compute the sums in the numerator and denominator of formula (9.2.5). Finally, the code addresses a peculiar feature of (9.2.5): if $x = t_i$ for some value of $i$, the formula calls for division by zero. Analytically, L'Hôpital's rule applies, and the interpolant takes the prescribed data value for node $t_i$ (see Exercise 9.2.3). Because division by zero numerically causes a NaN (not a number) value to be produced, however, we look for such cases and revise them accordingly.

> **Example 9.2.2**
>
> We use $n = 3$ and $n = 6$ with equally spaced nodes for the function $\sin(e^{2x})$ over $[0, 1]$.
>
> ```
> f = @(x) sin(exp(2*x));
> fplot(f,[0,1]), hold on
> t = linspace(0,1,4)';
> p = polyinterp(t,f(t));
> fplot(p,[0,1])
> ```
>
>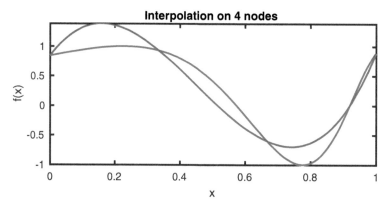
>
> ```
> cla
> fplot(f,[0,1]), hold on
> t = linspace(0,1,7)';
> p = polyinterp(t,f(t));
> fplot(p,[0,1]);
> ```
>
>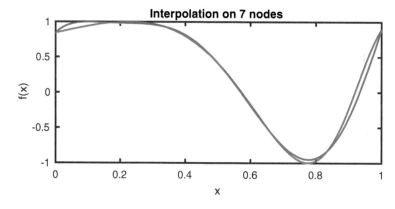
>
> The curves always intersect at the interpolation nodes.

## Stability

You might expect that as the evaluation point $x$ approaches a node $t_i$, subtractive cancellation error will creep into the barycentric formula because of the term $1/(x-t_i)$. While such errors do occur, they turn out not to cause trouble, though, because the *same* cancellation happens in the numerator and denominator. In fact the barycentric formula is a stable way to evaluate the interpolating polynomial, a statement which we do not try to prove here.

## Exercises

9.2.1. ✍ Complete the following:

(a) Find the barycentric weights for the nodes $t_0 = 0$, $t_1 = 1$, $t_2 = 3$.

(b) Compute the interpolant at $x = 2$ for the data $y_0 = -1$, $y_1 = 1$, $y_2 = -1$.

9.2.2. ✍ For each case of Exercise 9.1.1, write out the barycentric form of the interpolating polynomial.

9.2.3. ✍ Show using L'Hôpital's rule on (9.2.5) that $p(t_i) = y_i$ for all $i = 0, \ldots, n$.

9.2.4. ⌨ In each part, use Function 9.2.1 to interpolate the given function using $n+1$ evenly spaced nodes in the given interval. Plot each interpolant together with the exact function.

(a) $f(x) = \ln(x)$, $\quad n = 2, 3, 4$, $\quad x \in [1, 2]$.

(b) $f(x) = \tanh(x)$, $\quad n = 2, 3, 4$, $\quad x \in [0.5, 2]$.

(c) $f(x) = \cosh(x)$, $\quad n = 2, 3, 4$, $\quad x \in [0, 2]$.

(d) $f(x) = |x|$, $\quad n = 2, 3, 4$, $\quad x \in [-2, 1]$.

9.2.5. ⌨ Using code from Function 9.2.1 if you like, compute the barycentric weights numerically using $n+1$ equally spaced nodes in $[-1, 1]$ for $n = 30$, $n = 60$, and $n = 90$. On a single graph, plot their absolute values as a function of $t_i$ (which always ranges between $-1$ and 1) using a log–linear scale. (The resulting graphs are an indication of the trouble with equally spaced nodes that is explored in Section 9.3.)

9.2.6. ✍

(a) Use (9.2.5) to find an expression for the Lagrange cardinal function $\ell_k(x)$.

(b) Suppose $j \neq k$. Multiply both sides of the equation in part (a) by $(x - x_j)$ and use the result to show that $\ell'_k(x_j) = w_k / [w_j(x_j - x_k)]$.

## 9.3 ▪ Stability of polynomial interpolation

With barycentric interpolation available in the form of Function 9.2.1, we can explore polynomial interpolation using a numerically stable algorithm. Any remaining sensitivity to error is due to the interpolation process itself.

## Example 9.3.1

We choose a function over the interval [0, 1].

```
f = @(x) sin(exp(2*x));
fplot(f,[0 1])
```

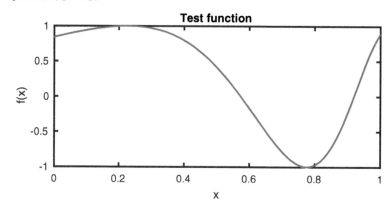

We interpolate it at equally spaced nodes for increasing values of $n$. We will sample the interpolant at a large number of points in order to estimate the interpolation error.

```
n = (5:5:60)';      err = 0*n;
x = linspace(0,1,1001)';          % for measuring error
for k = 1:length(n)
  t = linspace(0,1,n(k)+1)';      % equally spaced nodes
  y = f(t);                       % interpolation data
  p = polyinterp(t,y);
  err(k) = norm( f(x)-p(x), inf );
end
semilogy(n,err,'.-')
```

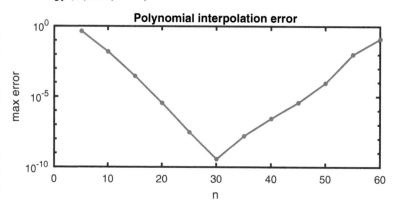

Initially the error decreases exponentially, i.e., as $O(K^{-n})$ for some $K > 1$. However, around $n = 30$ the error starts to *grow* exponentially.

## 9.3. Stability of polynomial interpolation

### Runge phenomenon

The disappointing loss of convergence in Example 9.3.1 is due to the use of equally spaced nodes. We will examine this effect using the error formula (9.1.4) as a guide:

$$f(x) - p(x) = \frac{f^{(n+1)}(\xi)}{(n+1)!} \Phi(x), \qquad \Phi(x) = \prod_{i=0}^{n} (x - t_i).$$

The $\Phi(x)$ term can be studied as a function of the nodes only.

**Example 9.3.2**

We plot $|\Phi(x)|$ over the interval $[-1, 1]$ with equispaced nodes for different values of $n$.

```
x = linspace(-1,1,1601)';
Phi = zeros(size(x));
for n = 10:10:50
    t = linspace(-1,1,n+1)';
    for k = 1:length(x)
        Phi(k) = prod(x(k)-t);
    end
    semilogy(x,abs(Phi)), hold on
end
```

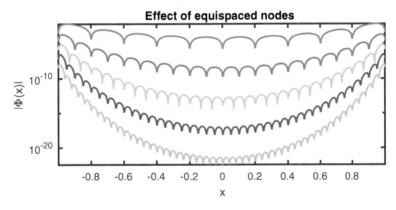

(Each time $\Phi$ passes through zero at an interpolation node, the value on the log scale should go to $-\infty$, which explains the numerous cusps on the curves.) Two observations are important: First, the size of $|\Phi|$ decreases exponentially at each fixed location in the interval (because the spacing between curves is constant for constant increments of $n$). Second, $|\Phi|$ is larger at the ends of the interval than in the middle, by an exponentially growing factor.

Even though $\Phi \to 0$ at every point in the interval, the exponentially growing gap between the ends and the middle of the interval can ruin the convergence of polynomial interpolation for many choices of $f$.

### Example 9.3.3

This function has infinitely many continuous derivatives on the entire real line and looks very easy to approximate over $[-1, 1]$.

```
f = @(x) 1./(x.^2 + 16);
fplot(f,[-1 1])
```

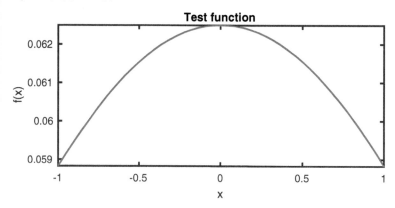

We start by doing polynomial interpolation for some rather small values of $n$.

```
x = linspace(-1,1,1601)';
n = (4:4:12)';
for k = 1:length(n)
    t = linspace(-1,1,n(k)+1)';        % equally spaced
        nodes
    p = polyinterp(t,f(t));
    semilogy( x, abs(f(x)-p(x)) );
end
```

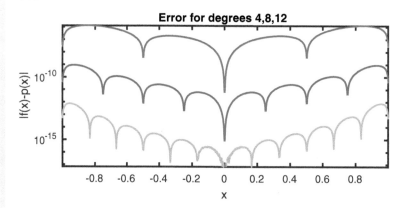

The convergence so far appears rather good, though not uniformly so. Now watch what happens as we continue to increase the degree.

```
n = 12 + 15*(1:3)';   clf
for k = 1:length(n)
    t = linspace(-1,1,n(k)+1)';        % equally spaced
        nodes
    p = polyinterp(t,f(t));
```

## 9.3. Stability of polynomial interpolation

```
        semilogy( x, abs(f(x)-p(x)) );
end
```

**Error for degrees 27,42,57**

[plot showing |f(x)-p(x)| on log scale vs x from -0.8 to 0.8, with error minimized near center around $10^{-10}$ and growing toward the ends]

The convergence in the middle can't get any better than machine precision. So maintaining the growing gap between the center and the ends pushes the error curves upward exponentially fast at the ends, wrecking the convergence.

The observation of instability in Example 9.3.3 is known as the **Runge phenomenon**. *The Runge phenomenon is an instability in the abstract mapping from a function to its polynomial interpolant, manifested when the nodes of the interpolant are equally spaced and the degree of the polynomial increases.* We reiterate that the phenomenon is rooted in the convergence theory and not a consequence of the algorithm chosen to implement polynomial interpolation.

Significantly, the convergence observed in Example 9.3.3 is stable within a middle portion of the interval. By redistributing the interpolation nodes, we will next sacrifice a little of the convergence in the middle portion in order to improve it enough near the ends to rescue the process globally.

### Chebyshev nodes

The observations above suggest that we might find success by having more nodes near the ends of the interval than in the middle. Though we will not give the details, it turns out that there is a precise asymptotic sense in which this must be done to make polynomial interpolation work over the entire interval. *One especially important node family that gives stable convergence for polynomial interpolation is the **Chebyshev points** of the second kind* (also known as Chebyshev extreme points) defined by

$$t_k = -\cos\left(\frac{k\pi}{n}\right), \qquad k = 0, \ldots, n. \tag{9.3.1}$$

These are the projections onto the $x$-axis of $n$ equally spaced points on a unit circle. As such they are densely clustered near the ends of $[-1, 1]$, and this turns out to overcome the Runge phenomenon.

### Example 9.3.4

We repeat Example 9.3.2 but replace equally spaced nodes with Chebyshev points.

```
x = linspace(-1,1,1601)';
Phi = zeros(size(x));
for n = 10:10:50
    theta = linspace(0,pi,n+1)';
    t = -cos(theta);
    for k = 1:length(x)
        Phi(k) = prod(x(k)-t);
    end
    semilogy(x,abs(Phi))
    hold on
end
```

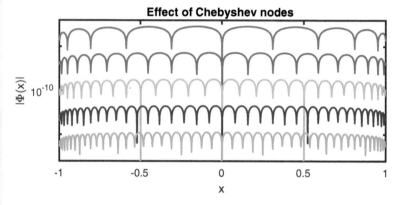

The convergence is a bit slower in the middle than with equally spaced points, but far more uniform over the entire interval, which is the key to global convergence.

### Example 9.3.5

We repeat Example 9.3.3, replacing equally spaced nodes with Chebyshev nodes.

This function has infinitely many continuous derivatives on the entire real line and looks very easy to approximate over $[-1, 1]$.

```
f = @(x) 1./(x.^2 + 16);
x = linspace(-1,1,1601)';
n = [4 10 16 40];
for k = 1:length(n)
    theta = linspace(0,pi,n(k)+1)';
    t = -cos(theta);
    p = polyinterp(t,f(t));
    semilogy( x, abs(f(x)-p(x)) );
end
```

## 9.3. Stability of polynomial interpolation

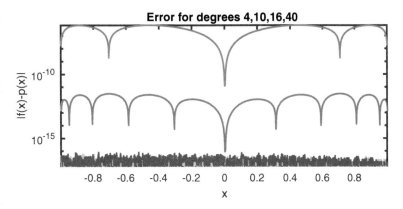

By degree 16 the error is uniformly within machine epsilon. Note that even as the degree continues to increase, the error near the ends does not grow as with the Runge phenomenon for equally spaced nodes.

As a bonus, for Chebyshev nodes the barycentric weights are simple:

$$w_k = (-1)^k d_k, \qquad d_k = \begin{cases} 1/2 & \text{if } k=0 \text{ or } k=n, \\ 1 & \text{otherwise.} \end{cases} \qquad (9.3.2)$$

### Spectral convergence

*If we take $n \to \infty$ and use polynomial interpolation on Chebyshev nodes, the convergence rate is exponential in $n$.* The following is typical of the results that can be proved.

> **Theorem 9.3.1**
>
> Suppose $f(x)$ is analytic in an open real interval containing $[-1,1]$. Then there exist constants $C > 0$ and $K > 1$ such that
>
> $$\max_{x \in [-1,1]} |f(x) - p(x)| \leq CK^{-n}, \qquad (9.3.3)$$
>
> where $p$ is the unique polynomial of degree $n$ or less defined by interpolation on $n+1$ Chebyshev second-kind points.

The condition "$f$ is analytic" means that the Taylor series of $f$ converges to $f(x)$ in an open interval containing $[-1,1]$.[39] A necessary condition of analyticity is that $f$ is infinitely differentiable.

In some contexts we refer to (9.3.3) as linear convergence, but here it is typical to say that the rate is exponential, geometric, or **spectral convergence**. One achieves constant reduction factors in the error by constant increments of $n$. By contrast, algebraic convergence in the form $O(n^{-p})$ for some $p > 0$ requires *multiplying $n$ by a constant*

---

[39] Alternatively it means the function is extensible to one that is differentiable in the complex plane.

factor in order to reduce error by a constant factor. Graphically, spectral error is linear on a log–linear scale, while algebraic convergence is a straight line on a log–log scale.

## Exercises

9.3.1. ▰ Revisit Example 9.3.1 and determine a good approximate value for the constant $K$ mentioned in the comments there.

9.3.2. ▰ For each case, compute the polynomial interpolant using $n$ second-kind Chebyshev nodes in $[-1,1]$ for $n = 4, 8, 12, \ldots, 60$. At each value of $n$, compute the infinity-norm error (that is, $\max|p(x) - f(x)|$ evaluated for at least 4000 values of $x$). Using a log–linear scale, plot the error as a function of $n$, then determine a good approximation to the constant $K$ in (9.3.3).

(a) $f(x) = 1/(25x^2 + 1)$.

(b) $f(x) = \tanh(5x + 2)$.

(c) $f(x) = \cosh(\sin x)$.

(d) $f(x) = \sin(\cosh x)$.

9.3.3. ▰ Write a function

```
function p = chebinterp(f,n)
```

that returns a function representing the polynomial interpolant of the function f using $n+1$ Chebyshev second-kind nodes (assuming $-1 \le x \le 1$). You should use (9.3.2) to compute the barycentric weights directly, rather than using the method in Function 9.2.1. Test your function by revisiting Example 9.3.3 to use Chebyshev rather than equally spaced nodes.

9.3.4. Theorem 9.3.1 assumes that the function being approximated has infinitely many derivatives over $[-1,1]$. But now consider the family of functions $f_m(x) = |x|^m$.

(a) ✎ How many continuous derivatives over $[-1,1]$ does $f_m$ possess?

(b) ▰ Compute the polynomial interpolant using $n$ second-kind Chebyshev nodes in $[-1,1]$ for $n = 10, 20, 30, \ldots, 100$. At each value of $n$, compute the infinity-norm error (that is, $\max|p(x) - f_m(x)|$ evaluated for at least 41000 values of $x$). Using a single log–log (not log–linear!) graph, plot the error as a function of $n$ for all 6 values $m = 1, 3, 5, 7, 9, 11$.

(c) ✎ Based on the results of parts (a) and (b), form a hypothesis about the asymptotic behavior of the error for fixed $m$ as $n \to \infty$.

9.3.5. The Chebyshev points can be used when the interval of interpolation is $[a, b]$ rather than $[-1, 1]$ by means of the change of variable

$$z = \psi(x) = a + (b-a)\frac{(x+1)}{2}. \qquad (9.3.4)$$

(a) ✎ Show that $\psi(-1) = a$, $\psi(1) = b$, and $\psi$ is strictly increasing on $[-1,1]$.

(b) ✎ Invert the relation (9.3.4) to solve for $x$ in terms of $\psi^{-1}(z)$.

## 9.4. Orthogonal polynomials

(c) ⚠ Let $t_0, \ldots, t_n$ be standard second-kind Chebyshev points. Then a polynomial in $x$ can be used to interpolate the function values $f(\psi(t_i))$. This in turn implies an interpolating function $\widetilde{P}(z) = P(\psi^{-1}(z))$. Show that $\widetilde{P}$ is a polynomial in $z$.

(d) 🖥 Implement the idea of part (c) to plot a polynomial interpolant of $f(x) = \cosh(\sin x)$ over $[0, 2\pi]$ using $n+1$ Chebyshev nodes with $n = 40$.

9.3.6. The Chebyshev points can be used for interpolation of functions defined on the entire real line by using the change of variable

$$z = \phi(x) = \frac{2x}{1-x^2}, \qquad (9.3.5)$$

which maps the interval $(-1, 1)$ in one-to-one fashion to the entire real line.

(a) ⚠ Find $\lim_{x \to 1^-} \phi(x)$ and $\lim_{x \to -1^+} \phi(x)$.

(b) ⚠ Invert (9.3.5) to express $x = \phi^{-1}(z)$. (Be sure to enforce $-1 \le x \le 1$.)

(c) 🖥 Let $t_0, \ldots, t_n$ be standard second-kind Chebyshev points. These map to the $z$ variable as $\zeta_i = \phi(t_i)$ for all $i$. Suppose that $f(z)$ is a given function whose domain is the entire real line. Then the function values $y_i = f(\zeta_i)$ can be associated with the Chebyshev nodes $t_i$, leading to a polynomial interpolant $p(x)$. This in turn implies an interpolating function on the real line, defined as

$$q(z) = p(\phi^{-1}(z)) = p(x).$$

Implement this idea to plot an interpolant of $f(z) = \tanh(3z - 2)$ using $n = 30$. Your plot should show $q(z)$ evaluated at 1000 evenly spaced points in $[-5, 5]$, with markers at the nodal values (those lying within the $[-5, 5]$ window).

## 9.4 ▪ Orthogonal polynomials

Interpolation is not the only way to use polynomials for global approximation of functions. We have also seen how to find least squares polynomial fits to data, by solving linear least squares matrix problems. This idea can be extended from fitting data directly to fitting functions.

### Example 9.4.1

Let's approximate $e^x$ over the interval $[-1, 1]$. We can sample it at, say, 40 points, and find the best-fitting straight line to that data.

```
t = linspace(-1,1,40)';
y = exp(t);
plot(t,y,'.')
V = [t.^0 t];
c = V\y
p = @(t) c(1) + c(2)*t;
hold on, fplot(p,[-1 1])
```

```
c =
    1.1846
    1.1091
```

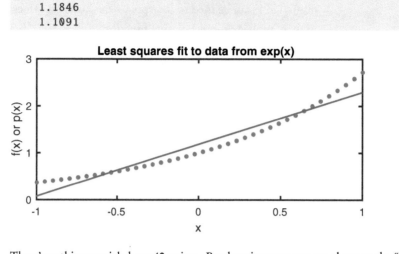

There's nothing special about 40 points. By choosing more we get closer to the "true" $e^x$.

```
t = linspace(-1,1,200)';
y = exp(t);
V = [t.^0 t];
c = V\y
```

```
c =
    1.1771
    1.1047
```

```
t = linspace(-1,1,1000)';
y = exp(t);
V = [t.^0 t];
c = V\y
```

```
c =
    1.1756
    1.1039
```

It's quite plausible that the coefficients of the best-fit line are approaching a limit as the number of nodes goes to infinity.

*We can extend least squares fitting from data to functions by extending several familiar finite-dimensional definitions to functions.* Let $S$ be the set of continuous real-valued functions on the interval $[-1, 1]$. By analogy with the inner product of two vectors, $u^T v = \sum u_i v_i$, we can define the inner product of any functions $f$ and $g$ in $S$:

$$\langle f, g \rangle = \int_{-1}^{1} f(x)g(x)\,dx. \qquad (9.4.1)$$

The inner product assigns a real number to any pair of functions in $S$. In abstract usage we refer to $S$ as an **inner product space**. This new definition of inner product allows

## 9.4. Orthogonal polynomials

us to extend to $S$ familiar vector concepts such as the 2-norm,

$$\|f\|_2^2 = \langle f, f \rangle, \tag{9.4.2}$$

and orthogonality,

$$\langle f, g \rangle = 0.$$

## Quasimatrices

If we are extending our notion of vectors to include continuous functions, what should serve as an extension of a matrix? One of our most important interpretations of a matrix is as a collection of its columns. Consider, for instance, the Vandermonde system $Vc = y$ from (2.1.2). We can connect it to the polynomial interpolation problem by writing the product as a linear combination of columns:

$$y = Vc = c_0 v_0 + c_1 v_1 + \cdots + c_n v_n,$$

in which the $i$th row of $v_j$ is $t_i^j$. We can interpret the columns $v_0, \ldots, v_n$ as discrete versions of the monomial functions $x^j$ for $j = 0, \ldots, n$. The identical linear combination expansion holds in the case of a linear least squares fit, except that there are more rows (evaluation nodes) than columns (monomial terms), and the equality is replaced by approximation.

Now consider the linear combination

$$c_0 \underline{1} + c_1 \underline{x} + \cdots + c_n \underline{x^n}, \tag{9.4.3}$$

where the "column vectors" are actually the monomial functions in the function space $S$. We underline the monomials here just to emphasize the connection to abstract vector spaces. We are motivated to define the **quasimatrix**

$$V = \begin{bmatrix} 1 & x & \cdots & x^n \end{bmatrix}. \tag{9.4.4}$$

The quasimatrix-vector product $Vc$ is to be interpreted as (9.4.3). Moreover, given a function $g(x)$, we define

$$V^T g = \begin{bmatrix} \langle 1, g \rangle \\ \vdots \\ \langle x^n, g \rangle \end{bmatrix}. \tag{9.4.5}$$

Finally, we define the operation $W = V^T V$ as the $(n+1) \times (n+1)$ matrix having $W_{ij} = \langle x^i, x^j \rangle$.

We are not limited to monomial functions as the columns of a quasimatrix. If $f_1(x), \ldots, f_k(x)$ are any $k$ functions in the space $S$, then we can define

$$F(x) = \begin{bmatrix} f_1 & f_2 & \cdots & f_k \end{bmatrix}.$$

Then $Fc$ is a linear combination of the functions, $F^T g$ is a vector of inner products, and $F^T F$ is a matrix of inner products. We consider any other expressions involving $F$ to be undefined. It might help to think of $F$ as an "$\infty \times k$" matrix, which is consistent with the definitions that $Fc$ is a function ($\infty \times 1$), $F^T g$ is a vector ($k \times 1$), and $F^T F$ is a matrix ($k \times k$).

## Normal equations

Let us return to the general discrete linear least squares problem (see (3.1.4)) of minimizing $\|f - Vc\|_2$ over all possible $c$, given matrix $V$ and vector $f$. Its solution is given by the normal equations (3.2.1),

$$c = [V^T V]^{-1} V^T f. \quad (9.4.6)$$

Now suppose $V$ is the quasimatrix (9.4.4) and $f \in S$. Thanks to our extended definitions, we can interpret $Vc$ as a polynomial, and $\|f - Vc\|_2$ has a precise meaning. Minimizing this norm gives the best polynomial approximation to $f$ in the least squares sense. Remarkably, equation (9.4.6) still provides the solution in this context! The underlying reason is that with our new definitions, $S$ captures the only ingredients necessary to prove the optimality of the normal equations: linear combination and inner product.

---

**Example 9.4.2**

We revisit approximation of $e^x$ as suggested in Example 9.4.1. With $V = \begin{bmatrix} 1 & x \end{bmatrix}$, we get

$$V^T e^x = \begin{bmatrix} \langle 1, e^x \rangle \\ \langle x, e^x \rangle \end{bmatrix} = \begin{bmatrix} \int_{-1}^{1} e^x \, dx \\ \int_{-1}^{1} x e^x \, dx \end{bmatrix} = \begin{bmatrix} e - e^{-1} \\ 2e^{-1} \end{bmatrix}$$

and

$$V^T V = \begin{bmatrix} \langle 1, 1 \rangle & \langle 1, x \rangle \\ \langle x, 1 \rangle & \langle x, x \rangle \end{bmatrix} = \begin{bmatrix} 2 & 0 \\ 0 & 2/3 \end{bmatrix}.$$

The normal equations (9.4.6) therefore have solution

$$c = \begin{bmatrix} 2 & 0 \\ 0 & 2/3 \end{bmatrix}^{-1} \begin{bmatrix} e - e^{-1} \\ 2e^{-1} \end{bmatrix} = \begin{bmatrix} \sinh(1) \\ 3e^{-1} \end{bmatrix} \approx \begin{bmatrix} 1.175201 \\ 1.103638 \end{bmatrix},$$

which is well in line with the values found in Example 9.4.1.

---

Clearly (9.4.6) becomes much simpler if $V^T V$ is diagonal. By our definitions, this would imply that the columns of $V$ are mutually orthogonal in the sense of the function inner product. Such is not true of the monomial functions $x^j$. But there are **orthogonal polynomials** which do satisfy this property.

## Legendre polynomials

Let $\mathcal{P}_n \subset S$ be the set of polynomials of degree $n$ or less. Define a sequence of polynomials by

$$\begin{aligned} P_0(x) &= 1, \\ P_1(x) &= x, \\ P_k(x) &= \frac{2k-1}{k} x P_{k-1}(x) - \frac{k-1}{k} P_{k-2}(x), \qquad k = 2, 3, \ldots. \end{aligned} \quad (9.4.7)$$

## 9.4. Orthogonal polynomials

These are known as the **Legendre polynomials**. One can show that $P_0, \ldots, P_n$ form a basis for $\mathscr{P}_n$. Most significantly, *the Legendre polynomials are* orthogonal:

$$\langle P_i, P_j \rangle = \begin{cases} 0, & i \neq j, \\ \alpha_i^2 = (i + \tfrac{1}{2})^{-1}, & i = j. \end{cases} \tag{9.4.8}$$

If we define the quasimatrix

$$\boldsymbol{L}_n(x) = \begin{bmatrix} \alpha_0^{-1} \underline{P_0} & \alpha_1^{-1} \underline{P_1} & \cdots & \alpha_n^{-1} \underline{P_n} \end{bmatrix}, \tag{9.4.9}$$

then $\boldsymbol{L}_n^T \boldsymbol{L}_n = \boldsymbol{I}$. The normal equations (9.4.6) thus simplify accordingly. Unraveling the definitions, we find the least squares solution

$$\boldsymbol{L}_n(\boldsymbol{L}_n^T f) = \sum_{k=0}^{n} c_k P_k(x), \quad \text{where } c_k = \frac{1}{\alpha_k^2} \langle P_k, f \rangle. \tag{9.4.10}$$

### Roots of orthogonal polynomials

Interesting properties can be deduced from the orthogonality conditions. The following result will be relevant in Section 9.6.

> **Theorem 9.4.1**
>
> All $n$ roots of the Legendre polynomial $P_n(x)$ are simple and real, and they lie in the open interval $(-1, 1)$.

*Proof.* Let $x_1, \ldots, x_m$ be all of the distinct roots of $P_n(x)$ between $-1$ and $1$ at which $P_n(x)$ changes sign (in other words, all roots of odd multiplicity). Define

$$r(x) = \prod_{i=1}^{m} (x - x_i).$$

By definition, $r(x)P_n(x)$ does not change sign over $(-1, 1)$. Therefore

$$\int_{-1}^{1} r(x) P_n(x)\, dx \neq 0. \tag{9.4.11}$$

Because $r$ is a degree-$m$ polynomial, we can express it as a combination of $P_0, \ldots, P_m$. If $m < n$, the integral (9.4.11) would be zero, by the orthogonality property of Legendre polynomials. So $m \geq n$. Since $P_n(x)$ has at most $n$ real roots, $m = n$. All of the roots must therefore be simple, and this completes the proof. $\square$

### Chebyshev polynomials

Equation (9.4.1) is not the only reasonable way to define an inner product on a function space. It can be generalized to

$$\langle f, g \rangle = \int_{-1}^{1} f(x) g(x) w(x)\, dx \tag{9.4.12}$$

for a positive function $w(x)$ called the **weight function** of the inner product. An important special case is

$$\langle f, g \rangle = \int_{-1}^{1} \frac{f(x)g(x)}{\sqrt{1-x^2}} dx. \qquad (9.4.13)$$

*The polynomials that are orthogonal with respect to this weight function are the Chebyshev polynomials*, which have their own recursive definition:

$$\begin{aligned} T_0(x) &= 1, \\ T_1(x) &= x, \\ T_k(x) &= 2x T_{k-1}(x) - T_{k-2}(x), \qquad k = 2, 3, \ldots. \end{aligned} \qquad (9.4.14)$$

As in the Legendre case, each $T_k$ is a polynomial of degree exactly $k$. Chebyshev polynomials also have a surprising alternate form,

$$T_k(x) = \cos(k \arccos(x)). \qquad (9.4.15)$$

In the weighted inner product, $\langle T_i, T_j \rangle = 0$ for $i \neq j$, and $\langle T_i, T_i \rangle = \gamma_i^2$, where $\gamma_0^2 = \pi$ and $\gamma_i^2 = \pi/2$ for $i > 0$. The rest of the definitions relating to least squares problems remain the same under the new inner product if we replace Legendre polynomials with Chebyshev polynomials. Note that the least squares solution is not the same in the Legendre and Chebyshev cases; both find the closest approximation to a given $f(x)$, but the norm used to measure the residual is not the same.

Theorem 9.4.1 also applies to Chebyshev polynomials. In fact, thanks to (9.4.15), the roots of $T_n$ are known explicitly:

$$t_k = \cos\left(\frac{2k-1}{2n}\pi\right), \qquad k = 1, \ldots, n. \qquad (9.4.16)$$

These are known as the **Chebyshev points of the first kind**. The chief difference between first-kind and second-kind points is that the latter type include the endpoints $\pm 1$. Both work well for polynomial interpolation and give spectral convergence.

## Least squares versus interpolation

Both interpolation and the solution of a linear least squares problem produce a **projection** of the original function $f$ onto the space of polynomials $\mathcal{P}_n$. A projection is a linear operator or matrix satisfying the identity $\boldsymbol{P}^2 = \boldsymbol{P}$. In words, applying a projection a second time has no effect.

The least squares case is special in that it produces an **orthogonal projection**. Minimization of the residual over the range of a matrix makes it orthogonal to the range, as shown in Theorem 3.2.1. The close connection with inner products and orthogonality makes the 2-norm (the norm established by (9.4.2), perhaps with a weight function) a natural setting for analysis. Because the unit weight function is the simplest choice for the inner product, Legendre polynomials are commonly used for least squares.

Interpolation has no easy connection to inner products or the 2-norm. With interpolants it's more fruitful to perform a different kind of approximation analysis, often involving the complex plane, in which the infinity norm (max norm) is the natural choice. For reasons beyond the scope of this text, Chebyshev polynomials are typically the most convenient to work with in this context.

There are many other families of orthogonal polynomials defined by the inner product (9.4.12) for different weight functions that may be favorable in particular problems. While Legendre and Chebyshev polynomials are the most prominent cases, any orthogonal family can be used with either least squares or interpolation.

## Exercises

9.4.1. Let $F$ be the quasimatrix $\begin{bmatrix} 1 & \cos(\pi x) & \sin(\pi x) \end{bmatrix}$ (all on $x \in [-1,1]$). Find $F^T x$ and $F^T F$.

9.4.2. Find the best linear approximation (in the least squares sense) to the function $\sin(x)$ on $[-1,1]$.

9.4.3. (a) Use (9.4.7) to write out $P_2(x)$ and $P_3(x)$. Plot $P_0, P_1, P_2, P_3$ on one graph for $-1 \le x \le 1$.

(b) Use (9.4.14) to write out $T_2(x)$ and $T_3(x)$. Plot $T_0, T_1, T_2, T_3$ on one graph for $-1 \le x \le 1$.

9.4.4. Use (9.4.7) to show that $P_n(x)$ is an odd function if $n$ is odd and an even function if $n$ is even.

9.4.5. Using (9.4.7), write a function

function P = legpoly(x,n)

such that the columns of the output matrix are the Legendre polynomials $P_0, P_1, \ldots, P_n$ evaluated at all the points in the vector $x$. Then use your function to plot $P_0, P_1, P_2, P_3$ on one graph.

9.4.6. (Continuation of previous problem.) Choose 1600 evenly spaced points in $[-1,1]$. For $n = 1, 2, \ldots, 16$, use this vector of points and the function legpoly to construct a $1600 \times (n+1)$ matrix that discretizes the quasimatrix

$$A_n = \begin{bmatrix} P_0 & P_1 & \cdots & P_n \end{bmatrix}.$$

Make a table of the matrix condition number $\kappa(A_n)$ as a function of $n$. (They will not be much larger than one thanks to the connection to the quasimatrix $L_n$ in (9.4.9).)

9.4.7. Using (9.4.15), write a function

function T = chebpoly(x,n)

such that the columns of the output matrix are the Chebyshev polynomials $T_0, T_1, \ldots, T_n$ evaluated at all the points in the vector $x$. Then use your function to plot $T_0, T_1, T_2, T_3$ on one graph.

9.4.8. (a) Use (9.4.15) to show that the first-kind points (9.4.16) are roots of $T_n$.

(b) Use (9.4.15) to show that the second-kind points (9.3.1) are local extreme points of $T_n$.

9.4.9. ✎ Show that the definition (9.4.15) satisfies the recursion relation in (9.4.14).

9.4.10. ✎ Use (9.4.15) to show that $\langle T_0, T_0 \rangle = \pi$ and $\langle T_k, T_k \rangle = \pi/2$ for $k > 0$ in the Chebyshev-weighted inner product.

## 9.5 ▪ Trigonometric interpolation

Up to this point all of our global approximating functions have been polynomials. While they are versatile and easy to work with, they are not always the best choice.

Suppose we want to approximate a function $f$ that is periodic, with one period represented by the standard interval $[-1, 1]$. Mathematically, $f(x+2) = f(x)$ for all real $x$. We could use polynomials to interpolate or project $f$. However, it seems reasonable to replace polynomials by functions that are also periodic, i.e., trigonometric functions.

Doing so leads to **trigonometric interpolation**. As it happens, *trigonometric interpolation allows us to return to equally spaced nodes without any problems*. We therefore define $N = 2n + 1$ equally spaced nodes inside the interval $[-1, 1]$ by

$$t_k = \frac{2k}{N}, \quad k = -n, \ldots, n. \tag{9.5.1}$$

The formulas in this section require some minor but important adjustments if $N$ is even instead. We have modified our standard indexing scheme here to make the symmetry within $[-1, 1]$ about $x = 0$ more transparent. Note that the endpoints $\pm 1$ are *not* among the nodes.

As usual, we have sample values $y_{-n}, \ldots, y_n$, perhaps representing values of a function $f(x)$ at the nodes. We also now assume that the sample values can be extended periodically forever in both directions, so that $y_{k+mN} = y_k$ for any integer $m$.

### Cardinal functions

We can explicitly state the cardinal function basis for equispaced trigonometric interpolation. It starts with

$$\tau(x) = \frac{2}{N} \left( \frac{1}{2} + \cos \pi x + \cos 2\pi x + \cdots \cos n\pi x \right) = \frac{\sin(N\pi x/2)}{N \sin(\pi x/2)}. \tag{9.5.2}$$

You can directly check (see Exercise 9.5.3) that this is 2-periodic, that $\tau(t_k) = 0$ for $k \neq 0$, and that $\tau(t_0) = \tau(0) = 1$ in the limiting sense.

Because any shift of a periodic function is also periodic, the cardinal basis for trigonometric interpolation is defined by $\tau_k(x) = \tau(x - t_k)$. Because the functions $\tau_{-n}, \ldots, \tau_n$ form a cardinal basis, the coefficients of the interpolant are just the sampled function values:

$$p(x) = \sum_{k=-n}^{n} y_k \tau_k(x). \tag{9.5.3}$$

## 9.5. Trigonometric interpolation

**Function 9.5.1** (`triginterp`) Trigonometric interpolation.

```
function p = triginterp(t,y)
% TRIGINTERP Trigonometric interpolation.
% Input:
%   t    equispaced interpolation nodes (vector, length N)
%   y    interpolation values (vector, length N)
% Output:
%   p    trigonometric interpolant (function)

N = length(t);
p = @value;

    function f = value(x)
        f = zeros(size(x));
        for k = 1:N
            f = f + y(k)*trigcardinal(x-t(k));
        end
    end

    function tau = trigcardinal(x)
        if rem(N,2)==1   % odd
            tau = sin(N*pi*x/2) ./ (N*sin(pi*x/2));
        else             % even
            tau = sin(N*pi*x/2) ./ (N*tan(pi*x/2));
        end
        tau(isnan(tau)) = 1;
    end

end
```

Function 9.5.1 is an implementation of trigonometric interpolation based on (9.5.3). The function accepts an $N$-vector of equally spaced nodes. (The case of even $N$ is included in the code.) Note too that evaluation of $\tau(0) = 1$ from (9.5.2) properly requires an application of L'Hôpital's rule, but we patch it after the fact by replacing NaN values.

**Example 9.5.1**

We get a cardinal function if we use data that is one at a node and zero at the others.

```
N = 7;   n = (N-1)/2;
t = 2*(-n:n)'/N;
y = zeros(N,1);   y(n+1) = 1;
plot(t,y,'.'), hold on
p = triginterp(t,y);
fplot(p,[-1 1])
```

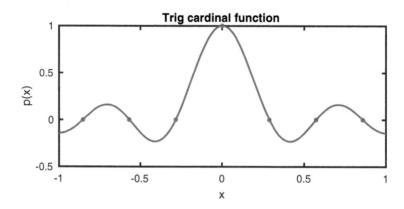

Here is a 2-periodic function and one of its interpolants.

```
clf
f = @(x) exp(sin(pi*x)-2*cos(pi*x));
fplot(f,[-1 1]), hold on
y = f(t);   plot(t,y,'.')
fplot(triginterp(t,y),[-1 1])
```

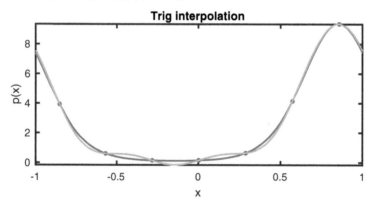

The convergence of the interpolant is exponential (spectral). We let $N$ go needlessly large here in order to demonstrate that, unlike polynomials, trigonometric interpolation is stable on equally spaced nodes.

```
N = (3:3:90)';
err = 0*N;
x = linspace(-1,1,1601)';    % for measuring error
for k = 1:length(N)
    n = (N(k)-1)/2;    t = 2*(-n:n)'/N(k);
    p = triginterp(t,f(t));
    err(k) = norm(f(x)-p(x),inf);
end
clf, semilogy(N,err,'.-')
```

## 9.5. Trigonometric interpolation

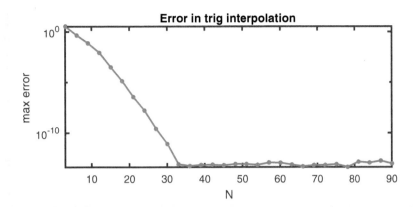

## Fast Fourier transform

Although the cardinal form of the interpolant is useful and stable, there is also a lot of importance attached to the equivalent forms

$$p(x) = \sum_{k=-n}^{n} c_k e^{ik\pi x} = \frac{a_0}{2} + \sum_{k=1}^{n} \Big[ a_k \cos(k\pi x) + b_k \sin(k\pi x) \Big], \tag{9.5.4}$$

where the constants $c_k$, or alternatively the constants $a_k$ and $b_k$, are determined by interpolation conditions. The connection between real and complex versions are Euler's formula,

$$e^{i\theta} = \cos(\theta) + i \sin(\theta), \tag{9.5.5}$$

and the consequential

$$\cos\theta = \frac{e^{i\theta} + e^{-i\theta}}{2}, \qquad \sin\theta = \frac{e^{i\theta} - e^{-i\theta}}{2i}. \tag{9.5.6}$$

While working with an all-real formulation seems natural when the data are real, the complex-valued version leads to more elegant formulas and standard usages, so we adopt it exclusively.

The $N = 2n+1$ unknown coefficients are determined by interpolation nodes at the $N$ nodes within $[-1, 1]$. By evaluating the complex exponential functions at these nodes, we get the $N \times N$ linear system

$$\begin{bmatrix} e^{-in\pi t_{-n}} & \cdots & 1 & e^{i\pi t_{-n}} & \cdots & e^{in\pi t_{-n}} \\ e^{-in\pi t_{-n+1}} & \cdots & 1 & e^{i\pi t_{-n+1}} & \cdots & e^{in\pi t_{-n+1}} \\ \vdots & & \vdots & \vdots & & \vdots \\ e^{-in\pi t_0} & \cdots & 1 & e^{i\pi t_0} & \cdots & e^{in\pi t_0} \\ \vdots & & \vdots & \vdots & & \vdots \\ e^{-in\pi t_n} & \cdots & 1 & e^{i\pi t_n} & \cdots & e^{in\pi t_n} \end{bmatrix} c = y$$

to be solved for the coefficients $c$. One of the most important (though not entirely original) observations of the 20th century was that *the linear system for the interpolation* *coefficients can be solved in just $O(N \log N)$ operations by an algorithm now known as the fast Fourier transform or FFT.* MATLAB has a function **fft** to perform this transform, but its conventions are a little different from ours. Instead of nodes in $(-1, 1)$, it expects the nodes to be defined in $[0, 2)$, and it returns the complex coefficients

$$N \begin{bmatrix} c_0 & c_1 & \cdots & c_n & c_{-n} & \cdots & c_{-1} \end{bmatrix}^T.$$

### Example 9.5.2

This function has two distinct frequencies.
```
f = @(x) 3*cos(5*pi*x) - exp(2i*pi*x);
```

We set up to use the built-in **fft**. Note how the definition of the nodes has changed.
```
n = 10;
N = 2*n+1;
t = 2*(0:N-1)'/N;     % nodes in $[0,2)$
y = f(t);
```

We perform Fourier analysis using **fft** and then examine the coefficients.
```
c = fft(y)/N;
k = [0:n -n:-1]';     % frequency ordering for MATLAB
plot(k,real(c),'.')
axis([-n n -2 2]), grid on
```

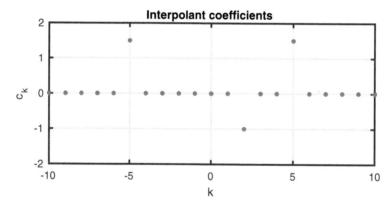

Note that $1.5e^{5i\pi x} + 1.5e^{-5i\pi x} = 3\cos(5\pi x)$ by Euler's formula, so this result is sensible.

Fourier's greatest contribution to mathematics was to point out that every periodic function is just a combination of frequencies—infinitely many of them in general, but truncated for computational use.

```
f = @(x) exp( sin(pi*x) );
c = fft(f(t))/N;
semilogy(k,abs(c),'.')
```

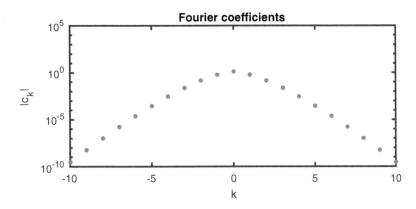

The Fourier coefficients of smooth functions decay exponentially in magnitude as a function of the frequency. This decay rate is directly linked to the convergence of the interpolation error.

The theoretical and computational aspects of Fourier analysis are vast and far-reaching. We have given only the briefest of introductions.

## Exercises

9.5.1.  Compute the trigonometric interpolant of the following functions on $-1 \leq x \leq 1$, using $N = 2n+1$ nodes with $n = 2, 4, 8, 16$. Plot the interpolants and the original function on the same plot using 500 evenly spaced points. Compute and plot the absolute error of the interpolants on 500 evenly spaced points; it may be helpful to use a semi-log scale.

(a) $f(x) = e^{\sin(\pi x)}$.

(b) $f(x) = \log[2 + \cos(3\pi x)]$.

(c) $f(x) = \cos^9(\pi x)$.

9.5.2. (a)  Show that the functions $\sin(r\pi x)$ and $\sin(s\pi x)$ are identical at all of the nodes given in (9.5.1) if $r - s = mN$ for an integer $m$. This important fact is called *aliasing*, and it implies that only finitely many frequencies can be distinguished on a fixed node set.

(b)  Demonstrate part (a) with a graph for the case $N = 15$, $s = 2$, $r = 17$. (Plot the two functions on one graph, and show that they intersect at all of the interpolation nodes.)

9.5.3.  Verify that the cardinal function given in equation (9.5.2)

(a) is 2-periodic,

(b) satisfies $\tau(t_k) = 0$ for $k \neq 0$ at the nodes (9.5.1), and

(c) satisfies $\lim_{x \to 0} \tau(x) = 1$.

9.5.4.  Prove the equality of the two expressions in (9.5.2). (Hint: Set $z = e^{i\pi x/2}$ and rewrite the sum using $z$ by applying Euler's identity.)

9.5.5. ▣ As always, spectral convergence is predicated on having infinitely many continuous derivatives. At the other extreme is a function with a jump discontinuity. Trigonometric interpolation across a jump leads to a lack of convergence altogether, a fact famously known as the *Gibbs phenomenon*.

(a) Define `f=@(x)sign(x+eps)`. This function jumps from $-1$ to $1$ at $x = -\varepsilon_{\text{mach}}$. Plot the function over $-0.05 \leq x \leq 0.15$.

(b) Let $n = 30$ and $N = 2n+1$. Using Function 9.5.1, add a plot of the trigonometric interpolant to $f$ to the graph from (a).

(c) Repeat (b) for $n = 80$ and $n = 180$.

(d) You should see that the interpolants overshoot and oscillate near the step. The widths of the overshoots decrease with $n$, but the heights approach a limiting value. By zooming in to the graph, find the height of the overshoot to two decimal places.

9.5.6. ▣ Let $f(x) = x$. Plot $f$ and its trigonometric interpolants of length $N = 2n+1$ for $n = 6, 20, 50$ over $-1 \leq x \leq 1$. Why is the convergence not spectral?

## 9.6 • Spectrally accurate integration

In Chapter 5 we derived methods of order 2, 4, and higher for numerical integration. (Recall that *quadrature* is another common term for numerical integration.) These all started with an expression of the form

$$\int_{-1}^{1} f(x)\,dx \approx \sum_{k=0}^{n} c_k f(t_k) \qquad (9.6.1)$$

for a collection of nodes $t_0, \ldots, t_n$ in $[-1, 1]$ and weights $c_0, \ldots, c_n$. (Throughout this section we use $[-1, 1]$ as the domain of the integral; for a general interval $[a, b]$, see Exercise 9.6.4.) The nodes and weights are independent of the integrand $f(x)$ and determine the implementation and properties of the formula.

The process for deriving a specific method was to interpolate the integrand, then integrate the interpolant. Piecewise linear interpolation at equally spaced nodes, for instance, leads to the trapezoid formula. *When the integrand is approximated by a spectrally accurate global function, the integration formulas are also spectrally accurate.*

### Periodic functions

For a function periodic on $[-1, 1]$, the most natural interpolant is the trigonometric polynomial (9.5.3). However, from (9.5.2) one finds that

$$\int_{-1}^{1} \sum_{k=-n}^{n} y_k \tau_k(x)\,dx = \sum_{k=-n}^{n} y_k \left[\int_{-1}^{1} \tau_k(x)\,dx\right] = \frac{1}{2n+1} \sum_{k=-n}^{n} y_k. \qquad (9.6.2)$$

In Exercise 9.6.1 you are asked to verify that this result is identical to the value of the trapezoid formula on $2n+1$ nodes. That is, *the trapezoid integration formula is exponentially accurate for periodic functions.*

## 9.6. Spectrally accurate integration

### Example 9.6.1

We use the trapezoidal integration formula to compute the perimeter of an ellipse with semi-axes 1 and 1/2. Parameterizing the ellipse as $x = \cos \pi t$, $y = \frac{1}{2}\sin \pi t$ leads to the integral

$$\int_{-1}^{1} \pi\sqrt{\cos^2(\pi t) + \tfrac{1}{4}\sin^2(\pi t)}\,dt.$$

```
f = @(t) pi*sqrt( cos(pi*t).^2+sin(pi*t).^2/4 );
N = (4:4:60)';
C = zeros(size(N));
for i = 1:length(N)
    h = 2/N(i);
    t = h*(0:N(i)-1);
    C(i) = h*sum(f(t));
end
format long
perimeter = C(end)
err = abs(C-C(end));
semilogy(N,err,'.-')

perimeter =
   4.844224110273837
```

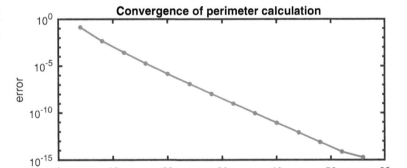

The approximations gain about one digit of accuracy for each constant increase in $N$, consistent with geometric (linear) convergence.

### Clenshaw–Curtis integration

Suppose $f$ is smooth but not periodic. If we use a global polynomial interpolating $f$ at the Chebyshev second-kind points from (9.3.1),

$$t_k = -\cos\left(\frac{k\pi}{n}\right), \qquad k = 0, \ldots, n,$$

and integrate the resulting polynomial interpolant, the method should have spectral accuracy for smooth integrands. The resulting algorithm is known as **Clenshaw–Curtis integration**.

Having specified the nodes in (9.6.1), all that remains is to find the weights. The Lagrange form of the interpolating polynomial is

$$p(x) = \sum_{k=0}^{n} f(x_k)\ell_k(x).$$

From this,

$$I = \int_{-1}^{1} f(x)\,dx \approx \int_{-1}^{1} p(x)\,dx = \int_{-1}^{1} \sum_{k=0}^{n} f(x_k)\ell_k(x)\,dx$$

$$= \sum_{k=0}^{n} f(x_k) \int_{-1}^{1} \ell_k(x)\,dx$$

$$= \sum_{k=0}^{n} c_k f(x_k), \qquad c_k = \int_{-1}^{1} \ell_k(x)\,dx.$$

For even values of $n$ the result is

$$c_k = \begin{cases} \dfrac{1}{n^2-1}, & k=0 \text{ or } k=n, \\ \dfrac{4}{n} \sum_{j=0}^{n/2} \dfrac{\cos(2\pi j k/n)}{\gamma_j(1-4j^2)}, & k=1,\ldots,n-1, \end{cases} \qquad \gamma_j = \begin{cases} 2, & j=0 \text{ or } n/2, \\ 1, & j=1,2,\ldots,n/2-1. \end{cases}$$

(9.6.3)

There are different formulas for odd values of $n$. Note that all of the weights depend on $n$; e.g., $c_2$ for $n=4$ is not the same as $c_2$ for $n=10$. Also note that the interpolant itself never needs to be computed. Function 9.6.1 performs Clenshaw–Curtis integration for even values of $n$.[40]

## Gauss–Legendre integration

Let us reconsider the generic numerical integration formula (9.6.1),

$$\int_{-1}^{1} f(x)\,dx \approx \sum_{k=1}^{n} c_k f(t_k) = Q_n[f],$$

where $Q_n[f]$ stands for the application of the formula to function $f$. (We now start the sum from $k=1$ instead of $k=0$ for notational convenience in what follows.) The interpolation approach spurred us to use Chebyshev nodes. But it's far from clear that these are the best nodes for the specific application of finding an integral. Instead, the formula can still be defined as the integral of a polynomial interpolant, but with the nodes and weights all chosen to satisfy an optimality criterion.

The definition of optimality that leads to a very useful result is to require that the formula be exact for all polynomial integrands of as high a degree as possible. Denote the set of all polynomials of degree at most $m$ by $\mathscr{P}_m$. Since there are $n$ nodes and $n$ weights to be determined, it seems plausible to expect $m=2n-1$, and this turns out

---

[40] This function is modeled after the function clencurt.m of [68].

## 9.6. Spectrally accurate integration

**Function 9.6.1** (`ccint`) Clenshaw–Curtis numerical integration.

```
function [I,x] = ccint(f,n)
% CCINT   Clenshaw-Curtis numerical integration.
% Input:
%   f      integrand (function)
%   n      one less than the number of nodes (even integer)
% Output:
%   I      estimate of integral(f,-1,1)
%   x      evaluation nodes of f (vector)

% Find Chebyshev extreme nodes.
theta = pi*(0:n)'/n;
x = -cos(theta);

% Compute the C-C weights.
c = zeros(1,n+1);
c([1 n+1]) = 1/(n^2-1);
theta = theta(2:n);
v = ones(n-1,1);
for k = 1:n/2-1
  v = v - 2*cos(2*k*theta)/(4*k^2-1);
end
v = v - cos(n*theta)/(n^2-1);
c(2:n) = 2*v/n;

% Evaluate integrand and integral.
I = c*f(x);    % use vector inner product
```

to be correct. Hence the goal is now to find nodes $t_k$ and weights $c_k$ such that

$$\int_{-1}^{1} p(x)\,dx = Q_n[p] = \sum_{k=1}^{n} c_k p(t_k) \qquad \text{for all } p \in \mathscr{P}_{2n-1}. \tag{9.6.4}$$

If these conditions are satisfied, the resulting method is called **Gauss–Legendre integration**, or often *Gaussian quadrature*.

### Example 9.6.2

As an example, consider the case $n = 2$, which should allow us to satisfy (9.6.4) for the polynomials 1, $x$, $x^2$, and $x^3$. Applying the integration formula in each case, we get the conditions

$$\begin{aligned} 2 &= c_1 + c_2, \\ 0 &= c_1 t_1 + c_2 t_2, \\ \frac{2}{3} &= c_1 t_1^2 + c_2 t_2^2, \\ 0 &= c_1 t_1^3 + c_2 t_2^3. \end{aligned} \tag{9.6.5}$$

These equations can be solved to obtain

$$c_1 = c_2 = 1, \quad x_1 = -\frac{1}{\sqrt{3}}, \quad x_2 = \frac{1}{\sqrt{3}},$$

which specifies the two-point Gaussian quadrature formula.

Note that in Example 9.6.2 it was sufficient for the formula to be exact for just the four polynomials $1, x, x^2$, and $x^3$. Because the formula is linear, i.e., $Q_n[\alpha p + q] = \alpha Q_n[p] + Q_n[q]$, exactness will hold for all of $\mathcal{P}_3$. This means that the conditions here are sufficient to make $Q_2$ exact for all polynomials in $\mathcal{P}_3$.

Generalizing the process above to general $n$ would be daunting, as the conditions on the nodes and weights are nonlinear. Fortunately, a more elegant approach is possible.

> **Theorem 9.6.1**
>
> The roots of the Legendre polynomial $P_n(x)$ are the nodes of an $n$-point Gaussian quadrature formula.

*Proof.* Choose an arbitrary $p \in \mathcal{P}_{2n-1}$, and let $I_n[p](x)$ be the interpolating polynomial for $p$ using the as-yet unknown nodes $t_1, \ldots, t_n$. By definition,

$$Q_n[p] = \int_{-1}^{1} I_n[p](x)\,dx.$$

Since $I_n[p](x)$ has degree $n-1$, it is exactly equal to $p$ if $p \in \mathcal{P}_{n-1}$, and (9.6.4) is trivially satisfied. Otherwise, the error formula (9.1.4) implies

$$p(x) - I_n[p](x) = \frac{p^{(n)}(\xi(x))}{n!}\Phi(x) = \frac{p^{(n)}(\xi(x))}{n!}(x-t_1)\ldots(x-t_n).$$

Trivially, the left-hand side is a polynomial in $\mathcal{P}_{2n-1}$ of degree at least $n$, so the right-hand side must be too. Thus, we can write

$$p(x) - I_n[p](x) = \Psi(x)\Phi(x),$$

where as above, $\Phi(x) = \prod(x-t_i)$, and $\Psi(x) \in \mathcal{P}_{n-1}$ is unknown. The optimality requirement (9.6.4) becomes

$$0 = \int_{-1}^{1} p(x)\,dx - Q_n[p] = \int_{-1}^{1}\left[p(x) - I_n[p](x)\right]dx = \int_{-1}^{1}\Psi(x)\Phi(x)\,dx.$$

Given that $\Psi(x) \in \mathcal{P}_{n-1}$, we can ensure that this condition is satisfied if

$$\int_{-1}^{1} q(x)\Phi(x)\,dx = 0 \quad \text{for all } q \in \mathcal{P}_{n-1}. \tag{9.6.6}$$

Hence satisfaction of (9.6.6) implies satisfaction of (9.6.4). But by the orthogonality property of Legendre polynomials, satisfaction of (9.6.6) is guaranteed if $\Phi(x) = cP_n(x)$ for a constant $c$. Thus $\Phi$ and $P_n$ have the same roots. $\square$

From Theorem 9.4.1 we know that the roots of $P_n$ are distinct and all within $(-1, 1)$. (Conversely, it would be strange to have the integral of a function depend on some of its values outside the integration interval!) There is no explicit formula for the roots. However, they are widely available to high precision, and there are fast algorithms to compute them on demand. Function 9.6.2 uses one of the oldest methods for computing the roots and is practical up to a hundred nodes or so.

## 9.6. Spectrally accurate integration

**Function 9.6.2** (`glint`) Gauss–Legendre numerical integration.

```
function [I,x] = glint(f,n)
% GLINT   Gauss-Legendre numerical integration.
% Input:
%   f       integrand (function)
%   n       number of nodes (integer)
% Output:
%   I       estimate of integral(f,-1,1)
%   x       evaluation nodes of f (vector)

% Nodes and weights are found via a tridiagonal eigenvalue problem.
beta = 0.5./sqrt(1-(2*(1:n-1)).^(-2));
T = diag(beta,1) + diag(beta,-1);
[V,D] = eig(T);
x = diag(D); [x,idx] = sort(x);      % nodes
c = 2*V(1,idx).^2;                   % weights

% Evaluate the integrand and compute the integral.
I = c*f(x);       % vector inner product
```

### Convergence

Both Clenshaw–Curtis and Gauss–Legendre integration are based on the integration of a global polynomial interpolant, and both are spectrally accurate. The Clenshaw–Curtis method on $n+1$ points is exact for polynomials in $\mathcal{P}_n$, whereas the Gauss–Legendre method with $n$ points is exact on all of $\mathcal{P}_{2n-1}$. For this reason, it is possible for Gauss–Legendre to converge at a rate that is "twice as fast," i.e., with roughly the square of the error of Clenshaw–Curtis. But the full story is not simple.

**Example 9.6.3**

First consider the integral

$$\int_{-1}^{1} \frac{1}{1+4x^2}\, dx = \arctan(2).$$

```
f = @(x) 1./(1+4*x.^2);
exact = atan(2);
```

We compare the two spectral integration methods for a range of $n$ values.

```
n = (8:4:96)';
errCC = 0*n;
errGL = 0*n;
for k = 1:length(n)
   errCC(k) = exact - ccint(f,n(k));
   errGL(k) = exact - glint(f,n(k));
end
semilogy( n, abs(errCC), '.-'), hold on
semilogy( n, abs(errGL), '.-')
```

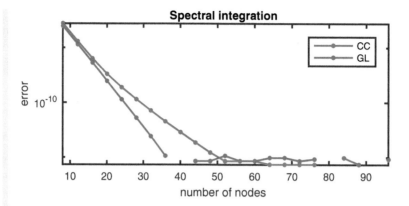

(The missing dots are where the error is exactly zero.) Gauss–Legendre does converge faster here, but at something less than twice the rate. Now we try a more sharply peaked integrand:

$$\int_{-1}^{1} \frac{1}{1+16x^2}\, dx = \frac{1}{2}\arctan(4).$$

```
f = @(x) 1./(1+16*x.^2);
exact = atan(4)/2;

clf
n = (8:4:96)';
errCC = 0*n;
errGL = 0*n;
for k = 1:length(n)
   errCC(k) = exact - ccint(f,n(k));
   errGL(k) = exact - glint(f,n(k));
end
semilogy( n, abs(errCC), '.-'), hold on
semilogy( n, abs(errGL), '.-')
```

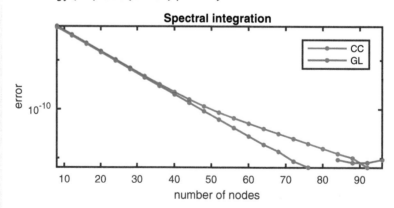

The two are very close until about $n = 40$, when the Clenshaw–Curtis method slows down.

Now let's compare the spectral performance to that of our earlier adaptive method in adaptquad. We will specify varying error tolerances and record the error as well as the total number of evaluations of $f$.

```
tol = 10.^(-2:-2:-14)';
n = 0*tol;   errAdapt = 0*tol;
for k = 1:length(n)
  [Q,t] = intadapt(f,-1,1,tol(k));
  errAdapt(k) = exact - Q;
  n(k) = length(t);
end
plot(n,abs(errAdapt),'.-')
plot(n,n.^(-4),'--')         % 4th order error
title('Spectral vs 4th order')
```

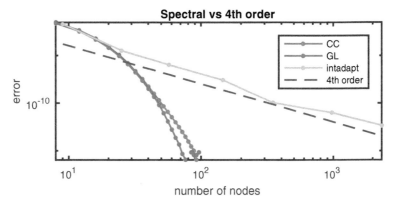

At the core of intadapt is a fourth-order formula, and the results track that rate closely. For all but the most relaxed error tolerances, both spectral methods are far more efficient than the low-order counterpart.

---

The difference between Clenshaw–Curtis and Gauss–Legendre is often modest—and far less than the difference between spectral and algebraic convergence. It is possible, though, to construct integrands for which adaptivity is critical. Choosing a method is highly problem-dependent, but a rule of thumb is that for large error tolerances, an adaptive low-order method is likely to be a good choice, while for high accuracy the spectral methods often dominate.

---

## Exercises

9.6.1. ✍ Suppose $f$ is periodic on $[-1, 1]$. Show that the result of applying the trapezoid formula on $2n + 1$ points is identical to (9.6.2).

9.6.2. ✍ For each integral, use Gauss–Legendre quadrature with $n = 2$ to write out the terms $c_1 f(t_1)$ and $c_2 f(t_2)$ explicitly.

(a) $\int_{-1}^{1} e^{-x} dx = 2\sinh(1)$.

(b) $\int_{-1}^{1} e^{-x^2} dx$.

(c) $\displaystyle\int_{-1}^{1} (2+x)^{-1}\,dx.$

9.6.3. ▭ For each integral, compute approximations using Functions 9.6.1 and 9.6.2 with $n = 8, 12, 16,\ldots, 64$. Plot the errors of both methods together as functions of $n$ on a semi-log scale.

(a) $\displaystyle\int_{-1}^{1} e^{-x}\,dx = 2\sinh(1).$

(b) $\displaystyle\int_{-1}^{1} e^{-x^2}\,dx = \sqrt{\pi}\,\operatorname{erf}(1).$

(c) $\displaystyle\int_{-1}^{1} \operatorname{sech}(x)\,dx = 2\tan^{-1}[\sinh(1)].$

(d) $\displaystyle\int_{-1}^{1} \frac{1}{1+25x^2}\,dx = \frac{2}{5}\tan^{-1}(5).$

9.6.4. (a) ▱ (See also Exercise 9.3.5.) Using the change of variable

$$z = \phi(x) = a + (b-a)\frac{(x+1)}{2},$$

show that

$$\int_a^b f(z)\,dz = \frac{b-a}{2}\int_{-1}^{1} f(\phi(x))\,dx.$$

(b) ▭ Rewrite Functions 9.6.1 and 9.6.2 to accept additional inputs for $a$ and $b$ and compute integrals over $[a,b]$.

(c) ▭ Repeat the steps of Exercise 9.6.3 for the integral

$$\int_{\pi/2}^{\pi} x^2 \sin 8x\,dx = -\frac{3\pi^2}{32}.$$

9.6.5. ▱ Prove the claim about linearity of the Gauss–Legendre integration formula alluded to in the derivation of Theorem 9.6.1. Namely, show that condition (9.6.4) is true if and only if

$$\int_{-1}^{1} x^j\,dx = \sum_{k=1}^{n} c_k x_k^j$$

for all $j = 0,\ldots,2n-1$.

## 9.7 • Improper integrals

When the interval of integration or the integrand function is unbounded, we say an integral is *improper*. Improper integrals present particular challenges to numerical computation.

In the case of an infinite interval, infinitely many nodes are needed to represent the integrand everywhere, which is absurd. However, in order for the integral to be finite,

## 9.7. Improper integrals

the integrand has to decay at infinity, which brings up the possibility of truncating the interval:

$$\int_{-\infty}^{\infty} f(x)\,dx \approx \int_{-M}^{M} f(x)\,dx. \qquad (9.7.1)$$

This integral can be discretized finitely by, say, the trapezoid formula. Yet this approach might not be realistic.

### Example 9.7.1

Consider the case with $f(x) = 1/(1+x^2)$,

$$\int_{-\infty}^{\infty} \frac{1}{1+x^2}\,dx = \pi.$$

Although $f$ decays fast enough for the integral to be finite, the decay is not fast enough for simple truncation. Note that $\int_M^\infty f\,dx \approx \int_M^\infty x^{-2}\,dx = M^{-1}$ when $M$ is large compared to 1. To get 6 digits of accuracy, then, we need to truncate with $M > 10^6$. A trapezoid discretization of (9.7.1) with $h < 1$ would need millions of nodes.

We can improve the situation a great deal by introducing a variable transformation $x(t)$. In the new variable $t$, the integrand becomes $f(x(t))x'(t)$, and the combination can be made to decay much more rapidly than $f(x)$ does.

## Review of hyperbolic functions

The variable transformations we will use are most naturally stated in terms of the hyperbolic functions, so it's worth a quick refresher on those. Recall the definitions

$$\sinh(t) = \frac{e^t - e^{-t}}{2}, \quad \cosh(t) = \frac{e^t + e^{-t}}{2}. \qquad (9.7.2)$$

Away from $t \approx 0$, these functions behave essentially like exponentials:

$$\sinh(t) \approx \pm\frac{1}{2}e^{|t|}, \quad \cosh(t) \approx \frac{1}{2}e^{|t|}, \quad \text{as } t \to \pm\infty. \qquad (9.7.3)$$

The identities

$$\frac{d}{dt}\sinh(t) = \cosh(t), \quad \frac{d}{dt}\cosh(t) = \sinh(t), \quad \cosh^2(t) - \sinh^2(t) = 1 \qquad (9.7.4)$$

frequently are handy. Finally, we recall the hyperbolic tangent,

$$\tanh(t) = \frac{\sinh(t)}{\cosh(t)} \to \pm 1, \quad \text{as } t \to \pm\infty. \qquad (9.7.5)$$

## Doubly exponential transformation

Returning to the integral (9.7.1), a particularly useful way to change the integration variable is

$$x = \sinh\left(\frac{\pi}{2}\sinh t\right). \qquad (9.7.6)$$

Note that $x = 0$ when $t = 0$, and $x \to \pm\infty$ as $t \to \pm\infty$. More specifically,

$$x \approx \pm\frac{1}{2}e^{\frac{\pi}{4}e^{|t|}}, \qquad \text{as } t \to \pm\infty,$$

and (9.7.6) is often referred to as a **doubly exponential** transformation.

By the chain rule,

$$\int_{-\infty}^{\infty} f(x)\,dx = \int_{-\infty}^{\infty} f(x(t))\frac{dx}{dt}\,dt = \frac{\pi}{2}\int_{-\infty}^{\infty} f(x(t))\cosh\left(\frac{\pi}{2}\sinh t\right)\cosh t\,dt. \qquad (9.7.7)$$

The exponential terms introduced by the chain rule grow doubly exponentially, so we seem to have made the integral much more difficult! But the decay of $f$ in the new variable more than makes up for the new terms, and *doubly exponential transformation makes truncating an infinite interval easy.*

---

**Example 9.7.2**

Consider again the case $f(x) = 1/(1+x^2)$ from Example 9.7.1. Suppose $x = x(t)$ as in (9.7.6). Although the chain rule term is doubly exponential in $t$, $x$ itself is doubly exponential, and since it's squared and in the denominator, the integrand in (9.7.7) *decays* doubly exponentially.

```
f = @(x) 1./(1+x.^2);
x = @(t) sinh(pi*sinh(t)/2);
chain = @(t) pi/2*cosh(t).*cosh(pi*sinh(t)/2);
integrand = @(t) f(x(t)).*chain(t);
fplot(integrand,[-4 4])
set(gca,'yscale','log')
```

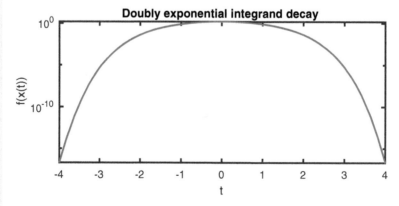

This graph suggests that we may integrate $t$ from $-4$ to $4$ and capture all of the integrand values that are larger than machine epsilon.

## 9.7. Improper integrals

**Function 9.7.1 (intde)** Doubly exponential integration over $(-\infty, \infty)$.

```
function [I,x] = intde(f,h,M)
% INTDE   Doubly exponential integration over (-inf,inf).
% Input:
%   f    integrand (function)
%   h    discretization size (positive scalar)
%   M    truncation point of integral (positive scalar)
% Output:
%   I    approximation to intergal(f,-M,M)
%   x    evaluation nodes (vector)

% Find where to truncate the trapezoid sum.
K = ceil( log(4/pi*log(2*M))/h );

% Integrate in a transformed variable t.
t = h*(-K:K)';
x = sinh(pi/2*sinh(t));
dxdt = pi/2*cosh(t).*cosh(pi/2*sinh(t));

I = h*sum( f(x).*dxdt );
```

We apply the trapezoid rule to the truncated integral in the new variable $t$:

$$\int_{-\infty}^{\infty} f(x)\,dx = \int_{-\infty}^{\infty} g(t)\,dt \approx h \sum_{k=-K}^{K} g(kh), \tag{9.7.8}$$

where $g$ consists of $f$ and the chain rule terms from (9.7.7). This implies that we truncate the integral in $t$ at $t = \pm Kh$. If we interpret this in terms of the original variable, the truncation point is $\pm M = \pm x(Kh)$. As $M \to \infty$ we get the rule of thumb

$$M \approx \frac{1}{2} e^{\frac{\pi}{4} e^{Kh}}, \quad \text{or} \quad K \approx \frac{1}{h} \log\left(\frac{4}{\pi} \log 2M\right). \tag{9.7.9}$$

Function 9.7.1 implements doubly exponential integration using $h$ and $M$ as input parameters of the discretization. The `ceil` function helps determine the truncation point.

### Example 9.7.3

We use doubly exponential transformation to integrate $\pi = \int_{-\infty}^{\infty} 1/(1+x^2)\,dx$. We let $h$ and $M$ vary over orders of magnitude and measure the effect on the accuracy of the result from Function 9.7.1.

```
f = @(x) 1./(1+x.^2);
h = logspace(-3,-1,60);
M = logspace(3,16,60);
for i = 1:length(h)
    for j = 1:length(M)
        I(i,j) = intde(f,h(i),M(j));
    end
```

```
end
err = abs(I-pi);
contourf(h,M,-log10(err)')
```

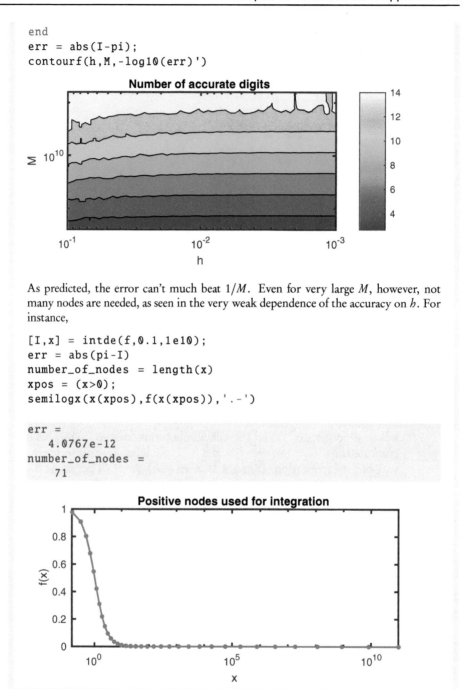

As predicted, the error can't much beat $1/M$. Even for very large $M$, however, not many nodes are needed, as seen in the very weak dependence of the accuracy on $h$. For instance,

```
[I,x] = intde(f,0.1,1e10);
err = abs(pi-I)
number_of_nodes = length(x)
xpos = (x>0);
semilogx(x(xpos),f(x(xpos)),'.-')
```

```
err =
    4.0767e-12
number_of_nodes =
    71
```

## Integrand singularities

If $f$ asymptotically approaches infinity as $x$ approaches an integration interval endpoint, its exact integral may or may not be finite. Even if $f$ is integrable, however, the methods we have used so far will generally not work well.

## 9.7. Improper integrals

### Example 9.7.4

Let's use Function 5.7.1 to try to integrate the function $\sqrt{x}$ over $[0,1]$. The exact result is $2/3$.

```
[I,x] = intadapt(@sqrt,0,1,1e-10);
err = I - 2/3
number_of_nodes = length(x)

err =
  -1.9074e-09
number_of_nodes =
   221
```

The adaptive integrator was reasonably successful. But if we integrate $1/\sqrt{x}$, which is unbounded at the origin, the number of nodes goes up dramatically.

```
[I,x] = intadapt(@(x) 1./sqrt(x),eps,1,1e-10);
err = I - 2
number_of_nodes = length(x)

err =
  -2.4282e-08
number_of_nodes =
   973
```

The nodes are packed in very closely at the origin; in fact they are placed with exponential spacing.

```
loglog(x(1:end-1),diff(x),'.')
xlim([1e-16 1]), ylim([1e-16 1])
```

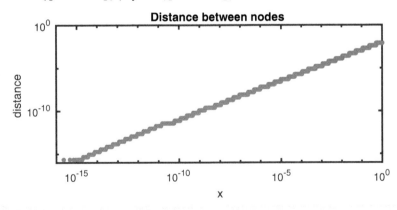

Doubly exponential transformations can be helpful for integrals with such singularities. The change of variable

$$x = \tanh\left(\frac{\pi}{2}\sinh t\right) \tag{9.7.10}$$

**Function 9.7.2** (`intsing`) Integrate function with endpoint singularities.

```
function [I,x] = intsing(f,h,delta)
% INTSING    Integrate function with endpoint singularities.
% Input:
%   f       integrand  (function)
%   h       discretization node spacing (positive scalar)
%   delta   distance away from endpoints (positive scalar)
% Output:
%   I       approximation to integral(f,-1+delta,1-delta)
%   x       evaluation nodes (vector)

% Find where to truncate the trapezoid sum.
K = ceil(log(-2/pi*log(delta/2))/h);

% Integrate over a transformed variable.
t = h*(-K:K)';
x = tanh(pi/2*sinh(t));
dxdt = pi/2*cosh(t) ./ (cosh(pi/2*sinh(t)).^2);

I = h*sum( f(x).*dxdt );
```

is a transformation between $x \in (-1,1)$ and $t \in (-\infty, \infty)$. One finds now that

$$\int_{-1}^{1} f(x)\,dx = \int_{-\infty}^{\infty} f(t)\frac{dx}{dt}\,dt = \frac{\pi}{2}\int_{-\infty}^{\infty} f(t)\cosh t \left[\cosh\left(\frac{\pi}{2}\sinh t\right)\right]^{-2}\,dt. \tag{9.7.11}$$

Now there is a squared doubly exponential term in the denominator, so that *the doubly exponential transformation is able to cancel out even unbounded growth in $f$ as one approaches an endpoint.*

Again in the new $t$ variable we have an unbounded domain that is to be truncated before applying the trapezoid rule, as in (9.7.8), defined by the discretization size $h$ and the upper sum limit $K$. Since the parameter $K$ is not simple to choose, we will swap it for another one that is easier to understand. Let us truncate the original integration interval in $x$ to $[-1+\delta, 1-\delta]$ for $\delta \ll 1$. One can show that for large positive values of $s$,

$$\tanh(s) \approx 1 - 2e^{-2s}, \tag{9.7.12}$$

and hence the truncated endpoints in $x$ map to $\log\!\left[-(2/\pi)\log(\delta/2)\right]$ in $t$. If this equals the rightmost trapezoid node at $t = hK$, we have

$$K \approx \frac{1}{h}\log\!\left[-\frac{2}{\pi}\log\!\left(\frac{\delta}{2}\right)\right].$$

This formula is used in Function 9.7.2, which accepts $h$ and $\delta$ to define the discretization.

### Example 9.7.5

We return to the problem of computing $\int_0^1 \sqrt{x}\,dx$. In order to apply `intsing`, we first have to transform the interval of integration to $[-1, 1]$. We can do this through $z = 2x - 1$. Note that

$$\int_0^1 \sqrt{x}\,dx = \int_{-1}^1 \sqrt{\tfrac{1}{2}(z+1)} \cdot \tfrac{1}{2}\,dz.$$

```
f = @(z) sqrt((1+z)/2);
[I,z] = intsing(f,0.1,1e-12);
err = I/2 - 2/3
number_of_nodes = length(z)

err =
  -7.0499e-14
number_of_nodes =
    59
```

The integration required very few nodes. For the more difficult integral of $1/\sqrt{x}$, the results are limited by how accurately we can represent $-1 + \delta$.

```
f = @(z) 1./sqrt((1+z)/2);
[I,z] = intsing(f,0.1,1e-14);
err = I/2 - 2
number_of_nodes = length(z)

err =
  -2.0378e-08
number_of_nodes =
    63
```

If we make $\delta$ any smaller, the outermost trapezoid nodes will be indistinguishable from $z = \pm 1$, i.e., the exact endpoints of the interval. We would need to use special code to evaluate $f$ indirectly in the limits $t \to \pm\infty$.

Doubly exponential mapped integration is a remarkably effective technique for a variety of integrals on unbounded domains or with singular integrands. It can also be beneficial for easier problems. Choosing the parameters well can be a nontrivial matter in the general case, however.

## Exercises

9.7.1. ▨ Use Function 9.7.1 to estimate the given integral. Experiment with the parameters to get the fewest number of integrand evaluations while achieving at least 12 digits of accuracy.

(a) $\displaystyle\int_{-\infty}^{\infty} \frac{1}{1+x^2+x^4}\,dx = \frac{\pi}{\sqrt{3}}.$

(b) $\displaystyle\int_{-\infty}^{\infty} e^{-x^2} \cos(x)\,dx = \frac{\sqrt{\pi}}{e^{1/4}}.$

(c) $\displaystyle\int_{-\infty}^{\infty} (1+x^2)^{-2/3}\,dx = \frac{\sqrt{\pi}\,\Gamma(1/6)}{\Gamma(2/3)}.$

9.7.2. ▣ Use Function 9.7.2 to estimate the given integral, possibly after transforming the given integration interval. Experiment with the parameters to get the fewest number of integrand evaluations while achieving at least 10 digits of accuracy. If you cannot achieve 10 digits, report the best accuracy you can get.

(a) $\displaystyle\int_{-1}^{1} \sqrt{1-x^2}\,dx = \frac{\pi}{2}.$

(b) $\displaystyle\int_{0}^{1} (\log x)^2\,dx = 2.$

(c) $\displaystyle\int_{0}^{\pi/2} \sqrt{\tan(x)}\,dx = \frac{\pi}{\sqrt{2}}.$

9.7.3. ✍ Derive the approximation (9.7.12) for large $s$.

9.7.4. For integration on a "half-infinite" interval such as $x \in [0,\infty)$, a third doubly exponential transformation is useful. Let $x(t) = \exp\left(\frac{\pi}{2}\sinh(t)\right)$.

(a) ✍ Show that $t \in (-\infty,\infty)$ is mapped to $x \in (0,\infty)$.

(b) ✍ Derive an analog of (9.7.7) for the chain rule on $\int_0^\infty f\,dx$.

(c) ✍ Show that truncation of $t$ to $[-Kh, Kh]$ will truncate $x$ to $[1/M, M]$. Derive an approximation for $K$ (similar to (9.7.9)) in terms of $h$ and $M$ when $M$ is large.

(d) ▣ Write a function inthalf(f,h,M) for the half-infinite integration problem. Test it on the integral

$$\int_0^\infty \frac{e^{-x}}{\sqrt{x}}\,dx = \sqrt{\pi}.$$

## Key ideas in this chapter

1. There is a unique constructible polynomial of degree less than $n+1$ interpolating $n+1$ points (page 359).

2. The Lagrange cardinal polynomials give a simple (though unstable) expression for the interpolating polynomial (page 361).

3. There is a useful formula for the error in a polynomial interpolant, when the data are samples of a smooth function (page 362).

4. The barycentric formula is the key to efficient and stable evaluation of a polynomial interpolant (page 366).

5. The Runge phenomenon is an instability in the abstract mapping from a function to its polynomial interpolant, manifested when the nodes of the interpolant are equally spaced and the degree of the polynomial increases (page 373).

6. One especially important node family that gives stable convergence for polynomial interpolation is the Chebyshev points of the second kind (page 373).

7. If we let the degree $n \to \infty$ and use polynomial interpolation on Chebyshev nodes, the convergence rate is exponential in $n$ (page 375).

8. We can extend least squares fitting from data to functions by extending several familiar finite-dimensional definitions to functions (page 378).

9. The Legendre polynomials are orthogonal with respect to function inner products (page 381).

10. The Chebyshev polynomials are orthogonal with respect to an inner product having a nonunit weight function (page 382).

11. Trigonometric interpolation can use equally spaced nodes without any difficulty (page 384).

12. The FFT solves for $N$ trigonometric interpolation coefficients in $O(N \log N)$ flops (page 387).

13. When an integrand is interpolated by a spectrally accurate method, the resulting integration formula (e.g., Clenshaw–Curtis or Gauss–Legendre) is also spectrally accurate (page 390).

14. The trapezoid integration formula is exponentially accurate for periodic functions (page 390).

15. A doubly exponential variable transformation makes truncating an infinite integration interval easy (page 400).

16. A doubly exponential transformation is able to cancel out a singularity in $f$ as one approaches an integral endpoint (page 404).

## Where to learn more

The topics in this chapter come mainly under the heading of *approximation theory*, on which there are many good references. A thorough introduction to polynomial interpolation and approximation, emphasizing the complex plane and going well beyond the basics given here, is [71]. A more thorough treatment of the least squares case is given in [22].

A thorough comparison of Clenshaw–Curtis and Gauss–Legendre integration is given in [69].

The literature on the FFT is vast; a good place to start is with the brief and clear original paper by Cooley and Tukey [20]. A historical perspective by Cooley on the acceptance and spread of the method can be found at the SIAM History Project at history.siam.org/pdf/jcooley.pdf (reprinted from Nash [48]). The FFT has a long and interesting history.

Doubly exponential integration, by contrast, is not often included in books. The original idea is presented in the readable paper [65], and the method is compared to Gaussian quadrature in [8], which is the source of some of the integration exercises in Section 9.7.

# Chapter 10
# Boundary-value problems

> Don't shoot! Don't shoot!
> —C3PO, *Star Wars: A New Hope*

In Chapter 6 we examined how to solve initial-value problems (IVPs) for ordinary differential equations (ODEs). In an IVP the supplemental conditions give complete information about the state of the system at one value of the independent variable. However, not all ODE problems come in this form. Instead we might have only partial information about the solution at two different points. Such ODE problems are called *boundary-value problems* (BVPs).

The difference between IVP and BVP may seem minor, but in fact the problems are quite different both theoretically and numerically. Conceptually the difference is like that between time and space. In an IVP the independent variable is often time, which flows naturally from the past into the future; our numerical methods start from the known initial state and propagate it forward incrementally. In a BVP the independent variable is more typically space, not time. *Information in a BVP flows both forward and backward, and hence the state at all points in the domain depend on what happens at both boundaries.*

We begin with a numerical method that attempts to treat a BVP using IVP methods. It's unstable, so we turn to numerical methods that find the solution everywhere simultaneously. One of these directly uses a discrete form of the ODE to express the computational problem, while another first restates the ODE in terms of integration. Both methods lead to algebraic systems of equations to be solved for the entire solution all at once, using the methods of Chapters 2 and 4.

## 10.1 • Shooting

The specific mathematical problem of this chapter is the **boundary-value problem** (BVP)

$$u''(x) = \varphi(x, u, u'), \qquad a \leq x \leq b,$$
$$u(a) = \alpha \text{ or } u'(a) = \alpha, \qquad (10.1.1)$$
$$u(b) = \beta \text{ or } u'(b) = \beta.$$

This BVP is said to be **linear** if the dependence of $\varphi$ on the solution $u(x)$ is purely linear, i.e., if we can write $\varphi(x, u, u') = p(x)u' + q(x)u + r(x)$ for some coefficient functions $p$, $q$, and $r$.

---

**Example 10.1.1**

A micromechanical electrically driven actuator consists of two flat disk-shaped surfaces in parallel, one at $z = 0$ and the other at $z = 1$. The surface at $z = 0$ is a rigid metal plate. The surface at $z = 1$ is an elastic membrane fixed only at its boundary. When the surfaces are given different electric potentials, the membrane deflects in response to the induced electric field, and the field is also a position of the deflection. Assuming circular symmetry, one may derive the ordinary differential equation (ODE) [54]

$$\frac{d^2 w}{dr^2} + \frac{1}{r}\frac{dw}{dr} = \frac{\lambda}{w^2}, \qquad (10.1.2)$$

where $w(r)$ is the vertical position of the membrane at radius $r$, and $\lambda$ is proportional to the applied electric potential. We will use $0 \leq r \leq 1$. There are also the supplemental physical conditions

$$w'(0) = 0, \qquad w(1) = 1, \qquad (10.1.3)$$

derived from the circular symmetry and fixing the edge of the membrane, respectively. This is a nonlinear BVP. The solution should respect $0 < w(r) \leq 1$ in order to be physically meaningful.

---

In the example above we can try to apply an IVP solver. In order to implement this idea, we rewrite (10.1.2) as a first-order system using $v_1 = w$, $v_2 = w'$:

$$v_1' = v_2,$$
$$v_2' = \frac{\lambda}{v_1^2} - \frac{v_2}{r}. \qquad (10.1.4)$$

The boundary conditions are now $v_2(0) = 0$ and $v_1(1) = 1$. Suppose we supplement the known value $v_2(0)$ with a guess for the unknown $v_1(0)$. We then compute the numerical solution of this IVP at $r = 1$ and compare the computed solution $v_1$ at $r = 1$ to its required value of 1. Depending on the result, we can adjust our guess for $v_1(0)$ and iterate.

## 10.1. Shooting

### Example 10.1.2

We'll examine the guessing/IVP approach for the case $\lambda = 0.6$.

```
lambda = 0.6;
phi = @(r,w,dwdr) lambda./w.^2 - dwdr./r;
```

We have to avoid $r = 0$, or we would divide by zero.

```
a = eps;   b = 1;
```

We convert the ODE to a first-order system in $v_1$, $v_2$.

```
f = @(r,v) [ v(2); phi(r,v(1),v(2)) ];
```

Now we try multiple guesses for the unknown $w(0)$ and plot the solutions.

```
for w0 = 0.4:0.1:0.9
    [r,v] = ode45( f, [a,b], [w0;0] );
    plot(r,v(:,1)), hold on
end
```

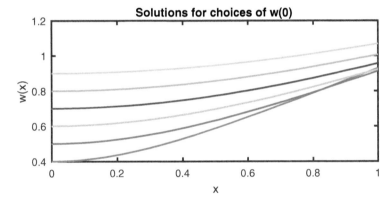

On the graph, it's the curve starting at $w(0) = 0.8$ that comes closest to the required condition $w(1) = 1$.

Following up on Example 10.1.2, we could manually try different starting values and get closer to the correct $w(0)$ by inspection. But it's much better to cast the problem into a familiar type for which we already have algorithms. Each time we pick a guessed value $v_1(0) = s$, we obtain some numerical value for $v_1$ at $r = 1$. Our goal is to adjust $s$ until $v_1(1) = 1$. Define $F(s)$ as the implied map from $s$ to the value $v_1(1) - 1$. Then we are just seeking a root of the function $F$! We don't have an explicit formula for $F(s)$, but all that a rootfinder such as `fzero` requires is that we provide a procedure for computing $F$ at any given $s$. The method that results from this description is called **shooting**.

Let's state the idea using notation for our more general problem (10.1.1). For the sake of argument, say that (as in the example) $u'(a) = \alpha$ and $u(b) = \beta$ are given. We introduce a parameter $s$ such that

$$u(a;s) = s, \qquad u'(a;s) = \alpha.$$

Then we define the objective function $F$ as $F(s) = u(b;s) - \beta$. If a root $s_0$ is found such that $F(s_0) = 0$, then $u(x;s_0)$ is the solution of the BVP. This description is straightforward to modify depending on which information is given at each endpoint.

**Function 10.1.1** (shoot) Shooting method for a two-point boundary-value problem.

```
function [x,u,dudx] = shoot(phi,xspan,lval,lder,rval,rder,init)
%SHOOT    Shooting method for a two-point boundary-value problem.
% Input:
%   phi       defines u'' = phi(x,u,u') (function)
%   xspan     endpoints of the domain (vector)
%   lval      prescribed value for u(a) (use [] if unknown)
%   lder      prescribed value for u'(a) (use [] if unknown)
%   rval      prescribed value for u(b) (use [] if unknown)
%   rder      prescribed value for u'(b) (use [] if unknown)
%   init      initial guess for lval or lder (scalar)
% Output:
%   x         nodes in x (length n+1)
%   u         values of u(x)  (length n+1)
%   dudx      values of u'(x) (length n+1)

% Tolerances for IVP solver and rootfinder.
ivp_opt = odeset('reltol',1e-6,'abstol',1e-6);
optim_opt = optimset('tolx',1e-5);

% Find the unknown quantity at x=a by rootfinding.
s = fzero(@objective,init,optim_opt);

% Don't need to solve the IVP again. It was done within the
% objective function already.
u = v(:,1);              % solution
dudx = v(:,2);           % derivative

    % Difference between computed and target values at x=b.
    function F = objective(s)
        if isempty(lder)      % u(a) given
            v_init = [ lval; s ];
        else                  % u'(a) given
            v_init = [ s; lder ];
        end

        [x,v] = ode45(@shootivp,xspan,v_init,ivp_opt);

        if isempty(rder)      % u(b) given
            F = v(end,1) - rval;
        else                  % u'(b) given
            F = v(end,2) - rder;
        end
    end

    % ODE posed as a first-order equation in 2 variables.
    function f = shootivp(x,v)
        f = [ v(2); phi(x,v(1),v(2)) ];
    end

end
```

## 10.1. Shooting

Our implementation of shooting is given as Function 10.1.1. The top-level function calls fzero in order to find the root of $F(s)$. It's difficult to know in general what initial guess for $s$ ought to be given to fzero. Function 10.1.1 could be modified to accept this guess as an additional input argument.

We have to supply fzero with the function objective, which computes $F$ when given $s$. This function in turn calls the IVP solver ode45, which solves the first-order system equivalent to (10.1.1):

$$\begin{aligned} v_1' &= v_2, \\ v_2' &= \varphi(x, v_1, v_2). \end{aligned} \qquad (10.1.5)$$

We provide ode45 with the function shootivp that encodes this ODE system.

Note that at the end of the rootfinding process, the most recent call made to objective caused the correct solution of the BVP to be computed over the whole domain. Because objective is nested within the top-level scope, those values are available without requiring another IVP solution.

### Example 10.1.3

We eliminate the guesswork for $w(0)$ and let shoot do the work for us.

```
lambda = 0.6;
phi = @(r,w,dwdr) lambda./w.^2 - dwdr./r;
a = eps;   b = 1;    % avoid r=0 in denominator
```

We specify the given and unknown endpoint values.

```
lval = [];   lder = 0;    % w(a)=?, w'(a)=0
rval = 1;    rder = [];   % w(b)=1, w'(b)=?

[r,w,dwdx] = shoot(phi,[a,b],lval,lder,rval,rder,0.8);
plot(r,w)
```

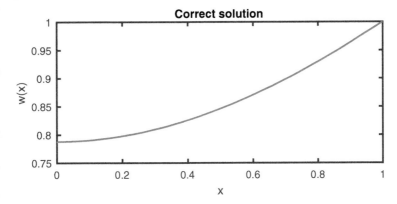

The value of $w$ at $r = 1$, meant to be exactly one, was computed to be

```
format long
```

```
w(end)

ans =
    0.999989011251061
```

The accuracy is consistent with the error tolerance used for `ode45` by `shoot`. The initial value $w(0)$ that gave this solution is

```
w(1)

ans =
    0.787737737464385
```

## Instability of shooting

The accuracy of the shooting method should be comparable to those of the component pieces, the rootfinder and the IVP solver. One has to take some care that the error goal used by the IVP solver is somewhat smaller than that of the rootfinder, or else the noise caused by the lack of precision in $F$ might make the rootfinding process impossible.

But more importantly, *the shooting method for BVPs is unstable.* An example illustrates the problem.

### Example 10.1.4

We solve the problem

$$u'' = \lambda^2 u + \lambda^2, \quad 0 \leq x \leq 1, \quad u(0) = -1, \; u(1) = 0. \qquad (10.1.6)$$

The exact solution is easily confirmed to be

$$u(x) = \frac{\sinh(\lambda x)}{\sinh(\lambda)} - 1.$$

This solution satisfies $-1 \leq u(x) \leq 0$ for all $x \in [0, 1]$.

We compute shooting solutions for several values of $\lambda$.

```
lambda = (6:4:18)';
for k = 1:4
    lam = lambda(k);
    lval = -1;   rval = 0;
    phi = @(x,u,dudx) lam^2*u + lam^2;
    [x,u] = shoot(phi,[0,1],lval,[],rval,[],0);
    plot(x,u,'-'), hold on
end
```

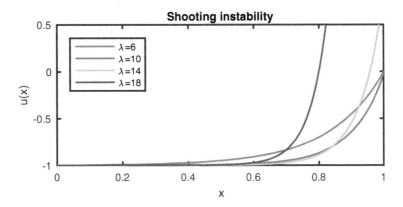

The solutions do not stay negative on the right half of the domain as $\lambda$ increases.

The behavior is readily explained. The solution to (10.1.6) with $u(0)=-1$ and $u'(0)=s$ is

$$u(x,s) = \frac{s}{\lambda} \sinh(\lambda x) - 1. \tag{10.1.7}$$

If $x$ is a fixed value in $[0,1]$, we note that the absolute condition number of (10.1.7) with respect to $s$ is

$$\left| \frac{\partial}{\partial s} [u(x,s)] \right| = \left| \frac{\sinh \lambda x}{\lambda} \right|,$$

which for $\lambda \gg 1$ is approximately $e^{\lambda x}/2\lambda$. Thus, any error in the initial slope $s$ will be multiplied by a value that grows exponentially larger in $x$ and nearly so in $\lambda$.

The essence of the difficulty in Example 10.1.4 is that errors (including roundoff) can grow exponentially away from the boundary at $x = a$. Using shooting, acceptable accuracy near $x = b$ therefore means requiring extraordinarily high accuracy near $x = a$.

The instability of shooting can be circumvented by breaking the interval into smaller pieces and thus limiting the potential for error growth, but in doing so the algorithm starts to lose its appeal. BVPs are fundamentally different from IVPs, and the methods in the rest of this chapter treat both ends of the domain symmetrically and solve over the whole domain at once.

## Exercises

10.1.1. ✎ Write each BVP in the form (10.1.1). (In other words, identify $\varphi$, $a$, $b$, $\alpha$, $\beta$.) State whether the problem is linear.

(a) $x^2 u'' + x u' + (x^2 - 1)u = 0$, $0 < x < 1$ (Bessel's equation).

(b) $u'' - e^{u+1/2} = 0$, $0 < x < 1$.

(c) $\epsilon u'' + 2(1-x^2)u + u^2 = 1$, $-1 < x < 1$. Here $0 < \epsilon \ll 1$ is a constant (Carrier equation [16]).

10.1.2. ✎ For each BVP, verify that the given answer solves the problem—that is, check the differential equations and the boundary conditions.

(a) A *Hermite equation* is

$$u'' - 2xu' + 8u = 0, \quad 0 < x < 1,$$

subject to $u(0) = 1$, $u(1) = -5/3$. The solution is $u(x) = (4/3)x^4 - 4x^2 + 1$; this answer is a multiple of the Hermite polynomial $H_4(x)$.

(b) A *Laguerre equation* is

$$xu'' + (1-x)u' + 3u = 0, \quad 0 < x < 2,$$

subject to $u(0) = 1$, $u(1) = -5/3$. The solution is $u(x) = (6 - 18x + 9x^2 - x^3)/6$; this answer is the Laguerre polynomial $L_3(x)$.

(c) A *Chebyshev equation* is

$$(1-x^2)u'' - xu' + 25u = 0, \quad -1 < x < 1,$$

subject to $u(1) = -1$, $u(1) = 1$. The solution is $u(x) = 16x^5 - 20x^3 + 5x$ (the Chebyshev polynomial $T_5(x)$).

(d) A *Legendre equation* is

$$(1-u^2)u'' - 2xu' + 12u = 0, \quad 0 < x < 1,$$

subject to $u(0) = 0$, $u(1) = 1$. The solution is $u(x) = (5x^3 - 3x)/2$ (the Legendre polynomial $P_3(x)$).

10.1.3. For each BVP, use Function 10.1.1 to compute the solution. The domain of $x$ is implied by the boundary conditions given. Plot the solution and its error as functions of $x$. (Make two graphs using `subplot`, or use **plotyy** to get two $y$-axes.)

(a)
$$u'' - \frac{3}{x}u' + \frac{4}{x^2}u = 0, \quad u(0) = 0, \quad u(2) = 4\ln 2.$$

Exact solution: $u(x) = x^2 \ln x$.

(b)
$$u'' - \frac{1}{x}u' + \frac{2}{x^2}u = 2(1 + \ln x), \quad u(1) = 1, \quad u(\pi) = \pi.$$

Exact solution: $u(x) = x^2 + (1 - \pi)x^2 \sin(\ln(x))/\sin(\ln(\pi))$.

(c)
$$u'' - \frac{1}{(x+1/2)}u' + \frac{2}{(x+1/2)^2}u = \frac{10}{(x+1/2)^4},$$
$$u(1/2) = 1, \quad u(5/2) = 1/9.$$

Exact solution: $u(x) = (x + 1/2)^{-2}$.

(d)
$$u'' + u = 0, \quad u(0) = 0, \quad u(3) = \sin 3.$$

Exact solution: $u(x) = \sin x$.

(e)
$$uu'' = 3(u')^2, \qquad u(-1) = 1, \quad u(2) = 1/2.$$

Exact solution: $u(x) = (x+2)^{-1/2}$.

10.1.4. 📖 The *stationary Allen–Cahn equation* is a model of phase changes, e.g., the change from liquid to solid. In one spatial dimension it can be written as

$$\epsilon u'' = u^3 - u, \qquad 0 \le x \le 1, \qquad u(0) = -1, \quad u(1) = 1.$$

As $\epsilon \to 0$, the solution tends toward a step function transition between $-1$ and $1$. By symmetry, $u'(x) = -u'(1-x)$.

(a) Use Function 10.1.1 to solve (10.4.7) using $\epsilon = 0.2$. Plot the solution and compute the numerical value of $u'(0) - u'(1)$.

(b) Repeat for $\epsilon = 0.02$.

(c) Repeat for $\epsilon = 0.002$. What is clearly wrong about the graph of this solution?

10.1.5. 📖 An ideal pendulum in the absence of energy dissipation satisfies the nonlinear equation $\theta'' + \sin \theta = 0$ for the angle $\theta$ with respect to the vertical direction.

(a) We want to start the pendulum from rest at angle $\alpha$ and have it come to the position $\theta = \pi/2$ after two seconds. What value of $\alpha$ should be chosen?

(b) Now we want the position to be $\theta = \pi/2$ at $t = 5$ instead of $t = 2$. Find one solution that passes through $\theta = 0$ and another that does not, and plot the solutions on one graph.

10.1.6. ✎ Suppose one wants to use shooting to solve (10.1.1) with $u(a) = \alpha$ and $u(b) = \beta$ given. As in the text, we introduce dependence on a parameter $s$ by the definition $u'(a,s) = s$, and the shooting objective function is $F(s) = u(b,s) - \beta$. A Newton iteration on $F(s)$ requires computing $F'(s) = \partial [u(b,s)]/\partial s = u_s(x,s)$; subscripts denote partial differentiation here. One way to do so is described in the following.

(a) Define $z(x,s) = u_s(x,s)$. Derive the IVP

$$z'' = \varphi_u(x,u,u')z + \varphi_{u'}(x,y,u')z', \quad z(a,s) = 0, \ z'(a,s) = 1, \quad (10.1.8)$$

where primes denote differentiation with respect to $x$. The terms $\varphi_u$ and $\varphi_{u'}$ refer to the partial derivatives of $\varphi(x,u,u')$ with respect to its second and third arguments, respectively. (Hint: Start by differentiating (10.1.1) with respect to $s$, applying the chain rule and treating $x$ and $s$ as independent variables.) Hence $F'(s) = z(b,s)$ can be found by solving this IVP numerically up to $x = b$.

(b) Show that if $\varphi(x,u,u') = p(x)u' + q(x)u + r(x)$ (i.e., the original BVP is linear), then $z(b,s)$ is independent of $s$. What does this imply about $F(s)$?

## 10.2 • Differentiation matrices

In Section 5.4 we used finite differences to turn a discrete collection of function values into an estimate of the derivative of the function at a point. Just as in elementary calculus, we can generalize differences at a point into an operation that maps discretized functions to discretized functions.

Corresponding to any function $f(x)$ defined over $[a,b]$ and a set of nodes $x_0,\ldots,x_n$ in $[a,b]$, there is the vector of nodal values,

$$\mathbf{f} = \begin{bmatrix} f(x_0) \\ f(x_1) \\ \vdots \\ f(x_{n-1}) \\ f(x_n) \end{bmatrix}.$$

Recall that our node indexing scheme starts at zero, which requires adding one to any indexing operation in our implementations in MATLAB. Also recall that we always define vectors with a column shape—this convention becomes crucial for what follows.

### Finite difference matrix

We first discretize the interval $x \in [a,b]$ into equal pieces of length $h = (b-a)/n$, leading to the nodes

$$x_i = a + ih, \qquad i = 0,\ldots,n.$$

Our goal is to find a vector $\mathbf{g}$ such that $g_i \approx f'(x_i)$ for $i = 0,\ldots,n$. Our first try is the forward difference formula (5.4.3),

$$g_i = \frac{f_{i+1} - f_i}{h}, \qquad i = 0,\ldots,n-1.$$

However, this leaves $g_n$ undefined, because the formula would refer to the unavailable value $f_{n+1}$. For $g_n$ we could resort to the backward difference

$$g_n = \frac{f_n - f_{n-1}}{h}.$$

We can summarize the entire set of formulas using linear algebra:

$$\begin{bmatrix} f'(x_0) \\ f'(x_1) \\ \vdots \\ f'(x_{n-1}) \\ f'(x_n) \end{bmatrix} \approx \frac{1}{h} \begin{bmatrix} -1 & 1 & & & \\ & -1 & 1 & & \\ & & \ddots & \ddots & \\ & & & -1 & 1 \\ & & & -1 & 1 \end{bmatrix} \begin{bmatrix} f(x_0) \\ f(x_1) \\ \vdots \\ f(x_{n-1}) \\ f(x_n) \end{bmatrix}, \quad \text{or} \quad \mathbf{f}' = \mathbf{D}_x \mathbf{f}. \qquad (10.2.1)$$

## 10.2. Differentiation matrices

Here, as elsewhere, elements of $D_x$ that are not shown are zero. We call $D_x$ a **differentiation matrix**. Each row of $D_x$ gives the weights of the finite difference formula being used at one of the nodes. *Thus, left-multiplication by $D_x$ of the function values at the nodes gives values for the desired derivative approximations at all of the nodes.*

We are free to choose whatever finite difference formulas we like in each row. However, it makes sense to use rows that are as similar as possible. Using second-order centered differences where possible and one-sided formulas (see Table 5.2) at the boundary points leads to

$$D_x = \frac{1}{h} \begin{bmatrix} -\frac{3}{2} & 2 & -\frac{1}{2} & & & & \\ -\frac{1}{2} & 0 & \frac{1}{2} & & & & \\ & -\frac{1}{2} & 0 & \frac{1}{2} & & & \\ & & \ddots & \ddots & \ddots & & \\ & & & & -\frac{1}{2} & 0 & \frac{1}{2} \\ & & & & \frac{1}{2} & -2 & \frac{3}{2} \end{bmatrix}. \quad (10.2.2)$$

Observe that the differentiation matrices so far are banded matrices—i.e., all the nonzero values are along diagonals close to the main diagonal.[41]

### Second derivative

Similarly, we can define differentiation matrices for second derivatives. For example,

$$\begin{bmatrix} f''(x_0) \\ f''(x_1) \\ f''(x_2) \\ \vdots \\ f''(x_{n-1}) \\ f''(x_n) \end{bmatrix} \approx \frac{1}{h^2} \begin{bmatrix} 2 & -5 & 4 & -1 & & & \\ 1 & -2 & 1 & & & & \\ & 1 & -2 & 1 & & & \\ & & \ddots & \ddots & \ddots & & \\ & & & & 1 & -2 & 1 \\ & & & -1 & 4 & -5 & 2 \end{bmatrix} \begin{bmatrix} f(x_0) \\ f(x_1) \\ f(x_2) \\ \vdots \\ f(x_{n-1}) \\ f(x_n) \end{bmatrix} = D_{xx} f. \quad (10.2.3)$$

Note that $D_{xx}$ need not be the square of any particular $D_x$. As pointed out in Section 5.4, squaring the first derivative is a valid approach but would place entries in $D_{xx}$ farther from the diagonal than is necessary.

Together the matrices (10.2.2) and (10.2.3) give second-order approximations of the first and second derivatives at all nodes. These matrices, as well as the nodes $x_0, \ldots, x_n$, are returned by Function 10.2.1.

---

[41] In order to exploit this structure efficiently in MATLAB, these matrices first need to be converted to sparse form by the `sparse` command. See Chapter 8.

**Function 10.2.1 (diffmat2)** Second-order accurate differentiation matrices.

```
function [x,Dx,Dxx] = diffmat2(n,xspan)
%DIFFMAT2    Second-order accurate differentiation matrices.
% Input:
%   n       number of subintervals (one less than the number of nodes)
%   xspan   interval endpoints
% Output:
%   x       equispaced nodes
%   Dx      matrix for first derivative
%   Dxx     matrix for second derivative

a = xspan(1);   b = xspan(2);
h = (b-a)/n;
x = a + h*(0:n)';   % nodes

% Define most of Dx by its diagonals.
dp = 0.5*ones(n,1)/h;       % superdiagonal
dm = -0.5*ones(n,1)/h;      % subdiagonal
Dx = diag(dm,-1) + diag(dp,1);

% Fix first and last rows.
Dx(1,1:3) = [-1.5,2,-0.5]/h;
Dx(n+1,n-1:n+1) = [0.5,-2,1.5]/h;

% Define most of Dxx by its diagonals.
d0 =  -2*ones(n+1,1)/h^2;   % main diagonal
dp =  ones(n,1)/h^2;        % superdiagonal and subdiagonal
Dxx = diag(dp,-1) + diag(d0) + diag(dp,1);

% Fix first and last rows.
Dxx(1,1:4) = [2,-5,4,-1]/h^2;
Dxx(n+1,n-2:n+1) = [-1,4,-5,2]/h^2;

end
```

### Example 10.2.1

We test first-order and second-order differentiation matrices for the function $x + e^{\sin 4x}$ over $[-1, 1]$.

```
f = @(x) x + exp(sin(4*x));
```

For reference, here are the exact first and second derivatives.

```
dfdx  = @(x) 1 + 4*exp(sin(4*x)).*cos(4*x);
d2fdx2 = @(x) 4*exp(sin(4*x)).*(4*cos(4*x).^2-4*sin(4*x));
```

We discretize on equally spaced nodes and evaluate $f$ at the nodes.

```
[t,Dx,Dxx] = diffmat2(12,[-1 1]);
y = f(t);
```

Then the first two derivatives of $f$ each require one matrix-vector multiplication.

```
yx = Dx*y;
```

## 10.2. Differentiation matrices

```
yxx = Dxx*y;
```

The results are fair but not very accurate for this small value of $n$.

```
subplot(2,1,1)
fplot(dfdx,[-1 1]), hold on
plot(t,yx,'.')
subplot(2,1,2)
fplot(d2fdx2,[-1 1]), hold on
plot(t,yxx,'.')
```

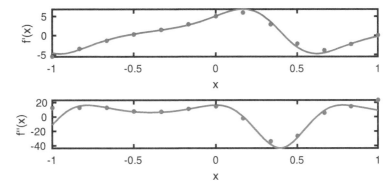

An experiment confirms the order of accuracy.

```
n = round( 2.^(4:.5:11)' );
err1 = 0*n;
err2 = 0*n;
for k = 1:length(n)
    [t,Dx,Dxx] = diffmat2(n(k),[-1 1]);
    y = f(t);
    err1(k) = norm( dfdx(t) - Dx*y, inf );
    err2(k) = norm( d2fdx2(t) - Dxx*y, inf );
end
```

For $O(n^{-p})$ convergence, we use a log–log plot of the errors.

```
loglog(n,[err1,err2],'.-'), hold on
loglog(n,10*10*n.^(-2),'--')
```

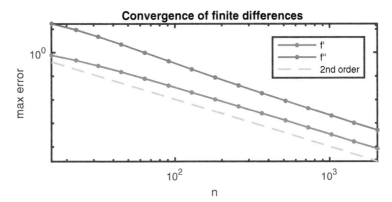

## Spectral differentiation

Recall that finite difference formulas are derived in three steps:

1. Choose a node index set $S$ near node $i$.
2. Interpolate with a polynomial using the nodes in $S$.
3. Differentiate the interpolant and evaluate at node $i$.

We can modify this process by using a global interpolant, either polynomial or trigonometric, as in Chapter 9. Rather than choosing a different index set for each node, we use all of the nodes each time. When this is done using Chebyshev second-kind points,

$$x_k = -\cos\left(\frac{k\pi}{n}\right), \qquad k = 0, \ldots, n,$$

**Function 10.2.2** (diffcheb) Chebyshev differentiation matrices.

```
function [x,Dx,Dxx] = diffcheb(n,xspan)
%DIFFCHEB   Chebyshev differentiation matrices.
% Input:
%   n       number of subintervals (integer)
%   xspan   interval endpoints (vector)
% Output:
%   x       Chebyshev nodes in domain (length n+1)
%   Dx      matrix for first derivative (n+1 by n+1)
%   Dxx     matrix for second derivative (n+1 by n+1)

x = -cos( (0:n)'*pi/n );    % nodes in [-1,1]
Dx = zeros(n+1);
c = [2; ones(n-1,1); 2];    % endpoint factors
i = (0:n)';                 % row indices

% Off-diagonal entries
for j = 0:n
    num = c(i+1).*(-1).^(i+j);
    den = c(j+1)*(x-x(j+1));
    Dx(:,j+1) = num./den;
end

% Diagonal entries
Dx(isinf(Dx)) = 0;           % fix divisions by zero on diagonal
Dx = Dx - diag(sum(Dx,2));   % "negative sum trick"

% Transplant to [a,b]
a = xspan(1);  b = xspan(2);
x = a + (b-a)*(x+1)/2;
Dx = 2*Dx/(b-a);

% Second derivative
Dxx = Dx^2;

end
```

## 10.2. Differentiation matrices

then the resulting **Chebyshev differentiation matrix** has entries

$$D_{00} = \frac{2n^2+1}{6}, \qquad D_{nn} = -\frac{2n^2+1}{6},$$

$$D_{ij} = \begin{cases} -\dfrac{x_i}{2(1-x_i^2)}, & i = j, \\ \dfrac{c_i}{c_j} \dfrac{(-1)^{i+j}}{x_i - x_j}, & i \neq j, \end{cases} \qquad (10.2.4)$$

where $c_0 = c_n = 2$ and $c_i = 1$ for $i = 1,\ldots,n-1$. Note that this is a full matrix (no sparsity). The simplest way to compute a second derivative is by squaring $\boldsymbol{D}_x$, as there is no longer any concern about the bandwidth of the result.

Function 10.2.2 returns these two matrices. The function uses a change of variables to transplant the standard $[-1,1]$ for Chebyshev nodes to any $[a,b]$. It also takes a different approach to computing the diagonal elements of $\boldsymbol{D}_x$ than the formulas in (10.2.4) (see Exercise 10.2.5).

### Example 10.2.2

Here is a $4 \times 4$ Chebyshev differentiation matrix.

```
[t,Dx] = diffcheb(3,[-1 1]);
format rat, Dx

Dx =
       -19/6            4         -4/3          1/2
          -1          1/3            1         -1/3
         1/3           -1         -1/3            1
        -1/2          4/3           -4         19/6
```

We again test the convergence rate.

```
f = @(x) x + exp(sin(4*x));
dfdx = @(x) 1 + 4*exp(sin(4*x)).*cos(4*x);
d2fdx2 = @(x) 4*exp(sin(4*x)).*(4*cos(4*x).^2-4*sin(4*x));
n = (5:5:70)';
err1 = 0*n;
err2 = 0*n;
for k = 1:length(n)
    [t,Dx,Dxx] = diffcheb(n(k),[-1 1]);
    y = f(t);
    err1(k) = norm( dfdx(t) - Dx*y, inf );
    err2(k) = norm( d2fdx2(t) - Dxx*y, inf );
end
semilogy(n,[err1,err2],'.-'), hold on
```

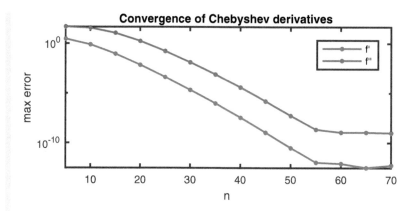

Notice that the graph has a log–linear scale, not log–log. Hence the convergence is exponential, as we expect for a spectral method on a smooth function.

According to Theorem 9.3.1, the convergence of polynomial interpolation to $f$ using Chebyshev nodes is spectral if $f$ is analytic (at least having infinitely many derivatives) on the interval. The derivative—in fact, all derivatives—of $f$ are also approximated with spectral accuracy.

## Exercises

10.2.1. (a) ✍ Calculate $D_x^2$ using (10.2.1) to define $D_x$.

(b) ⌨ Repeat the experiment of Example 10.2.1, but using this version of $D_x^2$ to estimate $f''$.

10.2.2. (a) ✍ Find the first and second derivatives of $f(x) = \exp(|4\sin(x)|)$ on the interval $[-1, 1]$. (Hint: It's best to treat it like a piecewise-defined function.)

(b) ⌨ Adapt Example 10.2.1 to operate on the function from part (a). You will need to adjust the observed orders of accuracy.

(c) ✍ Write a short paragraph discussing why the level of accuracy changed.

10.2.3. ⌨ To get a fourth-order accurate version of $D_x$, five points per row are needed, including two special rows at each boundary. For a fourth-order $D_{xx}$, five symmetric points per row are needed for interior rows, and six points are needed for the rows near a boundary.

(a) Modify Function 10.2.1 to a function `diffmat4`, which outputs fourth-order accurate differentiation matrices. You may want to use Function 5.4.1.

(b) Repeat the experiment of Example 10.2.1 using `diffmat4` in place of Function 10.2.1.

10.2.4. ✍ What is the purpose of line 30 in Function 10.2.2? Be precise with your answer.

10.2.5. (a) ✍ What is the derivative of a constant function?

(b) 🔍 Explain why, for any reasonable differentiation matrix $D$, we should find $\sum_{j=0}^{n} D_{ij} = 0$ for all $i$.

(c) 🔍 What does this have to do with Function 10.2.2? Refer to specific line(s) in the function for your answer.

10.2.6. Let
$$T = \begin{bmatrix} 1 & & & \\ 1 & 1 & & \\ \vdots & & \ddots & \\ 1 & 1 & \cdots & 1 \end{bmatrix}.$$

(a) 🔍 What is $Tu$ for any vector $u$? How is this like integration?

(b) 🔍 Find the inverse of $T$ in general. (Use MATLAB to find the answer if you like, but argue why your answer is correct.) What does this have to do with the inverse of integration?

## 10.3 • Collocation for linear problems

Let us now devise a numerical method for the second-order linear BVP

$$u'' + p(x)u' + q(x)u = r(x), \quad a \le x \le b, \quad u(a) = \alpha, \quad u(b) = \beta. \quad (10.3.1)$$

The first step is to select nodes $x_0 = a < x_1 < \cdots < x_n = b$. For finite differences these will most likely be equally spaced, but for spectral differentiation they will be Chebyshev points. Rather than solving for a function, we will solve for a vector of its values (or, more accurately, approximations of its values) at the nodes:

$$u = \begin{bmatrix} u_0 \\ u_1 \\ \vdots \\ u_{n-1} \\ u_n \end{bmatrix} \approx \begin{bmatrix} \hat{u}(x_0) \\ \hat{u}(x_1) \\ \vdots \\ \hat{u}(x_{n-1}) \\ \hat{u}(x_n) \end{bmatrix}, \quad (10.3.2)$$

where $\hat{u}$ is the exact solution of (10.3.1). If we so desire, we can use interpolation to convert the values $(x_i, u_i)$ into a function after the solution is found.

### Collocation

Having defined values at the nodes as our unknowns, *we impose approximations to the ODE at the same nodes*. This approach is known as **collocation**. Recall from Section 10.2 that approximate differentiation can be accomplished using differentiation matrices. For example,

$$\begin{bmatrix} \hat{u}'(x_0) \\ \hat{u}'(x_1) \\ \vdots \\ \hat{u}'(x_n) \end{bmatrix} \approx u' = D_x u \quad (10.3.3)$$

with an appropriately chosen differentiation matrix $D_x$. Similarly, we define

$$\begin{bmatrix} \hat{u}''(x_0) \\ \hat{u}''(x_1) \\ \vdots \\ \hat{u}''(x_n) \end{bmatrix} \approx u'' = D_{xx} u, \qquad (10.3.4)$$

with $D_{xx}$ chosen in accordance with the node set.

The discrete form of (10.3.1) at the $n+1$ chosen nodes is

$$u'' + Pu' + Qu = r, \qquad (10.3.5)$$

where

$$P = \begin{bmatrix} p(x_0) & & \\ & \ddots & \\ & & p(x_n) \end{bmatrix}, \quad Q = \begin{bmatrix} q(x_0) & & \\ & \ddots & \\ & & q(x_n) \end{bmatrix}, \qquad (10.3.6)$$

$$r = \begin{bmatrix} r(x_0) \\ \vdots \\ r(x_n) \end{bmatrix}. \qquad (10.3.7)$$

If we apply the definitions of $u'$ and $u''$ and rearrange, we obtain

$$Lu = r, \qquad L = D_{xx} + PD_x + Q. \qquad (10.3.8)$$

which is a linear system of $n+1$ equations in $n+1$ unknowns.

We have not yet incorporated the boundary conditions. Those take the form of the additional linear conditions $u_0 = \alpha$ and $u_n = \beta$. We might regard this situation as an overdetermined system, suitable for linear least squares. However, it's usually preferred to impose the boundary conditions and collocation (differential equation) conditions exactly, so we need to discard two of the collocation equations to keep the system square. The obvious candidates for deletion are the collocation conditions at the two endpoints. We may express these deletions by means of a matrix that is an $(n+1) \times (n+1)$ identity with the first and last rows deleted:

$$E = \begin{bmatrix} 0 & 1 & 0 & \cdots & 0 \\ \vdots & & \ddots & & \vdots \\ 0 & \cdots & 0 & 1 & 0 \end{bmatrix} = \begin{bmatrix} e_1^T \\ \vdots \\ e_{n-1}^T \end{bmatrix}, \qquad (10.3.9)$$

where as always $e_k$ is the $k$th column (here starting from $k=0$) of an identity matrix. The product $EA$ deletes the first and last rows of $A$, leaving a matrix that is $(n-1) \times (n+1)$. Similarly, $Er$ deletes the first and last rows of $r$.

Finally, we note that $\hat{u}(a) = e_0^T u$ and $\hat{u}(b) = e_n^T u$, so the linear system including both the ODE and the boundary condition collocations is

$$\begin{bmatrix} e_0^T \\ EL \\ e_n^T \end{bmatrix} u = \begin{bmatrix} \alpha \\ Er \\ \beta \end{bmatrix}, \quad \text{or} \quad Au = b. \qquad (10.3.10)$$

## 10.3. Collocation for linear problems

**Function 10.3.1** (bvplin) Solve a linear boundary-value problem.

```
function [x,u] = bvplin(p,q,r,xspan,lval,rval,n)
% BVPLIN    Solve a linear boundary-value problem.
% Input:
%   p,q,r    u'' + pu' + qu = r  (functions)
%   xspan    endpoints of problem domain (vector)
%   lval     value at left boundary (scalar)
%   rval     value at right boundary (scalar)
%   n        number of subintervals (integer)
% Output:
%   x        collocation nodes (vector, length n+1)
%   u        solution at nodes (vector, length n+1)

[x,Dx,Dxx] = diffmat2(n,xspan);

P = diag(p(x));
Q = diag(q(x));
L = Dxx + P*Dx + Q;    % ODE expressed at the nodes
r = r(x);

% Replace first and last rows using boundary conditions.
I = speye(n+1);
A = [ I(:,1)'; L(2:n,:); I(:,n+1)' ];
b = [ lval; r(2:n); rval ];

% Solve the system.
u = A\b;
```

### Implementation

Our implementation of linear collocation is Function 10.3.1. It uses second-order finite differences but makes no attempt to exploit the sparsity of the matrices. It would be trivial to change the function to use spectral differentiation. Note that there is no need to explicitly form the "row deletion" matrix $E$ from (10.3.9). Since it only appears as left-multiplying $L$ or $r$, we simply perform the row deletions as needed using indexing.

**Example 10.3.1**

Consider the linear BVP

$$u'' - (\cos x)u' + (\sin x)u = 0, \quad 0 \le x \le \pi/2, \quad u(0) = 1, \, u(\pi/2) = e.$$

Its exact solution is simple.

```
exact = @(x) exp(sin(x));
```

The problem is presented in our standard form, so we can identify the coefficient functions in the ODE. Each should be coded so that they can accept either scalar or vector inputs.

```
p = @(x) -cos(x);
q = @(x) sin(x);
r = @(x) 0*x;      % not a scalar value!
```

We solve the BVP and compare the result to the exact solution.

```
[x,u] = bvplin(p,q,r,[0 pi/2],1,exp(1),25);
subplot(2,1,1),
plot(x,u,'-')
subplot(2,1,2)
plot(x,exact(x)-u,'.-')
```

## Accuracy and stability

We revisit Example 10.1.4 on page 414 in order to verify second-order convergence (i.e., error that is $O(n^{-2})$).

**Example 10.3.2**

The BVP is
$$u'' - \lambda^2 u = \lambda^2, \quad 0 \leq x \leq 1, \quad u(0) = -1, \; u(1) = 0.$$

```
lambda = 10;
exact = @(x) sinh(lambda*x)/sinh(lambda) - 1;
```

These functions must accept and return vectors.

```
p = @(x) 0*x;
q = @(x) -lambda^2*x.^0;
r = @(x) lambda^2*x.^0;
```

We compare the computed solution to the exact one for increasing $n$.

## 10.3. Collocation for linear problems

```
n = [32 64 128 256 512]';
error = 0*n;
for k = 1:length(n)
    [x,u] = bvplin(p,q,r,[0 1],-1,0,n(k));
    error(k) = norm(exact(x)-u,inf);
end
table(n,error)

ans =
  5x2 table
     n        error
    ___    _____
    32     0.0014805
    64     0.00037262
    128    9.3493e-05
    256    2.3383e-05
    512    5.847e-06
```

Each time $n$ is doubled, the error is reduced by a factor very close to 4, which is indicative of second-order convergence.

```
loglog(n,error,'.-')
hold on, loglog(n,n.^(-2),'--')
```

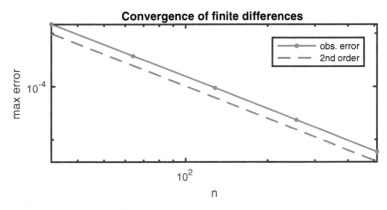

If we write the solution $u$ of equation (10.3.10) as the exact solution minus an error vector $\varepsilon$, i.e., $u = \hat{u} - \varepsilon$, we obtain

$$A\hat{u} - A\varepsilon = b,$$
$$\varepsilon = A^{-1}[A\hat{u} - b] = A^{-1}\tau(h),$$

where $\tau$ is the truncation error of the finite differences (except at the boundary rows, where it is zero). It follows that $\|\varepsilon\|$ vanishes at the same rate as the truncation error if $\|A^{-1}\|$ is bounded above as $h \to 0$. In the present context, this property is known as **stability**. Proving stability is too technical to walk through here, but stability is guaranteed under some reasonable conditions on the BVP.

## Exercises

10.3.1. ✍ For each BVP, verify that the given solution is correct. Then write out by hand for $n = 3$ the matrices $D_{xx}$, $D_x$, $P$, and $Q$, and the vector $r$.

(a)
$$u'' + u = 0, \quad u(0) = 0, \quad u(3) = \sin 3.$$

Exact solution: $u(x) = \sin x$.

(b)
$$u'' - \frac{3}{x}u' + \frac{4}{x^2}u = 0, \quad u(1) = 0, \quad u(4) = 32\ln 2.$$

Exact solution: $u(x) = x^2 \ln x$.

(c)
$$u'' - \frac{1}{x}u' + \frac{2}{x^2}u = 2(1 + \ln x), \quad u(1) = 1, \quad u(\pi) = \pi.$$

Exact solution: $u(x) = x^2 + (1-\pi)x^2 \sin(\ln(x))/\sin(\ln(\pi))$.

(d)
$$u'' - \frac{1}{(x+1/2)}u' + \frac{2}{(x+1/2)^2}u = \frac{10}{(x+1/2)^4},$$
$$u(1/2) = 1, \quad u(5/2) = 1/9.$$

Exact solution: $u(x) = (x + 1/2)^{-2}$.

10.3.2. 💻 For each of the cases in Exercise 10.3.1, use Function 10.3.1 to solve the problem with $n = 60$ and make a plot of its error as a function of $x$. Then, for each $n = 10, 20, 40, \ldots, 640$, find the infinity norm of the error. Make a log-log plot of error versus $n$ and include a graphical comparison to second-order convergence.

10.3.3. 💻 Modify Function 10.3.1 to use spectral differentiation rather than second-order finite differences. For each of the cases in Exercise 10.3.1, solve the problem with $n = 10, 15, 20, \ldots, 60$, finding the infinity norm of the error in each case. Make a log–linear plot of error versus $n$.

10.3.4. 💻 Use Function 10.3.1 to solve *Bessel's equation*,
$$x^2 u'' + xu' + x^2 y = 0, \quad 0 < x < 8, \quad u(0) = 1, \quad u(8) = 0.$$

Plot the solution for $n = 200$.

10.3.5. 💻 The *Airy equation* is $u'' = xu$. For $x > 0$ its solution is exponential and for $x < 0$ it is oscillatory. The exact solution is given by $u = c_1 \text{Ai}(x) + c_2 \text{Bi}(x)$, where Ai and Bi are Airy functions. In MATLAB they are computed by `airy(0,x)` and `airy(2,x)`, respectively.

(a) Suppose that $u(-10) = -1$, $u(2) = 1$. By setting up and solving a $2 \times 2$ linear system in MATLAB, find numerical values for $c_1$ and $c_2$. Plot the resulting exact solution.

(b) Use Function 10.3.1 with $n = 120$ to find the solution with the boundary conditions in part (a). In a $2 \times 1$ subplot array, plot the finite difference solution and its error.

(c) Repeat part (b) with $n = 800$.

10.3.6. ▣ Let $[a, b] = [0, 1]$ and $p(x) = q(x) = 1$ in (10.3.1). Using $n = 20$, show that the matrix $L$ in (10.3.8) is numerically singular. (This shows that the boundary conditions are an essential part of making the problem have a unique solution, computationally as well as mathematically.)

10.3.7. Consider the BVP $\epsilon u'' + (1+\epsilon)u' + u = 0$, $x \in (0, 1)$, $u(0) = 0$, $u(1) = 1$. You are to investigate how many grid points are required to reach an acceptable level of error as the parameter $\epsilon$ is decreased. The rapid variation of the solution near $x = 0$ is often called a boundary layer.

(a) ✎ Show that the exact solution to the problem is $u(x) = (e^{-x} - e^{-x/\epsilon})/(e^{-1} - e^{-1/\epsilon})$.

(b) ▣ For the values $\epsilon = 0.5, 0.25, 0.1, 0.05, 0.025, 0.01$, find the largest value of $n$ for which the error is below $10^{-4}$. Start your search at $n = 50$, and test $n$ in increments of 25.

(c) ▣ Plot the solution and the error as a function of $x$ for the smallest value of $\epsilon$. Where is the error the largest, and where does the solution change fastest?

(d) ▣ Develop a hypothesis for the value of $n$ needed as a function of $\epsilon$, and plot the actual $n$ and your hypothesis together on a log–log plot.

(e) ✎ Can this error increase with decreasing $\epsilon$ be consistent with $O(n^{-2})$ error for the finite difference method? Justify your answer.

## 10.4 ▪ Nonlinearity and boundary conditions

Collocation for nonlinear BVPs operates on the same principle as for linear problems: replace functions by vectors and replace derivatives by differentiation matrices. But because the differential equation is nonlinear, the resulting algebraic equations are as well. We will need to use an appropriate method for such nonlinear algebraic systems as part of the solution process.

We continue to focus on second-order scalar problems, now in the more general explicit form

$$u'' = \varphi(x, u, u'), \quad a \le x \le b, \quad u(a) = \alpha, \ u(b) = \beta, \tag{10.4.1}$$

where $\varphi$ may be nonlinear with respect to any of its arguments. As in Section 10.3, the function $u(x)$ is replaced by a vector $\boldsymbol{u}$ of its (approximated) values at a set of nodes $x_0, x_1, \ldots, x_n$ (see equation (10.3.2)). We define derivatives of the sampled function as in (10.3.3) and (10.3.4), using suitable differentiation matrices $\boldsymbol{D}_x$ and $\boldsymbol{D}_{xx}$. Our collocation equations, ignoring the boundary conditions for now, are

$$\boldsymbol{D}_{xx}\boldsymbol{u} - r(\boldsymbol{u}) = \boldsymbol{0},$$

where
$$r_i(u) = \varphi(x_i, u_i, u_i'), \qquad i = 0, \ldots, n, \tag{10.4.2}$$

and
$$u_i' = e_i^T(D_x u) = (e_i^T D_x) u, \tag{10.4.3}$$

which equals the $i$th row of $u'$, i.e., the numerical approximation to $u'(x_i)$ in the BVP.

We impose the boundary conditions in much the same way as in Section 10.3. Again define the rectangular "boundary removal" matrix $E$ as in (10.3.9), and replace the equations in those two rows by the boundary conditions in the same manner as (10.3.10):

$$f(u) = \begin{bmatrix} u_0 - \alpha \\ E(D_{xx} u - r(u)) \\ u_n - \beta \end{bmatrix} = 0. \tag{10.4.4}$$

The left-hand side of (10.4.4) is a nonlinear function of the unknowns in the vector $u$, with $n + 1$ components. *The original problem has been converted to a set of nonlinear equations, amenable to solution by the techniques of Sections 4.5 and 4.6.*

### Example 10.4.1

Given the BVP
$$u'' - \sin(xu) + \exp(xu') = 0, \qquad u(0) = -2, \quad u(3/2) = 1,$$
we compare to the standard form (10.4.1) and recognize
$$\varphi(x, u, u') = \sin(xu) - \exp(xu').$$

Suppose $n = 3$ on an equispaced grid, so that $h = 1/2$, $x_0 = 0$, $x_1 = 1/2$, $x_2 = 1$, and $x_3 = 3/2$. There are four unknowns. We compute

$$ED_{xx} = \frac{1}{1/4}\begin{bmatrix} 1 & -2 & 1 & 0 \\ 0 & 1 & -2 & 1 \end{bmatrix}, \quad ED_x = \frac{1}{1}\begin{bmatrix} -1/2 & 0 & 1/2 & 0 \\ 0 & -1/2 & 0 & 1/2 \end{bmatrix},$$

$$Er(u) = \begin{bmatrix} \sin(u_1/2) - \exp[1/2(u_2/2 - u_0/2)] \\ \sin(u_2) - \exp(u_3/2 - u_1/2) \end{bmatrix},$$

$$f(u) = \begin{bmatrix} u_0 + 2 \\ (4u_0 - 8u_1 + 4u_2) - \sin(u_1/2) + \exp(u_2/4 - u_0/4) \\ (4u_1 - 8u_2 + 4u_3) - \sin(u_2) + \exp(u_3/2 - u_1/2) \\ u_3 - 1 \end{bmatrix}.$$

In order to complete the solution, we would continue by applying a quasi-Newton method to solve this nonlinear algebraic system. That step is discussed below.

## Neumann and Robin conditions

A condition on the value of the solution, like $u(a) = \alpha$ and $u(b) = \beta$, is typically called a **Dirichlet condition**. An equally important type of boundary condition is a **Neumann condition**, which is a prescribed value of $u'(a)$ or $u'(b)$. If the prescribed value (of either type) is zero, the condition is said to be homogeneous. A **Robin condition** generalizes both cases to $c_1 u(a) + c_2 u'(a) = \alpha$.

Within our framework, replacing a Dirichlet condition with a more general one is simple. Suppose that the boundary conditions are given in the generic form

$$\gamma(u(a), u'(a)) = 0, \quad \delta(u(b), u'(b)) = 0. \tag{10.4.5}$$

Both $\gamma$ and $\delta$ are functions of two scalar values. In the discrete nonlinear problem (10.4.4), we substitute the equations

$$\sigma \beta(u_0, e_0^T D_x u) = 0, \tag{10.4.6a}$$

$$\sigma \delta(u_n, e_n^T D_x u) = 0 \tag{10.4.6b}$$

for the first and last rows. The reason for the extra factor of $\sigma$ in these equations is to scale the rows similarly to how they are scaled by $D_{xx}$ in (10.4.4). Doing so improves the condition numbers of the linear systems solved by the rootfinder. For a Dirichlet condition we choose $\sigma = h^{-2}$, while for any condition that includes the derivative of the solution, we use $\sigma = h^{-1}$.

## Quasi-Newton implementation

Our implementation using second-order finite differences is Function 10.4.1. It's surprisingly short, considering how general it is, thanks to `levenberg` (Function 4.6.2) from Chapter 4 and how MATLAB makes vectors and matrices easy to work with.

The heart of Function 10.4.1 is the computation of $f(u)$ for a given $u$, as done by the nested `residual` function. It follows the definitions leading up to (10.4.4), except that it does not construct $E$; instead, wherever $E$ multiplies a vector, the code simply removes the first and last entries of the vector. Note also that in line 27 the vector $D_x u$ is calculated, so its endpoint values are available for the boundary conditions.

In order to define a particular problem, we must write a function that computes $\varphi$ for vector-valued inputs $x$, $u$, and $u'$, and the two boundary condition functions (10.4.5). Note that these functions do *not* require the explicit use of differentiation matrices; arguments depending on $u'$ are supplied from within Function 10.4.1 itself.

**Function 10.4.1 (bvp)** Solve a nonlinear boundary-value problem.

```
function [x,u] = bvp(phi,xspan,lval,lder,rval,rder,init)
%BVP     Solve a boundary-value problem by finite differences
%        with either Dirichlet or Neumann BCs.
% Input:
%   phi       defines u'' = phi(x,u,u') (function)
%   xspan     endpoints of the domain
%   lval      prescribed value for u(a) (use [] if unknown)
%   lder      prescribed value for u'(a) (use [] if unknown)
%   rval      prescribed value for u(b) (use [] if unknown)
%   rder      prescribed value for u'(b) (use [] if unknown)
%   init      initial guess for the solution (length n+1 vector)
% Output:
%   x         nodes in x (vector, length n+1)
%   u         values of u(x)  (vector, length n+1)
%   res       function for computing the residual

n = length(init) - 1;
[x,Dx,Dxx] = diffmat2(n,xspan);
h = x(2)-x(1);

u = levenberg(@residual,init);
u = u(:,end);

    function f = residual(u)
        % Computes the difference between u'' and phi(x,u,u') at the
        % interior nodes and appends the error at the boundaries.
        dudx = Dx*u;           % discrete u'
        d2udx2 = Dxx*u;        % discrete u''
        f = d2udx2 - phi(x,u,dudx);

        % Replace first and last values by boundary conditions.
        if isempty(lder)       % u(a) given
            f(1) = (u(1) - lval)/h^2;
        else                   % u'(a) given
            f(1) = (dudx(1) - lder)/h;
        end
        if isempty(rder)       % u(b) given
            f(n+1) = (u(n+1) - rval)/h^2;
        else                   % u'(b) given
            f(n+1) = (dudx(n+1) - rder)/h;
        end
    end

end
```

**Example 10.4.2**

Suppose a damped pendulum satisfies the nonlinear equation $\theta'' + 0.05\theta' + \sin\theta = 0$. We want to start the pendulum at $\theta = 2.5$ and give it the right initial velocity so that it reaches $\theta = -2$ at exactly $t = 5$. This is a BVP, because $\theta(0) = 2.5$ and $\theta(5) = -2$.

```
phi = @(t,theta,omega) -0.05*omega - sin(theta);
init = linspace(2.5,-2,101)';
[t,theta] = bvp(phi,[0,5],2.5,[],-2,[],init);
plot(t,theta)
```

## 10.4. Nonlinearity and boundary conditions

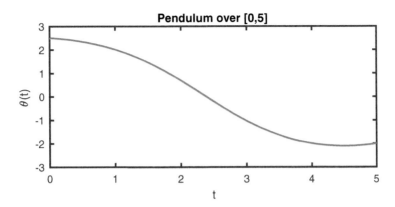

Note how the angle decreases throughout the motion until near the end. If we extend the time interval longer, then we find a rather different solution.

```
[t,theta] = bvp(phi,[0,8],2.5,[],-2,[],init);
plot(t,theta)
```

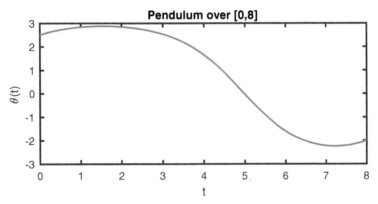

This time the angle initially *increases*, and the pendulum hovers near the vertical position before swinging back down.

As always with solving nonlinear equations, there may be multiple solutions, and the solution that is found can depend on the initial guess to the solver.

### Example 10.4.3

We use bvp to find a solution to the membrane deflection problem of (10.1.2)–(10.1.3): $w'' + w'/r = \lambda/w^2$, $w'(0) = 0$, $w(1) = 1$. To avoid division by zero at $r = 0$, we modify the domain to $[\varepsilon_{\text{mach}}, 1]$.

```
lambda = 0.5;
phi = @(r,w,dwdr) lambda./w.^2 - dwdr./r;
init = ones(301,1);
[r,w1] = bvp(phi,[0,1],[],0,1,[],init);
plot(r,w1)
```

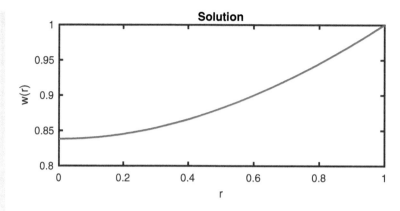

By choosing a different initial guess, we arrive at another solution.

```
init = 0.5*ones(301,1);
[r,w2] = bvp(phi,[0,1],[],0,1,[],init);
hold on, plot(r,w2)
```

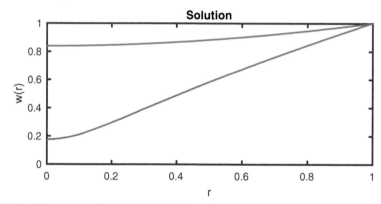

Sometimes the best way to get a useful starting guess for the iteration is to use the solution of a related easier problem, a technique known as **parameter continuation**. In this approach, one solves the problem at an easy parameter value and gradually changes the parameter value to the desired and more difficult value. The solution at the previous parameter value may be used as the initial guess for the current parameter value. The next example illustrates this technique.

### Example 10.4.4

We solve the stationary **Allen–Cahn equation**,

$$\epsilon u'' = u^3 - u, \quad 0 \leq x \leq 1, \quad u'(0) = 0, \; u(1) = 1. \tag{10.4.7}$$

Finding a solution is easy at some values of $\epsilon$.

```
epsilon = 0.05;
phi = @(x,u,dudx) (u.^3 - u) / epsilon;
init = linspace(-1,1,141)';
```

## 10.4. Nonlinearity and boundary conditions

```
[x,u1] = bvp(phi,[0,1],-1,[],1,[],init);
plot(x,u1)
```

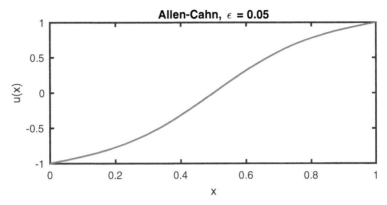

However, finding a good starting guess is not trivial for smaller values of $\epsilon$. Note below that the iteration stops without converging to a solution.

```
epsilon = 0.005;
phi = @(x,u,dudx) (u.^3 - u) / epsilon;
[x,u] = bvp(phi,[0,1],-1,[],1,[],init);
```

`Warning: Iteration did not find a root.`

A simple way around this problem is to use the result of a solved version as the starting guess for a more difficult version.

```
phi = @(x,u,dudx) (u.^3 - u) / 0.005;
[x,u2] = bvp(phi,[0,1],-1,[],1,[],u1);
hold on, plot(x,u2)
```

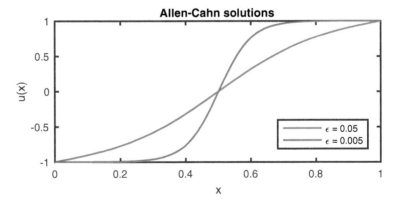

In this case we can continue further.

```
phi = @(x,u,dudx) (u.^3 - u) / 0.0005;
[x,u3] = bvp(phi,[0,1],-1,[],1,[],u2);
plot(x,u3)
```

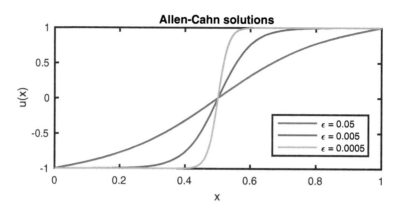

This approach eases solution of a wide variety of difficult problems.

## Exercises

10.4.1. This exercise is on the nonlinear BVP

$$u'' = \frac{3(u')^2}{u}, \quad -1 < x < 2, \quad u(-1) = 1, \quad u(2) = 1/2.$$

(a) ✍ Verify that the exact solution is $u(x) = (x+2)^{-1/2}$.

(b) ✍ Write out the finite difference approximation (10.4.4) with a single interior point (that is, with $n = 2$).

(c) ✍ Solve by hand the equation of part (b) for the lone interior value $u_1$.

(d) ⌨ Use Function 10.4.1 to solve the problem for $n = 80$. In a $2 \times 1$ subplot array, plot the finite difference solution and its error.

(e) ⌨ For each $n = 10, 20, 40, \ldots, 640$, find the infinity norm of the error. Make a log–log plot of error versus $n$ and include a graphical comparison to second-order convergence.

10.4.2. ⌨ (Adapted from [3].) Use Function 10.4.1 twice with $n = 200$ to solve

$$u'' + e^{u+0.5} = 0, \quad 0 < x < 1, \quad y(0) = y(1) = 0,$$

with initial guesses of $7 \sin x$ and $0.25 \sin x$. Plot the solutions together on one graph.

10.4.3. ⌨ Is the solution in Example 10.4.4 antisymmetric around the line $x = 0.5$? In other words, does $u(1-x) = -u(x)$? Check the condition numerically.

10.4.4. ⌨ This problem concerns the BVP

$$u'' = x \, \text{sign}(1-x) u, \quad u(-6) = 1, \quad u'(6) = 0.$$

If $u$ is continuous, then $u''$ is discontinuous at $x = 1$, so $u$ has a singularity there and finite differences may not converge at their natural order of accuracy.

(a) Solve the problem using Function 10.4.1 with $n = 1600$, and make a plot of the solution. (Note: MATLAB has a built-in sign function.)

(b) Assign the last entry of the solution vector, which corresponds to the node $x = 6$, to another variable name. This will serve as an "exact" solution at this point.

(c) For each $n = 100, 200, 300, \ldots, 1200$, apply Function 10.4.1 and compute the error at $x = 6$. Compare the convergence graphically to second order.

10.4.5. ▪ Example 10.4.3 found two solutions of (10.1.2)–(10.1.3) at $\lambda = 0.5$. Continue both solutions for $\lambda$ slowly increasing up to 0.79. Make a plot with $\lambda$ on the horizontal axis and $w(0)$ on the vertical axis, with one point to represent each solution found. You should get two paths that converge as $\lambda$ approaches 0.79 from below.

## 10.5 ▪ The Galerkin method

Using finite differences we defined a collocation method, in which an approximation of the differential equation is required to hold at a finite set of nodes. In this section we present an alternative based on integration rather than differentiation. Our presentation will be limited to the linear BVP

$$u'' = p(x)u' + q(x)u + r(x), \quad a \leq x \leq b, \quad u(a) = 0, \, u(b) = 0.$$

However, we will assume that the linear problem is presented in a special, but equivalent, form:

$$-\frac{d}{dx}\left[c(x)u'(x)\right] + s(x)u(x) = f(x), \quad a \leq x \leq b, \quad u(a) = 0, \, u(b) = 0. \quad (10.5.1)$$

Such a transformation is always possible, at least in principle (see Exercise 10.5.3), and the case when $u(a)$ and $u(b)$ are nonzero can also be incorporated (see Exercise 10.5.2). As with finite differences, a nonlinear problem is typically solved by using a Newton iteration to create a sequence of linear problems.

### Weak formulation

Let (10.5.1) be multiplied by a generic function $\psi(x)$ (called the **test function**), and then integrate both sides in $x$:

$$\int_a^b f(x)\psi(x)\,dx = \int_a^b \left[-(c(x)u'(x))'\psi(x) + s(x)u(x)\psi(x)\right]dx$$

$$= \left[u'(x)\psi(x)\right]_a^b + \int_a^b \left[c(x)u'(x)\psi'(x) + s(x)u(x)\psi(x)\right]dx.$$

(10.5.2)

The last line above used an integration by parts.

We now make an important and convenient assumption about the test function. The first term in (10.5.2), consisting of boundary evaluations, disappears if we require that

$\psi(a) = \psi(b) = 0$. Doing so leads to

$$\int_a^b \left[c(x)u'(x)\psi'(x) + s(x)u(x)\psi(x)\right]dx = \int_a^b f(x)\psi(x)\,dx, \qquad (10.5.3)$$

which is known as the **weak form** of the differential equation (10.5.1). If $u(x)$ is a function such that (10.5.3) is satisfied for all valid choices of $\psi$, we say that $u$ is a **weak solution** of the BVP. While the weak form might look odd, in many mathematical models it could be considered more fundamental than the differential form (10.5.1). Every solution of (10.5.1) (what we might now call the strong form of the problem) is a weak solution, but the converse is not always true.

## Galerkin conditions

Our goal is to solve a finite-dimensional problem that approximates the weak form of the BVP. Let $\phi_0, \phi_1, \ldots, \phi_m$ be linearly independent functions satisfying $\phi_i(a) = \phi_i(b) = 0$. If we require

$$\psi(x) = \sum_{i=1}^m z_i \phi_i(x),$$

then (10.5.3) becomes, after some rearrangement,

$$\sum_{i=1}^m z_i \left[\int_a^b \left[c(x)u'(x)\phi_i'(x)\,dx + s(x)u(x)\phi_i(x) - f(x)\phi_i(x)\right]dx\right] = 0.$$

One way to satisfy this condition is to ensure that the term inside the brackets is zero for each possible value of $i$, that is,

$$\int_a^b \left[c(x)u'(x)\phi_i'(x) + s(x)u(x)\phi_i(x)\right]dx = \int_a^b f(x)\phi_i(x)\,dx, \quad i = 1,\ldots,m.$$
(10.5.4)

The independence of the $\phi_i$ furthermore guarantees that this is the only possibility. We no longer need to consider the $z_i$.

Now that we have approximated the weak form of the BVP by a finite set of constraints, the next step is to represent the approximate solution by a finite set as well. A natural choice is to approximate $u(x)$ the same way as we did the test function $\psi$, where the $\phi_j$ form a basis for representing the solution:

$$u(x) = \sum_{j=1}^m w_j \phi_j(x). \qquad (10.5.5)$$

Substituting (10.5.5) into (10.5.4) implies

$$\int_a^b \left\{ c(x)\left[\sum_{j=1}^m w_j \phi_j'(x)\right]\phi_i'(x) + s(x)\left[\sum_{j=1}^m w_j \phi_j(x)\right]\phi_i(x) \right\} dx = \int_a^b f(x)\phi_i(x)\,dx$$

for $i = 1,\ldots,m$. This rearranges easily into

$$\sum_{j=1}^m w_j \left[\int_a^b c(x)\phi_i'(x)\phi_j'(x)\,dx + \int_a^b s(x)\phi_i(x)\phi_j(x)\,dx\right] = \int_a^b f(x)\phi_i(x)\,dx,$$
(10.5.6)

## 10.5. The Galerkin method

still for each $i = 1,\ldots,m$. These are the **Galerkin conditions** defining a numerical solution. They follow entirely from the BVP and the choice of the $\phi_i$.

*The conditions (10.5.6) are a linear system of equations for the unknown coefficients $w_j$.*
Define $m \times m$ matrices $\boldsymbol{K}$ and $\boldsymbol{M}$, and the vector $\boldsymbol{f}$, by

$$K_{ij} = \int_a^b c(x)\phi_i'(x)\phi_j'(x)\,dx, \tag{10.5.7a}$$

$$M_{ij} = \int_a^b s(x)\phi_i(x)\phi_j(x)\,dx, \qquad i,j = 0,\ldots,m, \tag{10.5.7b}$$

$$f_i = \int_a^b f(x)\phi_i(x)\,dx. \tag{10.5.7c}$$

Then (10.5.6) is simply
$$(\boldsymbol{K} + \boldsymbol{M})w = \boldsymbol{f}. \tag{10.5.8}$$

The matrix $\boldsymbol{K}$ is called the **stiffness matrix**, and $\boldsymbol{M}$ is called the **mass matrix**. By their definitions, they are symmetric. The last piece of the puzzle is to make some selection for $\phi_1,\ldots,\phi_m$ and obtain a fully specified algorithm.

### Example 10.5.1

Suppose we are given $-u'' + 4u = x$, with $u(0) = u(\pi) = 0$. We could choose the basis functions $\phi_k = \sin(kx)$ for $k = 1,2,3$. Then

$$M_{ij} = 4\int_0^\pi \sin(ix)\sin(jx)\,dx,$$

$$K_{ij} = ij\int_0^\pi \cos(ix)\cos(jx)\,dx,$$

$$f_i = \int_0^\pi x\sin(ix)\,dx.$$

With some calculation (or computer algebra), we find

$$M = 2\pi\begin{bmatrix} 1 & 0 & 0 \\ 0 & 1 & 0 \\ 0 & 0 & 1 \end{bmatrix}, \qquad K = \frac{\pi}{2}\begin{bmatrix} 1 & 0 & 0 \\ 0 & 4 & 0 \\ 0 & 0 & 9 \end{bmatrix}, \qquad f = \pi\begin{bmatrix} 1 \\ -1/2 \\ 1/3 \end{bmatrix}.$$

Upon solving the resulting diagonal linear system, the approximate solution is

$$\frac{2}{5}\sin(x) - \frac{1}{8}\sin(2x) + \frac{2}{39}\sin(3x).$$

### Finite elements

One of the most useful and general choices for the $\phi_i$ is the piecewise linear hat functions constructed in Section 5.2. As usual, we select nodes $a = t_0 < t_1 < \cdots < t_n = b$. Also define
$$h_i = t_i - t_{i-1}, \qquad i = 1,\ldots,n.$$

Then we set $m = n-1$, and the $\phi_i$ in (10.5.5) are

$$\phi_i(x) = H_i(x) = \begin{cases} \dfrac{x - t_{i-1}}{h_i} & \text{if } x \in [t_{i-1}, t_i], \\ \dfrac{t_{i+1} - x}{h_{i+1}} & \text{if } x \in [t_i, t_{i+1}], \\ 0 & \text{otherwise.} \end{cases} \qquad (10.5.9)$$

Recall that these functions are cardinal, i.e., $H_i(t_i) = 1$ and $H_i(t_j) = 0$ if $i \neq j$. Hence

$$u(x) = \sum_{j=1}^{m} w_j \phi_j(x) = \sum_{j=1}^{n-1} u_j H_j(x), \qquad (10.5.10)$$

where as usual $u_j$ is the value of the numerical solution at $t_j$. Note that we omit $H_0$ and $H_n$, which equal one at the endpoints of the interval, because of the boundary conditions on $u$.

The importance of the hat function basis in the Galerkin method is that each one is nonzero in only two adjacent intervals. As a result, we shift the focus from integrations over the entire interval in (10.5.7) to integrations over each subinterval, $I_k = [t_{k-1}, t_k]$. Specifically, we use

$$K_{ij} = \sum_{k=1}^{n} \left[ \int_{I_k} c(x) H_i'(x) H_j'(x)\,dx \right], \qquad i,j = 1, \ldots, n-1, \qquad (10.5.11\text{a})$$

$$M_{ij} = \sum_{k=1}^{n} \left[ \int_{I_k} s(x) H_i(x) H_j(x)\,dx \right], \qquad i,j = 1, \ldots, n-1, \qquad (10.5.11\text{b})$$

$$f_i = \sum_{k=1}^{n} \left[ \int_{I_k} f(x) H_i(x)\,dx \right], \qquad i = 1, \ldots, n-1. \qquad (10.5.11\text{c})$$

Start with the first subinterval, $I_1$. The only hat function that is nonzero over $I_1$ is $H_1(x)$. Thus the only integrals we need to consider over $I_1$ have $i = j = 1$:

$$\int_{I_1} c(x) H_1'(x) H_1'(x)\,dx, \qquad \int_{I_1} s(x) H_1(x) H_1(x)\,dx, \qquad \int_{I_1} f(x) H_1(x)\,dx,$$

which contribute to the sums for $K_{11}$, $M_{11}$, and $f_1$, respectively.

Before writing more formulas, we make one more very useful simplification. Unless the coefficient functions $c(x)$, $s(x)$, and $f(x)$ specified in the problem are especially simple functions, the natural choice for evaluating all of the required integrals is numerical integration, say by the trapezoid formula. As it turns out, though, such integration is not really necessary. The fact that we have approximated the solution of the BVP by a piecewise linear interpolant makes the numerical method second-order accurate overall (assuming a uniform grid). It can be proven that the error is still second order if we replace each of the coefficient functions by a constant over $I_k$, namely, the average of the endpoint values:

$$c(x) \approx \bar{c}_k = \frac{c(t_{k-1}) + c(t_k)}{2} \qquad \text{for } x \in I_k.$$

## 10.5. The Galerkin method

Thus the integrals in (10.5.11) can be evaluated solely from the node locations. For instance,

$$\int_{I_1} c(x) H_1'(x) H_1'(x)\,dx \approx \bar{c}_1 \int_{t_0}^{t_1} h_1^{-2}\,dx = \frac{\bar{c}_1}{h_1}.$$

Now consider interval $I_2 = [t_1, t_2]$. Here both $H_1$ and $H_2$ are nonzero, so there are contributions to all of the matrix elements $K_{11}, K_{12} = K_{21}, K_{22}$, to $M_{11}, M_{12} = M_{21}, M_{22}$, and to $f_1$ and $f_2$. Over $I_2$ we have $H_2' \equiv h_2^{-1}$ and $H_1' \equiv -h_2^{-1}$. Hence the contributions to $K_{11}$ and $K_{22}$ in (10.5.11a) are $\bar{c}_2/h_2$, and the contributions to $K_{12} = K_{21}$ are $-\bar{c}_2/h_2$. We summarize the relationship by

$$\frac{\bar{c}_k}{h_k}\begin{bmatrix} 1 & -1 \\ -1 & 1 \end{bmatrix} \rightsquigarrow \begin{bmatrix} K_{11} & K_{12} \\ K_{21} & K_{22} \end{bmatrix},$$

where the squiggly arrow is meant to show that the values of the 2 × 2 matrix on the left are added to the submatrix of $K$ on the right. Similar expressions are obtained for contributions to $M$ in (10.5.11b) and in (10.5.11c); see below.

In general, over $I_k$ for $1 < k < n$, we have $H_k' \equiv h_k^{-1}$ and $H_{k-1}' \equiv -h_k^{-1}$. The stiffness matrix contributions over $I_k$ become

$$\frac{\bar{c}_k}{h_k}\begin{bmatrix} 1 & -1 \\ -1 & 1 \end{bmatrix} \rightsquigarrow \begin{bmatrix} K_{k-1,k-1} & K_{k-1,k} \\ K_{k,k-1} & K_{k,k} \end{bmatrix}, \qquad k = 2, \ldots, n-1. \tag{10.5.12}$$

One finds the contributions to the other structures by similar computations:

$$\frac{\bar{s}_k h_k}{6}\begin{bmatrix} 2 & 1 \\ 1 & 2 \end{bmatrix} \rightsquigarrow \begin{bmatrix} M_{k-1,k-1} & M_{k-1,k} \\ M_{k,k-1} & M_{k,k} \end{bmatrix}, \qquad k = 2, \ldots, n-1, \tag{10.5.13}$$

and

$$\frac{\bar{f}_k h_k}{2}\begin{bmatrix} 1 \\ 1 \end{bmatrix} \rightsquigarrow \begin{bmatrix} f_{k-1} \\ f_k \end{bmatrix}, \qquad k = 2, \ldots, n-1. \tag{10.5.14}$$

The contribution from $I_n$ affects just $K_{n-1,n-1}, M_{n-1,n-1}$, and $f_{n-1}$, and it produces formulas similar to those for $I_1$.

Each $I_k$ contributes to four elements of each matrix and two of the vector $f$, except for $I_1$ and $I_n$, which each contribute to just one element of each matrix and $f$. The spatially localized contributions to the matrices characterize a **finite element method** (FEM). Putting together all of the contributions to (10.5.8) to form the complete algebraic system is often referred to as the *assembly* process.

### Implementation and convergence

Function 10.5.1 implements the piecewise linear FEM on the linear problem as posed in (10.5.3), using an equispaced grid. The code closely follows the description above.

**Function 10.5.1 (fem)** Piecewise linear finite elements for a linear BVP.

```
function [x,u] = fem(c,s,f,a,b,n)
%FEM     Piecewise linear finite elements for a linear BVP.
% Input:
%   c,s,f    coefficient functions of x describing the ODE (functions)
%   a,b      domain of the independent variable (scalars)
%   n        number of grid subintervals (scalar)
% Output:
%   x        grid points (vector, length n+1)
%   u        solution values at x (vector, length n+1)

% Define the grid.
h = (b-a)/n;
x = a + h*(0:n)';

% Templates for the subinterval matrix and vector contributions.
Ke = [1 -1; -1 1];
Me = (1/6)*[2 1; 1 2];
fe = (1/2)*[1; 1];

% Evaluate coefficent functions and find average values.
cval = c(x);    cbar = (cval(1:n)+cval(2:n+1)) / 2;
sval = s(x);    sbar = (sval(1:n)+sval(2:n+1)) / 2;
fval = f(x);    fbar = (fval(1:n)+fval(2:n+1)) / 2;

% Assemble global system, one interval at a time.
K = zeros(n-1,n-1);   M = zeros(n-1,n-1);   f = zeros(n-1,1);
K(1,1) = cbar(1)/h;   M(1,1) = sbar(1)*h/3;   f(1) = fbar(1)*h/2;
K(n-1,n-1) = cbar(n)/h;  M(n-1,n-1) = sbar(n)*h/3;  f(n-1) = fbar(n)*h/2;
for k = 2:n-1
  K(k-1:k,k-1:k) = K(k-1:k,k-1:k) + (cbar(k)/h) * Ke;
  M(k-1:k,k-1:k) = M(k-1:k,k-1:k) + (sbar(k)*h) * Me;
  f(k-1:k) = f(k-1:k) + (fbar(k)*h) * fe;
end

% Solve system for the interior values.
u = (K+M) \ f;
u = [0; u; 0];        % put the boundary values into the result
```

**Example 10.5.2**

We solve the equation

$$-(x^2 u')' + 4y = \sin(\pi x), \qquad u(0) = u(1) = 0,$$

in which

$$c(x) = x^2, \qquad s(x) \equiv 4, \qquad f(x) = \sin(\pi x).$$

The coefficient functions need to accept a vector input and return a vector of the same size.

```
c = @(x) x.^2;
q = @(x) 4*ones(size(x));
f = @(x) sin(pi*x);

[x,u] = fem(c,q,f,0,1,50);
plot(x,u), xlabel('x'), ylabel('u')
```

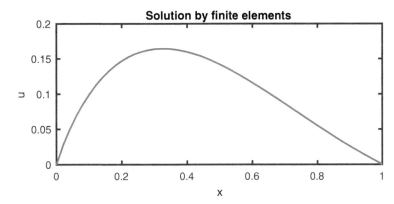

Because piecewise linear interpolation on a uniform grid of size $h$ is $O(h^2)$ accurate, the accuracy of the FEM based on linear interpolation as implemented here is similar to the second-order finite difference method.

## Exercises

10.5.1. For each linear BVP, use Function 10.5.1 to solve the problem and plot the solution for $n = 40$. Then for each $n = 10, 20, 40, \ldots, 640$, compute the norm of the error. Make a log–log convergence plot of error versus $n$ and compare graphically to second-order convergence.

    (a) $-u'' + u = -8 + 16x^2 - x^4$, $u(0) = u(2) = 0$, exact solution $x^2(4-x^2)$.

    (b) $[(2+x)u']' + 11xu = -e^x(12x^3 + 7x^2 + 1)$, $u(-1) = u(1) = 0$, exact solution $e^x(1-x^2)$.

    (c) $u'' + x(u' + u) = -x[4\sin(x) + 5x\cos(x)]$, $u(0) = u(2\pi) = 0$, exact solution $-x^2 \sin(x)$.

10.5.2. Suppose you want to solve the differential equation $-(c(x)u')' + d(x)u = f(x)$, as in equation (10.5.1), except with the boundary conditions $u(a) = \alpha$, $u(b) = \beta$. Find constants $p$ and $q$ such that if $v(x) = u(x) + px + q$, then $v$ satisfies the same BVP, except that $v(a) = v(b) = 0$ and $f$ is replaced by a different function.

10.5.3. Suppose $p(x)u''(x) + q(x)u'(x) + r(x) = 0$, and assume that $p(x) \neq 0$ for all $x$ in $[a, b]$. Let $z(x)$ be any function satisfying $z' = q/p$. Show that the differential equation is equivalent to one in the form (10.5.1), and find the functions $c(x)$, $d(x)$, and $f(x)$ in that equation. (Hint: Start by multiplying through the equation with $\exp(z)$.)

10.5.4. ✍ Derive (10.5.13), starting from (10.5.11b). You should replace $s(x)$ by the constant $\bar{s}_k$ within interval $I_k$.

10.5.5. Suppose the Dirichlet boundary conditions $u(a) = u(b) = 0$ are replaced by the homogeneous Neumann conditions $u'(a) = u'(b) = 0$.

(a) ✍ Explain why the weak form (10.5.3) can be derived without any boundary conditions on the test function $\psi$.

(b) 💻 The result of (a) suggests replacing (10.5.10) with

$$u(x) = \sum_{j=0}^{n} u_j H_j(x)$$

and making (10.5.11) hold for all $i, j$ from 0 to $n$. Modify Function 10.5.1 to do this and thereby solve the Neumann problem. (Note that $I_1$ and $I_n$ now each make multiple contributions, like all the other integration subintervals.)

(c) 💻 Test your function on the problem

$$u'' + u = -2\sin(x), \quad 0 \le x \le 1, \quad u'(0) = u'(1) = 0,$$

whose exact solution is $(x-1)\cos x - \sin x$. Show second-order convergence.

## Key ideas in this chapter

1. The solution of a BVP depends everywhere on information from both ends of the domain (page 409).
2. The shooting method for BVPs is unstable (page 414).
3. Left-multiplication by a differentiation matrix can generate derivative approximations at all nodes in an interval (page 419).
4. Approximating the solution values at nodes, and finding them by approximating the BVP at the same nodes, is called the collocation method (page 425).
5. Nonlinear BVPs, after discretization, require the solution of nonlinear algebraic systems (page 432).
6. The Galerkin method uses a set of basis functions to approximate the solution, and the process solves for the coefficients of those basis functions through an integral formulation of the problem (page 441).

## Where to learn more

A text a bit above the level of this text is by Ascher and Petzold [3], which covers shooting and finite difference collocation methods for linear and nonlinear BVPs, with a number of theoretical and applications problems. A graduate-level text solely on numerical solution of BVPs is by Ascher, Mattheij, and Russell [2]. Besides the shooting and finite difference methods, it briefly discusses Galerkin and spline-based methods, and it goes into more depth on theoretical issues. A more detailed treatment of the

Galerkin method can be found in Quarteroni, Sacco, and Saleri [55]. An older and accessible treatment of Galerkin and FEMs can be found in Strang and Fix [64].

For spectral methods, an introduction to BVPs may be found in Trefethen's book [68], and a more theoretical take is in Quarteroni et al. [55].

In this chapter, a number of linear variable-coefficient BVPs for so-called special functions were mentioned: Bessel's equation, Laguerre's equation, etc. These ODEs and their solutions arise in the solution of partial differential equations of mathematical physics and were extensively characterized prior to wide use of computers [1]. These special functions have come to be used for many things, and a modern summary in digital form has recently appeared (dlmf.nist.gov [51]).

# Chapter 11
# Diffusion equations

> That's impossible, even for a computer.
> — Wedge Antilles, *Star Wars: A New Hope*

To this point we have considered only ordinary differential equations (ODEs), those having only one independent variable. In the rest of this book, we introduce the huge topic of solving partial differential equations (PDEs). We begin by pairing time with space.

As we have seen with initial- and boundary-value problems, the crucial difference between time and space is that information can flow only forward in time. PDEs that include both time and space variables are sometimes referred to as initial-boundary-value problems (IBVPs), because they bear characteristics of both types. As with IVP methods, our IBVP methods advance a solution in time. Like BVP methods, the boundary values must be respected, and all values in space are represented simultaneously.

Finally, while we usually refer to "time" and "space," the independent variables can be more abstract. (One such example is introduced in the next section.) However, they are all time-like or space-like. Mathematical analysis presented in courses on PDEs can be used to determine definitively what type of PDE one is dealing with.

## 11.1 • Black–Scholes equation

Suppose that at time $t = 0$ you buy a stock whose share price is $S(t)$. At a later time, if $S(t) > S(0)$, you can sell the stock and make money. But if $S(t) < S(0)$, you stand to lose money—potentially, your entire investment. You might prefer to find a way to mitigate this risk. One way to do this is to buy a *call option* instead of the stock. This is a contract with a fixed *strike time* $T$ and *strike price* $K$ that gives you the right to buy the stock from the contract issuer at cost $K$ at time $T$. (We discuss only so-called European options; American options are more difficult computationally.)

Suppose $S(T) > K$. Then you could buy the stock at price $K$ and instantly resell it at price $S(T)$, so you make profit $S(T) - K$. On the other hand, if $S(T) \le K$, there is no advantage to exercising the option, because the stock is less valuable than the guaranteed purchase price. However, you have lost only what you paid for the option. These observations are summarized by the *payoff function*

$$H(S) = \max\{S - K, 0\} = (S - K)_+. \qquad (11.1.1)$$

The question now is, what is a fair (to both seller and buyer) price for you to pay for the option contract?

This question is answered to close approximation by the famous **Black–Scholes equation**,

$$\frac{\partial v}{\partial t} + \frac{1}{2}\sigma^2 S^2 \frac{\partial^2 v}{\partial S^2} + rS\frac{\partial v}{\partial S} - rv = 0, \qquad (11.1.2)$$

where $r$ is the *risk-free interest rate* (what you could earn with a very safe investment), and $\sigma$ is the *volatility* of the stock—essentially, the standard deviation in the rate of return for the stock. In the Black–Scholes model, both $r$ and $\sigma$ are assumed to be known and constant. The value $v(S, t)$ of the option depends on the time $t$ and on the stock price $S$, which may be varied independently. At the strike time, the payoff condition $v(S, T) = H(S)$ is imposed, and the goal is to solve the equation backward in time to find $v(S, 0)$.

## Evolutionary PDEs

Equation (11.1.2) is a *time-dependent partial differential equation* or **evolutionary PDE**. As with our first-order IVPs, we require an initial value for the dependent variable. In order to follow the usual convention of having time flow forward instead of backward, we define a new variable $\eta = T - t$, in which case (11.1.2) becomes

$$-v_\eta + \frac{1}{2}\sigma^2 S^2 v_{SS} + rS v_S - rv = 0, \qquad (11.1.3)$$

now defined for $0 \le \eta \le T$. Here we have adopted the common notation of using subscripts for partial derivatives. In the new time variable we have the **initial condition**

$$v(S, 0) = H(S), \qquad (11.1.4)$$

and the goal is to find $v(S, \eta)$ for $\eta > 0$. Henceforth we will simply rename the $\eta$ variable as $t$ again, with the understanding that it runs in the opposite direction to actual time in this application.

Stock prices cannot be negative, so we require $S \ge 0$. This is a left boundary for the domain of the independent variable $S$, and at this boundary we must have $v = 0$ (the option is worthless). In principle $S$ is not bounded from above. There are different ways to handle that fact computationally; we choose to truncate the domain at some positive value $S_{\max}$. In light of the payoff function $H$, initially the value $v$ has slope equal to one at this boundary. We summarize the **boundary conditions** as

$$v(0, t) = 0, \qquad (11.1.5)$$

$$\frac{\partial v}{\partial S}(S_{\max}, t) = 1. \qquad (11.1.6)$$

## 11.1. Black–Scholes equation

The complete problem consists of (11.1.3), (11.1.4), (11.1.5), and (11.1.6). Note that an evolutionary PDE has both initial and boundary conditions (and is sometimes called an initial-boundary-value problem as a result). Solving it numerically requires aspects of methods for both IVPs and BVPs.

The Black–Scholes equation can be transformed by a change of variables into the simpler PDE. Here again we would have to redefine the time variable, but we keep using $t$ as the name. The resulting equation is

$$u_t = u_{xx}, \quad t \geq 0, \quad a \leq x \leq b. \tag{11.1.7}$$

This is a centrally important PDE known widely as the **heat equation** or the *diffusion equation*. As the second name implies, it models quantities that spread by diffusion—in particular, with a velocity proportional to the gradient. Steep changes in the solution flatten out quickly, and in the long run the solution becomes as flat as is allowed by the boundary conditions. The diffusion equation is the prototype for the class of PDEs known as **parabolic equations**.

Note that with two derivatives in space present in the PDE, one boundary condition at each end of the domain is appropriate. We classify these in the same way we did for ODEs. A boundary condition that prescribes a value for $u$ at a boundary is called a **Dirichlet condition**, while a prescribed value for $\partial u / \partial x$ is called a **Neumann condition**. In either case a prescribed value of zero is called a **homogeneous** condition.

### Example 11.1.1

Consider the following diffusion problem:

$$\text{PDE:} \quad u_t = u_{xx}, \ 0 < x < 1, \ 0 < t < \infty,$$
$$\text{BC:} \quad u(0,t) = u(1,t) = 0, \ 0 < t < \infty,$$
$$\text{IC:} \quad u(x,0) = \sin(\pi x), \ 0 \leq x \leq 1.$$

We wish to show that the solution to the problem is $\hat{u}(x,t) = e^{-\pi^2 t} \sin(\pi x)$. To do this we compute the partial derivatives:

$$\frac{\partial \hat{u}}{\partial t} = -\pi^2 e^{-\pi^2 t} \sin(\pi x),$$
$$\frac{\partial^2 \hat{u}}{\partial x^2} = -\pi^2 e^{-\pi^2 t} \sin(\pi x).$$

They are clearly equal, and so the PDE is satisfied. Turning to the boundary conditions, substituting either $x = 0$ or $x = 1$ causes $\hat{u} = 0$. Substitution of $t = 0$ into $u$ recovers the initial condition $\hat{u}(x,0) = \sin(\pi x)$. Thus $\hat{u}$ is a solution to the problem. *Solutions to the problem must satisfy all of the conditions: the PDE itself, the boundary conditions, and the initial conditions.*

In certain cases there are explicit formulas describing the solution of (11.1.7) (and hence (11.1.2)). These provide a lot of insight and can lead to practical solutions. But the exact solutions are fragile in the sense that minor changes to the model can make them impractical or invalid. By contrast, numerical methods are more robust to such changes.

## A naive numerical solution

Let's return to the Black–Scholes equation (11.1.3), subject to the initial condition (11.1.4) and the boundary conditions (11.1.5) and (11.1.6). We now try to solve it numerically, without any transformation tricks or other insight. Define

$$x_i = ih, \quad i = 0, \ldots, m, \quad h = S_{\max}/m,$$
$$t_j = j\tau, \quad j = 0, \ldots, n, \quad \tau = T/m. \qquad (11.1.8)$$

Observe that we have used the more conventional $x$ and $t$ in place of $S$ and $\eta$. The result is a grid function $V$ whose entries are $V_{ij} \approx v(x_i, t_j)$. By the initial condition (11.1.4), we set $V_{i,0} = H(x_i)$ for all $i$.

For the moment, let us pretend that $i$ is unbounded in both directions. Replacing the derivatives in (11.1.3) with some simple finite difference formulas, we get

$$-\frac{V_{i,j+1} - V_{i,j}}{\tau} + \frac{\sigma^2 x_i^2}{2} \frac{V_{i+1,j} - 2V_{i,j} + V_{i-1,j}}{h^2} + rx_i \frac{V_{i+1,j} - V_{i-1,j}}{2h} - rV_{i,j} = 0. \qquad (11.1.9)$$

We can rearrange (11.1.9) into

$$V_{i,j+1} = V_{i,j} + \frac{\lambda \sigma^2 x_i^2}{2}(V_{i+1,j} - 2V_{i,j} + V_{i-1,j}) + \frac{rx_i \mu}{2}(V_{i+1,j} - V_{i-1,j}) - r\tau V_{i,j}, \qquad (11.1.10)$$

where $\lambda = \tau/h^2$ and $\mu = \tau/h$. If we put $j = 0$ into (11.1.10), then everything on the right-hand side is known for each value of $i$, and thus we can get values for $V_{i,1}$. We can then use $j = 1$ and get all of $V_{i,2}$, and so on.

Now we must take the boundaries into account. The value $V_{0,j+1}$ is zero, so we don't even need to compute the solution using (11.1.10) at $i = 0$. At $i = m$, or $x_i = S_{\max}$, things are trickier. We don't know $V_{m,j+1}$ explicitly and would like to solve for it, yet formula (11.1.10) refers to the fictitious value $V_{m+1,j}$ when $i = m$. This is where the condition $\partial v/\partial x = 1$ must be applied. If the value $V_{m+1,j}$ did exist, then we could discretize the Neumann condition as

$$\frac{V_{m+1,j} - V_{m-1,j}}{2h} = 1. \qquad (11.1.11)$$

We can therefore solve (11.1.11) for the fictitious $V_{m+1,j}$ and use it when called for in the right-hand side of (11.1.10). We consider the problem with $\sigma = 0.06$, $r = 0.08$, $K = 3$, and $S_{\max} = 8$.

---

**Example 11.1.2**

Here are the parameters for a particular simulation of the Black–Scholes equation.

```
Smax = 8;   T = 6;
K = 3;   sigma = 0.06;   r = 0.08;
```

We discretize space and time.

## 11.1. Black–Scholes equation

```
m = 200;  h = Smax / m;
x = h*(0:m)';
n = 1000;   tau = T / n;
t = tau*(0:n)';
lambda = tau / h^2;   mu = tau / h;
```

Set the initial condition, then march forward in time.

```
V = zeros(m+1,n+1);
V(:,1) = max( 0, x-K );
for j = 1:n
    % Fictitious value from Neumann condition.
    Vfict = 2*h + V(m,j);
    Vj = [ V(:,j); Vfict ];
    % First row is zero by the Dirichlet condition.
    for i = 2:m+1
        diff1 = (Vj(i+1) - Vj(i-1));
        diff2 = (Vj(i+1) - 2*Vj(i) + Vj(i-1));
        V(i,j+1) = Vj(i) ...
            + (lambda*sigma^2*x(i)^2/2)*diff2 ...
            + (r*x(i)*mu)/2*diff1 - r*tau*Vj(i);
    end
end
```

We plot at a few times.

```
select_times = 1 + 250*(0:4);
show_times = t(select_times)
plot(x,V(:,select_times))
```

```
show_times =
         0
    1.5000
    3.0000
    4.5000
    6.0000
```

The lowest curve is the initial condition, and the highest curve is the last time. The results are easy to interpret, recalling that the time variable really means "time before strike." Say you are close to the option's strike time. If the current stock price is, say, $S = 2$, then it's not likely that the stock will end up over the strike price $K = 3$ and therefore the option has little value. On the other hand, if presently $S = 3$, then there are good odds that the option will be exercised at the strike time, and you will need to pay a substantial portion of the stock price in order to take advantage.

Let's try to extend the simulation time to $T = 8$, keeping everything else the same.

```
T = 8;
n = 1000;   tau = T / n;
t = tau*(0:n)';
lambda = tau / h^2;   mu = tau / h;
for j = 1:n
    % Fictitious value from Neumann condition.
    Vfict = 2*h + V(m,j);
    Vj = [ V(:,j); Vfict ];
    % First row is zero by the Dirichlet condition.
    for i = 2:m+1
        diff1 = (Vj(i+1) - Vj(i-1));
        diff2 = (Vj(i+1) - 2*Vj(i) + Vj(i-1));
        V(i,j+1) = Vj(i) ...
            + (lambda*sigma^2*x(i)^2/2)*diff2 ...
            + (r*x(i)*mu)/2*diff1 - r*tau*Vj(i);
    end
end
plot(x,V(:,select_times))
```

This "solution" is nonsensical. Look at the scale of the ordinate!

Something has gone very wrong with the last simulation—specifically, instability. Understanding the source of that instability comes later in this chapter. First, though, we consider a general and robust strategy for solving evolutionary PDEs.

## Exercises

11.1.1. ✎ Show that $u(x,t) = e^{-4\pi^2 t}\cos(2\pi x)$ is a solution to the IBVP

PDE: $\dfrac{\partial u}{\partial t} = \dfrac{\partial^2 u}{\partial x^2}$, $0 < x < 1$, $0 < t < \infty$,

BC: $\dfrac{\partial u}{\partial x}(0,t) = \dfrac{\partial u}{\partial x}(1,t) = 0$, $0 < t < \infty$,

IC: $u(x,0) = \cos(2\pi x)$, $0 \le x \le 1$.

11.1.2. ✎ Show that $u(x,t) = \sin[2\pi(x-2t)]$ is a solution to the IBVP

PDE: $\dfrac{\partial^2 u}{\partial t^2} - 4\dfrac{\partial^2 u}{\partial x^2} = 0$, $0 < x < 1$, $0 < t < \infty$,

BC: $\dfrac{\partial u}{\partial x}(0,t) = \dfrac{\partial u}{\partial x}(1,t) = \sin(4\pi t)$, $0 < t < \infty$,

IC: $u(x,0) = \sin(2\pi x)$, $\dfrac{\partial u}{\partial t}(x,0) = -4\pi\cos(2\pi x)$, $0 \le x \le 1$.

11.1.3. ✎ Show that $u(x,t) = \exp(-x^2/4t)/\sqrt{t}$ solves the heat equation (11.1.7) at any value of $t > 0$.

11.1.4. ✎ Equation (11.1.9) results from applying finite differences to the derivatives in (11.1.3), including a forward difference for the term $-v_\eta$.

   (a) Write out the method that results if a backward difference is used for $-v_\eta$ instead.

   (b) Explain why modifying the code from Example 11.1.2 to implement this formula is not straightforward.

11.1.5. ⌨ In this problem you are asked to revisit Example 11.1.2 in order to examine the instability phenomenon more closely.

   (a) Leaving other parameters alone, let $m = 100$. To the nearest ten, find the minimum value of $n$ that leads to a stable (i.e., not exponentially growing) solution.

   (b) Repeat (a) for $m = 120, 140, \ldots, 200$. Make a table of the minimum stable $n$ for each $m$. Is the relationship $n = O(m)$, or something else?

## 11.2 • The method of lines

Our strategy in Section 11.1 was to discretize both the time and space derivatives using finite differences, then rearrange so that we could march the solution forward through time. It was partially effective but, as the end of Example 11.1.2 shows, not a sure thing.

Let's use the simpler heat equation, $u_t = u_{xx}$, as a model. As always, we use $\hat{u}$ when we specifically refer to the exact solution of the PDE. Because boundaries always complicate things, we will start by doing the next best thing to having no boundaries at

**Figure 11.1.** *Left: A function whose values are the same at the endpoints of an interval does not necessarily extend to a smooth periodic function. Right: For a truly periodic function, the function values and all derivatives match at the endpoints of one period.*

all: **periodic end conditions**. Specifically, we will solve the PDE over $0 \le x < 1$ and require that

$$u(x+1,t) = u(x,t) \quad \text{for all } x.$$

This is a little different from simply $u(1,t) = u(0,t)$, as Figure 11.1 illustrates.

## Semidiscretization

In order to avoid carrying along redundant information about the function, we use $x_i = ih$ only for $i = 0,\ldots,m-1$, where $h = 1/m$, and it's understood that a reference to $x_m$ is translated to one at $x_0$. In fact, we can apply the identity

$$\hat{u}(x_i, t) = \hat{u}(x_{(i \bmod m)}, t) \tag{11.2.1}$$

to the exact solution $\hat{u}$ for any value of $i$.

Next we define a vector $\boldsymbol{u}(t)$ by $u_i(t) \approx \hat{u}(x_i, t)$. This is called **semidiscretization**, since at this stage space is discretized but time is not. As in Chapter 10, we will replace $u_{xx}$ with multiplication of $\boldsymbol{u}$ by a differentiation matrix $\boldsymbol{D}_{xx}$. The canonical choice is the three-point finite difference formula (5.4.8), which after applying the periodicity (11.2.1) leads to

$$\boldsymbol{D}_{xx} = \frac{1}{h^2} \begin{bmatrix} -2 & 1 & & & & 1 \\ 1 & -2 & 1 & & & \\ & \ddots & \ddots & \ddots & & \\ & & & 1 & -2 & 1 \\ 1 & & & & 1 & -2 \end{bmatrix}. \tag{11.2.2}$$

Because we will be using this matrix quite a lot, we create Function 11.2.1 to compute it, as well as the corresponding second-order $\boldsymbol{D}_x$ for periodic end conditions.

The PDE $u_t = u_{xx}$ is now a semidiscrete problem,

$$\frac{d\boldsymbol{u}(t)}{dt} = \boldsymbol{D}_{xx}\boldsymbol{u}(t), \tag{11.2.3}$$

## 11.2. The method of lines

**Function 11.2.1** (`diffper`) Differentiation matrices for periodic end conditions.

```
function [x,Dx,Dxx] = diffper(n,xspan)
%DIFFPER    Differentiation matrices for periodic end conditions.
% Input:
%   n       number of subintervals (integer)
%   xspan   endpoints of domain (vector)
% Output:
%   x       equispaced nodes (length n)
%   Dx      matrix for first derivative (n by n)
%   Dxx     matrix for second derivative (n by n)

a = xspan(1);  b = xspan(2);
h = (b-a)/n;
x = a + h*(0:n-1)';      % nodes, omitting the repeated data

% Construct Dx by diagonals, then correct the corners.
dp = 0.5*ones(n-1,1)/h;    % superdiagonal
dm = -0.5*ones(n-1,1)/h;   % subdiagonal
Dx = diag(dm,-1) + diag(dp,1);
Dx(1,n) = -1/(2*h);
Dx(n,1) = 1/(2*h);

% Construct Dxx by diagonals, then correct the corners.
d0 = -2*ones(n,1)/h^2;     % main diagonal
dp = ones(n-1,1)/h^2;      % superdiagonal
dm = dp;                   % subdiagonal
Dxx = diag(dm,-1) + diag(d0) + diag(dp,1);
Dxx(1,n) = 1/(h^2);
Dxx(n,1) = 1/(h^2);

end
```

which is simply a linear, constant-coefficient system of *ordinary* differential equations. Given the initial values $u(0)$ obtained from $u(x_i,0)$, we have an IVP that we already know how to solve!

Semidiscretization is more often called the **method of lines**. Despite the name, it is not exactly a single method, because both space and time discretizations have to be specified in order to get an algorithm. *The key concept is the separation of those two discretizations.* In that way, it's related to separation of variables in analytic methods for the heat equation.

Suppose we solve (11.2.3) using the Euler integrator (6.2.1) (that is, AB1 from Chapter 6). We select a time step $\tau$ and discrete times $t_j = j\tau$, $j = 0, 1, \ldots, n$. We now discretize the vector $u$ in time as well to get a sequence $u_j \approx u(t_j)$ for varying $j$. (Remember the distinction in notation between $u_j$, which is a vector, and $u_j$, which is a single element of a vector.) Then a fully discrete method for the heat equation is

$$u_{j+1} = u_j + \tau(D_{xx}u_j) = (I + \tau D_{xx})u_j. \qquad (11.2.4)$$

Or, we could use backward Euler (AM1) for time stepping, resulting in

$$\begin{aligned}u_{j+1} &= u_j + \tau(D_{xx}u_{j+1}),\\ (I - \tau D_{xx})u_{j+1} &= u_j.\end{aligned} \qquad (11.2.5)$$

In this case a linear system must be solved for $u_{j+1}$ at each time step, because backward Euler is an implicit method. But the system is tridiagonal, so it can be made about as fast as the forward Euler method anyway.

### Example 11.2.1

Let's try out the Euler and backward Euler time stepping methods using the second-order semidiscretization:

```
m = 100;   [x,Dx,Dxx] = diffper(m,[0,1]);
Ix = eye(m);
```

First we apply the Euler discretization.

```
tfinal = 0.05;   n = 500;
tau = tfinal/n;   t = tau*(0:n)';
U = zeros(m,n+1);
```

This is where we set the initial condition. It isn't mathematically periodic, but the end values and derivatives are so small that for numerical purposes it may as well be.

```
U(:,1) = exp( -60*(x-0.5).^2 );
```

The Euler time stepping simply multiplies by a constant matrix for each time step.

```
A = Ix + tau*Dxx;
for j = 1:n
    U(:,j+1) = A*U(:,j);
end
```

Things seem to start well.

```
plot(x,U(:,1:3:7))
```

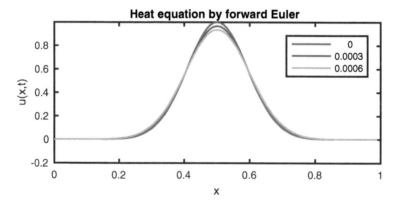

## 11.2. The method of lines

Shortly thereafter, though, there is nonphysical growth.

```
cla
plot(x,U(:,13:3:19))
```

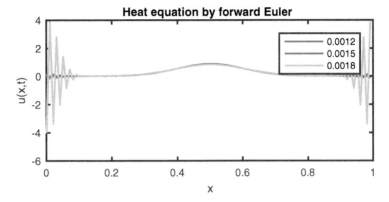

The growth is exponential in time.

```
M = max( abs(U), [], 1 );      % max in each column
semilogy(t,M)
```

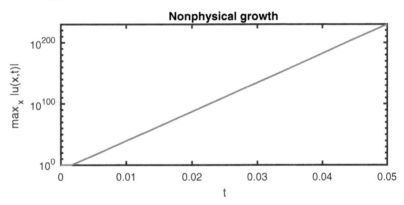

Now we try backward Euler. In this case there is a tridiagonal linear system to solve at each time step. We will use a sparse matrix to get sparse LU factorization, although the time savings at this size are negligible.

```
B = sparse(Ix - tau*Dxx);
for j = 1:n
    U(:,j+1) = B\U(:,j);
end
plot(x,U(:,1:100:501))
```

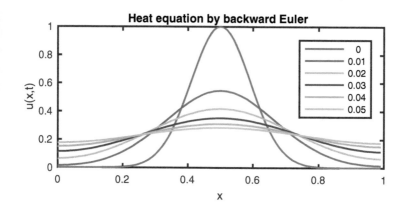

This solution looks physically realistic, as the large concentration in the center diffuses outward. Observe that the solution remains periodic in space.

Example 11.2.1 is evidence that implicit time stepping methods have an important role in diffusion. We will analyze the reason in the next few sections.

## Using black-box IVP solvers

Instead of applying one of the Runge–Kutta or multistep formulas manually for a method of lines discretization, we could try one of the built-in integrators directly on the ODE IVP (11.2.3).

> **Example 11.2.2**
>
> We set up the semidiscretization and initial condition in $x$ just as before.
>
> ```
> m = 100;   [x,Dx,Dxx] = diffper(m,[0,1]);
> u0 = exp( -60*(x-0.5).^2 );
> ```
>
> Now, though, we apply ode45 to the IVP $\mathbf{u}' = \mathbf{D}_{xx}\mathbf{u}$.
>
> ```
> tfinal = 0.05;
> ODE = @(t,u) Dxx*u;
> [t,U] = ode45(ODE,[0,tfinal],u0);
> U = U';    % rows for space, columns for time
> plot(x,U(:,1:100:501))
> ```

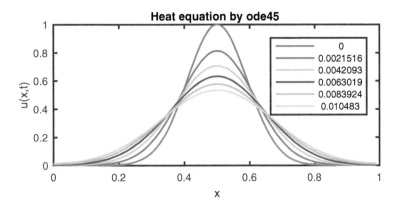

The solution is at least plausible. But the number of time steps that were selected automatically is surprisingly large, considering how smoothly the solution changes.

```
num_steps_ode45 = length(t)-1

num_steps_ode45 =
        2404
```

Now we apply another solver, ode15s.

```
[t,U] = ode15s(ODE,[0,tfinal],u0);
U = U';    % rows for space, columns for time
```

The number of steps selected is reduced by a factor of more than 50!

```
num_steps_ode15s = length(t)-1

num_steps_ode15s =
    41
```

The adaptive time integrators perform as designed. But they are certainly not interchangeable in every sense. Whether we choose to implement a method directly with a fixed step size, or automatically with adaptation, there is something to understand about the semidiscrete problem (11.2.3) that will occupy our attention in the next two sections.

## Exercises

11.2.1. ▣ Revisit Example 11.2.1 with forward Euler time stepping. For each $m = 20, 30, \ldots, 120$ points in space, let $n = 20, 40, 60, \ldots$ in time until you reach the smallest $n$ such that the numerical solution remains bounded above by 1.05 for all time; call this value $n_b$. Make a table and a log–log plot of $n_b$ as a function of $m$. If you suppose that $n_b = O(m^p)$ for a rational number $p$, what is a reasonable guess for $p$?

**11.2.2.** In Example 11.2.2, as $t \to \infty$ the solution $u(x, t)$ approaches a constant.

(a) Apply `ode15s` to find this constant value as accurately as you can.

(b) Show that $\int_0^1 u(x,t)\,dx$ is constant in time.

(c) Use (b) to relate the result of (a) to the initial condition.

**11.2.3.** Apply the trapezoid rule (AM2) to the semidiscretization (11.2.3) and derive what is known as the *Crank–Nicolson* method:

$$(I - \tfrac{1}{2}\tau D_{xx})u_{j+1} = (I + \tfrac{1}{2}\tau D_{xx})u_j. \tag{11.2.6}$$

Note that each side of the method is evaluated at a different time level.

**11.2.4.** Repeat Example 11.2.1 using the Crank–Nicolson method (11.2.6). What does the time step criterion for stability appear to be?

**11.2.5.** The PDE $u_t = 2u + u_{xx}$ combines growth with diffusion. Referring to Example 11.2.1 as a guide, solve this PDE with periodic boundary conditions for the same initial condition by means of backward Euler time stepping. Use $m = 200$ points in space, set $\tau = 1 \times 10^{-4}$, and plot the solution on one graph at times $t = 0, 0.02, 0.04, \ldots, 0.1$.

**11.2.6.** In this problem, you will analyze the convergence of the explicit method given by (11.2.4). Recall that the discrete approximation $u_{i,j}$ approximates the solution at $x_i$ and $t_j$.

(a) Write the method in scalar form as

$$u_{i,j+1} = (1 - 2\lambda)u_{i,j} + \lambda u_{i+1,j} + \lambda u_{i-1,j},$$

where $\lambda = \tau/h^2 > 0$.

(b) Taylor series of the exact solution $\hat{u}$ imply that

$$\hat{u}_{i,j+1} = \hat{u}_{i,j} + \frac{\partial \hat{u}}{\partial t}(x_i, t_j)\tau + O(\tau^2),$$

$$\hat{u}_{i\pm 1,j} = \hat{u}_{i,j} \pm \frac{\partial \hat{u}}{\partial x}(x_i, t_j)h + \frac{\partial^2 \hat{u}}{\partial x^2}(x_i, t_j)\frac{h^2}{2} \pm \frac{\partial^3 \hat{u}}{\partial x^3}(x_i, t_j)\frac{h^3}{6} + O(h^4).$$

Use these to show that

$$\hat{u}_{i,j+1} = \left[(1 - 2\lambda)\hat{u}_{i,j} + \lambda \hat{u}_{i+1,j} + \lambda \hat{u}_{i-1,j}\right] + O(\tau^2 + h^2)$$

$$= F(\lambda, \hat{u}_{i,j}, \hat{u}_{i+1,j}, \hat{u}_{i-1,j}) + O(\tau^2 + h^2).$$

(The last line should be considered a definition of the function $F$.)

(c) The numerical solution satisfies $u_{i,j+1} = F(\lambda, u_{i,j}, u_{i+1,j}, u_{i-1,j})$ exactly. Using this fact, subtract $u_{i,j+1}$ from both sides of the last line in part (b) to show that

$$e_{i,j+1} = F(\lambda, e_{i,j}, e_{i+1,j}, e_{i-1,j}) + O(\tau^2 + h^2),$$

where $e_{i,j} = \hat{u}_{i,j} - u_{i,j}$ is the error in the numerical solution for all $i$ and $j$.

(d) Define $E_j$ as the maximum of $|e_{i,j}|$ over all values of $i$, and use the result of part (c) to show that if $\lambda < 1/2$ is kept fixed as $h$ and $\tau$ approach zero, then for sufficiently small $\tau$ and $h$,

$$E_{j+1} = E_j + O\left(\tau^2 + h^2\right) \leq E_j + K_j\left(\tau^2 + h^2\right)$$

for a positive $K_j$ independent of $\tau$ and $h$.

(e) If the initial conditions are exact, then $E_0 = 0$. Use this to show finally that if the $K_j$ are bounded above and $\lambda < 1/2$ is kept fixed, then $E_n = O(\tau)$ as $\tau \to 0$.

## 11.3 ▪ Absolute stability

In Section 11.2 we applied several different time stepping methods to a linear, constant coefficient problem in the form $\boldsymbol{u}'(t) = \boldsymbol{A}\boldsymbol{u}(t)$. All of these methods are zero-stable, in the sense of Section 6.8, in the limit as the time step[42] $\tau \to 0$. Yet for experiments with small but finite sizes of $\tau$, we observed what seems to be exponential instability in some cases as $n \to \infty$.

First observe that if $\boldsymbol{A}$ has the eigenvalue decomposition $\boldsymbol{A} = \boldsymbol{V}\boldsymbol{D}\boldsymbol{V}^{-1}$, then

$$\boldsymbol{u}' = (\boldsymbol{V}\boldsymbol{D}\boldsymbol{V}^{-1})\boldsymbol{u},$$
$$(\boldsymbol{V}^{-1}\boldsymbol{u}') = \boldsymbol{D}(\boldsymbol{V}^{-1}\boldsymbol{u}),$$
$$\boldsymbol{y}' = \boldsymbol{D}\boldsymbol{y},$$

where $\boldsymbol{y}(t) = \boldsymbol{V}^{-1}\boldsymbol{u}(t)$. Because $\boldsymbol{D}$ is diagonal, the dynamics of the components of $\boldsymbol{y}$ are completely decoupled; each row is a self-contained equation of the form $y'_j = \lambda_j y_j$, where $\lambda_j$ is an eigenvalue of $\boldsymbol{A}$. Therefore we focus on the model problem

$$y' = \lambda y, \qquad y(0) = 1. \tag{11.3.1}$$

Because eigenvalues of real matrices may not be real, we let $\lambda$ be a complex number, and therefore $y(t)$ may be complex as well.[43]

The solution of (11.3.1) is trivially

$$y(t) = e^{t\lambda}.$$

If we write $\lambda$ in real and imaginary parts as $\lambda = \alpha + i\beta$, then by Euler's identity,

$$\left|e^{t(\alpha+i\beta)}\right| = \left|e^{t\alpha}\right| \cdot \left|e^{it\beta}\right| = e^{t\alpha}.$$

---

[42]In Chapter 6 we used $h$ rather than $\tau$, but now we reserve $h$ for spacing in the $x$ direction.
[43]In this section, $i$ is the imaginary unit, not an integer index.

Thus, solutions of (11.3.1) are bounded as $t \to \infty$ if and only if $\alpha = \operatorname{Re}\lambda \le 0$. Geometrically, we describe this as the left half of the complex plane.

## Stability regions

When an IVP method with step size $\tau$ is applied to the model equation (11.3.1), we find that the solution is bounded as $t \to \infty$ only for some values of $\lambda$ and $\tau$.

> **Example 11.3.1**
>
> Consider an Euler discretization of $y' = \lambda y$:
>
> $$y_{k+1} = y_k + \tau(\lambda y_k) = (1 + \tau\lambda)y_k.$$
>
> In order to simplify the notation, we define $\zeta = \lambda\tau$. We easily deduce that $y_k = (1+\zeta)^k$, and therefore
>
> $$|y_k| = |1+\zeta|^k.$$
>
> We see that $|y_k|$ remains bounded above as $k \to \infty$ if and only if $|1+\zeta| \le 1$. Because $\zeta$ is a complex number, it's easiest to interpret this condition geometrically:
>
> $$|\zeta + 1| = |\zeta - (-1)| \le 1,$$
>
> which reads that the distance in the plane from $\zeta$ to the point $-1$ is less than or equal to one. This description defines a closed disk of radius 1 centered at $(-1, 0)$.
>
> If we repeat the steps above for the backward Euler method, we obtain the numerical solution $y_k = (1-\zeta)^{-k}$. Absolute stability requires $|1-\zeta|^{-1} \le 1$, or $|\zeta - 1| \ge 1$. This describes the region *outside* of the open disk of radius 1 centered at $(1,0)$ in the plane.

It turns out that in all Runge–Kutta and multistep methods, just as in Example 11.3.1, the only parameter that matters is $\zeta = \lambda\tau$. The numerical solution is asymptotically bounded if and only if $\zeta$ lies in a subset of $\mathbb{C}$ called the *region of absolute stability*, or more simply the **stability region**, of the method. Stability regions for the most common IVP integrators are given in Figures 11.2–11.5. One can see that the implicit Adams–Moulton methods of Figure 11.3 are larger than those for the explicit Adams–Bashforth methods of Figure 11.2. For the implicit backward differentiation methods of Figure 11.4, the exteriors of the curves provide large regions of stability, but significant portions of the imaginary axis may be excluded. While the single-step Runge–Kutta methods have smaller regions of stability, some do include significant portions of the imaginary axis.

For any particular method and value of $\lambda$ in (11.3.1), we can use the stability region to deduce which, if any, values of the time step $\tau$ will give bounded solutions. Both the magnitude and the argument (angle) of $\lambda$ play a role in determining such constraints.

## 11.3. Absolute stability

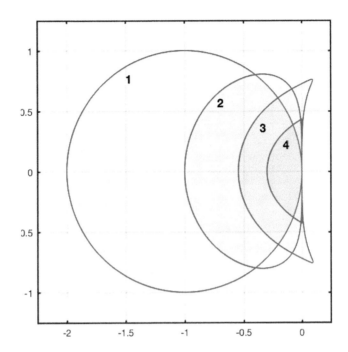

**Figure 11.2.** *Stability regions for Adams–Bashforth methods of order 1–4.*

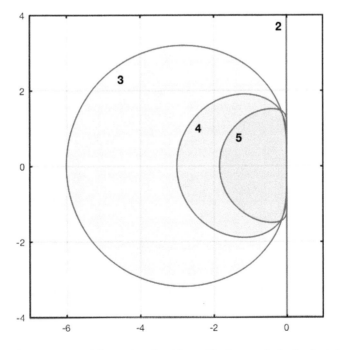

**Figure 11.3.** *Stability regions for Adams–Moulton methods of order 2–5.*

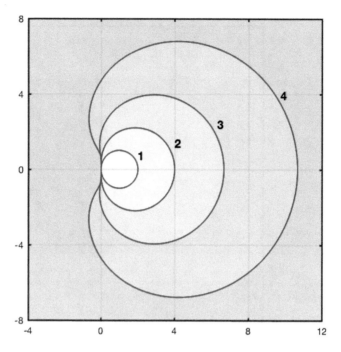

**Figure 11.4.** *Stability regions for backward differentiation methods of order 1–4 (exteriors of curves).*

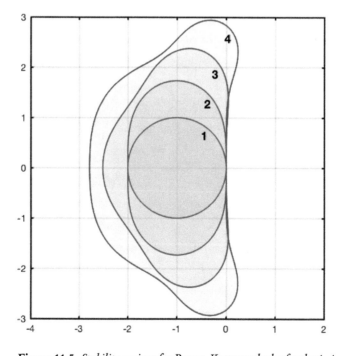

**Figure 11.5.** *Stability regions for Runge–Kutta methods of order 1–4.*

## 11.3. Absolute stability

**Example 11.3.2**

Suppose $\lambda = -4$ and Euler's method is applied. Since the time step is always positive, $\zeta = -4\tau$ is always on the negative real axis. The only part of that line that lies within the stability region of Euler as derived in Example 11.3.1 is the real interval $[-2, 0]$. Hence we require $\zeta \geq -2$, or $\tau \leq 1/2$. But the stability region of backward Euler includes the entire negative real axis, so absolute stability is assured regardless of $\tau$.

Now suppose instead that $\lambda = i$, so that $\zeta = i\tau$. Clearly $\zeta$ is always on the positive imaginary axis. But no part of this axis, aside from the origin, lies in the stability region of Euler's method, so is not absolutely stable for any nonzero value of $\tau$. The conclusion for backward Euler is the opposite: any value of $\tau$ will do, because the entire imaginary axis is within the stability region.

Example 11.3.2 does not contradict our earlier statements about the stability and convergence of Euler's method in general, even for the case $\lambda = i$. But those statements were based on the limit $\tau \to 0$ for $t$ in a finite interval $[a, b]$. Both this limit and the limit $t \to \infty$ imply the number of steps $n$ goes to infinity, but the limits behave differently.

*Two important trends are clear from the stability regions: as the order of accuracy increases, the stability regions get smaller, and implicit methods have larger stability regions than their explicit counterparts.* This second fact is the most important justification for implicit methods. While they have larger work requirements per step, they sometimes can take steps that are orders of magnitude larger than explicit methods and still remain stable.

When adaptive time stepping methods are used, as in most software for IVPs, the automatically determined time step is chosen to satisfy absolute stability requirements (otherwise errors grow exponentially). This phenomenon was manifested in Example 11.2.2. Because `ode45` is based on explicit IVP methods, the error control forced tiny step sizes compared to those in `ode15s`, which is based on implicit methods.

### Time stability in the heat equation

Now we return to the semidiscretization (11.2.3) of the heat equation, which was solved by Euler and backward Euler time stepping methods in Example 11.2.1.

**Example 11.3.3**

Both time stepping methods solved $\mathbf{u}' = \mathbf{D}_{xx}\mathbf{u}$ for the matrix

```
m = 40;  [~,~,Dxx] = diffper(m,[0,1]);
```

The eigenvalues of this matrix are real and negative.

```
lambda = eig(Dxx);
```

```
plot(real(lambda),imag(lambda),'.')
```

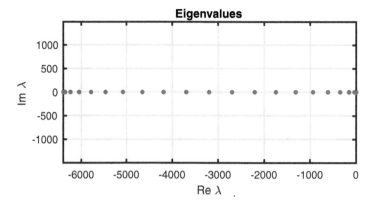

The Euler method is absolutely stable in the region $|\zeta + 1| \leq 1$ in the complex plane:

```
phi = linspace(0,2*pi,361);
z = exp(1i*phi) - 1;    % unit circle shifted to the left by 1
fill(real(z),imag(z),[.8 .8 1])
```

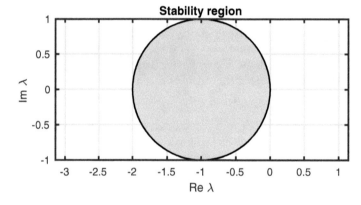

In order to get inside this region, we have to find $\tau$ such that $\lambda\tau > -2$ for all eigenvalues $\lambda$. This is an *upper* bound on $\tau$.

```
lambda_min = min(lambda)
max_tau = -2 / lambda_min
```

```
lambda_min =
     -6400
max_tau =
   3.1250e-04
```

Here we plot the resulting values of $\zeta = \lambda\tau$.

```
zeta = lambda*max_tau;
hold on
plot(real(zeta),imag(zeta),'.')
```

## 11.3. Absolute stability

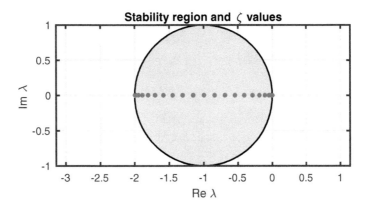

In backward Euler, the region is $|\zeta - 1| \geq 1$. Because they are all on the negative real axis, all of the $\zeta$ values will fit no matter what $\tau$ is chosen.

```
clf,   fill([-6 6 6 -6],[-6 -6 6 6],[.8 .8 1]),    hold on
z = exp(1i*phi) + 1;     % unit circle shifted right by 1
fill(real(z),imag(z),'w')
plot(real(zeta),imag(zeta),'.')
```

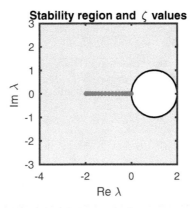

The matrix $D_{xx}$ occurring in (11.2.3) for semidiscretization of the periodic heat equation has eigenvalues that can be found explicitly. Assuming that $x \in [0, 1)$ (with periodic boundary conditions), for which $h = 1/m$, then the eigenvalues are

$$\lambda_j = -4m^2 \sin^2 \frac{j\pi}{m}, \qquad j = 0, \ldots, m-1. \tag{11.3.2}$$

This result agrees with the observation in Example 11.3.3 that the eigenvalues are real and negative. Furthermore, they lie within the interval $[-4m^2, 0]$. In Euler time integration, this implies that $-4\tau m^2 \geq -2$, or $\tau \geq 1/(2m^2) = O(m^{-2})$. For backward Euler, there is no time step restriction. We say that backward Euler is **unconditionally stable** for this problem.

The time step restrictions are independent of accuracy—both methods are first-order accurate, so both in principle converge to the correct solution at the same rate as $\tau \to 0$ at any particular time. However, the results for forward Euler will be exponentially large nonsense until the absolute stability restriction is satisfied.

## Exercises

11.3.1. ⚠ Use an eigenvalue decomposition to write the system
$$u'(t) = \begin{bmatrix} 0 & 4 \\ -4 & 0 \end{bmatrix} u(t)$$
as an equivalent diagonal system.

11.3.2. ⚠ For each system, state whether its solutions are bounded as $t \to \infty$.

(a) $u'(t) = \begin{bmatrix} 1 & 3 \\ 3 & 1 \end{bmatrix} u(t)$.

(b) $u'(t) = \begin{bmatrix} -1 & 3 \\ -3 & -1 \end{bmatrix} u(t)$.

(c) $u'(t) = \begin{bmatrix} 0 & 4 \\ -4 & 0 \end{bmatrix} u(t)$.

11.3.3. ⚠ Using Figures 11.2–11.5 and a ruler, find the time step restriction (if any) for the system
$$u'(t) = \begin{bmatrix} -4 & 0 & 0 \\ 0 & -2 & 0 \\ 0 & 0 & -0.5 \end{bmatrix} u(t)$$
for the following IVP methods:

(a) RK4.  (b) AM4.  (c) AB2.

11.3.4. ⚠ Using Figures 11.2–11.5 and a ruler, find the time step restriction (if any) for the system
$$u'(t) = \begin{bmatrix} -1 & 0 & 0 \\ 0 & 0 & 4 \\ 0 & -4 & 0 \end{bmatrix} u(t)$$
for the following IVP methods:

(a) RK4.  (b) AM4.  (c) AB3.

11.3.5. ⚠ Of the following methods, which would be unsuitable for a problem having eigenvalues on the imaginary axis? Justify your answer(s).

(a) AM2.  (b) AB2.  (c) RK2.  (d) RK3.

11.3.6. ⚠ Of the following methods, which would have a time step restriction for a problem with real, negative eigenvalues? Justify your answer(s).

(a) AM2.  (b) AM4.  (c) BD4.  (d) RK4.

11.3.7. ⚠ Let $D_{xx}$ be $m \times m$ and given by (11.2.2). Show that for each $k = 0, \ldots, m-1$, the pair
$$\lambda = -4m^2 \sin^2 \frac{k\pi}{m}, \quad v = \begin{bmatrix} 1 \\ e^{2i\pi/m} \\ e^{4i\pi/m} \\ \vdots \\ e^{2i(m-1)\pi/m} \end{bmatrix}$$
are an eigenvalue and eigenvector of $D_{xx}$.

**11.3.8.** ✍ Complete the following:

(a) Derive an algebraic inequality equivalent to absolute stability for the AM2 (trapezoid) formula.

(b) Argue that the inequality is equivalent to the restriction $\operatorname{Re}(\zeta) \leq 0$. (Hint: Complex magnitude is equivalent to distance in the plane.)

## 11.4 ▪ Stiffness

In Section 11.3 we analyzed time step constraints for the semidiscrete heat equation, $u' = D_{xx}u$, in terms of stability regions and the eigenvalues $\lambda_j$ of the matrix (given in (11.3.2)). Since all the eigenvalues are negative and real, the one farthest from the origin, at about $-4/h^2$, determines the specific time step restriction. In particular, any explicit time stepping method (such as Euler) must have a bounded stability region, so the step size requirement is $\tau = O(h^2)$ as $h \to 0$.

The solution to each $y' = \lambda_j y$ decays as $e^{-t|\lambda_j|}$. This exponential will be a factor of $1/e$ smaller when $t = 1/|\lambda_j|$. We may regard $|\lambda_j|^{-1}$, which has units of time, as a characteristic time scale of the scalar problem. Recall that $u' = D_{xx}u$ can be diagonalized into an equivalent collection of scalar problems over all $\lambda_0, \ldots, \lambda_{m-1}$. Thus, the (discretized) heat equation has phenomena at multiple time scales. Referring again to (11.3.2), those time scales range from $O(1)$ to $O(m^{-2}) = O(h^2)$. The exception is $\lambda_0 = 0$, which has an infinite time scale because it refers to a solution that is constant in time, i.e., a steady-state solution.

If we are interested in a numerical simulation that captures the longest finite time scale, which is $O(1)$, three things happen as $h \to 0$: the spatial discretization becomes more accurate like $O(h^2)$; the size of the matrix increases like $O(h^{-1})$; and if we use an explicit time stepping method, then absolute stability requires $O(h^{-2})$ steps. This restriction becomes more burdensome as $h \to 0$, i.e., as we improve the space discretization. Backward Euler, on the other hand, has no restriction on the time step for stability. These observations explain why implicit methods are preferred for the heat equation.

In general, problems that incorporate phenomena over time scales spanning orders of magnitude (or powers of $h$) are referred to as **stiff**. *Stiffness is not a binary condition but a spectrum.* It can arise in nonlinear problems and in many problems having nothing to do with diffusion. Stiff problems usually require implicit time stepping methods to avoid taking unreasonably small steps.

### Linearization

Why should the model equation $y' = \lambda y$ of absolute stability have wide relevance? Through diagonalization, it is easily generalized to $u' = Au$ for a constant matrix $A$. But that is still a severely limited type of problem. The emphasis on the boundedness of solutions also may seem mysterious.

Consider now a general vector nonlinear problem of the form $u' = f(t, u)$. The key to making a connection with absolute stability is to look not at an exact solution but to *perturbations* of one. Such perturbations always exist in real numerical solutions, such as those due to roundoff error, for example. But if we assume the perturbations are tiny, then we can use linear approximations. And if we conclude from such an approximation that the perturbation may grow without bound, then we must seriously question the value of the numerical solution.

Let's introduce more precision into the discussion. Suppose that $\hat{u}(t)$ is an exact solution that we wish to track, but that a perturbation has pushed us to a nearby solution curve $\hat{u}(t) + v(t)$. Substituting this solution into the governing ODE and appealing to a multidimensional Taylor series, we derive

$$[\hat{u}(t) + v(t)]' = f(t, \hat{u}(t) + v(t)),$$
$$\hat{u}'(t) + v'(t) = f(t, \hat{u}(t)) + J(t)v(t) + O(\|v(t)\|^2), \quad (11.4.1)$$
$$v'(t) \approx J(t)v(t).$$

We have used the Jacobian matrix $J$, with entries

$$J_{ij} = \left. \frac{\partial f_i}{\partial u_j} \right|_{(t,\hat{u}(t))}. \quad (11.4.2)$$

The result is called the **linearization** of the ODE about the exact solution $\hat{u}(t)$. Linearization governs the fate of infinitesimally small perturbations to the exact solution.

We make a further simplification in order to make the analysis tractable. If a perturbation begins at a moment $t_*$, we freeze the Jacobian there and let $A = J(t_*)$. The result is the equation $v' = Av$ for a constant matrix $A$.

---

**Example 11.4.1**

The **Oregonator** is a well-known ODE system modeling a chemical oscillator and is given by

$$\begin{aligned}
u_1' &= s[u_2(u_2 - u_1) + u_1(1 - qu_1)], \\
u_2' &= s^{-1}(u_3 - u_2 - u_1 u_2), \quad (11.4.3) \\
u_3' &= w(u_1 - u_3),
\end{aligned}$$

where $s$, $q$, and $w$ are constants. Linearization about an exact (but unknown) solution $\hat{u}(t)$ leads to the Jacobian

$$J(t) = \begin{bmatrix} -s(\hat{u}_2 - 2q\hat{u}_1 + 1) & s(1 - \hat{u}_1) & 0 \\ -\hat{u}_2/s & -(1 + \hat{u}_1)/s & 1/s \\ w & 0 & -w \end{bmatrix}. \quad (11.4.4)$$

Freezing coefficients at any time $t_*$ leads to the constant-coefficient problem $v' = J(t_*)v$ for the infinitesimal 3-component perturbation $v$.

## 11.4. Stiffness

### Multiple time scales

The equation $v' = Av$ is of the type we used to discuss the absolute stability of IVP solvers. What we know about stability regions suggests the following:

> **Rule of thumb for absolute stability.** The eigenvalues of the Jacobian appearing in the linearization about an exact solution, after scaling by the time step $\tau$, must lie in the stability region of the IVP solver.

It's a rule of thumb rather than a theorem here; we made several approximations and assumptions along the way. Nevertheless, if the rule of thumb is violated, we should expect perturbations to the exact solution to grow significantly with time, eventually rendering the numerical solution useless.

We already designated $|\lambda|^{-1}$ as the time scale of a negative real $\lambda$ in $y' = \lambda y$. If $\lambda$ is imaginary, so that $\lambda = i\omega$, then the solution $e^{i\omega t}$ has sines and cosines of frequency $\omega$, so $|\lambda|^{-1}$ again seems like a measure of the time scale. We adopt that convention for all complex $\lambda$.

A problem that models phenomena occurring over very different time scales simultaneously will have a Jacobian matrix with eigenvalues at different orders of magnitude. Say $|\lambda_1| \gg |\lambda_2|$. Any explicit integrator will have a bounded stability region and therefore impose a time step restriction proportional to $|\lambda_1|^{-1}$. Any good adaptive integrator will obey such a restriction naturally to control the error. But to observe the "slow" part of the solution, the simulation must go on for a time on the order of $|\lambda_2|^{-1}$, which is much longer.

**Example 11.4.2**

We simulate the Oregonator given by (11.4.3) and whose linearization is given by (11.4.4).

```
q = 8.375e-6;   s = 77.27;   w = 0.161;
f = @(t,u) [ s*(u(2)-u(1)*u(2)+u(1)-q*u(1)^2);...
    (-u(2)-u(1)*u(2)+u(3))/s; ...
    w*(u(1)-u(3)) ];
```

The ode15s solver is fast.

```
tic, [t,u] = ode15s(f,[0 6],[1,1,4]); toc
plot(t,u,'.-')
num_steps_ode15s = length(t)
```

```
Elapsed time is 0.031792 seconds.
num_steps_ode15s =
   180
```

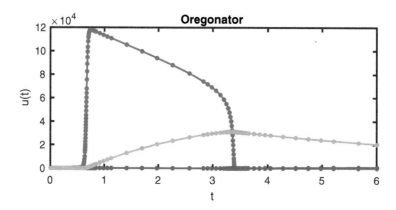

You can see that `ode15s` takes small time steps only when some part of the solution is changing rapidly. However, the `ode45` solver, based on an explicit RK4 method, is much slower and takes more steps compared to `ode15s`.

```
tic, [t,u] = ode45(f,[0 6],[1,1,4]); toc
num_steps_ode45 = length(t)
```

```
Elapsed time is 1.037573 seconds.
num_steps_ode45 =
      113661
```

Here is the Jacobian matrix at any value of $u$.

```
J = @(u) [ -s*(u(2)+1-2*q*u(1)), s*(1-u(1)), 0; ...
    -u(2)/s, (-1-u(1))/s, 1/s; ...
    w,0,-w];
```

During the early phase, the time steps seem fairly large. The eigenvalues around $t = 1/2$ are

```
i1 = find(t>0.5,1);
lambda1 = eig( J(u(i1,:)) )
```

```
lambda1 =
 -139.8308
   -1.1319
   -0.3824
```

These are real and negative. Checking the stability region of RK4 along the negative real axis, we see that stability requires a maximum time step:

```
maxstep1 = 2.8 / max(abs(lambda1))
```

```
maxstep1 =
    0.0200
```

The actual step size chosen by `ode45` was

```
step1 = t(i1+1) - t(i1)
```

## 11.4. Stiffness

```
step1 =
    0.0143
```

Later in the simulation, the steps seem quite small compared to the apparent rate of activity. We look near $t = 4$:

```
i2 = find(t>4,1);
lambda2 = eig( J(u(i2,:)) )

lambda2 =
   1.0e+04 *
   -1.8236
   -0.0000
   -0.0000
```

These are also real and negative. We compare the maximum and observed step sizes again:

```
maxstep2 = 2.8 / max(abs(lambda2))
step2 = t(i2+1) - t(i2)

maxstep2 =
   1.5354e-04
step2 =
   4.8870e-05
```

The phenomenon exhibited in Example 11.4.2 is stiffness, characterized by multiple time scales in the solution, or alternatively, by time step restrictions that are based on absolute stability, not local accuracy.

### A-stability

Since stability regions determine the fate of small perturbations to a solution, let's reconsider the stability regions given in Figures 11.2–11.5. Clearly, the larger the stability region, the more generous the time step restriction will be. However, in the context of a semidiscretization of the heat equation, we observed that the relevant eigenvalues of the finite-dimensional system matrix spanned $O(h^{-2})$. Consequently, any stability region that is bounded in the negative real direction will have a $\tau = O(h^2)$ restriction—only the leading constant will change.

Hence it is desirable in stiff problems generally, and diffusion problems in particular, to have a stability region that is unbounded. A stability region that includes a sector of angle $\alpha$ in either direction from the negative real axis is called **A($\alpha$)-stable**. For the heat equation, any $\alpha > 0$ will suffice for unconditional stability.

The ultimate stability region would include not only the negative real axis but also the entire left half of the complex plane. That set corresponds to all perturbations that can be expected to be bounded in the exact solution of the original ODE. A time stepping method whose stability region contains the entire left half-plane is called **A-stable**. When an A-stable method is used, time step size can be based on accuracy considerations alone. The backward Euler (AM1) and trapezoid (AM2) formulas are **A-stable**. Unfortunately, better A-stable methods are not easy to come by.

> **Theorem 11.4.1: Second Dahlquist stability barrier**
>
> An A-stable linear multistep method must be implicit and have order of accuracy no greater than 2.

The trapezoid formula is as accurate as we can hope for in the family of A-stable linear multistep methods. The situation with Runge–Kutta methods is a little different, but not a great deal more favorable; we will not go into the details.

## Exercises

11.4.1. ✍ Write the mechanical oscillator $x'' + cx' + kx = 0$ as a first-order linear system, $\boldsymbol{u}' = \boldsymbol{A}\boldsymbol{u}$. Show that if $c = k + 1$, this system is stiff as $k \to \infty$.

11.4.2. This exercise is about the IVP $y' = \cos(t) - 1000(y - \sin(t))$, $y(0) = 0$.

  (a) ✍ Show that $y(t) = \sin(t)$ is the exact solution, and find the linearization about this solution.

  (b) ✍ Find the lone eigenvalue of the Jacobian. What other time scale is also relevant in the solution?

  (c) ⌨ Use Function 6.7.1 (ab4) to solve the IVP over $t \in [0, \pi]$ with $n = 8000, 9000, \ldots, 12000$ steps. By comparing to the exact solution, show that the method gets either zero accurate digits or at least 13 accurate digits.

11.4.3. In Example 6.3.5 we derived the following system for two pendulums hanging from a rod:

$$u_1' = u_3,$$
$$u_2' = u_4,$$
$$u_3' = -\gamma u_3 - \frac{g}{L}\sin u_1 + k(u_2 - u_1),$$
$$u_4' = -\gamma u_4 - \frac{g}{L}\sin u_2 + k(u_1 - u_2).$$

  (a) ✍ Use the approximation $\sin x \approx x$ to write the problem as a linear system.

  (b) ⌨ Compute the eigenvalues of the system with $\gamma = 0.1$, $g/L = 1$, and $k = 10^d$ for $d = 0, 1, \ldots, 5$. Characterize the observed stiffness of the system as $O(k^p)$ for some rational $p$.

11.4.4. The equation $u' = u^2 - u^3$ is a simple model for combustion of a flame ball in microgravity. (This problem is adapted from Section 7.9 of [46].)

(a) 🖥 Solve the problem using Function 6.5.1, with $u(0) = 0.01$, for $0 \le t \le 200$ and with an error tolerance of $10^{-5}$. Plot the solution.

(b) ✏️ Find the $1 \times 1$ Jacobian "matrix" of this system and use it to derive an upper bound on the time step of RK2 when $u = 1$.

(c) 🖥 Find the size of the final time step taken by the numerical solution in part (a), and compare it to the prediction of part (b).

11.4.5. The *van der Pol equation* is a famous nonlinear oscillator given by

$$y'' - \mu(1-y^2)y' + y = 0, \tag{11.4.5}$$

where $\mu \ge 1$ is a constant.

(a) ✏️ Write the equation as a first-order system and find its Jacobian.

(b) ✏️ Find the eigenvalues of the Jacobian when $y = -2$ and $y' = 0$.

(c) 🖥 Solve the problem using `ode45` when $\mu = 50$, $y(0) = y'(0) = 1$, for $0 \le t \le 200$. Plot the solution.

(d) 🖥 Find the eigenvalues from part (b) at $\mu = 50$. Find a time when the numerical solution from part (b) is close to $y = -2$, $y' = 0$, and find the local step size that was taken to reach it. Multiply the eigenvalues by the time step size, and compare the result to the stability region of RK4.

## 11.5 ▪ Method of lines for parabolic PDEs

So far we have considered the method of lines for problems with periodic end conditions, which is much like having no boundary at all. How will boundary conditions be incorporated into this technique?

Suppose a nonlinear PDE of the form

$$u_t = \phi(t, x, u, u_x, u_{xx}), \qquad a \le x \le b, \tag{11.5.1}$$

is posed. Not all such PDEs are parabolic (essentially, diffusive), but we will assume this to be the case. Suppose also that the solution is subject to the boundary conditions

$$\alpha_\ell u(a,t) + \beta_\ell \frac{\partial u}{\partial x}(a,t) = \gamma_\ell,$$
$$\alpha_r u(b,t) + \beta_r \frac{\partial u}{\partial x}(b,t) = \gamma_r, \tag{11.5.2}$$

where $\alpha_i$, $\beta_i$, and $\gamma_i$ are known constants for $i = \ell, r$. These include all Dirichlet, Neumann, and Robin conditions.

### Boundary removal

As usual, we replace $u(x,t)$ by the semidiscretized $\mathbf{u}(t)$, where $u_i(t) \approx \hat{u}(x_i, t)$ and $i = 0, \ldots, m$. We require the endpoints of the interval to be included in the discretization, that is, $x_0 = a$ and $x_m = b$. This allows a division of the semidiscrete unknown $u(t)$ into interior and boundary nodes:

$$u = \begin{bmatrix} u_0 \\ v \\ u_m \end{bmatrix}, \qquad (11.5.3)$$

where $v$ are the solution values over the interior of the interval. The guiding principle is to let the interior unknowns $v$ be governed by a discrete form of the PDE, while the endpoint values are chosen to satisfy the boundary conditions. As a result, we will develop an IVP for the interior unknowns only:

$$\frac{dv}{dt} = f(t,v). \qquad (11.5.4)$$

The boundary conditions are used only in the definition of $f$.

After discretization using a differentiation matrix $D_x$, equation (11.5.2) takes the form

$$\begin{aligned} e_0^T[\alpha_\ell u + \beta_\ell(D_x u)] &= \gamma_\ell, \\ e_m^T[\alpha_r u + \beta_r(D_x u)] &= \gamma_r. \end{aligned} \qquad (11.5.5)$$

The rows of the identity appearing on the left side of these equations simply pick out the first and last elements of the vectors they multiply. Equation (11.5.5) can be rearranged into a $2 \times 2$ linear system for the end values $u_0$ and $u_m$:

$$B \begin{bmatrix} u_0 \\ u_m \end{bmatrix} = \gamma - Cv. \qquad (11.5.6)$$

For Dirichlet conditions we have $B = I$, but this need not be the case when derivatives of the solution are involved at the boundaries.

---

**Example 11.5.1**

Recall the Black–Scholes PDE (11.1.3) (rearranged into our more traditional notation),

$$-u_t + \frac{1}{2}\sigma^2 x^2 u_{xx} + rxu_x - rv = 0,$$

subject to $v(0) = 0$ and $u_x(S_{\max}) = 1$. These imply $u_0 = 0$ and

$$\frac{0.5 u_{m-2} - 2 u_{m-1} + 1.5 u_m}{h} = 1.$$

Hence

$$\begin{bmatrix} 1 & 0 \\ 0 & \frac{3}{2} \end{bmatrix} \begin{bmatrix} u_0 \\ u_m \end{bmatrix} = \begin{bmatrix} 0 \\ h \end{bmatrix} - \begin{bmatrix} 0 & \cdots & 0 & 0 \\ 0 & \cdots & -\frac{1}{2} & 2 \end{bmatrix} v.$$

We note here that when the indexing is started at zero, the first element of $v$ is $u_1$ and its last element is $u_{m-1}$; in MATLAB, the indexing needs to be taken into account.

---

If we use a spectrally accurate method for derivatives in space, the matrix $B$ in (11.5.6) need not even be diagonal.

## 11.5. Method of lines for parabolic PDEs

### Example 11.5.2

Returning to Example 11.5.1, suppose we use a global Chebyshev differentiation matrix for $D_x$ in (11.5.5). Then $u_0 = 0$ and

$$D_{m0} u_0 + D_{m1} u_1 + \cdots + D_{mm} u_m = 1.$$

Hence

$$\begin{bmatrix} 1 & 0 \\ D_{m0} & D_{mm} \end{bmatrix} \begin{bmatrix} u_0 \\ u_m \end{bmatrix} = \begin{bmatrix} 0 \\ 1 \end{bmatrix} - \begin{bmatrix} 0 & \cdots & 0 & 0 \\ D_{m1} & \cdots & D_{m,m-2} & D_{m,m-1} \end{bmatrix} v.$$

## Implementation

The steps to evaluate $f$ in (11.5.4) now go as follows:

Given a value of $t$ and $v$,
1. use (11.5.5) to solve for $u_0$ and $u_m$;
2. assemble the total vector $u$ from (11.5.3);
3. use the spatial semidiscretization to evaluate $\phi$ at all the nodes;
4. chop off the boundary nodes to get the value of $f(t, v)$.

In the examples that follow, we create a function called `extend` to perform the operations of extending a solution from $v$ to $u$ and another called `chop` to remove the boundary nodes from a full-length vector. The resulting ODE IVP is solved using `ode15s`, as is appropriate for a stiff system.

### Example 11.5.3

We solve $u_t = u_{xx}$ on $[-1, 1]$ subject to the Dirichlet conditions $u(-1, t) = 0$, $u(1, t) = 2$.

```
m = 100;   [x,Dx,Dxx] = diffcheb(m,[-1,1]);
```

Our next step is to write a function that defines $f$. Since the boundary values are given explicitly, there is no need to solve for them—we just append them to each end of the vector.

```
extend = @(v) [0;v;2];
```

We can also define the inverse operation of chopping off the boundary values from a full vector. Note that the indexing starts at 1 and ends at $m + 1$ for the extended vector.

```
chop = @(u) u(2:m);
```

All the pieces are now in place to define and solve the IVP.

```
    function f = ODE(t,v)
```

```
            u = extend(v);
            phi = Dxx*u;
            f = chop(phi);
        end
u0 = 1 + sin(pi/2*x) + 3*(1-x.^2).*exp(-4*x.^2);
t = linspace(0,0.15,5)';    % defines the output times
[t,V] = ode15s(@ODE,t,chop(u0));
V = V.';                    % rows for space, columns for time
```

We extend the solution to the boundaries at each time, then plot.

```
U = nan(m+1,length(t));
for j = 1:length(t)
    U(:,j) = extend(V(:,j));
end
plot(x,U)
```

[Heat equation plot showing curves for t = 0, t = 0.0375, t = 0.075, t = 0.1125, t = 0.15 over x from -1 to 1]

```
end
```

*There is little difference between solving a nonlinear PDE and a linear one from this perspective.* And because we have adopted a differentiation matrix approach, we are also free to use spectrally accurate methods, with little change to the code when the boundary conditions are of Dirichlet type.

### Example 11.5.4

We solve a diffusion equation with source term $u_t = u^2 + u_{xx}$ on $[-1, 1]$ subject to homogeneous Dirichlet conditions.

```
m = 100;    [x,Dx,Dxx] = diffcheb(m,[-1,1]);

extend = @(v) [0;v;0];      % extend to boundary
chop = @(u) u(2:m);         % discard boundary
function fI = ODE(t,v)
    u = extend(v);    uxx = Dxx*u;
    f = u.^2 + uxx;
    fI = chop(f);
```

## 11.5. Method of lines for parabolic PDEs

```
end
```

All the pieces are now in place to define and solve the IVP.

```
u0 = 6*(1-x.^2).*exp(-4*(x-.5).^2);
t = linspace(0,1.5,5)';    % defines the output times
[t,V] = ode15s(@ODE,t,chop(u0));
V = V.';                   % rows for space, columns for time
```

Extend the solution to the boundaries at each time, then plot.

```
U = nan(m+1,length(t));
for j = 1:length(t)
    U(:,j) = extend(V(:,j));
end
plot(x,U)
```

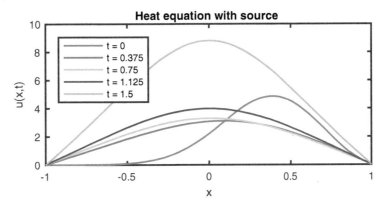

Finally, we return to the example of the Black–Scholes equation from Section 11.1. This time, we use the strategy outlined in Example 11.5.1 to handle the Neumann boundary condition at the right end.

### Example 11.5.5

```
Smax = 8;  T = 8;
K = 3;  sigma = 0.06;  r = 0.08;
m = 200;   [x,Dx,Dxx] = diffmat2(m,[0,Smax]);
h = x(2)-x(1);
```

Using the boundary conditions and defining the ODE follow next.

```
extend = @(v) [ 0; v; 2/3*(h-0.5*v(m-2)+2*v(m-1)) ];
chop = @(u) u(2:m);
function fI = ODE(t,v)
  u = extend(v);
  ux = Dx*u;   uxx = Dxx*u;
  f = sigma^2/2*x.^2.*uxx + r*x.*ux - r*u;
  fI = chop(f);
```

```
end
```

Now we define the initial conditions and solve the IVP.

```
v0 = max( 0, x-K );
t = linspace(0,T,5)';
[t,V] = ode15s(@ODE,t,chop(v0));
V = V';     % rows for space, columns for time
```

Extend the solution to the boundaries at each time, then plot.

```
U = nan(m+1,length(t));
for j = 1:size(V,2)
    U(:,j) = extend(V(:,j));
end
plot(x,U)
```

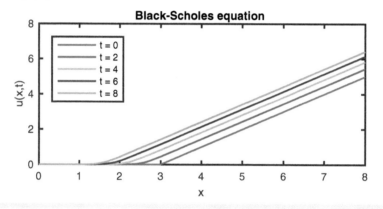

## Exercises

11.5.1. Consider Example 11.5.4 with the initial condition
$$u(x,0) = A(1-x^2)\exp(-4(x-0.5)^2).$$
The example uses $A = 6$.

(a) Solve numerically with $A = 4$ and provide graphical evidence that the solution decays to zero.

(b) For some $4 < A < 6$ the solution switches from decay to unbounded growth as $t \to \infty$. Find this transition value to at least two decimal places, and provide visual evidence supporting your answer.

11.5.2. The *Allen–Cahn equation* is used as a model for systems that prefer to be in one of two stable states. The governing PDE is
$$u_t = u(1-u^2) + \epsilon u_{xx}.$$
For this problem, assume $\epsilon = 10^{-3}$, $-1 \leq x \leq 1$, boundary conditions $u(\pm 1, t) = -1$, and initial condition
$$u(x,0) = -1 + \beta(1-x^2)e^{-20x^2},$$

where $\beta$ is a parameter. Use Chebyshev spectral discretization with $m = 200$.

(a) Simulate the problem with $\beta = 1.1$ up to time $t = 8$, plotting the solution at 6 equally spaced times. (You should see the solution decay down to the constant value $-1$.)

(b) Simulate again with $\beta = 1.6$. (This time the part of the bump will grow to just about reach $u = 1$ and stay there.)

11.5.3. ▣ The *Fisher equation* is $u_t = u_{xx} + u - u^2$. Assume that $0 \leq x \leq 5$, the boundary conditions are $u_x(0, t) = u(5, t) = 0$, and the initial condition is $u(x, 0) = \frac{1}{2}[1 + \cos(\pi x/2)]$. Use finite differences with $m = 300$ and plot the solution at times $t = 0, 0.5, \ldots, 3$. What is $u(0, 3)$?

11.5.4. ▣ Solve the heat equation for $0 \leq x \leq 1$ with initial condition $u(x, 0) = x(1-x)$ and subject to the boundary conditions $u(0, t) = 0$, $u(1, t) - u_x(1, t) = 1$. Plot the solution at $t = 1$ and report a value for $u(0.5, 1)$.

## Key ideas in this chapter

1. Solutions to the PDE problem must satisfy all of the conditions: the PDE itself, the boundary conditions, and the initial conditions (page 451).
2. The key concept in the method of lines is the separation of time and space discretizations (page 457).
3. The stability regions for time stepping methods have two important trends: (i) as the order of accuracy increases, the stability regions get smaller, and (ii) implicit methods have larger stability regions than their explicit counterparts (page 467).
4. Stiffness in a system of ODEs arises when the times scales of the solution are widely separated in size; such problems are usually best solved with implicit time stepping (page 471).
5. A distinct advantage of the method of lines is that there is little difference in the approach for solving linear or nonlinear problems (page 480).

## Where to learn more

There are many texts on PDEs; a fairly popular undergraduate-level text is by Haberman [28]. A more advanced treatment is by Ockendon et al. [50].

For a more traditional and analytical take on full discretization of the heat equation, one may consult Smith [59] or Morton and Mayers [47].

Several examples of using the method of lines with spectral method approximation maybe be found in Trefethen's text [68]. A classic text on using spectral methods on the PDEs from fluid mechanics is Canuto et al. [15]. The literature on computational fluid dynamics is vast, but one comprehensive monograph is by Roache [56].

For a first-hand account of the development of numerical methods for equations governing reservoir simulations (close to the heat equation), see the article by D. W. Peaceman (`history.siam.org/pdf/dpeaceman.pdf`, reprinted from [48]). Those were early days in computing!

# Chapter 12
# Advection equations

> Now, let's see if we can't figure out what you are, my little friend. And where you come from.
> — Obi-Wan Kenobi, *Star Wars: A New Hope*

Now that we have seen PDEs with both time and space variables, we have a new wrinkle to add. Some of these equations behave like those of the previous chapter, creating diffusive effects. Others, though, are about propagation or advection—generally, the behavior of waves.

Wave behavior is very different from diffusion. In idealized cases, waves travel with a finite speed and conserve energy, whereas diffusion smooths out features quickly and is associated with the dissipation of energy. There are many numerical methods that are specialized for purely advective problems, but we will consider only the method of lines approach from Chapter 11. Along the way we will see that wavelike behavior leads to some different conclusions about how to make these methods effective.

## 12.1 • Traffic flow

Have you ever been driving on a highway when you suddenly came upon a traffic jam? Maybe you had to brake harder than you would like to admit, endured a period of bumper-to-bumper progress, and experienced a much more gradual emergence from dense traffic than the abrupt entry into it. The mathematics of this phenomenon are well understood.

Consider a one-dimensional road extending in the $x$ direction. We represent the vehicles by a continuous density function $\rho(x)$. The flow rate or *flux* of vehicles, expressed as the number of cars per unit time crossing a fixed point on the road, is denoted by $q$. We assume that this flux depends on the local density of cars. It's reasonable to suppose that the flow will be zero when $\rho = 0$ (no cars), reach a maximum $q_m$ at some $\rho = \rho_m$, and approach zero again as the density approaches a critical density $\rho_c$. These conditions are met by the model

$$q = Q_0(\rho) = \frac{4q_m \rho_m \rho(\rho - \rho_c)(\rho_m - \rho_c)}{[\rho(\rho_c - 2\rho_m) + \rho_c \rho_m]^2}. \qquad (12.1.1)$$

Observations [76, Chapter 3] suggest that good values for a three-lane highway are $\rho_c = 1080$ vehicles per km, $\rho_m = 380$ vehicles per km, and $q_m = 4500$ vehicles per hour. In addition, we account for the fact that drivers anticipate slowing down or speeding up when they perceive changes in density, and therefore use $q = Q_0(\rho) - \epsilon \rho_x$ for a small $\epsilon > 0$.

## Conservation laws and hyperbolic equations

*Conservation laws* play a major role in science and engineering. They are typically statements that matter, energy, momentum, or some other meaningful quantity cannot be created or destroyed. In one dimension they take the form $\rho_t + q_x = 0$, where as above $\rho$ represents a density and $q$ represents a flux (flow rate). Using this in our traffic flow, we arrive at the evolutionary PDE

$$\rho_t + Q_0'(\rho)\rho_x = \epsilon \rho_{xx}. \qquad (12.1.2)$$

We recognize the first and last terms as indicative of diffusion, but the middle term has a different effect. (A similar term appeared in the Black–Scholes equation, but we did not discuss its interpretation.) Note that $Q_0'$ has the dimensions of [cars per time] over [cars per length], or length over time.

Let's momentarily consider the simplified version

$$u_t + c u_x = 0, \qquad (12.1.3)$$

where $c$ is constant. It's easy to produce a solution. Let $u(t,x) = \phi(x - ct)$, where $\phi$ is any differentiable function of one variable. Then the chain rule tells us that

$$u_t + c u_x = (-c)\phi'(x - ct) + c\phi'(x - ct) = 0,$$

and the PDE is satisfied. The form of the solution tells us that $u$ remains constant along any path with $x - ct = a$ for a constant $a$, or $x = a + ct$. So if $c > 0$, a fixed value of $u$ moves rightward with speed $|c|$, and if $c < 0$, it moves leftward with speed $|c|$. *We conclude that the solution to this linear advection equation propagates with constant speed and does not change shape.*

We call (12.1.3) the **advection equation**. It is the prototype of a **hyperbolic PDE**, just as the heat equation is canonical for parabolic PDEs. We can solve it by the method of lines as in Chapter 11. We need the first-derivative matrix for periodic end conditions:

$$D_x = \frac{1}{2h} \begin{bmatrix} 0 & 1 & & & -1 \\ -1 & 0 & 1 & & \\ & \ddots & \ddots & \ddots & \\ & & -1 & 0 & 1 \\ 1 & & & -1 & 0 \end{bmatrix}. \qquad (12.1.4)$$

This matrix is returned by Function 11.2.1.

### Example 12.1.1

We solve the advection equation on a domain with periodic end conditions. Our approach is the method of lines.

```
c = 2;
[x,Dx,Dxx] = diffper(300,[-4,4]);
f = @(t,u) -c*(Dx*u);
```

The following initial condition isn't mathematically periodic, but the deviation is less than machine precision.

```
u_init = 1 + exp(-3*x.^2);
t = linspace(0,3,41);
[t,U] = ode45(f,t,u_init);
```

A nice way to visualize solutions with moving features is by a **waterfall** plot.

```
waterfall(x,t,U)
view(-13,65)
```

The bump moves with speed 2 to the right, reentering on the left as it exits to the right because of the periodic conditions.

If you look carefully at Example 12.1.1, you'll notice that we used the time integrator ode45, a nonstiff method. As we will see later in this chapter, the pure advection equation is not stiff, unlike diffusion.

## Solutions for traffic flow

One interpretation of the left-hand side of (12.1.2), which is nonlinear in the unknown, is that constant traffic density $\rho$ moves with velocity $Q_0'(\rho)$. *This is different from the linear PDE (12.1.3): the velocity of a constant density value depends on the value of the density and is the key to the peculiar behavior of traffic jams.* In particular, the velocity is lower at high density, and higher at low density. The term on the right side of (12.1.2) provides a bit of diffusion, and the parameter $\epsilon$ determines the balance between the two effects—advection effects dominate if $\epsilon$ is small.

Exact solutions of (12.1.2) are much harder to come by than for the standard advection equation, but the method of lines is still effective. Because our model retains a little diffusion, we use the stiff solver `ode15s`.

### Example 12.1.2

We solve for traffic flow using periodic boundary conditions. The following are parameters and a function relevant to defining the problem.

```
rho_c = 1080;   rho_m = 380;   q_m = 10000;
Q0prime = @(rho) q_m*4*rho_c^2*(rho_c-rho_m)*rho_m ...
    *(rho_m-rho)./(rho*(rho_c-2*rho_m) + rho_c*rho_m).^3;
ep = 0.02;
```

Here we create a discretization on $m = 800$ points.

```
[x,Dx,Dxx] = diffper(800,[0,4]);
```

Next we define the ODE resulting from the method of lines.

```
odefun = @(t,rho) -Q0prime(rho).*(Dx*rho) + ep*(Dxx*rho);
```

Our first initial condition has moderate density with a small bump.

```
rho_init = 400 + 10*exp(-20*(x-3).^2);
t = linspace(0,1,6);
[t,RHO] = ode15s(odefun,t,rho_init);
plot(x,RHO)
```

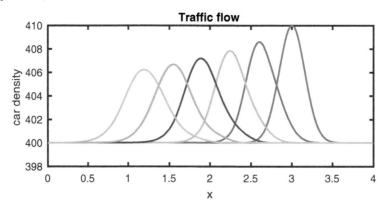

The bump slowly moves backward on the roadway, spreading out and gradually fading away due to the presence of diffusion.

Now we use an initial condition with a larger bump.

```
rho_init = 400 + 80*exp(-16*(x-3).^2);
t = linspace(0,0.5,6);
[t,RHO] = ode15s(odefun,t,rho_init);
plot(x,RHO)
```

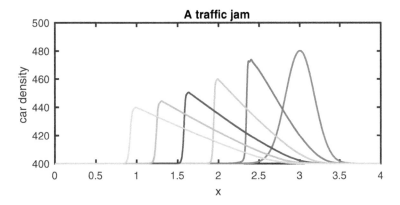

In this case the density bump travels backward along the road. It also steepens on the side facing the incoming traffic and decreases much more slowly on the other side. A motorist would experience this as an abrupt increase in density, followed by a much more gradual decrease in density and resulting gradual increase in speed.

The phenomenon in the second plot of Example 12.1.2 is called a *shock wave*. The underlying mathematics is not much different from the shock wave that comes off of the wing of a supersonic aircraft in the form of a sonic boom, or from cresting waves under certain conditions in the ocean. In the absence of diffusion ($\epsilon = 0$), the shock becomes a jump discontinuity in the solution, which breaks down both the finite differences and the original PDE, requiring different approaches. For many applications, the addition of a small amount of diffusion is appropriate and simple. However, we will first try to come to terms with pure advection in a linear problem.

## Exercises

12.1.1. By the analogy between (12.1.2) and (12.1.3), use (12.1.1) to confirm that constant traffic density moves backward (right to left) for $\rho_m < \rho < \rho_c$, as observed in Example 12.1.2. (Note that the derivative of $Q_0$ is given in the code for the example.)

12.1.2. Using as large a discretization and as small a dissipation parameter $\epsilon$ as you can get away with, perform experiments to estimate the speed of the shock wave in Example 12.1.2 in terms of the initial jump in the value of the solution at the shock.

12.1.3. The simplest model that includes both diffusion and nonlinear advection is the *viscous Burgers equation*, $u_t + u u_x = \epsilon u_{xx}$. Assume periodic end conditions on $-4 \leq x < 4$, let $\epsilon = 0.04$, and suppose $u(x,0) = e^{-2x^2}$. Solve the problem numerically with $m = 200$, plotting the solution on one graph at times $t = 0, 0.5, 1, \ldots, 3$. (You should see that the bump decays but also steepens as it moves to the right.)

12.1.4. The *Schrödinger equation* governs the wave function of quantum mechanics. In a special case we can write it as $i u_t = -u_{xx} + u$, where $|u(x,t)|^2$ is the probability density for a particle and $i^2 = -1$. This equation is primarily advective. Using periodic end conditions on $[-1,1]$ with $m = 200$, and letting $u(x,0) = \exp[-24x^2 + 20i\pi x]$, apply Function 6.7.2 (am2) with $n = 200$ steps to solve the Schrödinger equation. Make a waterfall plot of $|u|^2$.

12.1.5. ▥ The *Kuramoto–Sivashinsky equation*, $u_t + u u_x = -u_{xx} - \epsilon u_{xxxx}$, exhibits solutions that are considered chaotic. Assume periodic end conditions, $\epsilon = 0.006$, and initial condition $u(x,0) = 1 + e^{-2x^2}$. Using the method of lines, solve the problem numerically with $m = 200$ for $-4 \le x \le 4$, making a waterfall plot of the solution over $0 \le t \le 1$. (You may discretize the fourth derivative as two second derivatives.)

## 12.2 • Upwinding and stability

Let's focus on the constant-velocity linear advection equation,

$$u_t + c u_x = 0, \qquad (12.2.1)$$

with an initial condition $u(0,x) = u_0(x)$. For now, we suppose there are no boundaries. Keep in mind that $c$ is a velocity, not a speed: if $c > 0$, solutions travel rightward, and if $c < 0$, they travel leftward.

In Section 12.1 we argued that $u(x,t) = \phi(x - ct)$ is a solution of (12.2.1). By setting $t = 0$, we get $\phi(x) = u_0(x)$. As a result, we see that $u(x,t)$ depends on the initial condition at the single point $x - ct$. We call this point the **domain of dependence** at $(x,t)$. The direction in which the domain of dependence lies relative to $(x,t)$ is called the **upwind** direction. The upwind direction at $x$ is to the left or to the right of $x$ if $c$ is positive or negative, respectively.

Any numerical method we choose is going to have an analogous property called the **numerical domain of dependence**. Say we discretize $u_x$ by a centered difference,

$$u_x(x_i, t_j) \approx \frac{U_{i+1,j} - U_{i-1,j}}{2h}. \qquad (12.2.2)$$

Suppose also that we use the Euler time discretization method, so that

$$\mathbf{u}_{j+1} = (\mathbf{I} - c\tau \mathbf{D}_x)\mathbf{u}_j, \qquad (12.2.3)$$

where $\tau$ is the time step. Because the matrix in this time step is tridiagonal, the entry $U_{i,j}$ can depend directly only on $U_{i-1,j}$, $U_{i,j}$, and $U_{i+1,j}$. Going back another time step, the dependence extends to space positions $i - 2$ and $i + 2$, and so on. When we reach the initial time, the dependence of $U_{i,j}$ reaches from $x_{i-j}$ to $x_{i+j}$, or between $x_i - jh$ and $x_i + jh$. If we ignore boundaries, we have the situation illustrated in Figure 12.1. As $\tau, h \to 0$, the numerical domain of dependence fills in the shaded region in the figure.

### The CFL condition

We now state an important principle about a required relationship between the domains of dependence. It is a theorem, but we won't be making the precise definitions needed to make it rigorous.

## 12.2. Upwinding and stability

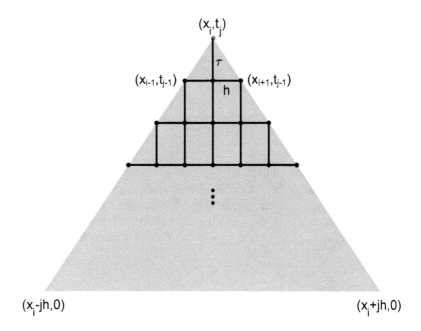

**Figure 12.1.** *Numerical domain of dependence for an explicit time stepping scheme. If $\tau$ and $h$ are infinitesimally small, the shaded region is filled in, and the exact domain of dependence must lie within it for convergence.*

---

**Theorem 12.2.1: Courant–Friedrichs–Lewy condition**

In order for a numerical method for an advection equation to converge to the correct solution, the limiting numerical domain of dependence must contain the exact domain of dependence.

---

To justify the statement of the theorem, consider that if the **CFL condition** of the theorem does not hold, the exact solution at $(x,t)$ could be affected by a change in the initial data while having no effect on the numerical solution. Hence there is no way for the method to get the solution correct for all problems. By contradiction, then, the CFL criterion is necessary for convergence.

Let's return to the explicit discretization method (12.2.3). In order for the numerical domain of dependence depicted in Figure 12.1 to contain the exact domain of dependence $\{x_i - ct_j\}$, it is necessary that $x_i - jh \leq x_i - ct_j \leq x_i + jh$, or $|cj\tau| \leq jh$. That is,

$$\frac{h}{\tau} \geq |c| \quad \text{as } \tau, h \to 0, \tag{12.2.4}$$

which is the implication of the CFL condition for this discretization. Notice that $h/\tau$ is the speed at which information moves in the numerical method; thus, a common restatement of the CFL condition is that *the maximum propagation speed in the numerical method must be at least the maximum speed in the original PDE problem.*

We can rearrange the CFL criterion to imply a time step restriction $\tau \leq h/|c|$. This condition must hold in order for the method to have a chance to converge, but it is only necessary, not sufficient. Compared to the $\tau = O(h^2)$ we derived for Euler on the heat equation in Section 11.4, the restriction for advection is much less severe. This is our first indication that advection is less stiff than diffusion.

Now consider what happens if we replace Euler by backward Euler, so that

$$(I + c\tau D_x)u_{j+1} = u_j,$$
$$u_{j+1} = (I + c\tau D_x)^{-1}u_j.$$

The inverse of a tridiagonal matrix is not necessarily tridiagonal, and in fact $U_{i,j+1}$ depends on *all* of the data at time level $j$. Thus the numerical domain of dependence includes the entire real line, and the CFL condition is always satisfied. Similar conclusions hold in general for explicit and implicit time discretizations: *An explicit method has to obey* $\tau = O(h)$, *while an implicit method is unrestricted by the CFL condition.*

### Example 12.2.1

We set up a test problem with velocity $c = 2$ and periodic end conditions.

```
[x,Dx] = diffper(200,[0 1]);
c = 2;
uinit = exp(-80*(x-0.5).^2);
```

For this problem we use ode113, which uses a combination of explicit multistep methods.

```
odefun = @(t,u) -c*(Dx*u);
[t,U] = ode113(odefun,[0,2],uinit);
pcolor(x,t(1:10:end),U(1:10:end,:))
```

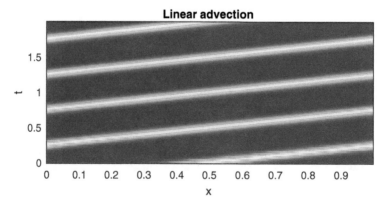

You can see the hump traveling rightward at constant speed, traversing the domain once for each integer multiple of $t = 1/2$. We note the average time step that was chosen:

```
avgtau1 = mean(diff(t))
```

```
avgtau1 =
    0.0015
```

## 12.2. Upwinding and stability

We cut $h$ by a factor of two and solve again.

```
[x,Dx] = diffper(400,[0 1]);
c = 2;
uinit = exp(-80*(x-0.5).^2);
odefun = @(t,u) -c*(Dx*u);
[t,U] = ode113(odefun,[0,2],uinit);
```

The CFL condition suggests that the time step should be cut by a factor of two also.

```
avgtau2 = mean(diff(t))
ratio = avgtau1 / avgtau2

avgtau2 =
    7.3801e-04
ratio =
    2.0975
```

### Upwinding and inflow

There are other ways to discretize the $u_x$ term. Specifically, we have the backward and forward differences,

$$u_x(x_i, t_j) \approx \frac{U_{i,j} - U_{i-1,j}}{h} \quad \text{(backward)}, \tag{12.2.5}$$

$$u_x(x_i, t_j) \approx \frac{U_{i+1,j} - U_{i,j}}{h} \quad \text{(forward)}. \tag{12.2.6}$$

Suppose an explicit time stepping method is used. In the backward difference case the solution $U_{i,j}$ depends only on points to the left of $x_i$, while in the forward difference case it depends only on points to the right. If $c > 0$, then the forward difference case can't possibly meet the CFL condition, because the exact domain of dependence is to the left of $x_i$; similarly, the backward difference case must fail if $c < 0$. The way to summarize these conclusions is that the discretization must choose at least one value from the upwind side, a technique called **upwinding**.

A parallel conclusion holds concerning boundaries. There is only a first-order derivative in $x$, so we should have only one boundary condition. If we impose it at the downwind side of the domain, there is no way for that boundary information to propagate into the interior of the domain as time advances. On the other hand, for points close to the upwind boundary, the natural domain of dependence eventually moves to the left of the boundary. This is impossible, so instead the domain of dependence has to stay at that boundary. In summary, we require an **inflow** condition on the PDE. This is true of the exact problem, not the discretization per se.

### Example 12.2.2

We set up advection over $[0,1]$ with velocity $c = -1$. This puts the right-side boundary in the upwind direction.

```
c = -1;   n = 100;
[x,Dx] = diffmat2(n,[0,1]);
uinit = exp(-100*(x-0.5).^2);
```

First we try imposing $u = 0$ at the right boundary, by appending that value to the end of the vector before multiplying by the differentiation matrix.

```
chop = @(u) u(1:n);    extend = @(v) [v;0];
odefun = @(t,v) -c*chop(Dx*extend(v));
[t,V] = ode113(odefun,[0,1],uinit(1:n));
pcolor(x(1:n),t,V)
```

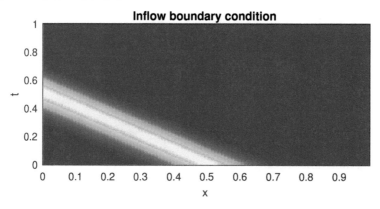

The data from the initial condition propagates out of the left edge. Because only zero is coming in from the upwind direction, the solution remains zero thereafter.

Now we try $u = 0$ imposed at the left boundary.

```
chop = @(u) u(2:n+1);    extend = @(v) [0;v];
odefun = @(t,v) -c*chop(Dx*extend(v));
[t,V] = ode113(odefun,[0,1],uinit(2:n+1));
pcolor(x(2:n+1),t,V)
```

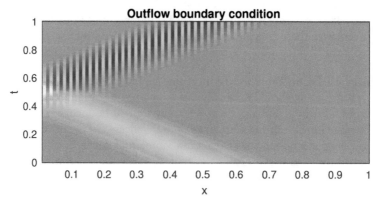

Everything seems OK until the data begins to interact with the inappropriate boundary condition. The resulting "reflection" is entirely wrong for advection from right to left.

## Exercises

12.2.1. ✍ Suppose you want to model the weather including wind speed up to 200 km/hr. If the shortest time step you can take is 4 hr, what is the CFL limit on the spatial resolution of the model? Is this a lower bound or an upper bound?

12.2.2. ✍ Suppose you want to model the traffic on a high-speed freeway. Derive a CFL condition on the allowable time step, stating your assumptions carefully.

12.2.3. ✍ For the heat equation, the domain of dependence at any $(x,t)$ with $t > 0$ is all of $x \in (-\infty, \infty)$. Show that the CFL condition implies that $\tau/h \to 0$ is required for convergence as $h \to 0$.

12.2.4. ✍ Suppose you wish to solve $u_t = u u_x$ for $x \in [-1, 1]$ and $u(x,0) = -2 + \sin(\pi x)$. Which is the inflow side of the domain, $x = -1$ or $x = 1$? Explain your answer.

## 12.3 • Absolute stability for advection

The CFL criterion gives a necessary condition for convergence. It suggests, but cannot confirm, that a step size of $O(h)$ may be adequate in the advection equation. More details emerge when we adopt the semidiscretization point of view.

Let the advection equation (12.2.1) over $[0, 1]$ be subjected to periodic end conditions for now. Suppose we use the central-difference matrix $\boldsymbol{D}_x$ defined in (12.1.4) to discretize the space derivative, leaving us with

$$\boldsymbol{u}' = -c \boldsymbol{D}_x \boldsymbol{u}.$$

To apply an IVP solver, we need to compare the stability region of the solver with the eigenvalues of $-c\boldsymbol{D}_x$, as in Section 11.3. You can verify (see Exercise 12.3.1) that for $m$ points in $[0, 1)$ these are

$$\lambda_k = -icm \sin\left(\frac{2\pi k}{m}\right), \qquad k = 0, \ldots, m-1. \tag{12.3.1}$$

Two things stand out about these eigenvalues: they are pure imaginary, and they extend no farther than $O(m) = O(h^{-1})$ away from the origin. Considering that the semidiscrete equations can be diagonalized to the uncoupled scalar equations $y'_j = \lambda_j y_j$, and $|e^{t\lambda}| \equiv 1$ for imaginary $\lambda$, we have found another expression of the conservative nature of the advection equation.

Both the shape and the size of the stability region of a time stepping method are relevant to conclusions about absolute stability.

### Example 12.3.1

For $c = 1$ we get the following eigenvalues:

```
[x,Dx] = diffper(40,[0,1]);
lambda = eig(Dx);
```

```
plot(real(lambda),imag(lambda),'.')
```

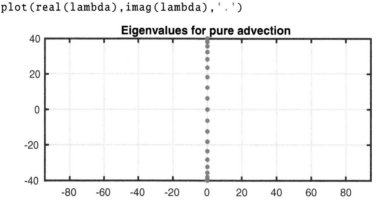

Let's choose a time step of $\tau = 0.1$ and compare to the stability regions of the Euler and backward Euler time steppers.

```
zc = exp(1i*linspace(0,2*pi,361)');    % points on |z|=1
subplot(1,2,1)
z = zc - 1;                             % shift left by 1
fill(real(z),imag(z),[.8 .8 1])
hold on, plot(real(0.1*lambda),imag(0.1*lambda),'.')
subplot(1,2,2)
fill([-6 6 6 -6],[-6 -6 6 6],[.8 .8 1])
z = zc + 1;                             % shift right by 1
hold on, plot(real(0.1*lambda),imag(0.1*lambda),'.')
fill(real(z),imag(z),'w')
```

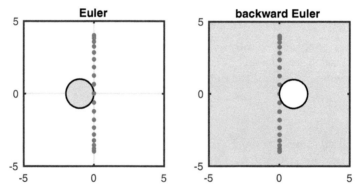

In the Euler case it's clear that *no* real value of $\tau > 0$ is going to make all (or even any) of the $\tau\lambda_j$ fit within the stability region. Hence Euler will never produce bounded solutions to this discretization of the advection equation. The A-stable backward Euler time stepping tells the exact opposite story; it will be absolutely stable regardless of $\tau$.

Many PDEs that conserve energy will have imaginary eigenvalues, causing Euler and some other IVP methods to fail regardless of step size. Diffusion problems, in which the eigenvalues are negative and real, are compatible with a wider range of integrators, though possibly with onerous stiff step size requirements.

The location of eigenvalues near $\pm ic/h$ also confirms what the CFL condition was suggesting. In order to use RK4, for example, whose stability region intersects the

## 12.3. Absolute stability for advection

imaginary axis at around $\pm 2.8i$, the time step stability restriction is $\tau c/h \leq 2.8$, or $\tau = O(h)$. This is much more favorable than for diffusion, whose eigenvalues were as large as $O(h^{-2})$, and it makes explicit IVP methods much more attractive for advection problems than for diffusion.

### Advection–diffusion equations

The traffic flow equation (12.1.2) combines a nonlinear advection with a diffusion term. The simplest linear problem with the same feature is $u_t + c u_x = \epsilon u_{xx}$, which is known as the **advection–diffusion equation**. The parameter $\epsilon$ controls the relative strength between the two behaviors, and the eigenvalues accordingly vary between the purely imaginary ones of advection and the negative real ones of diffusion.

**Example 12.3.2**

The eigenvalues of advection–diffusion are near-imaginary for $\epsilon \approx 0$ and more negative-real for increasing values of $\epsilon$.

```
[x,Dx,Dxx] = diffper(40,[0,1]);
tau = 0.1;
for ep = [0.001 0.01 0.05]
  lambda = eig(-Dx + ep*Dxx);
  plot(real(tau*lambda),imag(tau*lambda),'.'), hold on
end
```

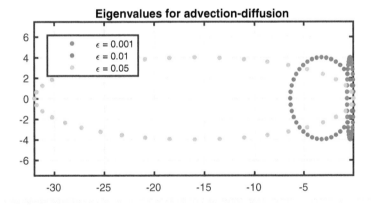

In a nonlinear problem, of course, the eigenvalues come from the linearization about an exact solution (Section 11.4).

### Boundary effects

Boundary conditions can have a dramatic effect on the eigenvalues of the semidiscretization. For instance, a homogeneous inflow condition is unable to provide additional energy to the solution as it passes out of the domain for good, and thus imaginary eigenvalues are inappropriate.

Suppose the linear advection equation $u_t + u_x = 0$ is solved on $[0, 1]$ with the homogeneous inflow condition $u(0, t) = 0$. As in Section 11.5, we need to derive an ODE for just the interior points, $v' = f(t, v)$, where $v$ has $m$ elements and excludes $x = 0$. Let $E$ be the $(m+1) \times (m+1)$ identity with the first row deleted. Then $v = Eu$, where $u$ includes the point $x = 0$, and $v' = (Eu)' = E(u')$. Furthermore,

$$u' = -D_x u = -D_x \begin{bmatrix} 0 \\ v \end{bmatrix} = -D_x(E^T v) = (-D_x E^T)v.$$

Putting this together, we derive

$$v' = E(u') = -(E D_x E^T)v.$$

As a result we conclude that $A = -(E D_x E^T)$ is the proper matrix for determining the eigenvalues of the semidiscretization. In other words, we simply delete the first row and first column from $-D_x$. This is the approach we take in the next example.

### Example 12.3.3

Deleting the first row and column places all the eigenvalues of the discretization into the left half of the complex plane.

```
[x,Dx,Dxx] = diffcheb(40,[0,1]);
A = -Dx(2:end,2:end);    % leave out first row and column
lambda = eig(A);
plot(real(lambda),imag(lambda),'.')
```

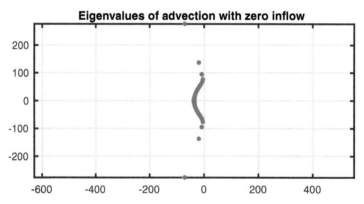

Note that the rightmost eigenvalues have real part at most

```
max( real(lambda) )
```

```
ans =
   -4.9320
```

Consequently all solutions decay exponentially to zero as $t \to \infty$. This matches the intuition of a flow with nothing at the inlet: eventually everything flows out of the domain.

## Exercises

12.3.1. ✍ Let $D_x$ be the $m \times m$ matrix (12.2.2), made for $0 \le x \le 1$. Show that for each $k = 0, \ldots, m-1$, the pair

$$\lambda = im \sin \frac{2k\pi}{m}, \qquad v = \begin{bmatrix} 1 \\ e^{2ki\pi/m} \\ e^{4ki\pi/m} \\ \vdots \\ e^{2ki(m-1)\pi/m} \end{bmatrix}$$

are an eigenvalue and eigenvector of $D_x$.

12.3.2. ⌨ (Adapted from [68].) Consider the problem $u_t + c(x)u_x = 0$ for $0 \le x \le 2\pi$, with periodic boundary conditions and speed $c(x) = 0.2 + \sin^2(x-1)$. This problem has a solution that is $T$-periodic in time for some $T \approx 13$.

(a) Find this value $T$ accurately to 10 digits by converting the integral $T = \int_0^T dt$ to an integral in $x$ through $dx/dt = c(x)$. (You should use numerical integration.)

(b) Using $u(x, 0) = e^{\sin x}$, solve the PDE numerically for $0 \le t \le T$ using the answer from part (a). Make a surface plot of the solution as a function of $x$ and $t$. (To keep the size of the graphic reasonable, have the solution evaluated at fewer than 100 points in the time interval.)

(c) By adjusting the accuracy in time and space, verify that

$$\|u(x, T) - u(x, 0)\|_\infty < 0.01.$$

12.3.3. ⌨ Refer to the semidiscretization in Example 12.3.3. Find an upper bound on $\tau$ that gives absolute stability for the Euler integration method.

12.3.4. ⌨ Modify Example 12.3.3 so that it produces the eigenvalues of the problem $u_t + u_x = 0$ with an outflow condition $u(1, t) = 0$. What is the behavior of solutions as $t \to \infty$?

12.3.5. It is possible to linearize a PDE about an exact solution $\hat{u}$.

(a) ✍ By substituting $u = \hat{u} + v$ into the *inviscid Burgers equation* $u_t + uu_x = 0$ and neglecting all terms quadratic and higher order in $v$, derive the perturbation evolution equation $v_t = -\hat{u}_x v - \hat{u} v_x$.

(b) ⌨ Using periodic end conditions and $m = 50$ points, plot the eigenvalues of the discretization of the linearized problem from part (a) when $\hat{u} \equiv 1$. (These are the same as the eigenvalues of the linearization of the semidiscretization.)

## 12.4 • The wave equation

Closely related to the advection equation is the **wave equation**,

$$u_{tt} - c^2 u_{xx} = 0, \qquad 0 \le x \le 1, \quad t > 0. \tag{12.4.1}$$

This is our first PDE having a second derivative in time. As in the advection equation, $u(x,t) = \phi(x-ct)$ is a solution of (12.4.1), but now so is $u(x,t) = \phi(x+ct)$ for any twice-differentiable $\phi$. Thus, *the wave equation supports advection in both directions simultaneously.* Because $u$ has two derivatives in $t$ and in $x$, we need two boundary conditions and two initial conditions. For example,

$$u(0,t) = u(1,t) = 0, \qquad t \geq 0, \qquad (12.4.2\text{a})$$
$$u(x,0) = f(x), \qquad 0 \leq x \leq 1, \qquad (12.4.2\text{b})$$
$$u_t(x,0) = g(x), \qquad 0 \leq x \leq 1. \qquad (12.4.2\text{c})$$

One approach is to discretize both the $u_{tt}$ and $u_{xx}$ terms using finite differences:

$$\frac{1}{\tau^2}(U_{i,j+1} - 2U_{i,j} + U_{i,j-1}) = \frac{c^2}{h^2}(U_{i+1,j} - 2U_{i,j} + U_{i-1,j}).$$

This equation can be rearranged to solve for $U_{i,j+1}$ in terms of values at time levels $j$ and $j-1$. Rather than pursue this method, however, we will turn to the method of lines.

## First-order system

In order to be compatible with the standard IVP solvers that we have encountered, we must recast (12.4.1) as a first-order system in time. Using our typical methodology, we would define $y = u_t$ and derive

$$\begin{aligned} u_t &= y, \\ y_t &= c^2 u_{xx}. \end{aligned} \qquad (12.4.3)$$

There is, however, another, less obvious option for reducing to a first-order system:

$$\begin{aligned} u_t &= z_x, \\ z_t &= c^2 u_x. \end{aligned} \qquad (12.4.4)$$

This second form is appealing because it's equivalent to Maxwell's equations for electromagnetism in one dimension. In the Maxwell form we typically replace (12.4.2c) with

$$z(x,0) = g(x),$$

which may be physically more relevant in many instances.

Because waves travel in both directions and there is no single upwind direction, a centered finite difference in space is most appropriate. Before application of the boundary conditions, semidiscretization of (12.4.4) leads to

$$\begin{bmatrix} u'(t) \\ z'(t) \end{bmatrix} = \begin{bmatrix} 0 & D_x \\ c^2 D_x & 0 \end{bmatrix} \begin{bmatrix} u(t) \\ z(t) \end{bmatrix}. \qquad (12.4.5)$$

The boundary conditions (12.4.2a) suggest that we should remove both of the end values of $u$ from the discretization, but retain all of the $z$ values. We use $w$ to denote the vector of all the unknowns to be solved by the method of lines. When computing $dw/dt$, we extract the $u$ and $z$ components, and we use dedicated functions for padding the $u$ vector with the zero end values or chopping off the zeros as necessary.

## 12.4. The wave equation

**Example 12.4.1**

We solve the wave equation (in Maxwell form) for speed $c = 2$, with homogeneous Dirichlet conditions on the first variable.

```
c = 2;  m = 200;
[x,Dx] = diffmat2(m,[-1,1]);
```

The boundary values of $u$ are given to be zero, so they are not unknowns in the ODEs we solve. Instead they are added or removed as necessary.

```
chop = @(u) u(2:m);
extend = @(v) [0;v;0];
```

The following function computes the time derivative of the system at interior points. (This entire example is written as a function, not a script, to make this definition possible.)

```
    function dwdt = odefun(t,w)
        u = extend(w(1:m-1));
        z = w(m:2*m);
        dudt = Dx*z;
        dzdt = c.^2 .* (Dx*u);
        dwdt = [ chop(dudt); dzdt ];
    end
```

Our initial condition is a single hump for $u$.

```
u_init = exp(-100*(x).^2);
z_init = -u_init;
w_init = [ chop(u_init); z_init ];
```

Because the wave equation is hyperbolic, we can use the nonstiff IVP solver ode45.

```
t = linspace(0,2,101);
[t,W] = ode45(@odefun,t,w_init);
W = W.';                % rows for x, columns for t
```

We extract the original $u$ and $z$ variables from the results, adding in the zeros at the boundaries for $u$.

```
n = length(t)-1;
U = [ zeros(1,n+1); W(1:m-1,:); zeros(1,n+1) ];
Z = W(m:2*m,:);
```

We plot the results for the original $u$ variable.

```
subplot(1,2,1)
waterfall(x,t,U.')
subplot(1,2,2)
pcolor(x,t,U.')
```

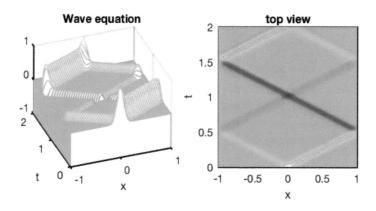

The original hump breaks into two pieces of different amplitudes. Each travels with speed $c = 2$, and they pass through one another without interference. When a hump encounters a boundary, it is perfectly reflected, but with inverted shape. At time $t = 2$ the exact solution looks just like the initial condition.

## Reflections

One interesting situation is when the wave speed $c$ changes discontinuously, as when light passes from one material into another. For this we must replace the term $c^2$ in (12.4.5) with the matrix $\text{diag}\big(c^2(x_0), \ldots, c^2(x_n)\big)$.

**Example 12.4.2**

We now use a wave speed that is discontinuous at $x = 0$.

```
m = 120;
[x,Dx] = diffcheb(m,[-1,1]);
c = 1 + (sign(x)+1)/2;
chop = @(u) u(2:m);
extend = @(v) [0;v;0];
```

This function computes the time derivative of the method of lines system.

```
function dwdt = odefun(t,w)
    u = extend(w(1:m-1));
    z = w(m:2*m);
    dudt = Dx*z;
    dzdt = c.^2 .* (Dx*u);
    dwdt = [ chop(dudt); dzdt ];
end
```

We set the initial conditions and solve using ode45.

```
u_init = exp(-100*(x+0.5).^2);
```

```
z_init = -u_init;
w_init = [ chop(u_init); z_init ];
t = linspace(0,2,101);
[t,W] = ode45(@odefun,t,w_init);
W = W.';                          % rows for x, columns
    for t
```

Now we extract the original $u$ and $z$ variables from the results, adding in the zeros at the boundaries for $u$.

```
n = length(t)-1;
U = [ zeros(1,n+1); W(1:m-1,:); zeros(1,n+1) ];
Z = W(m:2*m,:);
```

Finally, we plot the results.

```
pcolor(x,t,U.')
```

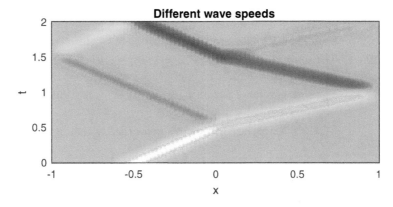

Each pass through the interface at $x = 0$ generates a reflected and transmitted wave. By conservation of energy, these are both smaller in amplitude than the incoming bump.

## Exercises

12.4.1. ✍ Consider the Maxwell equations (12.4.4) with smooth solution $u(x,t)$ and $z(x,t)$.

  (a) Show that $u_{tt} = c^2 u_{xx}$.

  (b) Show that $z_{tt} = c^2 z_{xx}$.

12.4.2. ✍ Suppose that $\phi(s)$ is any twice-differentiable function.

  (a) Show that $u(x,t) = \phi(x-ct)$ is a solution of $u_{tt} = c^2 u_{xx}$. (As in the advection equation, this is a traveling wave of velocity $c$.)

  (b) Show that $u(x,t) = \phi(x+ct)$ is another solution of $u_{tt} = c^2 u_{xx}$. (This is a traveling wave of velocity $-c$.)

12.4.3. ✍ Show that the following is a solution to the wave equation $u_{tt} = c^2 u_{xx}$ and its initial and boundary conditions (12.4.2a)–(12.4.2c):

$$u(x,t) = \frac{1}{2}[f(x-ct)+f(x+ct)] + \frac{1}{2c}\int_{x-ct}^{x+ct} g(\xi)\,d\xi.$$

This is *D'Almebert's solution* to the wave equation.

12.4.4. As mentioned in Example 12.4.1, the solution $u(x,2)$ there should be identical to $u(x,0)$.

  (a) Repeat the computation and plot $u(x,2)-u(x,0)$.

  (b) Repeat again with $m = 400$ and make the same plot. Is the result consistent with second-order accuracy?

  (c) Make the plot a third time using Chebyshev differentiation with $m = 100$, and comment on the accuracy compared to the finite difference versions.

12.4.5. Consider the wave equation with zero Neumann conditions on $u$ at each boundary. Using the Maxwell formulation (12.4.4), we have $z_t = c^2 u_x$, so $z$ is constant in time at each boundary. Use this fact with the solution parameters from Example 12.4.1 to solve the problem by the method of lines.

12.4.6. The equations $u_t = z_x - \sigma u$, $z_t = c^2 u_{xx}$ model electromagnetism in an imperfect conductor. Repeat Example 12.4.2 with $\sigma(x) = 1 + \text{sign}(x)$.

12.4.7. The *sine–Gordon equation* $u_{tt} - u_{xx} = \sin u$ has interesting nonlinear solutions.

  (a) Write the equation as a first-order system in the variables $u$ and $v = u_t$.

  (b) Assume periodic end conditions on $[-10, 10]$ and discretize at $m = 200$ points. Let $u(x,0) = \pi e^{-x^2}$ and $u_t(x,0) = 0$. Solve the system using `ode45` for 100 times between $t = 0$ and $t = 20$, and make a waterfall plot of the solution.

12.4.8. The deflections of a stiff beam, such as a ruler, are governed by the second-order PDE $u_{tt} = -u_{xxxx}$.

  (a) Show that the PDE is equivalent to the first-order system

  $$u_t = v_{xx},$$
  $$v_t = -u_{xx}.$$

  (b) Assuming periodic end conditions on $[-1,1]$, let $u(x,0) = \exp(-24x^2)$, $v(x,0) = 0$ and simulate the solution of the beam equation for $0 \le t \le 1$ using Function 6.7.2 with $n = 100$ time steps. Make a `pcolor` plot of the solution.

**Key ideas in this chapter**

1. The exact solutions to the advection equation can propagate in one direction with constant speed and do not change shape (page 486).
2. The solution to the nonlinear traffic example shows that for nonlinear equations, the speed of the solution can depend on the solution itself, and can lead to much richer solution behavior than for a linear problem (page 487).

3. A common restatement of the CFL condition is that the maximum propagation speed in the numerical method must be at least the maximum speed in the original PDE problem. As a consequence, an explicit method must have a time step of comparable size to the space step, while an implicit method is unrestricted by the CFL condition (page 491).
4. The wave equation, our first PDE with a second-order derivative in time, supports propagation in both directions in one space dimension. This led us to choose central differencing in space to achieve upwinding in both directions (page 500).

## Where to learn more

The numerical solution of advection and wave equations, particularly nonlinear conservation laws, is still an active area of research. A more detailed treatment on finite difference methods for scalar problems is in LeVeque [42]. An accessible treatment of nonlinear conservation laws can be found in the monograph by LeVeque [41]. Recent monographs aimed at hyperbolic PDEs, both scalar equations and systems, are from Trangenstein [67] and LeVeque [43].

A wide-ranging view of computational mathematics relating to conservation laws (in many cases, nonlinear versions of the advection equation) can be found in P. Lax's oral history (http://history.siam.org/oralhistories/lax.htm).

# Chapter 13
# Two-dimensional problems

> You have taken your first step into a larger world.
> — Obi-Wan Kenobi, *Star Wars: A New Hope*

We have graduated from ODEs, usually having either time or space as the independent variable, to PDEs in which both space and time are represented simultaneously. The final innovation in this book is to consider problems with two space dimensions. We will confine ourselves to the simplest possible class of two-dimensional regions, leaving aside the major issue of the geometry of the domain. Even so there are many important and valuable mathematical models within this constraint.

First we explore how to add another space dimension to the two classes of PDE encountered so far, the parabolic and hyperbolic equations. Last comes the third major class of PDE: *elliptic* equations. These problems omit time altogether and therefore are used to represent steady-state phenomena.

## 13.1 • Tensor-product discretizations

As you learned when starting double integration in vector calculus, the simplest extension of an interval to two dimensions is a rectangle. We will use a particular notation for rectangles:

$$[a,b] \times [c,d] = \{(x,y) \in \mathbb{R}^2 : a \leq x \leq b \text{ and } c \leq y \leq d\}. \tag{13.1.1}$$

The × in this notation is called a tensor product, and a rectangle is the fundamental example of a **tensor-product domain**. The implication of the tensor product is that each variable independently varies over a fixed set. The simplest three-dimensional tensor-product domain is the *cuboid* $[a,b] \times [c,d] \times [e,f]$. When the interval is the same in each dimension (that is, the region is a square or a cube), we may write $[a,b]^2$ or $[a,b]^3$. We will limit our discussion to two dimensions henceforth.

The discretization of a two-dimensional tensor-product domain is straightforward.

Given discretizations of the two intervals,

$$a \leq x_i \leq b, \quad i = 0, \ldots, m, \\ c \leq y_j \leq b, \quad j = 0, \ldots, n, \qquad (13.1.2)$$

then the natural discretization of $[a,b] \times [c,d]$ is the set

$$\{(x_i, y_j) : i = 0, \ldots, m \text{ and } j = 0, \ldots, n\}. \qquad (13.1.3)$$

It is conventional to require $x_0 < x_1 < \cdots < x_m$ and $y_0 < y_1 < \cdots < y_n$ and to regard the discrete points as a doubly indexed sequence rather than an unordered set. Such a discretization is often referred to as a **tensor-product grid**. *A two-dimensional tensor-product grid is precisely the tensor product of the individual x and y discretizations.*

## Functions on grids

The double indexing of a grid implies an irresistible connection to matrices. *Corresponding to any function $f(x,y)$ defined on the rectangle is an $(m+1) \times (n+1)$ matrix $F$ defined by collecting the values of $f$ at the points in the grid.* This correspondence of a function to a matrix is so important that we give it a formal name:

$$F = \text{mtx}(f) = \Big[ f(x_i, y_j) \Big]_{\substack{i=0,\ldots,m \\ j=0,\ldots,n}}. \qquad (13.1.4)$$

**Example 13.1.1**

Let the interval $[0, 2\pi]$ be divided into $m = 4$ equally sized pieces, and let $[1,3]$ be discretized in $n = 2$ equal pieces. Then the grid in the rectangle $[0, 2\pi] \times [1,3]$ is given by all points $(2i\pi/4, 1+j)$ for all choices $i = 0, 1, 2, 3, 4$ and $j = 0, 1, 2$. If $f(x,y) = \sin(xy)$, then

$$\text{mtx}(f) = \begin{bmatrix} \sin(0 \cdot 1) & \sin(0 \cdot 2) & \sin(0 \cdot 3) \\ \sin\left(\tfrac{1}{2}\pi \cdot 1\right) & \sin\left(\tfrac{1}{2}\pi \cdot 2\right) & \sin\left(\tfrac{1}{2}\pi \cdot 3\right) \\ \sin(\pi \cdot 1) & \sin(\pi \cdot 2) & \sin(\pi \cdot 3) \\ \sin\left(\tfrac{3}{2}\pi \cdot 1\right) & \sin\left(\tfrac{3}{2}\pi \cdot 2\right) & \sin\left(\tfrac{3}{2}\pi \cdot 3\right) \\ \sin(2\pi \cdot 1) & \sin(2\pi \cdot 2) & \sin(2\pi \cdot 3) \end{bmatrix} = \begin{bmatrix} 0 & 0 & 0 \\ 1 & 0 & -1 \\ 0 & 0 & 0 \\ -1 & 0 & 1 \\ 0 & 0 & 0 \end{bmatrix}.$$

To define the grid computationally, we create two **coordinate matrices** $X = \text{mtx}(x)$ and $Y = \text{mtx}(y)$ using the built-in **ndgrid** function.[44] Once the coordinate matrices are in hand, a given function $f(x,y)$ is evaluated at all of the grid points simultaneously using elementwise operations on these matrices. The function can then be visualized via a plotting command such as **surf**, **contourf**, or **pcolor**.

---

[44] Often one sees the older command meshgrid instead of ndgrid. It makes the $y$ variable, which is conventionally oriented vertically, correspond to the row dimension, and $x$ correspond to the column dimension. This can cause confusion because the standard ordering $(x,y)$ is reversed when indexing into the matrix. Most MATLAB functions, including the plotting functions, accept the grid defined either way, but at this writing there are still some exceptions, such as interp2d, that allow only meshgrid.

## 13.1. Tensor-product discretizations

### Example 13.1.2

Consider again the example $f(x,y) = \sin(xy)$ over $[0, 2\pi] \times [1, 3]$, with the dimensions discretized using $m = 4$ and $n = 2$ equal pieces, respectively.

```
m = 4;    x = (0:2*pi/m:2*pi)';
n = 2;    y = (1:2/n:3)';
```

We create a representation of the grid using two matrices created by the ndgrid function.

```
[X,Y] = ndgrid(x,y)
```

```
X =
         0          0          0
    1.5708     1.5708     1.5708
    3.1416     3.1416     3.1416
    4.7124     4.7124     4.7124
    6.2832     6.2832     6.2832
Y =
     1     2     3
     1     2     3
     1     2     3
     1     2     3
     1     2     3
```

As you see above, the entries of X vary in the first dimension (rows), while the entries of Y vary along the second dimension (columns).

We can also visualize this grid on the rectangle.

```
plot(X,Y,'bo','markersize',6)
set(gca,'xtick',x,'ytick',y), grid on
```

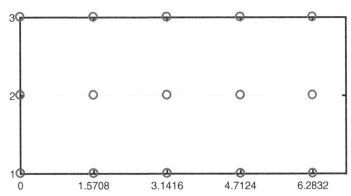

For a given definition of $f(x, y)$ we can find $\text{mtx}(f)$ by elementwise operations on the coordinate matrices X and Y.

```
f = @(x,y) sin(x.*y-y);
F = f(X,Y)
```

```
F =
    -0.8415   -0.9093   -0.1411
     0.5403    0.9093    0.9900
     0.8415   -0.9093    0.1411
    -0.5403    0.9093   -0.9900
    -0.8415   -0.9093   -0.1411
```

We can make nice plots of the function by first choosing a much finer grid.

```
m = 70;    x = (0:2*pi/m:2*pi)';
n = 50;    y = (1:2/n:3)';
[X,Y] = ndgrid(x,y);
F = f(X,Y);

subplot(1,2,1)
contourf(X,Y,F,10)
subplot(1,2,2)
surf(X,Y,F), shading flat
```

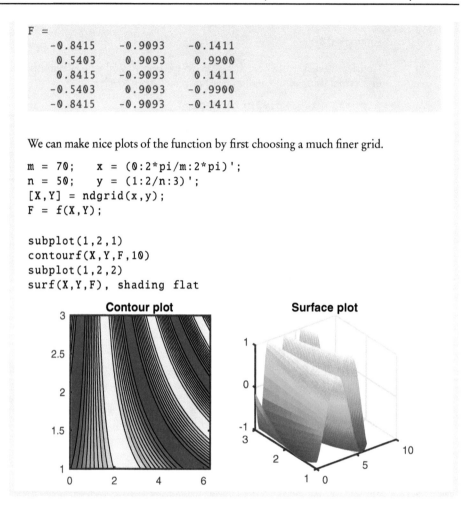

## Parameterized surfaces

We are not limited to rectangles by tensor products. Many regions and surfaces may be parameterized by means of $x(u,v)$, $y(u,v)$, and $z(u,v)$, where $u$ and $v$ lie in a rectangle. Such "logically rectangular" surfaces include the unit disk,

$$\begin{cases} x = u\cos v, \\ y = u\sin v, \\ z = 0, \end{cases} \quad \begin{aligned} 0 &\leq u < 1, \\ 0 &\leq v \leq 2\pi, \end{aligned} \tag{13.1.5}$$

and the unit sphere,

$$\begin{cases} x = \cos u \sin v, \\ y = \sin u \sin v, \\ z = \cos v, \end{cases} \quad \begin{aligned} 0 &\leq u < 2\pi, \\ 0 &\leq v \leq \pi. \end{aligned} \tag{13.1.6}$$

## 13.1. Tensor-product discretizations

**Example 13.1.3**

Construction of a function over the unit disk is straightforward. We define a grid in $(r,\theta)$ space and compute accordingly. For the function $f(r,\theta) = 1 - r^4$, for example,

```
r = linspace(0,1,41);
theta = linspace(0,2*pi,81);
[R,Theta] = ndgrid(r,theta);
F = 1-R.^4;
surf(R,Theta,F)
```

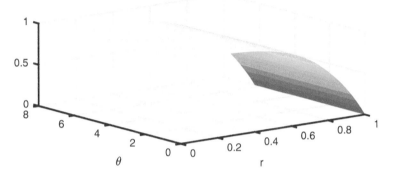

**A polar function**

Of course we are used to seeing such plots over the $(x,y)$ plane. This is easily accomplished.

```
X = R.*cos(Theta);   Y = R.*sin(Theta);
surf(X,Y,F)
```

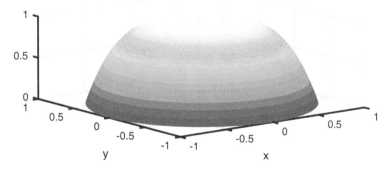

**Function over the unit disk**

In such functions it's up to us to ensure that the values along the line $r = 0$ are identical, and that the values on the line $\theta = 0$ are identical to those on $\theta = 2\pi$. Otherwise the interpretation of the domain as the unit disk is nonsensical.

On the unit sphere we can use color to indicate a function value. For the function $f(x,y,z) = xyz^3$, for instance,

```
theta = linspace(0,2*pi,61);
phi = linspace(0,pi,61);
[Theta,Phi] = ndgrid(theta,phi);
```

```
X = cos(Theta).*sin(Phi);  Y = sin(Theta).*sin(Phi);
Z = cos(Phi);
F = X.*Y.*Z.^3;
surf(X,Y,Z,F), colorbar
```

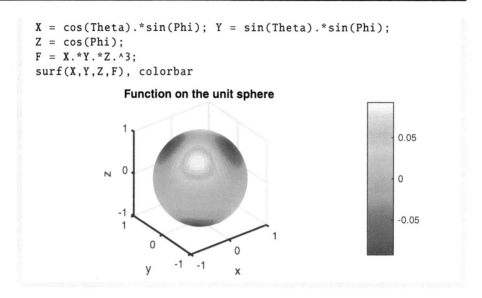

## Partial derivatives

In order to solve BVPs in one dimension by collocation, we replaced an unknown function $u(x)$ by a vector of its values at selected nodes and discretized the derivatives in the equation using differentiation matrices. In the two-dimensional case we convert a function $u(x,y)$ to its values on the grid (13.1.3) in the form of the matrix mtx($u$), as defined in (13.1.4). Now we will construct discrete analogs of the partial derivatives $\partial u/\partial x$ and $\partial u/\partial y$. *The partial derivatives are approximated through multiplication by differentiation matrices.*

Consider first $\partial u/\partial x$. Recall that in the definition of this partial derivative, the independent variable $y$ is held constant. Note that $y$ is constant within each column of $U = \text{mtx}(u)$. Thus, we may regard a single column $u_j$ as a (discretized) function of $x$. We know how to construct a (discrete) derivative of this column: left-multiply by a differentiation matrix $D_x$ such as (10.2.2). In the present context, we need to multiply by each column of $U$ by $D_x$ on the left independently. But this is exactly what is meant by $D_x U$! Symbolically we may say

$$\text{mtx}\left(\frac{\partial u}{\partial x}\right) \approx D_x \, \text{mtx}(u). \qquad (13.1.7)$$

Note that this relation is not an equality, because the left-hand side is a discretization of the exact partial derivative, while the right-hand side is a finite difference approximation. Yet it is a natural analog for what we mean by partial differentiation when we are given not $u(x,y)$ but only the value matrix $U$.

Now we tackle $\partial u/\partial y$. Here the inactive coordinate $x$ is held fixed within each *row* of $U$. Clearly we need a finite difference matrix $D_y$, but differencing of a row-oriented vector is not something we have encountered before. However, we can easily recover the familiar situation by transposing the row into a column, multiplying on the left by $D_y$, and finally transposing the result back into the row shape. Now observe that

## 13.1. Tensor-product discretizations

transposing the entire matrix $U$ turns *each* row into a column, multiplying on the left by $D_y$ applies finite differences to every column, and transposing that (matrix) result restores each column to a row. Stating this chain of operations as linear algebra leads to

$$\text{mtx}\left(\frac{\partial u}{\partial y}\right) \approx \left(D_y U^T\right)^T = \text{mtx}(u) D_y^T. \quad (13.1.8)$$

Keep in mind some hidden details within (13.1.7) and (13.1.8). The differentiation matrix $D_x$ is based on the discretization $x_0, \ldots, x_m$, and as such it must be $(m+1) \times (m+1)$. On the other hand, $D_y$ is based on $y_0, \ldots, y_n$ and is $(n+1) \times (n+1)$. This is exactly what is needed dimensionally to make the products consistent. More subtly, if the differentiation is based on equispaced grids in each variable, the value of $h$ in a formula such as (5.4.6) will be different for $D_x$ and $D_y$.

**Example 13.1.4**

We define a function and, for reference, its two exact partial derivatives.

```
u    = @(x,y) sin(x.*y-y);
dudx = @(x,y) y.*cos(x.*y-y);
dudy = @(x,y) (x-1).*cos(x.*y-y);
```

We use an equispaced grid and second-order finite differences as implemented by diffmat2.

```
m = 80;   [x,Dx] = diffmat2(m,[0,2*pi]);
n = 60;   [y,Dy] = diffmat2(n,[1,3]);
[X,Y] = ndgrid(x,y);
U = u(X,Y);
```

Here we compare the exact $\partial u / \partial x$ with the finite difference approximation.

```
subplot(1,2,1)
contourf(X,Y,dudx(X,Y),10)
subplot(1,2,2)
contourf(X,Y,Dx*U,10)
```

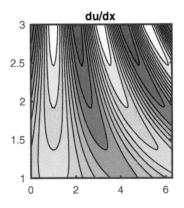
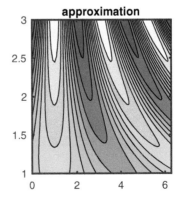

We now do the same for $\partial u/\partial y$,

```
subplot(1,2,1)
contourf(X,Y,dudy(X,Y),10)
subplot(1,2,2)
contourf(X,Y,U*Dy',10)
```

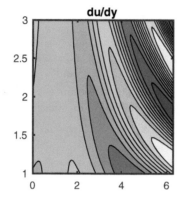
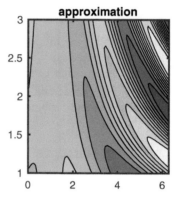

To the eye there is little difference to be seen, though we expect that the results probably have no more than 2–4 correct digits for these discretization sizes.

## Exercises

13.1.1. In each part, make side-by-side surface and contour plots of the given function over the given domain. Label the axes and give a title.

(a) $f(x,y) = 2y + e^{x-y}$, $[0,2] \times [-1,1]$.

(b) $f(x,y) = \tanh[5(x + xy - y^3)]$, $[-2,2] \times [-1,1]$.

(c) $f(x,y) = \exp[-6(x^2 + y^2 - 1)^2]$, $[-2,2] \times [-2,2]$.

13.1.2. For each function in Exercise 13.1.1, make side-by-side surface plots of $f_x$ and $f_y$ using Chebyshev spectral differentiation.

13.1.3. For each function in Exercise 13.1.1, make a contour plot of the mixed derivative $f_{xy}$ using Chebyshev spectral differentiation.

13.1.4. In each part, make a plot of the function (given in polar or cartesian coordinates) over the unit disk.

(a) $f(r,\theta) = r^2 - 2r\cos\theta$. (b) $f(r,\theta) = e^{-r^2}$. (c) $f(x,y) = xy + 2\sin(x)$.

13.1.5. Plot $f(x,y,z) = xy - xz - yz$ as a function on the unit sphere. (Use `axis equal` to get equal aspect ratios for the axes.)

13.1.6. Plot $f(x,y,z) = xy - xz - yz$ as a function on the unit cylinder $r = 1$ for $-1 \le z \le 2$. (Use `axis equal` to get equal aspect ratios for the axes.)

## 13.2 • Two-dimensional diffusion and advection

We next describe how to apply the method of lines to PDEs of the form

$$u_t = \phi(u, u_x, u_y, u_{xx}, u_{xy}, u_{yy}), \quad (x,y) \in [a,b] \times [c,d]. \tag{13.2.1}$$

The PDE may be of either parabolic or hyperbolic type, with the primary difference being potential restrictions on the time step size. To keep descriptions and implementations relatively simple, we will consider only periodic conditions or Dirichlet boundary conditions imposed on all the edges of the rectangular domain.

As described in Section 13.1, the rectangular domain is discretized by a grid $(x_i, y_j)$ for $i = 0, \ldots, m$ and $j = 0, \ldots, n$. The solution is semidiscretized as a matrix $U(t)$ such that $U_{ij} \approx u(x_i, y_j, t)$. Terms involving the spatial derivatives of $u$ are readily replaced by discrete counterparts: $D_x U$ for $u_x$, $U D_y^T$ for $u_y$, and so on.

### Matrix and vector shapes

Our destination is a set of ODEs that can be solved by a Runge–Kutta or multistep solver. These solvers (such as the ones built into MATLAB) are intended for vector problems in the form $w' = f(t, w)$. But our unknowns naturally have a matrix shape, which is the most convenient for the differentiation formulas (13.1.7) and (13.1.8).

Fortunately it's easy to *translate between a matrix and an equivalent vector*. We first define the **vec** operator, which stacks the columns of any $m \times n$ matrix into an $mn \times 1$ vector:

$$\mathrm{vec}(A) = \begin{bmatrix} A_{11} \\ \vdots \\ A_{m1} \\ \vdots \\ A_{1n} \\ \vdots \\ A_{mn} \end{bmatrix}. \tag{13.2.2}$$

The vec operation defines a linear ordering for the two-dimensional grid of points by taking all the $x$ entries for the first value of $y$, then all the $x$ entries for the second value of $y$, and so on. For completeness we also define unvec, the inverse mapping. Both are one-liners in MATLAB, using **reshape** or colon notation.

> **Example 13.2.1**
>
> ```
> m = 2;   n = 3;
> v = (1:6)';
> ```
>
> The unvec operation:
>
> ```
> V = reshape(v,2,3)
> ```
>
> ```
> V =
>      1     3     5
>      2     4     6
> ```
>
> Two equivalent syntaxes for the vec operation:
>
> ```
> reshape(V,6,1);
> v = V(:)
> ```
>
> ```
> v =
>      1
>      2
>      3
>      4
>      5
>      6
> ```

Suppose the end conditions are periodic (Dirichlet boundary conditions are considered below). The unknowns in the method of lines are elements of the matrix $U(t)$ representing grid values of the numerical solution. For the purposes of ODE solution, this matrix is equivalent to the vector $w(t)$ defined as $w = \text{vec}(U)$. All of the computations are done using the matrix shape, and reshaping is used to go back and forth between the two representations, in two functions we call **pack** and **unpack**.

> **Example 13.2.2**
>
> We solve a two-dimensional heat equation $u_t = 0.1(u_{xx} + u_{yy})$ on the square $[-1,1] \times [-1,1]$. We assume periodic behavior in both directions.
>
> ```
> m = 60;   n = 40;
> [x,Dx,Dxx] = diffper(m,[-1,1]);
> [y,Dy,Dyy] = diffper(n,[-1,1]);
> [X,Y] = ndgrid(x,y);
> ```
>
> Note that the initial condition must also be periodic on the domain.
>
> ```
> U0 = sin(4*pi*X).*exp(cos(pi*Y));
> pcolor(X,Y,U0)
> ```

## 13.2. Two-dimensional diffusion and advection

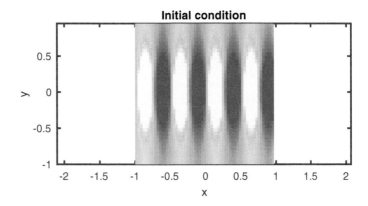

**Initial condition**

The next two functions map between the natural matrix shape of the unknowns and the vector shape demanded by the ODE solvers.

```
function U = unpack(u)
    U = reshape(u,m,n);
end
function u = pack(U)
    u = U(:);
end
```

This function computes the time derivative for the unknowns. The actual calculations take place using the matrix shape.

```
function dudt = timederiv(t,u)
    U = unpack(u);
    Uxx = Dxx*U;   Uyy = U*Dyy';      % 2nd partials
    dUdt = 0.1*(Uxx + Uyy);   % PDE
    dudt = pack(dUdt);
end
```

Since this problem is parabolic, a stiff integrator like `ode15s` is a good choice.

```
t = linspace(0,.2,3);
[t,W] = ode15s(@timederiv,t,pack(U0));
W = W.';      % each column is one time instant
```

Here we plot the solution at two different times. (The results are best viewed using an animation.)

```
for k = 1:2
    subplot(1,2,k)
    pcolor(X,Y,unpack(W(:,k+1)))
end
```

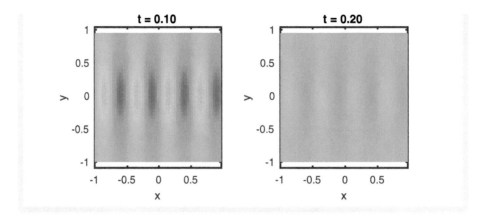

## Boundary conditions

For problems with boundary conditions, the utility function Function 13.2.1 encapsulates our rectangular discretization techniques for use in PDE solving. One of its outputs is an array of logical values, i.e., `true` or `false`. This array has the same size as a grid coordinate matrix and has `true` at the boundary point locations. Its use is for addressing only boundary or interior locations within a grid function, as explained below.

**Function 13.2.1** (`rectdisc`) Discretization on a rectangle.

```
function [X,Y,disc] = rectdisc(m,xspan,n,yspan)
% RECTDISC   Discretization on a rectangle.
% Input:
%   m        number of grid points in x (integer)
%   xspan    extent of domain in x direction (2-vector)
%   n        number of grid points in y (integer)
%   yspan    extent of domain in y direction (2-vector)
% Output:
%   X,Y      grid coordinate matrices (m+1 by n+1)
%   disc     discretization tools (structure)

% Initialize grid and finite differences.
[x,Dx,Dxx] = diffmat2(m,xspan);
[y,Dy,Dyy] = diffmat2(n,yspan);
[X,Y] = ndgrid(x,y);

% Get the diff. matrices recognized as sparse.
disc.Dx = sparse(Dx);    disc.Dxx = sparse(Dxx);
disc.Dy = sparse(Dy);    disc.Dyy = sparse(Dyy);
disc.Ix = speye(m+1);    disc.Iy = speye(n+1);

% Locate boundary points.
disc.isbndy = true(m+1,n+1);
disc.isbndy(2:m,2:n) = false;

disc.vec = @(U) U(:);
disc.unvec = @(u) reshape(u,m+1,n+1);

end
```

## 13.2. Two-dimensional diffusion and advection

Recall that in Section 11.5 we coped with boundary conditions by removing the boundary values from the vector of unknowns being solved in the semidiscretized ODE. Each evaluation of the time derivative required us to extend the values to include the boundaries before applying differentiation matrices in space.

We proceed similarly here. *The ODE unknowns in the vector $w(t)$ include grid values of the solution only at the interior grid points.* Given a value of $w$, it is not difficult to reconstruct the corresponding grid function $U$ that has the correct values at the boundary and the interior values described by $w$. The inverse operation is straightforward as well. All of these manipulations are compartmentalized in the functions **pack** and **unpack**, so that the PDE-defined computations can just use the convenient matrix shapes.

Note in the following example that we use the discretized initial condition $u(x,y,0)$ to supply the boundary values, since it must be consistent with the boundary conditions.

### Example 13.2.3

We solve an advection–diffusion problem, $u_t + u_x = 1 + 0.05(u_{xx} + u_{yy})$, where $u = 0$ on the boundary of the square $[-1,1]^2$.

```
m = 50;   n = 50;
[X,Y,d] = rectdisc(m,[-1,1],n,[-1,1]);
```

The initial condition we specify here is used to impose its boundary values on the solution at all times.

```
U0 = (1-X.^4).*(1-Y.^4);
pcolor(X,Y,U0)
```

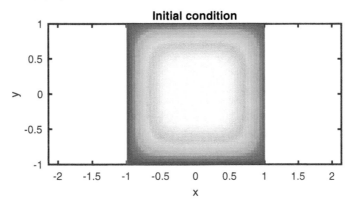

This next function maps the unknowns, given in a vector shape, to a matrix of values including the boundaries.

```
function U = unpack(w)
    U = U0;                  % get the boundary right
    U(~d.isbndy) = w;        % overwrite the interior
end
```

The next function drops the boundary values and returns a vector of the interior values. It's the inverse of the unpack function.

```
function w = pack(U)
    w = U(~d.isbndy);
end
```

This function computes the time derivative at the interior nodes only.

```
function dwdt = timederiv(t,w)
    U = unpack(w);
    Uxx = d.Dxx*U;  Uyy = U*d.Dyy';    % 2nd partials
    dUdt = 1 - d.Dx*U + 0.05*(Uxx + Uyy);   % PDE
    dwdt = pack(dUdt);
end
```

Since this problem is parabolic, a stiff integrator like ode15s is a good choice.

```
t = linspace(0,1,3);
[t,W] = ode15s(@timederiv,t,pack(U0));
W = W.';       % each column is one time instant
```

We plot the solution at two different times. Observe that the boundary values are identical to those of the initial condition.

```
for k = 1:2
    subplot(1,2,k)
    U = unpack(W(:,k+1));
    pcolor(X,Y,U)
end
```

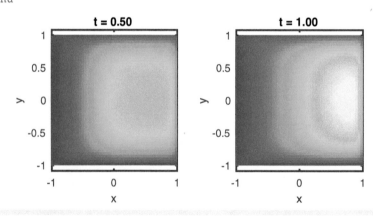

Full understanding of the pack and unpack functions in Example 13.2.3 hinges on logical matrix indexing in MATLAB. The variable d.isbndy is of logical type and the same size as U. Hence U(~d.isbndy) refers to a subset of the elements of U, automatically interpreted as a vector. The assignment statements using this idiom operate on vectors of identical size on the two sides of the equal sign.

Our last example of this section is the wave equation $u_{tt} = u_{xx} + u_{yy}$ on the square $[-2,2] \times [-2,2]$, where $u = 0$ on the boundary. We now have an additional complication of reducing the PDE to a first-order system, so that the grid unknowns are a *pair*

## 13.2. Two-dimensional diffusion and advection

of matrices $U(t)$ and $V(t)$. Furthermore, the boundary values of $U$ are prescribed, while those of $V$ are not. All of the unknowns still get packed into a single vector $w(t)$ for the ODE solver, and the recipe for doing so is still confined to the `pack` and `unpack` functions.

**Example 13.2.4**

Our first step is to write the wave equation as the equivalent first-order system

$$u_t = v,$$
$$v_t = u_{xx} + u_{yy}.$$
(13.2.3)

```
m = 60;   n = 60;
[X,Y,d] = rectdisc(m,[-2,2],n,[-2,2]);
```

Here is the initial condition. The boundary values of $u$ will remain constant.

```
U0 = (X+.2).*exp(-12*(X.^2+Y.^2));
V0 = 0*U0;
pcolor(X,Y,U0)
```

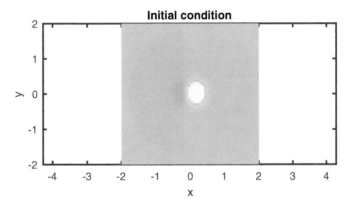

The `unpack` function separates the unknowns for $u$ and $v$, applies the boundary conditions on $u$, and returns two functions on the grid.

```
function [U,V] = unpack(w)
    numU = (m-1)*(n-1);    % number of unknowns for U
    U = U0;
    U(~d.isbndy) = w(1:numU);    % overwrite the interior
    V = d.unvec( w(numU+1:end) );    % use all values
end
```

The next function drops the boundary values of $u$ and returns a vector of all the unknowns for both components of the solution. It's the inverse of the `unpack` function.

```
function w = pack(U,V)
    w = U(~d.isbndy);
    w = [ w; V(:) ];
end
```

The following function computes the time derivative of the unknowns. Besides the translation between vector and matrix shapes, it's quite straightforward.

```
function dwdt = timederiv(t,w)
    [U,V] = unpack(w);
    dUdt = V;
    dWdt = d.Dxx*U + U*d.Dyy';
    dwdt = pack(dUdt,dWdt);
end
```

Since this problem is hyperbolic, not parabolic, a nonstiff integrator like `ode45` is fine and faster than a stiff integrator.

```
t = linspace(0,3,5);
v0 = pack(U0,V0);
[t,W] = ode45(@timederiv,t,v0);
W = W';   % each column is one time instant
```

We plot the solution at four different times.

```
for k = 1:2
    subplot(1,2,k)
    [U,~] = unpack(W(:,k+1));
    pcolor(X,Y,U)
end
```

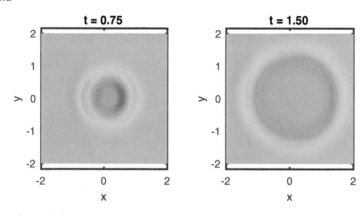

```
for k = 1:2
    subplot(1,2,k)
    [U,~] = unpack(W(:,k+3));
    pcolor(X,Y,U)
end
```

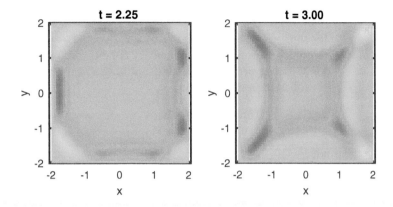

## Exercises

13.2.1. ◼ For the given $u(x,y)$, make a plot of the given quantity on the square $[-2,2]^2$ using appropriate differentiation matrices.

(a) $u(x,y) = \exp(x - y^2)$; plot $u_{xx} + u_{yy}$.

(b) $u(x,y) = \cos(\pi x) + \sin(\pi y)$; plot $u_x + u_y$.

(c) $u(x,y) = \exp(-x^2 - 4y^2)$; plot $xu_y$.

13.2.2. ◼ Following Example 13.2.2 as a model, solve the Allen–Cahn equation $u_t = u(1-u^2) + 0.001(u_{xx} + u_{yy})$ on the square $[-1,1]^2$ with periodic conditions, taking $u(x,y,0) = \sin(\pi x)\cos(2\pi y)$. Use $m = n = 60$ and solve for 60 times up to $t = 4$. You can use the following code to animate your solution:

```
for k = 1:length(t)
    U = unpack(W(:,k));
    surf(X,Y,U)
    axis([-1 1 -1 1 -1 1])
    shading flat
    shg, pause(0.05)
end
```

13.2.3. ◼ Following Example 13.2.3 as a model, solve $u_t = yu_x - u_y + 0.03(u_{xx} + u_{yy})$ on the square $[-1,1]^2$, with $u(x,y,0) = (1-x^2)(1-y^2)$ and homogeneous Dirichlet boundary conditions. Use $m = n = 50$ and solve for 50 times up to $t = 2$. You can use the following code to animate your solution:

```
for k = 1:length(t)
    U = unpack(W(:,k));
    surf(X,Y,U)
    axis([-1 1 -1 1 0 1])
    shading flat
    shg, pause(0.05)
end
```

13.2.4. ▣ Following Example 13.2.4 as a model, solve $u_{tt} = u_{xx} + u_{yy} + \cos(7t)$ on the square $[-1,1]^2$, with $u(x,y,0) = x(1-x^6)(1-y^2)$, $u_t(x,y,0) = 0$, subject to homogeneous Dirichlet boundary conditions. Take $m = n = 60$ and solve for 200 times between $t = 0$ and $t = 12$. You can animate the solution using the following template code:

```
for k = 1:length(t)
    [U,~] = unpack(W(:,k));
    surf(X,Y,U)
    axis([-1 1 -1 1 -1 1])
    shading flat
    shg, pause(0.05)
end
```

## 13.3 • Laplace and Poisson equations

Consider the heat equation in two dimensions, $u_t = u_{xx} + u_{yy}$. After a long time, the distribution of temperature will stop changing. This steady-state solution must satisfy the PDE $u_{xx} + u_{yy} = 0$, which is called the **Laplace equation**. A slight generalization is the **Poisson equation**,

$$u_{xx}(x,y) + u_{yy}(x,y) = f(x,y), \qquad (x,y) \in [a,b] \times [c,d], \qquad (13.3.1)$$

which might describe the steady state when sources or sinks of heat are present. Poisson's equation is often written compactly as $\Delta u = f$, where $\Delta$ is known as the **Laplacian operator**.

*The Laplace/Poisson equation is the prototype of the class of elliptic PDEs.* All linear PDEs with no higher than second derivatives are parabolic, hyperbolic, or elliptic. Elliptic problems are notable for having no time (or analogous) variable.

In order to get a fully specified problem, the Laplace or Poisson equations must be complemented with a boundary condition. We will consider only the Dirichlet condition $u(x,y) = g(x,y)$ along the entire boundary.

### Sylvester equation

With the unknown solution represented by its values $U = \text{mtx}(u)$ on a rectangular grid, and second-derivative finite difference or spectral differentiation matrices $D_{xx}$ and $D_{yy}$, the Poisson equation (13.3.1) becomes the discrete equation

$$D_{xx} U + U D_{yy}^T = F, \qquad (13.3.2)$$

where $F = \text{mtx}(f)$. Equation (13.3.2), with an unknown matrix $U$ multiplied on the left and right in different terms, is known as a **Sylvester equation**. We will solve it by casting it into another form, after defining a new matrix operation.

## Kronecker products

Let $A$ be $m \times n$ and $B$ be $p \times q$. The **Kronecker product** $A \otimes B$ is the $mp \times nq$ matrix given by

$$A \otimes B = \begin{bmatrix} A_{11}B & A_{12}B & \cdots & A_{1n}B \\ A_{21}B & A_{22}B & \cdots & A_{2n}B \\ \vdots & \vdots & & \vdots \\ A_{m1}B & A_{m2}B & \cdots & A_{mn}B \end{bmatrix}. \tag{13.3.3}$$

Unlike regular matrix products, Kronecker products place no restrictions on the sizes of the operand matrices.

**Example 13.3.1**
```
A = [1 2; -2 0]
B = [ 1 10 100; -5 5 3 ]

A =
     1     2
    -2     0
B =
     1    10   100
    -5     5     3
```

Applying the definition manually,
```
AkronB = [ A(1,1)*B, A(1,2)*B;
           A(2,1)*B, A(2,2)*B ]

AkronB =
     1    10   100     2    20   200
    -5     5     3   -10    10     6
    -2   -20  -200     0     0     0
    10   -10    -6     0     0     0
```

But it's simpler to use the built-in **kron**.
```
kron(A,B)

ans =
     1    10   100     2    20   200
    -5     5     3   -10    10     6
    -2   -20  -200     0     0     0
    10   -10    -6     0     0     0
```

The Kronecker product obeys several natural-looking identities:

$$A \otimes (B+C) = A \otimes B + A \otimes C, \qquad (13.3.4\text{a})$$
$$(A+B) \otimes C = A \otimes C + B \otimes C, \qquad (13.3.4\text{b})$$
$$(A \otimes B) \otimes C = A \otimes (B \otimes C), \qquad (13.3.4\text{c})$$
$$(A \otimes B)^T = A^T \otimes B^T, \qquad (13.3.4\text{d})$$
$$(A \otimes B)^{-1} = A^{-1} \otimes B^{-1} \quad \text{(if } A \text{ and } B \text{ are invertible)}, \qquad (13.3.4\text{e})$$
$$(A \otimes B)(C \otimes D) = (AC) \otimes (BD) \quad \text{(if } AC \text{ and } BD \text{ are defined)}. \qquad (13.3.4\text{f})$$

Finally, we come to an identity crucial for our Sylvester equation:

$$\text{vec}(ABC^T) = (C \otimes A)\text{vec}(B). \qquad (13.3.5)$$

This is valid for any compatibly sized matrices $A$, $B$, and $C$.

## Poisson as a linear system

*We can use Kronecker products to express the Sylvester form (13.3.2) of the discrete Poisson equation as an ordinary linear system.* First we pad (13.3.2) with "silent" identity matrices, i.e.,

$$D_{xx}UI_y + I_x U D_{yy}^T = F,$$

where $I_x$ and $I_y$ are, respectively, the $(m+1) \times (m+1)$ and $(n+1) \times (n+1)$ identities. Upon taking the vec of both sides and applying (13.3.5), we obtain

$$\left[(I_y \otimes D_{xx}) + (D_{yy} \otimes I_x)\right]\text{vec}(U) = \text{vec}(F),$$
$$Au = b. \qquad (13.3.6)$$

This is in the form of a standard linear system in $(m+1)(n+1)$ variables. However, boundary conditions of the PDE must yet be applied.

As has been our practice for one-dimensional BVPs, we will replace the collocation equation for the PDE at each boundary point with an equation that assigns that boundary point its prescribed value. The details are a bit harder to express algebraically in the two-dimensional geometry. Say that $N = (m+1)(n+1)$ is the number of entries in the unknown $U$, and let $B$ be a subset of $\{1, \ldots, N\}$ such that $i \in B$ if and only if $(x_i, y_i)$ is on the boundary. Then for each $i \in B$, we want to replace row $i$ of the system $Au = b$ with the equation

$$e_i^T u = g(x_i, y_i).$$

Hence from $A$ we should subtract away its $i$th row and add $e_i^T$ back in. When this is done for all $i \in B$, the result is replacement of the relevant rows of $A$ with relevant rows of the identity:

$$\tilde{A} = A - I_B A + I_B, \qquad (13.3.7)$$

where $I_B$ is a matrix with zeros everywhere except for ones at $(i, i)$ for all $i \in B$. A similar expression can be derived for the modification of $b$. However, rather than implementing a literal interpretation of (13.3.7), the changes to $A$ and $b$ are made more easily through logical indexing. A small example is more illustrative than further description.

## 13.3. Laplace and Poisson equations

**Example 13.3.2**

Here is a forcing function for Poisson's equation.

```
f = @(x,y) x.^2 - y + 2;
```

We pick a crude discretization for illustrative purposes.

```
m = 5;  n = 6;
[X,Y,d] = rectdisc(m,[0 3],n,[-1 1]);
```

Next, we evaluate $f$ on the grid.

```
F = f(X,Y);
```

Here are the equations for the PDE collocation, before any modifications are made for the boundary conditions.

```
A = kron(d.Iy,d.Dxx) + kron(d.Dyy,d.Ix);
spy(A)
b = d.vec(F);
N = length(b)
```

N =
   42

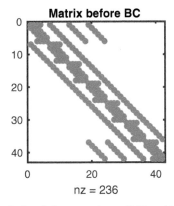

The array d.isbndy is logical and the same size as X, Y, and F.

```
spy(d.isbndy), title('Boundary points')
```

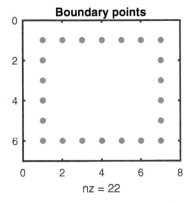

Next replace the boundary rows of the system by rows of the identity.

```
I = speye(size(A));
A(d.isbndy,:) = I(d.isbndy,:);      % Dirichlet conditions
spy(A)
```

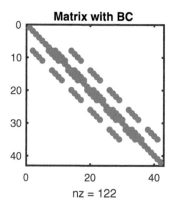

**Matrix with BC**

nz = 122

Finally, we must replace the rows in the vector $b$ by the boundary values being assigned to the boundary points. Here, we let the boundary values be zero everywhere.

```
b(d.isbndy) = 0;                    % Dirichlet values
```

Now we can solve for $u$ and reinterpret it as the matrix-shaped $U$, the solution on our grid. This grid is much too coarse for the result to look like a smooth function of two variables.

```
u = A\b;
U = d.unvec(u);
mesh(X,Y,U)
```

**Coarse solution**

## Implementation

Function 13.3.1 solves the Poisson equation in much the same way as Example 13.3.2: it creates the linear system, modifies it for the boundary conditions, and solves it using backslash. The matrix is $N = (m+1)(n+1)$ on each side and very sparse, so we take care to use **sparse** matrices to exploit that structure. There is a small but important

## 13.3. Laplace and Poisson equations

**Function 13.3.1** (`poissonfd`) Solve Poisson's equation by finite differences.

```
function [U,X,Y] = poissonfd(f,g,m,xspan,n,yspan)
%POISSONFD   Solve Poisson's equation by finite differences.
% Input:
%   f         forcing function (function of x,y)
%   g         boundary condition (function of x,y)
%   m         number of grid points in x (integer)
%   xspan     endpoints of the domain of x (2-vector)
%   n         number of grid points in y (integer)
%   yspan     endpoints of the domain of y (2-vector)
%
% Output:
%   U         solution (m+1 by n+1)
%   X,Y       grid matrices (m+1 by n+1)

% Initialize the rectangle discretization.
[X,Y,d] = rectdisc(m,xspan,n,yspan);

% Form the collocated PDE as a linear system.
A = kron(d.Iy,d.Dxx) + kron(d.Dyy,d.Ix);   % Laplacian matrix
b = d.vec(f(X,Y));

% Replace collocation equations on the boundary.
scale = max(abs(A(n+2,:)));
I = speye(size(A));
A(d.isbndy,:) = scale*I(d.isbndy,:);              % Dirichet assignment
b(d.isbndy) = scale*g( X(d.isbndy),Y(d.isbndy) ); % assigned values

% Solve the linear system and reshape the output.
u = A\b;
U = d.unvec(u);

end
```

change from Example 13.3.2: the boundary conditions are rescaled to read $\sigma u(x,y) = \sigma g(x,y)$, where $\sigma$ is the largest element of a row of $A$. This tweak improves the condition number of the final matrix.

### Example 13.3.3

We can engineer an example by choosing the solution first. Let $u(x,y) = \sin(3xy - 4y)$. Then one can derive $f = \Delta u = -\sin(3xy-4y)(9y^2 + (3x-4)^2)$ for the forcing function and use $g = u$ on the boundary.

First we define the problem on $[0,1] \times [0,2]$.

```
f = @(x,y) -sin(3.*x.*y-4*y).*(9*y.^2+(3*x-4).^2);
g = @(x,y) sin(3.*x.*y-4*y);
xspan = [0 1];   yspan = [0 2];
```

Here is the finite difference solution.

```
[U,X,Y] = poissonfd(f,g,50,xspan,80,yspan);
mesh(X,Y,U)
```

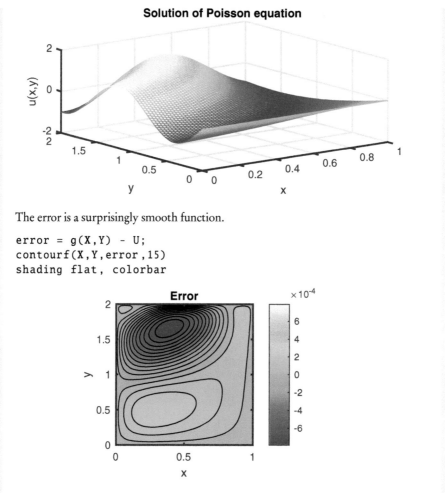

The error is a surprisingly smooth function.

```
error = g(X,Y) - U;
contourf(X,Y,error,15)
shading flat, colorbar
```

Because $u$ is specified on the boundary, we should observe that the error is zero on the boundary—else we have done something incorrectly.

## Accuracy and efficiency

Let's simplify the discussion by assuming that $m = n$. The discretization of the derivatives in Function 13.3.1 is at second-order accuracy, so the error in the solution can be expected to be $O(n^{-2})$. The matrix $\boldsymbol{A}$ has size $N = O(n^2)$. The upper and lower bandwidths of the matrix $\boldsymbol{A}$ are identical and $O(n)$, so the time taken by sparse Cholesky factorization can be no better than $O(n^2 N) = O(n^4)$. If instead we use running time $T$ as the free parameter, then $n = O(T^{1/4})$ and error is $O(1/\sqrt{T})$. Clearly, investing more time (or more computing power) into the computation provides additional benefits only at a slow rate.

## Exercises

13.3.1. ✍ Using general $2 \times 2$ matrices, verify the identities (13.3.4d) and (13.3.4f).

13.3.2. ✍ Prove that the matrix appearing in (13.3.6) is symmetric.

13.3.3. 💻 Use Function 13.3.1 to solve the following problems on $[0,1] \times [0,1]$ using $m = n = 50$. In each case, make a plot of the solution and a plot of the error.

   (a) $u_{xx} + u_{yy} = 2x^2[x^2(x-1)+(10x-6)(y^2-y)]$, with $u = 0$ on the boundary. Exact solution: $u(x,y) = x^4(1-x)y(1-y)$.

   (b) $u_{xx} + u_{yy} = \left(-16x^2 - (1-4y)^2\right)\sinh(x-4xy)$, with $u = \sinh(4xy - x)$ on the boundary. Exact solution: $u(x,y) = \sin(4\pi x))$.

   (c) $u_{xx} + u_{yy} = -(20\pi^2)\sin(4\pi x)\cos(2\pi y)$, with $u = \sin(4\pi x)\cos(2\pi y)$. Exact solution: $u(x,y) = \sin(4\pi x)\cos(2\pi y)$.

13.3.4. 💻 For each item in Exercise 13.3.3, solve the problem using Function 13.3.1 with $m = n = 20, 30, 40, \ldots, 120$. For each numerical solution compute the maximum absolute error on the grid. On a log-log plot, compare the convergence of the error as a function of $n$ to theoretical second-order accuracy.

13.3.5. 💻 Copy Function 13.3.1 to `poissoncheb.m`, then modify the new file to use a Chebyshev discretization rather than finite differences. For each item in Exercise 13.3.3, solve the problem using `poissoncheb.m` using $m = n = 20, 25, 30, \ldots, 60$. For each numerical solution compute the maximum absolute error on the grid. Show the convergence of the error as a function of $n$ on a log–linear plot.

13.3.6. 💻 Sometimes boundary conditions are specified using a piecewise definition, with a different formula for each side of the domain. Use Function 13.3.1 with $m = n = 60$ to solve the Laplace equation on $[0,1]^2$ with boundary conditions

$$u(0,y) = u(1,y) \equiv 0, \quad u(x,0) = \sin(3\pi x), \quad u(x,1) = e^{2x}(x-x^2).$$

Make a surface plot of your numerical solution.

## 13.4 · Nonlinear elliptic PDEs

Many nonlinear PDEs are elliptic and include specific references to the Laplacian operator.

### Example 13.4.1

Recall the micromechanical deflector modeled in a disk by (10.1.2). A fully two-dimensional equivalent is (see [54])

$$\Delta u - \frac{\lambda}{(u+1)^2} = 0. \qquad (13.4.1)$$

This may be posed on any region, with $u = 0$ specified everywhere on the boundary.

More generally, we want to solve the nonlinear equation

$$f(u, u_x, u_y, u_{xx}, u_{yy}) = 0 \qquad \text{in } R, \tag{13.4.2a}$$
$$u(x,y) = g(x,y) \qquad \text{on the boundary,} \tag{13.4.2b}$$

where $R$ is a rectangle. Most of the time we will just write $f(u)$, with the understanding that partial derivatives may be applied to $u$ as part of the definition. To keep the notation calmer, we use $m = n$ on the grid in $R$, so $N = (n+1)^2$ is the total number of scalar unknowns. After collocation, $u$ becomes $u = \text{vec}(U)$ and $f$ becomes a non-linear function $f$. That is, *the discretization of a nonlinear elliptic PDE is a nonlinear system of equations to be solved.*

Following (13.3.7), we modify the nonlinear function to incorporate boundary values:

$$\tilde{f}(u) = I_B(u - g) + (I - I_B) f(u), \tag{13.4.3}$$

where $g$ represents evaluation of $g$ on the grid (though only the values at boundary points are used and need to be calculated). Thus the discrete problem is to find a root, $\tilde{f}(u) = 0$.

Our prototype for Newton and quasi-Newton methods is the iteration $u_0, u_1, \ldots$ defined by

$$u_{k+1} = u_k - d_k, \qquad \tilde{A}_k d_k = \tilde{f}(u_k) \tag{13.4.4}$$

for a given $u_0$ and a sequence of matrices $\tilde{A}_k$. In the pure Newton case, $\tilde{A}_k$ is the Jacobian matrix of $\tilde{f}$, evaluated at $u_k$. This is the situation we consider in this section.

## Finding the Jacobian

We begin by observing that the boundary condition presents no new issues. Taking the Jacobian of both sides of (13.4.3), we find

$$\tilde{f}'(u) = I_B + (I - I_B) f'(u), \tag{13.4.5}$$

thanks to the linearity of differentiation. So the main task is to find $f'(u)$, the Jacobian of the PDE discretization.

We are motivated by one definition of the Jacobian,

$$Jv = \lim_{\epsilon \to 0} \frac{f(u + \epsilon v) - f(u)}{\epsilon},$$

except that it's more convenient to apply it to the continuous problem $f(u) = 0$ and discretize the result. The process is as follows:

1. Write out $f(u + \epsilon v)$.

## 13.4. Nonlinear elliptic PDEs

2. Expand it in a power series in $\epsilon$. The $O(1)$ term must equal $f(u)$. Calculate the $O(\epsilon)$ term explicitly and leave the others simply as $O(\epsilon^2)$.

3. Subtraction of $f(u)$ cancels the $O(1)$ term. Upon dividing by $\epsilon$ and taking $\epsilon \to 0$, the original $O(\epsilon^2)$ terms also vanish. What remains is a linear operator on the function $v$. Discretizing this operator gives $\boldsymbol{J} = \boldsymbol{f}'(\boldsymbol{u})$.

**Example 13.4.2**

Referring back to Example 13.4.1, we compute

$$f(u+\epsilon v) = \Delta(u+\epsilon v) - \frac{\lambda}{(u+\epsilon v+1)^2}$$

$$= \Delta u + \epsilon \Delta v - \left[\frac{\lambda}{(u+1)^2} - \frac{2\lambda}{(u+1)^3}\epsilon v + \cdots\right]$$

$$= f(u) + \left[\Delta v + \frac{2\lambda}{(u+1)^3}v\right]\epsilon + O(\epsilon^2).$$

Only the term in brackets remains after following the recipe above. It is a linear operator on $v$, in two terms. The first of these, $\Delta v$, is just the Laplacian operator on $v$. Discretizing it on the grid leaves $\boldsymbol{L}\boldsymbol{v}$, where $\boldsymbol{v} = \text{vec}(\text{mtx}(v))$ and $\boldsymbol{L}$ is the matrix $\boldsymbol{A}$ from (13.3.6). The first term, $\frac{2\lambda}{(u+1)^3}v$, leads to a vector whose $i$th element on the grid equals $\frac{2\lambda v_i}{(u_i+1)^3}$. But this is equal to

$$\text{diag}\left(\frac{2\lambda}{(u_i+1)^3}\right)\boldsymbol{v},$$

where by diag we mean to construct a diagonal $N \times N$ matrix with diagonal elements equal to the given expression. In summary, the Jacobian is

$$\boldsymbol{f}'(\boldsymbol{u}) = \boldsymbol{L} + \text{diag}\left(\frac{2\lambda}{(u_i+1)^3}\right). \tag{13.4.6}$$

Note that in Example 13.4.2, the Jacobian matrix $\boldsymbol{f}'(\boldsymbol{u})$ depends on $\boldsymbol{u}$, as the derivative of a nonlinear function should.

### Implementation

Function 13.4.1 implements the Newton iteration to solve the PDE (13.4.2). In order to help globalize the convergence, a damped Newton step is used (see Section 4.6). The input parameter **f** is a function of the discrete solution $\boldsymbol{U}$ and the first two differentiation matrices in each dimension, as computed by Function 13.2.1. This function must return the residual $\boldsymbol{f}(\boldsymbol{u})$ (in matrix shape) and the Jacobian matrix.

**Function 13.4.1** (`newtonpde`) Newton's method to solve an elliptic PDE.

```
function [U,X,Y] = newtonpde(f,g,m,xspan,n,yspan)
%NEWTONPDE   Newton's method to solve an elliptic PDE.
% Input:
%   f              defines the PDE, f(u)=0 (function)
%   g              boundary condition (function)
%   n              number of grid points in each dimension (integer)
% Output:
%   U              solution (n+1 by n+1)
%   X,Y            coordinate matrices (n+1 by n+1)

% Discretization.
[X,Y,d] = rectdisc(n,xspan,n,yspan);
I = speye((n+1)^2);    % used for row replacements

% This evaluates the discretized PDE and its Jacobian, with all the
% boundary condition modifications applied.
    function [r,J] = residual(U)
        [R,J] = f(U,X,Y,d);
        scale = max(abs(J(:)));
        J(d.isbndy,:) = scale*I(d.isbndy,:);
        XB = X(d.isbndy);   YB = Y(d.isbndy);
        R(d.isbndy) = scale*(U(d.isbndy) - g(XB,YB));
        r = d.vec(R);
    end

% Intialize the Newton iteration.
U = zeros(size(X));
[r,J] = residual(U);
tol = 1e-10;  itermax = 20;
s = Inf;  normr = norm(r);  k = 1;
I = speye(numel(U));

lambda = 1;
while (norm(s) > tol) && (normr > tol)
    s = -(J'*J + lambda*I) \ (J'*r);   % damped step
    Unew = U + d.unvec(s);
    [rnew,Jnew] = residual(Unew);

    if norm(rnew) < normr
        % Accept and update.
        lambda = lambda/6;   % dampen the Newton step less
        U = Unew;  r = rnew;  J = Jnew;
        normr = norm(r);
        k = k+1;
        disp(['Norm of residual = ',num2str(normr)])
    else
        % Reject.
        lambda = lambda*4;   % dampen the Newton step more
    end

    if k==itermax
        warning('Maximum number of Newton iterations reached.')
        break
    end
end

end
```

## 13.4. Nonlinear elliptic PDEs

### Example 13.4.3

We solve (13.4.1), using the Jacobian matrix derived in Example 13.4.2.

This function defines the PDE, by way of $f(u)$ and its derivative. Note that the function will have access to all of the properties of a discretization, as if they were returned by `rectdisc`.

```
lambda = 1.5;
    function [F,J] = pde(U,X,Y,d)
        LU = d.Dxx*U + U*d.Dyy';    % apply Laplacian
        F = LU - lambda./(U+1).^2;   % residual
        L = kron(d.Dyy,d.Ix) + kron(d.Iy,d.Dxx);
        u = d.vec(U);
        J = L + sparse( diag(2*lambda./(u+1).^3) );
    end
```

Now we solve and plot the result.

```
g = @(x,y) zeros(size(x));    % boundary condition
[U,X,Y] = newtonpde(@pde,g,100,[0 2.5],80,[0 1]);
surfl(X,Y,U), colormap copper
```

```
Norm of residual = 15.0971
Norm of residual = 1.9805
Norm of residual = 0.075294
Norm of residual = 0.00013857
Norm of residual = 6.8429e-09
Norm of residual = 1.9367e-11
```

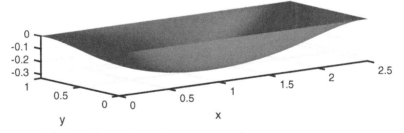

### Example 13.4.4

The stationary Allen–Cahn equation in two dimensions is $u(1-u^2)+\delta\Delta u = 0$. Using the same technique as in Example 13.4.2, we find that the Jacobian matrix for this problem (before boundary modifications) is

$$f'(u) = \mathrm{diag}(1-3u_i^2) + \delta L.$$

The following function defines the PDE, by way of $f(u)$ and its derivative:

```
function [F,J] = pde(U,X,Y,d)
```

```
        LU = d.Dxx*U + U*d.Dyy';      % apply Laplacian
        F = U.*(1-U.^2) + 0.05*LU;    % residual
        L = kron(d.Dyy,d.Ix) + kron(d.Iy,d.Dxx);
        u = d.vec(U);
        J = sparse( diag(1-3*u.^2) ) + 0.05*L;   % Jacobian
    end
```

Now we solve and plot the result.

```
g = @(x,y) tanh(5*(x+2*y-1));   % boundary condition
[U,X,Y] = newtonpde(@pde,g,100,[0 1],100,[0 1]);
surf(X,Y,U)
```

```
Norm of residual = 50.471
Norm of residual = 11.5011
Norm of residual = 0.91811
Norm of residual = 0.0072331
Norm of residual = 4.9834e-06
Norm of residual = 9.1686e-10
Norm of residual = 1.35e-11
```

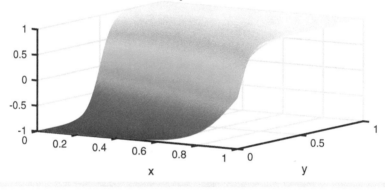

### Example 13.4.5

The steady-state limit of the advection–diffusion equation in Example 13.2.3 is $1 - u_x + \delta \Delta u = 0$, with a Dirichlet value of zero on the boundary. This problem is linear and thus is its own linearization. Upon discretization, the constant Jacobian matrix is found to be

$$f'(u) = -(I_y \otimes D_x) + \delta L.$$

The following function defines the PDE, by way of $f(u)$ and its Jacobian.

```
function [F,J] = pde(U,X,Y,d)
    LU = d.Dxx*U + U*d.Dyy';      % apply Laplacian
    F = 1 - d.Dx*U + 0.05*LU;     % residual
    L = kron(d.Dyy,d.Ix) + kron(d.Iy,d.Dxx);
    J = -kron(d.Iy,d.Dx) + 0.05*L;   % Jacobian
end
```

Now we solve and plot the result.

```
g = @(x,y) 0*x;      % boundary condition
[U,X,Y] = newtonpde(@pde,g,100,[-1,1],100,[-1,1]);
surf(X,Y,U)

Norm of residual = 45.0025
Norm of residual = 7.4088
Norm of residual = 0.23932
Norm of residual = 0.0013258
Norm of residual = 1.2299e-06
Norm of residual = 1.9035e-10
Norm of residual = 4.0724e-12
```

**Steady advection-diffusion**

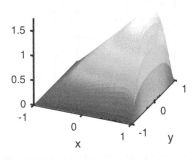

## Exercises

13.4.1. ✍ Use the steps in Example 13.4.2 to justify the Jacobian used in Example 13.4.4.

13.4.2. ⌨ Solve the advection–diffusion problem

$$u_{xx} + u_{yy} = u_y + x + 2 \quad \text{in } [-1,1] \times [-1,1],$$
$$u(x,y) = 0 \quad \text{on the boundary,}$$

for $m = n = 80$ points, and plot the solution.

13.4.3. ⌨ Consider

$$u_{xx} + u_{yy} = u_x + 4u^3 - u \quad \text{in } [0,1] \times [0,1],$$
$$u(x,y) = \begin{cases} 3 & \text{if } x > 0.5 \text{ and } y > 0.25, \\ -2 & \text{otherwise,} \end{cases} \quad \text{on the boundary.}$$

The solution $u(x,y)$ is not continuous on the boundary, but a reasonable solution may still be found using finite differences.

(a) Solve for $m = n = 20$ and make a surface plot of the solution.

(b) Solve for $m = n$ and $n = 4, 8, 12, \ldots, 32$, using `tic/toc` commands in each case to measure the execution time. Then make a log–log plot of execution time versus $n$.

13.4.4. A soap film stretched on a wire frame above the $(x, y)$ plane assumes a shape $u(x, y)$ of minimum area and is governed by

$$\nabla \cdot \left( \frac{\nabla u}{\sqrt{1 + u_x^2 + u_y^2}} \right) = 0 \quad \text{in region } R,$$

$$u(x, y) = g(x, y) \quad \text{on the boundary of } R.$$

(a) 🔺 Find the linearization of the PDE.

(b) 🔺 Write out the Jacobian matrix for a discretization of the linearization.

(c) 💻 Solve the equation for the domain and boundary data in Example 13.4.4.

## Key ideas in this chapter

1. A tensor-product grid is the doubly indexed set of points that represents all pairs of points in $x$ and $y$ discretizations (page 508).
2. Corresponding to a function defined on a rectangle is a matrix of its values on a tensor-product grid (page 508).
3. Partial derivatives of a grid function are computed by left- and right-multiplication by differentiation matrices (page 512).
4. Unknowns on a tensor-product grid must be reshaped to vectors for solutions by the method of lines (page 515).
5. The unknowns in the method of lines do not include grid points on the boundary (page 519).
6. The Poisson equation is the prototype of elliptic PDEs (page 524).
7. We can use Kronecker products as a compact and efficient way to express the discrete Poisson equation as an ordinary linear system (page 526).
8. The discretization of a nonlinear elliptic PDE is a nonlinear algebraic system of equations (page 532).

## Where to learn more

Trefethen [68] discusses techniques for tensor-product geometry, including eigenvalue problems and problems posed on a disk. Both Iserles [34] and Morton and Mayers [47] discuss finite differences at curved boundaries and present introductions to the finite element method in two dimensions.

# Appendix A
# Review of linear algebra

An $m \times n$ matrix $A$ is a rectangular $m \times n$ array of numbers called **elements** or **entries**. The numbers $m$ and $n$ are called the **row dimension** and the **column dimension**, respectively; collectively they describe the **size** of $A$. We say $A$ belongs to $\mathbb{R}^{m \times n}$ if its entries are real or $\mathbb{C}^{m \times n}$ if they are complex valued. A **square** matrix has equal row and column dimensions. A **row vector** has dimension $1 \times n$, while a **column vector** has dimension $m \times 1$. We use $\mathbb{R}^n$ or $\mathbb{C}^n$ to denote spaces of vectors. An ordinary number in $\mathbb{R}$ or $\mathbb{C}$ may be called a **scalar**.

We use capital letters in bold to refer to matrices, and lowercase bold letters for vectors. In this book, all vectors are column vectors—in other words, matrices with multiple rows and one column. The bold symbol $\mathbf{0}$ may refer to a vector of all zeros or to a zero matrix, depending on context; we use 0 as the scalar zero only.

To refer to a specific element of a matrix, we use the uppercase name of the matrix *without* boldface, as in $A_{24}$ to mean the $(2,4)$ element of $A$. To refer to an element of a vector, we use just one subscript, as in $x_3$. If you see a boldface character with one or more subscripts, then you know that it is a matrix or vector that belongs to a numbered collection.

We will have frequent need to refer to the individual columns of a matrix as vectors. Our convention is to use a lowercase bold version of the matrix name, with a subscript to represent the column number. Thus, $\boldsymbol{a}_1, \boldsymbol{a}_2, \ldots, \boldsymbol{a}_n$ are the columns of the $m \times n$ matrix $\boldsymbol{A}$. Conversely, whenever we define a sequence of vectors $\boldsymbol{v}_1, \ldots, \boldsymbol{v}_p$, we can implicitly consider them to be columns of a matrix $\boldsymbol{V}$. Sometimes we might write $\boldsymbol{V} = \begin{bmatrix} \boldsymbol{v}_j \end{bmatrix}$ to emphasize the connection.

The **diagonal** (main diagonal) of an $n \times n$ matrix $\boldsymbol{A}$ refers to the entries $A_{ii}$, $i = 1, \ldots, n$. The entries $A_{ij}$, where $j - i = k$, are on a **superdiagonal** if $k > 0$ and a subdiagonal if $k < 0$. The diagonals are numbered as suggested here:

$$\begin{bmatrix} 0 & 1 & 2 & \cdots & n-1 \\ -1 & 0 & 1 & \cdots & n-2 \\ \vdots & \ddots & \ddots & \ddots & \vdots \\ -n+2 & \cdots & -1 & 0 & 1 \\ -n+1 & \cdots & -2 & -1 & 0 \end{bmatrix}.$$

A **diagonal** matrix is one whose entries are all zero off the main diagonal. An **upper triangular** matrix $U$ has entries $U_{ij}$ with $U_{ij} = 0$ if $i > j$, and a **lower triangular** matrix $L$ has $L_{ij} = 0$ if $i < j$.

The **transpose** of $A \in \mathbb{C}^{m \times n}$ is the matrix $A^T \in \mathbb{C}^{n \times m}$ given by

$$A^T = \begin{bmatrix} A_{11} & A_{21} & \cdots & A_{m1} \\ \vdots & \vdots & & \vdots \\ A_{1n} & A_{2n} & \cdots & A_{mn} \end{bmatrix}.$$

The **conjugate transpose** or **hermitian** is given by $A^* = \overline{A^T}$, where the bar denotes taking a complex conjugate.[45] If $A$ is real, then $A^* = A^T$. A square matrix is **symmetric** if $A^T = A$ and **hermitian** if $A^* = A$.

## Algebra

Matrices of the same size may be added elementwise. Multiplication by a scalar is also defined elementwise. These operations obey the familiar laws of commutativity, associativity, and distributivity. The multiplication of two matrices, on the other hand, is much less straightforward.

In order for matrices $A$ and $B$ to be multiplied, it is necessary that their "inner" dimensions match—i.e., $A$ is $m \times p$ and $B$ is $p \times n$. Note that even if $AB$ is defined, $BA$ may not be, unless $m = n$. In terms of scalar components, the $(i, j)$ entry of $C = AB$ is given by

$$C_{ij} = \sum_{k=1}^{p} A_{ik} B_{kj}. \tag{A.1}$$

An important identity is that when $AB$ is defined,

$$(AB)^T = B^T A^T. \tag{A.2}$$

The latter product is always defined in this situation.

However, $AB = BA$ is *not* true in general, even when both products are defined. That is, *matrix multiplication is not commutative.* It is, however, associative: $ABC = (AB)C = A(BC)$. Changing the ordering of operations by associativity is a trick we use repeatedly.

---

[45] The conjugate of a complex number is found by replacing all references to the imaginary unit $i$ by $-i$. We do not do much with complex numbers until Chapter 7.

# Appendix A. Review of linear algebra

It is worth reinterpreting (A.1) at a vector level. If $\boldsymbol{A}$ has dimensions $m \times n$, it can be multiplied on the right by an $n \times 1$ column vector $\boldsymbol{v}$ to produce an $m \times 1$ column vector $\boldsymbol{Av}$, which satisfies

$$\boldsymbol{Av} = \begin{bmatrix} \sum_k A_{1k} v_k \\ \sum_k A_{2k} v_k \\ \vdots \\ \sum_k A_{mk} v_k \end{bmatrix} = v_1 \begin{bmatrix} A_{11} \\ A_{21} \\ \vdots \\ A_{m1} \end{bmatrix} + v_2 \begin{bmatrix} A_{12} \\ A_{22} \\ \vdots \\ A_{m2} \end{bmatrix} + \cdots + v_n \begin{bmatrix} A_{1n} \\ A_{2n} \\ \vdots \\ A_{mn} \end{bmatrix} = v_1 \boldsymbol{a}_1 + \cdots + v_n \boldsymbol{a}_n. \tag{A.3}$$

In words, we say that $\boldsymbol{Av}$ is a **linear combination** of the columns of $\boldsymbol{A}$. Equation (A.3) is very important: *Multiplying a matrix on the right by a column vector produces a linear combination of the columns of the matrix.* There is a similar interpretation of multiplying $\boldsymbol{A}$ on the left by a row vector. Keeping to our convention that boldface letters represent column vectors, we write, for $\boldsymbol{v} \in \mathbb{R}^m$,

$$\boldsymbol{v}^T \boldsymbol{A} = \begin{bmatrix} \sum_k v_k A_{k1} & \sum_k v_k A_{k2} & \cdots & \sum_k v_k A_{kn} \end{bmatrix}$$
$$= v_1 \begin{bmatrix} A_{11} & \cdots & A_{1n} \end{bmatrix} + v_2 \begin{bmatrix} A_{21} & \cdots & A_{2n} \end{bmatrix} + \cdots + v_m \begin{bmatrix} A_{m1} & \cdots & A_{mn} \end{bmatrix}. \tag{A.4}$$

Thus *multiplying a matrix on the left by a row vector produces a linear combination of the rows of the matrix.*

These two observations extend to more general matrix-matrix multiplications. One can show that (here again $\boldsymbol{A}$ is $m \times p$ and $\boldsymbol{B}$ is $p \times n$)

$$\boldsymbol{AB} = \boldsymbol{A} \begin{bmatrix} \boldsymbol{b}_1 & \boldsymbol{b}_2 & \cdots & \boldsymbol{b}_n \end{bmatrix} = \begin{bmatrix} \boldsymbol{Ab}_1 & \boldsymbol{Ab}_2 & \cdots & \boldsymbol{Ab}_n \end{bmatrix}. \tag{A.5}$$

In words, a matrix-matrix product is a horizontal concatenation of matrix-vector products involving the columns of the right-hand matrix. Equivalently, if we write $\boldsymbol{A}$ in terms of rows, then

$$\boldsymbol{A} = \begin{bmatrix} \boldsymbol{w}_1^T \\ \boldsymbol{w}_2^T \\ \vdots \\ \boldsymbol{w}_m^T \end{bmatrix} \quad \text{implies} \quad \boldsymbol{AB} = \begin{bmatrix} \boldsymbol{w}_1^T \boldsymbol{B} \\ \boldsymbol{w}_2^T \boldsymbol{B} \\ \vdots \\ \boldsymbol{w}_m^T \boldsymbol{B} \end{bmatrix}. \tag{A.6}$$

Thus, a matrix-matrix product is also a vertical concatenation of vector-matrix products involving the rows of the left-hand matrix. All of our representations of matrix multiplication are equivalent, so whichever one is most convenient at any moment can be used.

**Example A.1**

Let
$$\boldsymbol{A} = \begin{bmatrix} 1 & -1 \\ 0 & 2 \\ -3 & 1 \end{bmatrix}, \quad \boldsymbol{B} = \begin{bmatrix} 2 & -1 & 0 & 4 \\ 1 & 1 & 3 & 2 \end{bmatrix}.$$

Then, going by (A.1), we get

$$AB = \begin{bmatrix} (1)(2)+(-1)(1) & (1)(-1)+(-1)(1) & (1)(0)+(-1)(3) & (1)(4)+(-1)(2) \\ (0)(2)+(2)(1) & (0)(-1)+(2)(1) & (0)(0)+(2)(3) & (0)(4)+(2)(2) \\ (-3)(2)+(1)(1) & (-3)(-1)+(1)(1) & (-3)(0)+(1)(3) & (-3)(4)+(1)(2) \end{bmatrix}$$

$$= \begin{bmatrix} 1 & -2 & -3 & 2 \\ 2 & 2 & 6 & 4 \\ -5 & 4 & 3 & -10 \end{bmatrix}.$$

But note also, for instance, that

$$A\begin{bmatrix} 2 \\ 1 \end{bmatrix} = 2\begin{bmatrix} 1 \\ 0 \\ -3 \end{bmatrix} + 1\begin{bmatrix} -1 \\ 2 \\ 1 \end{bmatrix} = \begin{bmatrix} 1 \\ 2 \\ -5 \end{bmatrix},$$

and so on, as according to (A.5).

The **identity matrix of size** $n$, called $I$ (or sometimes $I_n$), is a diagonal $n \times n$ matrix with every diagonal entry equal to one. As can be seen from (A.5) and (A.6), it satisfies $AI = A$ for $A \in \mathbb{C}^{m \times n}$ and $IB = B$ for $B \in \mathbb{C}^{n \times p}$. It is therefore the matrix analog of the number 1.

### Example A.2

Let

$$B = \begin{bmatrix} 2 & 1 & 7 & 4 \\ 6 & 0 & -1 & 0 \\ -4 & -4 & 0 & 1 \end{bmatrix}.$$

Suppose we want to create a zero in the (2,1) entry by adding $-3$ times the first row to the second row, leaving the other rows unchanged. We can express this operation as a product $AB$ as follows. From dimensional considerations alone, $A$ will need to be $3 \times 3$. According to (A.4), we get "$-3$ times row one plus row two" from left-multiplying $B$ by the vector $\begin{bmatrix} -3 & 1 & 0 \end{bmatrix}$. Equation (A.6) tells us that this must be the second row of $A$. Since the first and third rows of $AB$ are the same as those of $B$, similar logic tells us that the first and third rows of $A$ are the same as the identity matrix:

$$\begin{bmatrix} 1 & 0 & 0 \\ -3 & 1 & 0 \\ 0 & 0 & 1 \end{bmatrix} B = \begin{bmatrix} 2 & 1 & 7 & 4 \\ 0 & -3 & -22 & -12 \\ -4 & -4 & 0 & 1 \end{bmatrix}.$$

This can be verified directly using (A.1).

Note that a square matrix $A$ can always be multiplied by itself to get a matrix of the same size. Hence we can define the integer powers $A^2 = (A)(A)$, $A^3 = (A^2)A = (A)A^2$ (by associativity), and so on. By definition, $A^0 = I$.

According to the rules for multiplying matrices, there are two ways for vectors to be multiplied together. If $v$ and $w$ are in $\mathbb{C}^n$, their **inner product** is

$$v^*w = \sum_{k=1}^{n} \overline{v_k} w_k.$$

Trivially, one finds that $w^*v = \overline{(v^*w)}$. Additionally, *any* two vectors $v \in \mathbb{C}^m$ and $w \in \mathbb{C}^n$ have an **outer product**, which is an $m \times n$ matrix:

$$vw^* = \begin{bmatrix} v_1\overline{w_1} & v_1\overline{w_2} & \cdots & v_1\overline{w_n} \\ v_2\overline{w_1} & v_2\overline{w_2} & \cdots & v_2\overline{w_n} \\ \vdots & \vdots & & \vdots \\ v_m\overline{w_1} & v_m\overline{w_2} & \cdots & v_m\overline{w_n} \end{bmatrix}.$$

## Linear systems and inverses

Given a square $n \times n$ matrix $A$ and $n$-vectors $x$ and $b$, the equation $Ax = b$ is equivalent to

$$a_{11}x_1 + a_{12}x_2 + \cdots + a_{1n}x_n = b_1,$$
$$a_{21}x_1 + a_{22}x_2 + \cdots + a_{2n}x_n = b_2,$$
$$\vdots$$
$$a_{n1}x_1 + a_{n2}x_2 + \cdots + a_{nn}x_n = b_n.$$

We say that $A$ is **nonsingular** or **invertible** if there exists another $n \times n$ matrix $A^{-1}$, the **inverse** of $A$, such that $AA^{-1} = A^{-1}A = I$, the identity matrix. Otherwise, $A$ is **singular**. If a matrix is invertible, its inverse is unique. This and the following facts are usually proved in an elementary text on linear algebra.

### Theorem A.1

The following statements are equivalent:

1. $A$ is nonsingular.
2. $(A^{-1})^{-1} = A$.
3. $Ax = 0$ implies that $x = 0$.
4. $Ax = b$ has a unique solution, $x = A^{-1}b$, for any $n$-vector $b$.

## Exercises

A.1. In racquetball, the winner of a rally serves the next rally. Generally, the server has an advantage. Suppose that when Ashley and Barbara are playing racquetball, Ashley wins 60% of the rallies she serves and Barbara wins 70% of the rallies she serves. If $x \in \mathbb{R}^2$ is such that $x_1$ is the probability that Ashley serves first and $x_2 = 1 - x_1$ is the probability that Barbara serves first, define a matrix $A$ such that $Ax$ is a vector of the probabilities that Ashley and Barbara each serve the second rally. What is the meaning of $A^{10}x$?

A.2. Suppose we have lists of $n$ terms and $m$ documents. We can define an $m \times n$ matrix $A$ such that $A_{ij} = 1$ if term $j$ appears in document $i$, and $A_{ij} = 0$ otherwise. Now suppose that the term list is

'numerical', 'analysis', 'more', 'cool', 'accounting'

and that $x = \begin{bmatrix} 1 & 1 & 0 & 1 & 0 \end{bmatrix}^T$. Give an interpretation of the product $Ax$.

A.3. Let
$$A = \begin{bmatrix} 0 & 1 & 0 & 0 \\ 0 & 0 & 0 & 1 \\ 0 & 0 & 0 & 0 \\ 0 & 0 & 1 & 0 \end{bmatrix}.$$

Show that $A^n = 0$ when $n \geq 4$.

A.4. Find two matrices $A$ and $B$, neither of which is the zero matrix, such that $AB = 0$.

A.5. Prove that when $AB$ is defined, $B^T A^T$ is defined too, and use equation (A.1) to show that $(AB)^T = B^T A^T$.

A.6. Show that if $A$ is invertible, then $(A^T)^{-1} = (A^{-1})^T$. (This matrix is often just written as $A^{-T}$.)

A.7. Prove true, or give a counterexample: The product of upper triangular square matrices is upper triangular.

# Bibliography

[1] M. J. Abramowitz and I. Stegun, *Handbook of Mathematical Functions*, Dover, 1972. Reprinted with corrections from NBS, Washington, DC, 1964; 10th printing 1972 (65-12253). (Cited on p. 447)

[2] U. M. Ascher, R. M. M. Mattheij, and R. D. Russell, *Numerical Solution of Boundary Value Problems for Ordinary Differential Equations*, SIAM, Philadelphia, 1995 (QA379 .A83 1995). (Cited on p. 446)

[3] U. M. Ascher and L. R. Petzold, *Computer Methods for Ordinary Differential Equations and Differential–Algebraic Equations*, SIAM, Philadelphia, 1998 (QA372 .A78 1998). (Cited on pp. 438, 446)

[4] K. E. Atkinson, *An Introduction to Numerical Analysis*, Wiley, New York, 1989, 2nd ed. (QA297 .A84 1989). (Cited on p. 278)

[5] K. H. Atkinson and W. Han, *Elementary Numerical Analysis*, Wiley, New York, 2004, 3rd ed. (QA297 .A83 2004). (Cited on p. 225)

[6] J. L. Aurentz, T. Mach, R. Vanderbril, and D. S. Watkins, "Fast and backward stable computation of roots of polynomials," *SIAM J. Matrix Anal. Appl.*, 36(3):942–973 (2015). (Cited on p. 173)

[7] S. M. Baer and T. Erneux, "Singular Hopf bifurcation to relaxation oscillations," *SIAM J. Appl. Math.*, 46:721–739 (1986). (Cited on p. 249)

[8] D. H. Bailey, K. Jeyabalan, and X. S. Li, "A comparison of three high-precision quadrature schemes," *Experimental Mathematics*, 14(3):317–329 (2005). (Cited on pp. 216, 223, 407)

[9] A. Berke and S. Mueller, "The kinetics of lid motion and its effects on the tear film," in *Lacrimal Gland, Tear Film, and Dry Eye Syndromes* 2, D. A. Sullivan et al., eds., Plenum Press, New York, 1998 (QP188.T4 L332 1998). (Cited on p. 171)

[10] Å. Björk, *Numerical Methods for Least Squares Problems*, SIAM, Philadelphia, 1996 (QA214 .B56 1996). (Cited on p. 119)

[11] K. E. Brenan, S. L. Campbell, and L. R. Petzold, *Numerical Solution of Initial-Value Problems in Differential–Algebraic Equations*, SIAM, Philadelphia, 1995 (QA379 .A83 1995). (Cited on p. 278)

[12] N. F. Britton, *Essential Mathematical Biology*, Springer, Berlin, 2003 (QH323.5 .B745 2003). (Cited on p. 248)

[13] J. K. Brosch, Z. Wu, C. G. Begley, T. A. Driscoll, and R. J. Braun, "Blink characterization using curve fitting and clustering algorithms," *J. Modeling Ophthalmol.*, 3:60–81 (2017). (Cited on pp. 171, 172)

[14] R. L. Burden and J. D. Faires, *Numerical Analysis*, Brooks/Cole, Pacific Grove, 2001, 7th ed. (QA297 .B84 2001). (Cited on p. 225)

[15] C. Canuto, M. Y. Hussaini, A. Quarteroni, and T. A. Zang, *Spectral Methods in Fluid Dynamics*, Springer, Berlin, 1993 (QA377 .S676 1987). (Cited on p. 483)

[16] G. F. Carrier, "Singular perturbation theory and geophysics," *SIAM Rev.*, 12:175-193 (1970). (Cited on p. 415)

[17] E. W. Cheney and D. R. Kincaid, *Numerical Mathematics and Computing*, Brooks/Cole, Pacific Grove, 2013, 7th ed. (QA297 .C426 2013). (Cited on p. 225)

[18] A. M. Cohen, "Is the polynomial so perfidious?," *Numer. Math.*, 68:225-238 (1994). (Cited on p. 29)

[19] A. R. Conn, K. Scheinberg, and L. N. Vicente, *Introduction to Derivative-Free Optimization*, SIAM and MPS, Philadelphia, 2009 (TA342 .C67 2009). (Cited on p. 173)

[20] J. W. Cooley and J. W. Tukey, "An algorithm for the machine calculation of complex Fourier series," *Mathematics of Computation*, 19(90):297-301, 1965. (Cited on p. 407)

[21] R. Corless and N. Fillon, *Graduate Introduction to Numerical Methods*, Springer, Berlin, 2013 (QA297 .C665 2013). (Cited on p. 278)

[22] P. J. Davis, *Interpolation and Approximation*, Dover, 2014. Reprint of 1963 edition (QA221.D33). (Cited on p. 407)

[23] P. J. Davis and P. Rabinowitz, *Methods of Numerical Integration*, Academic, Orlando, 1984, 2nd ed. (QA299.3 .D28 1984). (Cited on p. 225)

[24] C. de Boor, *A Practical Guide to Splines*, Springer, Berlin, 1978. Reprinted 2001 (QA224 .D43). (Cited on p. 225)

[25] B. Fornberg, *A Practical Guide to Pseudospectral Methods*, Cambridge University Press, 2003 (QA320 .F65 1996). (Cited on p. 225)

[26] G. H. Golub and D. P. O'Leary, "Some history of the conjugate gradient and Lanczos algorithms: 1948-1976," *SIAM Rev.*, 31:50-102 (1989). (Cited on p. 358)

[27] G. H. Golub and C. F. Van Loan, *Matrix Computations*, Johns Hopkins University Press, Baltimore, 1996, 3rd ed. (QA188 .G65 1996). (Cited on pp. 93, 119, 310)

[28] R. Haberman, *Elementary Applied Partial Differential Equations with Fourier Series and Boundary Value Problems*, Prentice-Hall, Upper Saddle River, 1998, 3rd ed. (QA377.H27 1998). (Cited on p. 483)

[29] G. J. Hahn, "A conversation with Donald Marquardt," *Stat. Sci.*, 10:377-393 (1995). (Cited on p. 173)

[30] E. Hairer, S. P. Nørsett, and G. Wanner, *Solving Ordinary Differential Equations I: Nonstiff Problems*, Springer, Berlin, 2009, 2nd rev. ed. (QA371 .H334 2009). (Cited on p. 278)

[31] P. C. Hansen, V. Pereyra, and G. Scherer, *Least Squares Data Fitting with Applications*, Johns Hopkins University Press, Baltimore, 2013 (QA275 .H26 2013). (Cited on p. 119)

[32] M. R. Hestenes and E. Stiefel, "Method of conjugate gradients for solving linear systems," *J. Res. Natl. Bur. Standards*, 49:409-438 (1952). (Cited on p. 358)

[33] N. J. Higham, *Accuracy and Stability of Numerical Algorithms*, SIAM, Philadelphia, 2002, 2nd ed. (QA297 .H53 2002). (Cited on pp. 28, 29, 93, 119)

[34] A. Iserles, *A First Course in the Numerical Analysis of Differential Equations*, Cambridge University Press, Cambridge, 1996 (QA371 .I813 1996). (Cited on pp. 278, 538)

[35] I. T. Jolliffe, *Principal Component Analysis*, Springer, Berlin, 2002, 2nd ed. (QA278.5 .J65 2002). (Cited on p. 310)

[36] C. T. Kelley, *Iterative Methods for Linear and Nonlinear Equations*, SIAM, Philadelphia, 1995 (QA297.8 .K45 1995). (Cited on p. 173)

[37] W. O. Kermack and A. G. McKendrick, "A contribution to the mathematical theory of epidemics," *Proc. Roy. Soc. A*, 115:700-721. (Cited on p. 170)

[38] D. C. Lay, *Linear Algebra and Its Applications*, Pearson, Boston, 2006 (QA184.2 .L39 2006). (Cited on p. 93)

[39] S. J. Leon, *Linear Algebra with Applications*, Prentice-Hall, Upper Saddle River, 2006 (QA184.2 .L46 2006). (Cited on p. 93)

[40] K. Levenberg, "A method for the solution of certain non-linear problems in least squares," *Quart. Appl. Math.*, 2:164-168 (1944). (Cited on p. 173)

[41] R. J. LeVeque, *Numerical Methods for Conservation Laws*, Birkhäuser, Basel, 1992 (QA377 .L4157 1992). (Cited on p. 505)

[42] R. J. LeVeque, *Finite Difference Methods for Ordinary and Partial Differential Equations*, SIAM, Philadelphia, 2007 (QA431.L548 2007). (Cited on p. 505)

[43] R. J. LeVeque, *Finite Volume Methods for Hyperbolic Problems*, Cambridge University Press, Cambridge, 2002 (QA377 .L41566 2002). (Cited on p. 505)

[44] D. W. Marquardt, "An algorithm for least-squares estimation of nonlinear parameters," *J. Soc. Ind. Appl. Math.*, 11(2):431-441 (1963). (Cited on p. 173)

[45] D. B. Meade and A. A. Struthers, "Differential equations in the new millennium: The parachute problem," *Int. J. of Eng. Education*, 15(6): 417-424 (1999). (Cited on p. 234)

[46] C. Moler, *Numerical Computing with MATLAB*, SIAM, Philadelphia, 2004 (QA297 .M625 2004). (Cited on p. 476)

[47] K. W. Morton and D. F. Mayers, *Numerical Solution of Partial Differential Equations*, Cambridge University, Cambridge, 2005, 2nd ed. (QA377 .M69 2005). (Cited on pp. 483, 538)

[48] S. G. Nash, *A History of Scientific Computing*, ACM, New York, 1990 (QA76.17.H59 1990). (Cited on pp. 278, 358, 407, 483)

[49] R. Newton, L. J. Broughton, M. J. Lind, P. J. Morrison, H. J. Rogers, and I. D. Bradbrook, "Plasma and salivary pharmacokinetics of caffeine in man," *Eur. J. Clin. Pharmacol.*, 21(1): 45-52 (1981). (Cited on p. 234)

[50] J. R. Ockendon, S. D. Howison, A. A. Lacey, and A. B. Movchan, *Applied Partial Differential Equations*, Oxford University Press, Oxford, 1999 (QA377 .A675 2003). (Cited on p. 483)

[51] F. W. J. Olver, D. W. Lozier, R. F. Boisvert, and C. W. Clark, *NIST Handbook of Mathematical Functions*, Cambridge University Press, Cambridge, 2010 (QA331 .N57 2010). (Cited on p. 447)

[52] J. M. Ortega and W. C. Rheinboldt, *Iterative Solution of Nonlinear Equations in Several Variables*, SIAM, Philadelphia, 2000 (QA297.8 .O77 2000). (Cited on p. 173)

[53] B. N. Parlett, *The Symmetric Eigenvalue Problem*, SIAM, Philadelphia, 1998. Reprint of the 1980 edition (QA188 .P37 1998). (Cited on p. 310)

[54] J. A. Pelesko and T. A. Driscoll, The effect of the small-aspect-ratio approximation on canonical electrostatic MEMS models, *J. Eng. Math.*, 53(3-4):239–252, 2005. (Cited on pp. 410, 531)

[55] A. Quarteroni, R. Sacco, and F. Saleri, *Numerical Mathematics*, Springer, Berlin, 2000 (QA297 .Q83 2000). (Cited on pp. 172, 173, 358, 447)

[56] P. J. Roache, *Fundamentals of Computational Fluid Dynamics*, Hermosa, Albuquerque, 1998 (QA911.R57 1998). (Cited on p. 483)

[57] Y. Saad, *Iterative Methods for Sparse Linear Systems*, SIAM, Philadelphia, 2003, 2nd ed. (QA188 .S17 2003). (Cited on p. 358)

[58] L. F. Shampine and M. W. Reichelt, "The MATLAB ODE suite," *SIAM J. Sci. Comput.*, 18(1):1–22 (1997). (Cited on p. 278)

[59] G. D. Smith, *Numerical Solution of Partial Differential Equations: Finite Difference Methods*, Clarendon, Oxford, 1992 (QA374 .S86 1992). (Cited on p. 483)

[60] G. W. Stewart, *Matrix Algorithms. Volume II: Eigensystems*, SIAM, Philadelphia, 2001 (QA188 .S714 1998 v. 2). (Cited on p. 310)

[61] G. S. Strang, *Computational Science and Engineering*, Wellesley-Cambridge Press, Wellesley, MA, 2007; 2nd printing, 2012 (TA330 .S78 2007). (Cited on p. 173)

[62] G. S. Strang, *Introduction to Linear Algebra*, Wellesley-Cambridge Press, Wellesley, MA, 2016 (QA184 .S78 2016). (Cited on p. 93)

[63] G. S. Strang and K. Borre, *Linear Algebra, Geodesy, and GPS*, Wellesley-Cambridge Press, Wellesley, MA, 1997 (TA347 .L5 S87 1997). (Cited on p. 119)

[64] G. S. Strang and G. J. Fix, *An Analysis of the Finite Element Method*, Wellesley-Cambridge Press, Wellesley, MA, 1998. (Cited on p. 447)

[65] H. Takahasi and M. Mori, "Double exponential formulas for numerical integration," *Publ. RIMS, Kyoto University*, 9:721–741, 1974. (Cited on p. 407)

[66] P. J. G. Teunissen, *Dynamic Data Processing: Recursive Least-Squares*, VSSD, Delft, 2009 (QA9.615). (Cited on p. 119)

[67] J. A. Trangenstein, *Numerical Solution of Hyperbolic Partial Differential Equations*, Cambridge University Press, Cambridge, 2009. (QA377.T62) (Cited on p. 505)

[68] L. N. Trefethen, *Spectral Methods in MATLAB*, SIAM, Philadelphia, 2000 (QA377 .T65 2000). (Cited on pp. 392, 447, 483, 499, 538)

[69] L. N. Trefethen, "Is Gauss quadrature better than Clenshaw–Curtis?," *SIAM Rev.*, 50(1):67–87 (2008). (Cited on p. 407)

[70] L. N. Trefethen, "Six myths of polynomial interpolation and quadrature," *Mathematics Today*, 184–188 (August 2011). (Cited on p. 29)

[71] L. N. Trefethen, *Approximation Theory and Approximation Practice*, SIAM, Philadelphia, 2013 (QA221 .T73 2013). (Cited on p. 407)

[72] L. N. Trefethen and D. Bau III, *Numerical Linear Algebra*, SIAM, Philadelphia, 1997 (QA184 .T74 1997). (Cited on pp. 93, 358)

# Bibliography

[73] L. N. Trefethen and M. Embree, *Spectra and Pseudospectra: The Behavior of Nonnormal Matrices and Operators*, Princeton University Press, 2005 (QA320 .T67 2005). (Cited on pp. 292, 310)

[74] H. A. van der Vorst, *Iterative Krylov Methods for Large Linear Systems*, Cambridge University Press, 2003 (QA297.8 .V67 2003). (Cited on p. 358)

[75] C. F. Van Loan, *Introduction to Scientific Computing: A Matrix-Vector Approach Using Matlab*, Prentice-Hall, Upper Saddle River, 2000 (QA76.95 .V35 1999). (Cited on p. 225)

[76] G. B. Whitham, *Linear and Nonlinear Waves*, Wiley, New York, 1974 (QA927 .W48). (Cited on p. 486)

[77] J. H. Wilkinson, "The perfidious polynomial," in *Studies in Numerical Analysis*, G. H. Golub, ed., *Studies in Mathematics*, vol. 24, Mathematical Association of America, Washington, DC, 1984, pp. 1–28. (Cited on p. 29)

[78] Z. Wu, C. G. Begley, P. Situ, T. Simpson and H. Liu, "The effects of mild ocular surface stimulation and concentration on spontaneous blink parameters," *Curr. Eye Res.*, 39(1):9–20 (2014). (Cited on p. 171)

# Index

A-stability, 476
adaptivity, 218
adjacency matrix, 283, 312
advection equation, 486, 492
advection-diffusion, 487, 497, 536, 537
Airy equation, 430
algorithm, 20
aliasing, 389
Allen–Cahn equation, 417, 436, 482, 535
Antilles, Wedge, 449
Arnoldi iteration, 335, 344
asymptotic notation, 61

\(backslash), 45, 104, 112, 156, 167
backward error, 24, 83, 125
barycentric interpolation formula, 366
barycentric weights, 365
Bauer–Fike theorem, 289
beam equation, 504
Bernoulli numbers, 211
Bessel's equation, 415, 430
Black–Scholes equation, 450
boundary conditions, numerical implementation of, 426, 433, 452, 477, 518, 532
boundary layer, 431
boundary-value problem, 410
Broyden update, 161
Burgers equation
 inviscid, 499
 viscous, 489

C3PO, 121, 409
cardinal functions, 180, 184, 360, 384
Carrier equation, 415

CFL condition, 491
characteristic polynomial, 287
Chebyshev equation, 416
Chebyshev points
 first kind, 382
 second kind, 373, 422
Chebyshev polynomials, 382
Clenshaw–Curtis integration, 391
collocation, 425
condition number, 16
 of a matrix, 80, 105
 of eigenvalues, 289
 of elementary functions, 18
 of finite differences, 205
 of initial-value problems, 231
 of interpolation, 179
 of least squares, 105
 of rootfinding, 123
conjugate gradients, 345
conservation law, 486
consistency, 239
contraction map, 134
convergence rate, *see also* order of accuracy
 linear, 132, 323, 340, 344
 quadratic, 139, 329
 spectral, 375, 390
 superlinear, 146
Crank–Nicolson, 462

D'Alembert's solution, 504
Dahlquist theorems
 equivalence, 276
 first stability barrier, 277
 second stability barrier, 476
deflation, 127
differentiation matrix, 456, 512
diffusion equation, 451
digits (significant digits), 11

dimension, 242
dimension reduction, 304, 333
Dirichlet condition, 433, 451
discretization, 175
domain of dependence, 490
double precision, 12
doubly exponential transformation, 400

eigenvalue, 286, 338
 conditioning of, 289
 decomposition, 287, 316, 463
 dominant, 320, 326
elliptic PDE, 524
eps, 12
error function, 224
Euler's method, 235, 246
Euler–Maclaurin formula, 211
evolutionary PDE, 450
explicit method
 Adams–Bashforth, 262
 Euler's (for a system), 246
 Euler's (for IVPs), 235
 Heun's, 252
 improved Euler, 250
 modified Euler, 252
 one-step, 238
 Runge–Kutta, 249
extrapolation, 212

FFT, 388
field of values, 303
fill-in, 313
finite difference
 parabolic PDE, 456
finite differences, 160, 196
 boundary value problems, 425
 elliptic PDE, 512
 matrix, 418
 parabolic PDE, 452

551

finite element method, 443
    assembling equations, 443
Fisher equation, 483
Fitzhugh–Nagumo equations, 249
fixed point, 127
floating point numbers, 9, 12
Francis QR iteration, 291

Galerkin conditions, 441
Gauss–Newton method, 167
Gaussian
    distribution, 14
    integration, 393
generating polynomials, 262
Gibbs phenomenon, 390
Givens rotation, 118
GMRES, 339
    and MINRES, 344
    preconditioning, 353
    restarting, 340
Gram–Schmidt, 110
graph nodes and edges, 282
Gregory quadrature formula, 216

hat functions, 183
heat equation, 451, 458, 460, 467, 516
Hermite
    equation, 416
    interpolant, 36
hermitian positive definite, 302
Horner's rule, 21
Householder reflection, 113
hyperbolic PDE, 486

IEEE 754, 12, 14
implicit method, 268
    Adams–Moulton, 262
    backward differentiation, 262
    backward Euler, 263
inflow condition, 493
initial-value problem, 227
inner product
    of functions, 378
    of vectors, 542
interpolation, 31, 148, 175
    by cubic splines, 189
    by piecewise linear polynomials, 182, 441
    by piecewise polynomials, 209
    by polynomials, 176, 198, 359

by trigonometric functions, 384
    condition number of, 179, 186
inverse interpolation, 148
inverse iteration, 326

Jacobian matrix, 153, 160, 472, 532

Kenobi, Obi-Wan, 1, 485, 507
Kronecker product, 525
Krylov subspace, 332
Kuramoto–Sivashinsky equation, 490

Lagrange interpolation formula, 361
Laguerre equation, 416
Lanczos iteration, 344
Laplace equation, 524
least squares
    linear, 95
Legendre equation, 416
Legendre polynomials, 381
Levenberg method, 162
linear combination, 541
linear system
    ODEs, 243
    overdetermined, 95
    square, 31
linearization
    of a PDE, 532
    of an ODE, 472
Lipschitz condition, 134
local extrapolation, 224
logistic equation, 228

machine epsilon, 10
    in double precision, 12
mantissa, 10
mass matrix, 441
matrix
    adjacency, see adjacency matrix
    as image, 283, 305, 349
    banded, 85, 314, 419
    condition number, 80, 345
    coordinate, 508
    degree, 318
    diagonal, 540
    diagonalizable, 287
    differentiation, 419, 423
    elementary, 55
    exponential, 243

factorization, see also eigenvalue decomposition, and singular value decomposition
    Cholesky, 90, 105
    eigenvalue, 296
    LU, 51, 314, 354
    PLU, 70
    QR, 109, 113
    thin QR, 110
    for a function of two variables, 508
    graph Laplacian, 318
    hermitian, 296, 540
    Hilbert, 81, 83
    inverse, 44, 45, 104, 543
    Krylov, 332
    nonsingular, 543
    normal, 290
    ONC, 108
    orthogonal, see orthogonal matrix
    positive definite, 89, 104, 302, 345
    projection, 113
    similar, 288
    skew-symmetric, 92
    sparse, see sparse matrix
    symmetric, 37, 87, 104, 299, 344, 540
    Toeplitz, 292
    triangular, 46, 51, 109, 540
    unitary, see unitary matrix
    upper Hessenberg, 336
    Vandermonde, 32, 84, 97
method of lines, 457, 467, 477, 495
MINRES, 344
misfit function, 167
multistage methods, 250
multistep method, 261, 266, 273

NaN (not a number), 12
Neumann condition, 433, 451
Newton's method, 3, 138, 154, 433, 532
Newton–Cotes formula, 209
norm
    matrix, 75, 295
    vector, 74
normal equations, 103, 111
numerical computing, 4

ONC, 297

# Index

order of accuracy, 187, 192, 203, 211, 240, 263, 363, 445
Oregonator, 472
orthogonal
    matrix, 109, *see also* unitary matrix
    polynomials, 380
    vectors, 107
orthogonal matrix, 109, 113
orthogonal projection, 382
outer product, 543
overflow, 12

parabolic PDE, 451
parameter continuation, 436
partition of unity, 189
pendulum, 245, 417
periodic boundary conditions, 456
Poisson equation, 524
power iteration, 321
preconditioning, 353
predator–prey model, 242

quasi-Newton methods
    Broyden update, 161
    finite difference Jacobian, 160
    Levenberg, 162
    Levenberg–Marquardt, 165
quasimatrix, 379

racquetball, 543
random walk, 14
Rayleigh quotient, 301
residual, 125
*Return of the Jedi*, 1, 311
Robin condition, 433
root multiplicity, 125, 141
rootfinding problem, 121, 268

roots, 24
Runge phenomenon, 373
Runge–Kutta method, 249, 257

Schrödinger equation, 489
secant method, 145
Self-Organized Networks Database, 318
semidiscretization, *see* method of lines
Sherman–Morrison formula, 165
shift operator, 263
shooting method, 411
Simpson's rule, 213, 219
sine integral function, 224
sine–Gordon equation, 504
singular value, 293
    principal, 294
singular value decomposition, 293, 304
    thin form, 297
Skywalker, Luke, 311
soap film, 538
Solo, Han, 31, 95, 227, 359
sparse matrix, 312, 319, 354
spline
    cubic, 190
    natural, 192
    not-a-knot, 192
stability, 22, 72
    of collocation, 429
    of IVP solvers, 275, 463
    of polynomial interpolation, 369
    region, 464
*Star Wars: A New Hope*, 31, 121, 227, 409, 449, 485, 507

steepest descent, 162
stencil, 263
step size, 235
stiff problem, 271, 471
stiffness matrix, 441
subdiagonal, 540
subtractive cancellation, 15, 16, 23, 72, 108
superdiagonal, 539
Sylvester equation, 524
symbolic computing, 4

tensor-product
    domain, 507
    grid, 508
*The Empire Strikes Back*, 9, 95, 175, 281, 359
trapezoid formula, 210, 390
truncation error, 202, 238, 263

U.S. census, 101
unconditionally stable, 469
underflow, 12
unit roundoff, *see* machine epsilon
unitary matrix, 286, 293, 299
upwind direction, 490, 493

van der Pol equation, 477
vec, 515
von Neumann, John, xi

wave equation, 499, 521
weak solution, 440

Yoda, 9, 175, 281

zero-stability, 275